Available online at www.sciencedirect.com

Earth-Science Reviews 73 (2005) v–vi

www.elsevier.com/locate/earscirev

Special Issue

Fifty Years of Death Valley Research
A volume in honor of Lauren A. Wright and Bennie Troxel

Edited by

J. P. Calzia (Editor)

U.S. Geological Survey, Menlo Park, CA, U.S.A.

CONTENTS

fty years of Death Valley research: A volume in honor of Lauren Wright and Bennie Troxel
   J.P. Calzia   1
cknowledgements of a professional lifetime
   L.A. Wright   3
y geological career in Death Valley
   B.W. Troxel 13
eological landscapes of the Death Valley region
   M.B. Miller 17
bliography L.A. Wright and B.W. Troxel 1944–present 31

Eolian deposits in the Neoproterozoic Big Bear Group, San Bernardino Mountains, California, USA
J.H. Stewart   47

The relationship between the Neoproterozoic Noonday Dolomite and the Ibex Formation: New observations and their bearing on
'snowball Earth'
F.A. Corsetti and A.J. Kaufman   63

Interpretation of the Last Chance thrust, Death Valley region, California, as an Early Permian décollement in a previously undeformed
shale basin
C.H. Stevens and P. Stone   79

Structure and regional significance of the Late Permian(?) Sierra Nevada–Death Valley thrust system, east-central California
C.H. Stevens and P. Stone 103

The Black Mountains turtlebacks: Rosetta stones of Death Valley tectonics
M.B. Miller and T.L. Pavlis 115

Are turtleback fault surfaces common structural elements of highly extended terranes?
I. Çemen, O. Tekeli, G. Seyitoğlu and V. Isik 139

Large-scale gravity sliding in the Miocene Shadow Valley Supradetachment Basin, Eastern Mojave Desert, California
G.A. Davis and S.J. Friedmann 149

Late Cenozoic sedimentation and volcanism during transtensional deformation in Wingate Wash and the Owlshead Mountains, Death Valley
H.G. Luckow, T.L. Pavlis, L.F. Serpa, B. Guest, D.L. Wagner, L. Snee, T.M. Hensley and
A. Korjenkov 177

Miocene rapakivi granites in the southern Death Valley region, California, USA
J.P. Calzia and O.T. Rämö 221

Upper Neogene stratigraphy and tectonics of Death Valley — a review
J.R. Knott, A.M. Sarna-Wojcicki, M.N. Machette and R.E. Klinger 245

Late Quaternary denudation, Death and Panamint Valleys, eastern California
A.S. Jayko 271

Holocene fluvial geomorphology of the Amargosa River through Amargosa Canyon, California
D.E. Anderson 291

Macropolygon morphology, development, and classification on North Panamint and Eureka playas, Death Valley National Park CA
P. Messina, P. Stoffer and W.C. Smith 309

Base- and precious-metal deposits in the Basin and Range of Southern California and Southern Nevada—Metallogenic implications
of lead isotope studies
S.E. Church, D.P. Cox, J.L. Wooden, J.V. Tingley and R.B. Vaughn 323

Fifty Years of Death Valley Research

Edited by **James P. Calzia**
U.S. Geological Survey,
345 Middlefield Road, Menlo Park, CA 94025 U.S.A.

Amsterdam ● Boston ● Heidelberg ● London ● New York ● Oxford
Paris ● San Diego ● San Francisco ● Singapore ● Sydney ● Tokyo

ELSEVIER

Elsevier
Radarweg 29, PO Box 211, 1000 AE Amsterdam, The Netherlands
The Boulevard, Langford Lane, Kidlington, Oxford OX5 1GB, UK

First edition 2006

Notice
No responsibility is assumed by the Publisher for any injury and/or damage to persons or property as a matter of products liability, negligence or otherwise, or from any use or operation of any methods, products, instructions or ideas contained in the material herein. Because of rapid advances in the medical sciences, in particular, independent verification of diagnoses and drug dosages should be made.

British Library Cataloguing in Publication Data
A catalogue record is available from the British Library.

Library of Congress Cataloging in Publication Data
A catalog record is available from the Library of Congress.

ISBN-13:978-0-444-52752-3
ISBN-10:0-444-52752-4

Reprinted from the Elsevier journal
EARTH SCIENCE REVIEWS, Volume 73, Nos. 1-4 (December 2005)

For information on all book publications
visit our website at books.elsevier.com

Printed and bound in The Netherlands

06 07 08 09 10   10 9 8 7 6 5 4 3 2 1

Working together to grow
libraries in developing countries

www.elsevier.com | www.bookaid.org | www.sabre.org

ELSEVIER   BOOK AID
International   Sabre Foundation

The paper used in this publication meets the requirements of ANSI/NISO Z39.48-1992 (Permanence of Paper).
Printed in The Netherlands.

Available online at www.sciencedirect.com

Earth-Science Reviews 73 (2005) 1

www.elsevier.com/locate/earscirev

Fifty years of Death Valley research:
A volume in honor of Lauren Wright and Bennie Troxel

J.P. Calzia

Dr. Lauren A. Wright (Professor Emeritus at Penn State University) and Bennie W. Troxel are internationally recognized experts on the geology of Death Valley, California. In November 2002, they celebrated 50 years of cooperative research together. This Special Issue of *Earth-Science Reviews* commemorates that special occasion.

Wright and Troxel's research in Death Valley covers a wide variety of subjects including stratigraphy, structure, regional tectonics, Quaternary geology, and mineral resources. Their diversity in research is reflected in this Special Issue of *Earth-Science Reviews*. The first two papers add stratigraphic and ^{13}C data to the constantly growing volume of literature on Neoproterozoic global glaciation and the Snowball Earth theory. The next seven papers are nearly equally divided between late Paleozoic thrust faulting, middle Cenozoic extensional tectonics, and magmatism. Note that the three papers on extensional tectonics represent the latest contributions in a continuum of research since Wright and Troxel published their landmark paper, *Shallow fault interpretation of Basin and Range structure*, in 1973. The next four papers describe the late Neogene to Holocene geology and geomorphology of Death Valley, research topics very dear to Wright and Troxel in the last 10 years. The last paper describes the lead–zinc deposits of the southern Basin and Ranges, including the Death Valley region. It is almost ironic that a paper on mineral resources is the last paper in this volume because Lauren began his research in Death Valley by mapping Proterozoic talc deposits in

1942. Please pause, however, before reading these 14 provocative papers and reminisce with Wright and Troxel as they recall their 50 years in Death Valley together. Then review Marli Miller's photographic essay on the geologic landscapes of Death Valley. Miller's photographs perfectly capture the mystery and beauty of this fantastic place, and show why Wright, Troxel and so many other scientists continually return to the challenges of Death Valley.

This is not the first volume written in honor of Lauren Wright and Bennie Troxel. Brian Wernicke dedicated Geological Society of America Memoir 176, *Basin and Range extensional tectonics near the latitude of Las Vegas, Nevada*, to Wright and Troxel in 1990. In the intervening 15 years, much has been done, and there is much to do. To this day, Wright and Troxel continue their research in Death Valley and continue to share their knowledge, data, and ideas with any and all earth scientists. Their dedication, keen observational skills, and ability to merge observations with theory are well documented in this and earlier volumes. Combine these characteristics, however, with their compatible yet different personalities and willingness to help the youngest student or the most experienced professional and you have world class scientists that are an inspiration to all; no one could ask for better mentors! The scientific and personal standards set by Lauren Wright and Bennie Troxel over the last 50 years, combined with the theories and data presented here, will make this Special Issue of *Earth-Science Reviews* the standard reference for geologic research in the Death Valley region for the next 50 years!

0012-8252/$ - see front matter © 2005 Published by Elsevier B.V.
doi:10.1016/j.earscirev.2005.07.007

Available online at www.sciencedirect.com

SCIENCE DIRECT

Earth-Science Reviews 73 (2005) 1–

www.elsevier.com/locate/earscirev

ELSEVIER

Fifty years of Death Valley research: A volume in honor of Lauren Wright and Bennie Troxel

J.P. Calzia

0012-8252/$ - see front matter © 2005 Published by Elsevier B.V.
doi:10.1016/j.earscirev.2005.07.001

Available online at www.sciencedirect.com

Earth-Science Reviews 73 (2005) 3–11

www.elsevier.com/locate/earscirev

Acknowledgements of a professional lifetime

Lauren A. Wright

I am indebted to Jim Calzia in many ways, including this opportunity for both Bennie and me to thank the creators of this volume. So "thank you indeed, Jim," for conceiving of, arranging for, and editing it. And to each of the authors, a "thank you" for your contributions which add much to an understanding of the evolution of this very special part of the Earth's crust. Because Jim was motivated by knowing that November 9, 2002, marked the 50th anniversary of the Troxel–Wright collaboration, need I add that my indebtedness to Bennie is beyond measure?

I am also pleased that Jim has asked both of us to include recollections of our own professional lives. In this way, I can address a long-held wish to acknowledge in writing indebtedness to all of my geological companions who have been helpful in my journey and especially to five men whose support made it possible and inexpressively rewarding. In chronological order of our meeting, they are Ben Page, Ian Campbell, Dick Jahns, Olaf Jenkins and Levi Noble. Each was a valued mentor for as long as he lived. Each also became a personal friend.

My interest in rocks and landscapes dates from my early childhood when my parents and I spent summers in a mountain cabin on the west side of Lookout Mountain west of Denver. My father, who built the cabin, was a commercial artist and was thus free to work either in our modest residence in the Los Angeles basin or in our Colorado retreat. As early as the late 1920s, having become fascinated by the crystalline rocks and flat irons of the Front Range, I knew that I must study geology.

In 1936 I enrolled as a geology major at the University of Southern California where Ben Page (Fig. 1), fresh from receiving his doctorate at Stanford, was the newest faculty member. A grandson of a governor of California, Ben was both aristocratic and thoroughly likeable. My first geologic map was constructed as part of Ben's field geology course and covered an area of the Santa Monica Mountains in the vicinity of Cahuenga Pass. For my Masters degree at USC, and under the supervision of Tom Clements, I mapped in the eastern part of the Ventura Basin where the Upper Miocene Modelo and Mint Canyon Formations are closely associated. I also learned that a certain Richard H. Jahns had mapped in this area several years earlier while working on a Cal Tech doctorate minor thesis under the supervision of Chester Stock.

In the Spring of 1941, Ian Campbell (Fig. 2), then the Associate Head of the Department of Geology and Geophysics at Cal Tech, graciously spent the better part of an afternoon escorting me and another prospective graduate student through the departmental facilities. I later learned that Ian was frequently described as "living in a world inhabited only by good guys" and for good reason. A kinder man I have never known.

I entered the Cal Tech program in the fall of 1941 with Ian as my academic advisor. It was soon obvious that Dick Jahns (Fig. 3) had already become legendary

E-mail address: law11@psu.ed.

0012-8252/$ - see front matter © 2005 Published by Elsevier B.V.
doi:10.1016/j.earscirev.2005.07.008

Fig. 1. Ben Page, my principal undergraduate instructor at USC, shown here as my USGS party chief beside a wind-blown tent when we were mapping talc deposits in 1942.

as an undergraduate on the Cal Tech campus. In his senior year, he had changed majors, had taken all of the required courses in geology with straight A's, and

Fig. 3. A characteristic salutation from Dick Jahns, my long-time friend, confident, and mentor.

had played on the varsity baseball team. Of the courses I took that year, I most remember John Peter Buwalda's structural geology seminar where we read and discussed Levi Noble's 1941 classic paper on the Virgin Spring chaos. I wondered if I could ever do geology in that fantastic terrane.

Late in the spring semester of 1942, Ben called, saying that he had taken academic wartime leave to work in the strategic minerals program of the U.S. Geological Survey, and would I be interested in working as his field assistant? So, with Ian's approval and to my unbelieving delight, I obtained my first professional job at an annual salary of $2000 plus per diem.

At that time, high quality steatite talc was an important ingredient in high frequency insulators used in military equipment. Ben's assignment was to assess the reserves of steatite-grade talc in the western states. So for the next 5 months, we mapped by plane table the larger of the known steatite-grade talc deposits in the southwest. We also sampled several talc deposits in the Death Valley region. Although of non-steatite grade, their geologic setting in the Proterozoic formations of Death Valley so impressed me that I

Fig. 2. Ian Campbell, my Cal Tech thesis advisor and, from 1959 to 1961, my boss while he was serving as Olaf Jenkins' successor as Chief of the California Division of Mines and Geology.

quickly and mentally claimed them as the subject of my delayed doctoral thesis. Recalling my wish in Dr. Buwalda's class, I must have asked "Is this serendipity in action?"

In December of 1942, the field-oriented part of the talc program ended. Being with Ben for those 5 months had been pure delight and a wonderful learning experience. We mostly camped out or slept in miner's quarters and we talked and laughed about many things. Now I was up for reassignment by George Mansfield, a much-respected geologist and Chief of the Nonmetallic Minerals Branch of the USGS, but also a man I had never met. I was amazed and a bit overwhelmed to learn that I was to report to the now legendary, but still young, Dick Jahns. If George Mansfield had sent me elsewhere, my life from then on would have been totally different from the one reported here.

Dick had been studying the contact-metamorphic, beryllium-bearing deposits at Iron Mountain in southern New Mexico, also as part of the USGS strategic minerals program. He had arranged with his thesis advisor, Ian Campbell (who else?), to be in residence at Cal Tech for 2 or 3 months in order to write his doctoral thesis on these deposits. At that time, they were being explored in a U.S. Bureau of Mines drilling program and someone was needed to log the cores. I arrived at the USBM camp at Christmas time and was greeted warmly by Dick and his wife Frances. Soon thereafter, they left and I stayed for the rest of that winter.

For the following 12 months, from the spring of 1943 until the spring of 1944, I assisted Dick in investigations of pegmatite deposits in northern New Mexico, first in a drilling program at the Harding lithium–tantalum–beryllium mine near Dixon and later at numerous mica mines and prospects in the Petaca district west of Taos. As the Harding pegmatite contained a spectacular and challenging array of minerals, another rewarding friendship was initiated with a visit by Waldemar Schaller, a distinguished mineralogist with the USGS. Schaller, I learned, was the pioneer investigator of the minerals of the borate deposits of Death Valley and Mojave Desert regions.

The Petaca assignment involved plane table mapping of the deposits. For much of this time, we were joined by USGS geologists Porter Irwin and E. William Heinrich. Toward the end of the Petaca

project, Dick encouraged us to work almost independently although under his close guidance. So he let me map and describe the pegmatite body at the Globe mine, which has an especially well defined paragenesis. This project, in turn, resulted in my first paper in a refereed journal (with Dick as a referee).

In Petaca, we lived in an abandoned adobe store and Francis cooked for the five of us. We were the only non-Spanish speaking inhabitants. In the spring of 1944, Dick and Frances left for the pegmatite districts of the southeastern U.S. and I was left alone. On Easter Sunday, in observing how my neighbors celebrated, I concluded that they were about equally divided between traditional Catholics, Penitents and Pentecostals.

In the following June, I was inducted into the Army Air Force. After basic training at Buckley Field, Denver, I surprisingly found myself in the Air Transport Command. At that time, military supplies were being flown to Russia via Alaska and the Aleutian Islands and I became an instructor in a group teaching cold weather survival methods to military personnel about to enter the Arctic for the first time. The fact that I had never been in the Arctic did not seem to matter. I, indeed, welcomed the assignment and inwardly thanked the personnel officer for his decision. During my first months with this group, our base camp was at the foot of Mount Evans in the Colorado Front Range and almost within sight of the cabin where I spent my first 21 summers. We later moved to another base camp about midway between Edmonton and Jasper, Alberta. On weekend and weeklong passes, it was possible to hike into the high country of the Canadian Rockies and connect with the geology there. And I was thankful again. I was released in February 1945 in time to enroll for the Spring semester at Cal Tech.

There, to my surprise and delight, I was greeted by Dick Jahns, the newest member of the geology faculty. I was also welcomed by Ian Campbell who readily agreed to supervise the talc-oriented thesis. Dick had retained part-time employment with the USGS and was spending weekends mapping the gem-bearing pegmatites of San Diego County. On my arrival, he arranged a similar USGS appointment for me and we were together again, traveling between Pasadena and the Pala pegmatite district. On other intervals, I began my doctoral research by constructing a plane table map of the Silver Lake talc deposits near Baker. Gene Schumaker was then a Cal Tech undergraduate, and, being curious about the plane table technique, volunteered as my rod man for 2 or 3 days. The fact that he had already been dubbed "Supergene" by his classmates hinted that he was destined for geological greatness.

In 1946, when the Pala pegmatite project was nearing completion, we were visited in the field by Olaf Jenkins (Fig. 4), who had been recently promoted to Chief of the California Division of Mines. He had already agreed to publish our report with the map and photos of the gemstones in color. More importantly, he was also involved in a post-war reorganization of his agency. At the end of the final day, he encouraged me to apply for an opening of Associate Geologist. When I told him that I must finish my long delayed doctorate before seeking permanent employment, he asked about my thesis. When told that it was a study of the talc deposits of the Death Valley region, he said "I would make that your first assignment." So I moved to San Francisco in the Spring of 1947 and spent the next 13 years employed by the California Division of Mines, later termed the California Division of Mines and Geology, and now designated as the California Geological Survey.

The first 3 years were spent working in and out of the Division's headquarters in the Ferry Building. From the window of my second floor office, I could look at the full length of Market Street to Mt.

Fig. 4. Olaf Jenkins, State Geologist for California 1927–59, and Chief of the California Division of Mines and Geology, 1947–59; a dedicated and effective advocate of geology in the service of mankind.

Davidson on the skyline. Foremost among my new associates was Charles Chesterman, who I soon learned was in close touch with Waldemar Schaller. In the following years, the three of us enjoyed numerous shared experiences, including a trip to view pegmatite deposits on the Isle of Elba. Our friendship with him was truly avuncular as he wished us to call him "Unc". I continue to keep in touch with Charles Jr., a Captain in the U.S. Navy.

In my frequent contacts with Olaf, I observed in him a professional style quite different from Ben's, Ian's and Dick's, but also admirable. Like them, he saw in geology a means for improving the lives of people, but he had to present his programs to others, especially to members of the State Mining Board and the Senate Finance Committee. He pushed these programs with vigor and tenacity. An equivocator he was not, but he was an opportunist. When he became aware that, in my Death Valley activities, I had developed a friendship with State Senator Charles Brown of Shoshone and Chairman of the Senate Finance Committee, Olaf made certain that I sat by his side at each of Charlie's budget hearings.

True to his commitment, Olaf permitted me to devote a reasonable part of my time to the talc project. The time spent was also credited to my Cal Tech record as "research in absentia" (credit Ian, too, for that courtesy!). When, in 1949, I was ready to write my thesis, he assigned me to a 3-month duty on the Cal Tech campus with pay. Among my other assignments in the San Francisco office was an investigation of California foundry sands, pursued in collaboration with Heinrich Ries of Cornell University a good friend of Olaf's and a lover of fine cigars.

In 1948, during one of my talc-related excursions in southern California, I attended a field trip led by Bob Wallace who was completing his doctorate also at Cal Tech. Bob had finished mapping a 16-mile segment of the San Andreas fault zone. It was adjacent to the Pearland Quadrangle which had been mapped by Levi Noble who lived in nearby Valyermo. Noble was also there, and between stops, I told him of my thesis research. He responded with much interest and became the fifth of my treasured mentors. When we met, Noble had retired from the USGS. He had received the Interior Department's Distinguished Service Gold Medal for his pioneer investigations of the Precambrian and Cambrian formations of the

Grand Canyon, for preparing the first detailed geologic map of the Grand Canyon, for detecting evidence for major strike-slip movement on the San Andreas fault and, most importantly, for his work in Death Valley.

Meanwhile, Olaf occasionally sought my opinions on matters of policy. Once, when I differed with him, his reaction was so strong that I suffered chest pains for several days. Soon thereafter, he promoted me and placed me in charge of the Los Angeles office. At first I felt that he simply wanted me out of sight and still suspect that that was a factor. But we both tacitly knew that, in addition to increasing my responsibility, he was placing me much closer geographically to Death Valley and to Dick, Ian, and now Levi. For each of these advantages, I continue to be grateful to Olaf, an assertive, but most kindly, man.

The 10 years in the Los Angeles office, beginning in 1950, still resonate pleasantly in my memory. The office was staffed by 6 to 8 geologists and two secretaries. Dick Jahns was our honorary member. We were an especially congenial group thanks largely to Bea Reynolds our secretary (Fig. 5). She set the tone with her natural cheerfulness, her people-orientation and her dedication to her job. Whatever form I needed to sign appeared on my desk at exactly the right time.

I also recall a phone call from Don Carlisle of the UCLA geology faculty. He had learned that there was an opening in our office and believed that he had exactly the right person to fill it—one of his students named Bennie Troxel. His evaluation was dead center. Bennie took the exam, was hired, spent a brief time in the San Francisco office, and joined us in November 1950.

Fig. 5. Secretary Bea Reynolds and my colleague Cliff Gray, hard at work in the Los Angeles office of the California Division of Mines and Geology, ca. 1954.

At least several pages would be needed to even begin to tell of the activities and interactions of the Los Angeles staff. But I can again say "thank you" to George Cleveland, Tom Gay, Cliff Gray, Ed Kiessling, Paul Morton, Dick Saul, Dick Stewart, Hal Weber, and of course Bennie Troxel for the pleasure of working with you. Sadly, George, Tom, Cliff, Dick Saul and Dick Stewart are no longer with us.

I also must mention the informal group that we called the "Geolly Fellows". Nearly every month, we spent an evening at someone's home hearing about and discussing ongoing geologic research, mostly in southern California. Frequent participants, in addition to the Los Angeles staff, included Clarence Allen, John Crowell (John still has "Geolly" on his license plate), Tom Dibblee, Mason Hill, Dick Jahns, Thane McCulloh, Foster Hewett, Mike Murphy, Lee Silver, Jerry Winterer, Phil King, and A. O. Woodford.

Soon after my arrival in Los Angeles, I learned that Dick Jahns had persuaded Olaf to publish the tome entitled" Geology of Southern California" to coincide with the 1954 national meeting of the Geological Society of America in Los Angeles. It eventually consisted of 92 articles and map sheets and involved about 125 authors. It appeared as Bulletin 170 of the California Division of Mines, went through two editions, and is now a collector's item. It includes my first solo effort at mapping in the Death Valley region—a map of the geology of the Alexander Hills at the southern end of the Nopah Range. Most of the Los Angeles staff of the Division of Mines were also participants. Bennie, in particular, contributed a map sheet, authored a field guide and coauthored two other field guides (at which point Olaf was heard to exclaim "Bennie scores again!").

As an Associate Editor of this volume, and also as an acquaintance of Levi Noble (Fig. 6), I was persuaded (for me Dick had brought new meaning to the term "persuasion") to obtain from Levi two very important unpublished geologic maps. One was Levi's map of a 60-mile segment of the San Andreas fault and related features along the north side of the San Gabriel Mountains. The other was his reconnaissance map of the Death Valley region, which he had assembled mainly in the 1930s. He readily and enthusiastically agreed to contribute the first, but agreed to the latter only if I would help him fill in the remaining blank areas. Levi, characteristically, was

Fig. 6. Levi Noble and a young Lauren Wright soon after the beginning of our association, ca.1951.

drawn to the areas of greatest geologic complexity, and almost ignored the areas where the geology seemed simple. This tendency amused Levi's USGS colleague, Foster Hewett, who once observed that Levi's field maps had a polka dot appearance and contained so many notations that the maps seemed to be shrouded in a gray haze. I found, however, that they were meticulously plotted and annotated. I spent 3 rewarding weeks in the field with Levi, filling in those blank areas, while Olaf placed his drafting department on hold. This arrangement worked. For me the time with Levi was a chance-of-a-life-time experience which I continue to relive.

The Troxel–Wright collaboration began when we prepared a field guide to the Death Valley region for Bulletin 170. Bennie met Levi in Valyermo when we were engaged in this project. Afterward, with Olaf's approval and Levi's encouragement, we began to spend 3-day weekends, and even vacations, mapping in the Death Valley region. It became routine for us to devote Monday through Thursday morning to office responsibilities, to visit the Nobles in Valyermo en route late Thursday afternoon, and to arrive in

Shoshone before 10 p.m. when Maury Sorrells, Charlie Brown's son-in-law, turned off the diesel powered generator and the town went dark. We would drive back on Sunday evening after a full day in the field. As often as possible, Levi would come along. His wife, Dorothy, would occasionally accompany him in their Jaguar sedan.

Levi encouraged us to devote as much of our time as possible to a reexamination of the Amargosa chaos— an undertaking I had wished for since Dr. Buwalda's seminar 12 years earlier (serendipity again at work?). As most readers of this volume already know, the chaos is arguably the most puzzling geologic feature of the Death Valley terrane. In his 1941 paper, Levi Noble had divided the Amargosa chaos into the Virgin Spring, Calico, and Jubilee phases. He also suggested that the chaos was part of a regional thrust fault that possibly extended from the Panamint Mountains to the Kingston Range. By the early 1950s, however, he had come to view only the Virgin Spring phase as a definable structural feature, and preferred to interpret it as confined to the southern part of the Black Mountains. To clarify a continuing confusion over Levi's later use of the term Bennie and I have recently proposed the following definition: "A structural term for a mosaic of fault-bounded, commonly gigantic blocks derived from a stratigraphic succession, and arranged in proper stratigraphic order, but occupying only a small fraction of the thickness of the original succession".

For his 1941 paper, Levi had mapped the chaos, as a single unit, on a two-time enlargement of a part of the Avawatz Mountains one-degree topographic map. In the mid-1950s, we could map the chaos in much greater detail, on a four-times enlargement the newly available Confidence Hills 15-min quadrangle. I had become familiar with the formations of the chaos by mapping them in their original thickness where exposed in the southern part of the Nopah Range and adjacent Alexander Hills.

So on many of our weekends, the three of us entered the Virgin Spring area, the type locality of the Virgin Spring chaos, where Bennie and I would follow predetermined traverses, while Levi, in his seventies, stayed close to our vehicle taking notes on critical features. To have been able to "stand on his shoulders" in this manner I continue to be grateful.

Mainly because it was a part-time project, the mapping of the chaos progressed slowly, continuing well after Levi's death in 1964. The map and accompanying text was finally published in 1984. In the text, we interpreted the attenuation exhibited in the chaos as produced by extreme extension of an originally much thicker succession of Proterozoic and Paleozoic rocks. We feel certain that Levi would have agreed. This work also permitted us, under Levi's guidance, to piece together much of the volcanic stratigraphy of the southern Black Mountains.

Thanks largely to our close association with Levi and the attraction of the Amargosa chaos, in the 1950s, Bennie and I became well acquainted with some of the foremost American geologists of that time—most of them, with Levi, in the inner circle of the Geological Survey. These included Charlie Anderson, Charles Denny, Henry Ferguson, Jim Gilluly, Foster Hewett, Charles Hunt, Phil King, Chester Longwell, and Bill Pecora. How fortunate we were!

Meanwhile, Olaf allowed Bennie and me to extend our mapping to other parts of the Death Valley region. Bennie wished to record features in the southern Death Valley area and along the adjacent part of the Garlock fault zone. I chose to map in the area that bordered his on the north including the Ibex Hills and Sheephead Peak. These activities also were pursued with Levi's encouragement.

In 1958, Olaf retired as State Geologist and Chief of the California Division of Mines and Geology. So the search had begun for his successor. A number of us were aware that Ian Campbell might be looking for new challenges and urged him to make himself available. He obliged and the Division was favored by another distinguished leader. For a brief time, he again helpfully supervised my geologic activities. For the rest of their lives, he and his wife Kitty lived in San Francisco, their high-rise apartment on Russian Hill providing a view of San Francisco Bay from the Golden Gate to the Bay Bridge.

In 1960, Dick Jahns, at the urging of his friends Elbert Osborn, Frank Tuttle, and Wayne Burnham, joined them on the faculty of the College of Mineral Industries at Penn State, first as Head of the Division of Earth Sciences and, a year later, as Dean of the College. Before leaving southern California, however, he said he needed my help in simplifying and harmonizing the geologically oriented part of the college. Persuaded again, I visited Penn State early in 1961 on a get-acquainted trip. (The first day of my visit proved to be the most important ever for me when Dick introduced me to his administrative assistant and my future wife, Myrtle Davies). Soon afterward, I received an offer to chair a newly formed Department of Geology and Geophysics which, along with the Department of Mineralogy and Geochemistry, was to cover the earth sciences at Penn State. I accepted, but only on what today would seem an outrageous condition of employment—that I spend several weeks in the middle of the school year in Death Valley.

When I arrived at Penn State, in the fall of 1961, Dick, with the best of intentions, presented me with two season tickets to the up-coming Penn State football games and suggested that I take Myrtle to each. This was one of Dick's few miscues as it embarrassed both Myrtle and me. We were a while in finding ourselves, but find ourselves we did. On the fourth of the following May, we were married in the home of Dick and Frances in the presence of a few close friends. Thanks to Myrtle, the next 27 years were the best of my life.

Here, I could write volumes on my years at Penn State and what it was like to be an academic neophyte in the presence of an eminent faculty in both departments, including P. D. Krynine, John Griffiths, Frank Tuttle, Elbert Osborn and Frank Swartz, plus a younger group of energetic and talented teachers and scientists. But everyone was helpful. Myrtle was exceptionally so as she knew the university well, having worked in the Departments of Physics, Mathematics, and Meteorology before being hand-picked by Dick to help him in his administrative responsibilities (he called her "Sis"). Predictably, she resigned her position on the day of our marriage to become a full-time wife, mother and helpmate.

Then, too, our department was very fortunate in the presence of a dedicated and caring secretary. Dorothy Duck and Myrtle had been close friends well before my arrival. Like Myrtle, she was well versed in University procedures and related helpfully and cheerfully to those around her. In many ways, she controlled the ambiance of our department.

My sessions in Death Valley became routine, thanks to Dick's support, the understanding and cooperation of my new colleagues, and the fact that, during these intervals, I supervised the field work of graduate students who could not have worked in

Death Valley in the summer. John Griffiths, who chaired the other department, also accepted my absences uncomplainingly.

Myrtle and our son, Tony, accompanied me to Shoshone and the Death Valley region whenever possible. They both became attached to this landscape and friends with its inhabitants. (Tony still joins me in Shoshone when he can). We also spent memorable evenings with Ian and Kitty, in their Russian Hill apartment, watching the fog roll across San Francisco Bay and the city illuminate.

When, in 1964, Dick announced that he was leaving Penn State to become Dean of the School of Earth Sciences at Stanford, Myrtle and I were devastated. But, by keeping in close touch with the Jahns' by phone and as guests in their Palo Alto home, we recovered. When, in 1983, Dick died in the aftermath of heart surgery, we were again overwhelmed, especially so when Frances asked that I write his memorial for the Geological Society. How does one verbalize the indebtedness of an entire profession to a person of his character and accomplishments?

In the interval from 1965 to 1986, eighteen Penn State graduate students completed field-based theses on topics related to the geologic evolution of the Death Valley region. They effectively investigated the depositional environments of various Proterozoic, Cambrian and Cenozoic formations, igneous events of the Central Death Valley plutonic–volcanic field, the genesis of the Mesozoic plutons, and aspects of Cenozoic tectonism in the Death Valley region. Of these students, nine became college or university teachers, two were recruited to the petroleum industry, two became members of the U.S. Geological Survey and five have been geologically employed in other ways. Most of them have returned, from time to time, to extend their research and/or to participate in field trips; several have remained actively involved in Death Valley projects.

The supervision of their thesis research was shared in various ways by my colleagues Gene Williams and Charlie Thornton. They also have returned to Death Valley through the years and Gene is still of immeasurable help in our continuing work on the Proterozoic formations. So I join the two of them in thanking former students Syed Almashure, Ibrahim Cemen, Ed DeWitt, Paul Diehl, Rick Haefner, Izat Heydari, Alfred Hu, Ihor Kunasz, Bob Mazurak, Matt McMackin, Randy Maud, Gary Novak, Jim Otton, Tony Prave, Mike Roberts, Bob Scott, Don Siegel and John Stamm for their contributions and friendship.

Marli Miller was not a Penn State graduate student. But I think of her as an honorary one, having directed her attention to the Badwater turtleback even before she enrolled at the University of Washington where it became the subject of her doctoral thesis. My association with her and her former adviser, Darrel Cowan, is continuing and rewarding.

I also remember Thanksgiving Eve, 1966, when Preston Cloud arrived in Shoshone unannounced. The Troxel's and Wright's had already made Thanksgiving plans, but he persuaded (that word again!) Bennie and me to instead help him begin a search for prokaryotic life forms in the strata of the Proterozoic Pahrump Group. Pres, as he preferred to be called, has since been described by William Schopf, as "probably the greatest biogeosynthesist the United States ever produced" and also as wiry, lean, and feisty. In addition, we found him to be both gracious and appreciative. Thus began another close association which lasted until his death in 1991.

The prokaryote microfossils were found in black chert in the upper part of the Beck Spring Dolomite. They were described by his graduate student Gerry Licari who named one of them after Bennie and another after me. Pres, commonly accompanied by his wife Jan, returned to the Death Valley area many times, his interest extended to include the variously shaped stromatolites and tubular structures formed in the microbial mats of the Proterozoic carbonate platforms. He commonly brought visitors, including M. R. Walter, Misha Semikhatov, Hans Hofmann and Don Elston. From him and also from his visitors, Bennie and I learned much about Proterozoic environments. When his book "Cosmos, Earth, and Man" was published in 1980, he sent Myrtle and me each a copy, each separately and thoughtfully inscribed. He was, indeed, a gentleman.

Before my departure for Penn State in 1961, Bennie and I agreed to keep our association intact by jointly mapping the geology of the central and northern parts of the Funeral Mountains whenever we could schedule time together. (The scheduling was aided by the fact that he had succeeded me as head of the Los Angeles office). We had been aware of the geological complexity of this area since I accompanied Levi on the reconnaissance for the 1954 map. We began in 1962 and, by working several days at a time, finished in the mid-1980s aware that we had documented the existence of a textbook example of a metamorphic core complex.

We could not anticipate, however, that in 1987, 2 years after my retirement, the U.S. Congress was to designate nearby Yucca Mountain, Nevada, as a potential site for the storage of high-level nuclear waste and authorized a major feasibility program. Soon thereafter, Mike Carr of the USGS arranged to have the Survey publish our map as part of that program. It appeared in 1993. Without Mike's support, the map might well have remained unpublished.

In the mid-80s, too, I was encouraged by Ren Thompson, of the USGS, to compile a tectonic-geologic map of most of the Death Valley region also as part of the Yucca Mountain program. For this and other contributions of interest to the Survey, Ren has arranged for contractual support for Bennie and me. The regional map remains a continuing project, but through Ren's role as coordinator and through the computer graphic skills of Jeremy Workman, a version of this map was incorporated on a still larger map to be used in ground water modeling related to the Yucca Mountain program.

Still another undertaking made possible by the support and understanding of a large group of earth scientists was initiated by Bennie and Jim Calzia when they organized a session entitled "Tertiary basins and volcanism of the Death Valley region" for the 1994 meeting of the Cordilleran Section of the Geological Society of America. The papers presented in this session marked the first stage in the organization of GSA Special Paper 333, entitled "Cenozoic

basins of the Death Valley region". Thirty-five authors contributed to sixteen chapters. It was published in 1999 and sold out in 14 months. So "thank you again" to everyone involved in this project.

Finally, Bennie joins me in recognizing our much-extended geological family who, with on-going or recent research interests in the Death Valley region, stay in or pass through Shoshone. These include Tom B. Anderson, Tom H. Anderson, Diana Anderson, Stan Awramik, Sue Beard, Terry Blair, Rich Blakely, Ibrahim Cemen, John Cooper, Frank Corsetti, John Crowell, Darrel Cowan, Nick Christie-Blick, Bill Chavez, Rachael Craig, Greg Davis, Ed DeWitt, Damon DeYoung, Dave Diodato, Mike Ellis, Rolfe Erickson, Cris Fedo, Bob Fleck, Cris Fridrich, Nick Hayman, Rich Hereford, Izat Heydari, Paul Hoffman, Angela Jayko, Fred Johnson, Martin Keller, Martin Kennedy, Joe Kirschvink, Jeff and Dianne Knott, Chuck Langston. Dan Larsen, Ollie Lehnert, John Louie, Mike Machette, Matt McMackin, Marli Miller, Cris Menges, Roger Morrison, Dave Mrofka, Steve Nelson, Dick Parizek, Terry Pavlis, Ryan Petterson, Tony Prave, Steve Rowland, Eban Rose, Andre Sarna-Wojcicki, Jim Sears, Laura Serpa, Bill Schoenborn, Janet Slate, Burt Slemmons, Cathy Summa, Karl Thompson, Ren Thompson, Charlie Thornton, Dave Topping, Dave and Lynn Wagner, Steve Wells, Brian Wernicke, Gene Williams, and, of course, Jim Calzia. We have observed together, dined together, laughed together and discussed the unsolved geological problems of the Death Valley region. I wish for space enough to describe in detail each of these associations.

And as I reflect on all of my professional experiences – beginning with the field class with Ben Page – and look forward to still others, I return to Bennie. It seems that we have accomplished in a partnership what neither could have done alone.

Available online at www.sciencedirect.com

Earth-Science Reviews 73 (2005) 13–16

www.elsevier.com/locate/earscirev

My geological career in Death Valley

Bennie W. Troxel

I owe many people a debt of gratitude for contributing to my career. I cannot acknowledge all of them in this brief account, partly from a faulty memory but mostly because there are too many to list herein. To those who I failed to identify, I ask your forgiveness.

Robert E. Stevenson created my interest in geology and started my education in the science. I am especially grateful to him. The faculty of the geology department at UCLA in the late 1940s provided me

with a geological education. Lauren Wright was especially fundamental in making me become a scientist and improving my ability to write scientific articles.

1. The beginning

Following service in the U.S. Army Air Corps during WWII, I returned to work at Northrop Aircraft in Hawthorne, California. I soon realized that I needed a college education so I enrolled as a student at Compton Junior College in 1947. While there I took college prep courses that I avoided in high school and went to UCLA during summers. For a college science course requirement, I took an introductory physical geology course. The instructor, Robert E. Stevenson, was such an interesting and skillful teacher that I became extremely curious about science. A course in mineralogy from him and a course in historical geology from Gordon Oakeshott convinced me that I wanted to become a geologist.

I then went to UCLA and was fortunate to get my Master's degree in geology from a superb faculty that included W.C. Putnam, J. Murdock, Don Carlisle, John Crowell, Clem Nelson, Cordell Durrell, U.S. Grant IV, Jim Gilluly, "Parky" Parkinson and others. I completed my course work in 1952 and at the recommendation of Don Carlisle, I went to the Los Angeles office of the California Division of Mines where I met and was interviewed by Lauren Wright for a position in the Division.

0012-8252/$ - see front matter © 2005 Published by Elsevier B.V.
doi:10.1016/j.earscirev.2005.07.009

Olaf Jenkins, Chief of Division, said to hire me and after a few weeks working at the laboratory and at the information desk in the San Francisco headquarters office of the Division I was transferred to the Los Angeles office. While in San Francisco, I learned much about California geology from Gordon Oakeshott and Charles Chesterman.

In Los Angeles, I soon became involved in several projects which resulted in publications on mineral commodities (wollastonite, uranium, thorium, and abrasive minerals) and several geologic road guides for Bulletin 170. Lauren also encouraged me to publish a map of my thesis area (Shadow Mountains near Victorville, CA) in Bulletin 170. Paul Morton and I next undertook a 2-year investigation of the mineral resources of Kern County. Several other staff members of the Division contributed to the comprehensive report on Kern County. All of the material published from these studies was edited by Lauren and was improved by his input.

Soon after I started working in the Los Angeles office, Lauren invited me to accompany him on a trip to the Death Valley region. At that time Lauren was studying talc deposits for the Division and for a PhD degree at California Institute of Technology. On that trip he suggested that I consider making a geologic map of the north half of the Avawatz Pass 15-min quadrangle. I liked the idea and thus began a long and extremely satisfactory career of working with Lauren in Death Valley.

2. The early Death Valley events

Almost immediately after I started fieldwork on the Avawatz Pass quadrangle, Lauren and I began the preparation of a geologic guide through the Mojave Desert and Death Valley. I also met Levi Noble during this time, because of our interests in mapping in nearby areas Levi invited Lauren and I to map with him the complicated geology of the Virgin Spring area, the area of extreme importance to Levi. Subsequently we divided our long and frequent winter weekend trips between mapping our individual quadrangles and mapping with Levi. We usually visited Levi and his wife Dorothy at Valyermo on our way to Shoshone. Levi rode with us on many weekends, but occasionally he and

Dorothy would drive to Shoshone in their walnut paneled 4-door Jaguar. It was a real treat to drive their Jaguar into Virgin Spring Canyon. In fact, it was a better field car than the 1954 Chevrolet Sedan that was our only field car at the time. The Division later acquired a Chevrolet panel truck that was converted to 4-wheel drive and we were able to traverse more terrain with it.

In the 1950s we were provided with a vehicle and $16.50 per diem for expenses. The motel room in Shoshone for two was $6.50 per day. We cooked breakfast in the motel, packed a lunch, than ate dinner at the café in Shoshone. We plotted the geology with 7 to 9 H pencils then using crow-quill pens "inked in" our maps after dinner. Evening events were terminated at 10 p.m. when the electric generator for Shoshone was turned off. We mapped independently in the Virgin Spring area and discussed what we saw as we returned to Shoshone. Levi listened carefully and early next morning he would make notes about our observations. I commonly made wild interpretations of our work and Lauren was kind enough not to laugh at them. Our detailed mapping proceeded very slowly. We each rarely mapped more than 1/10 of a square mile in a day.

During this time a group of Levi's peers would make an annual fieldtrip to review our progress. The group included Henry Ferguson, Jim Gilluly, Chester Longwell, Bill Pecora, Charles Anderson, Charles Denny, Charles Hunt, and others. If the group was too large, Levi would make an itinerary for Lauren and I; Levi would stay home. Before Levi died he had agreed with us that his Amargosa Chaos was formed by extensional activity rather than compression. Unfortunately, we never published a joint paper with him to document this fact.

Levi also had an interest in the Funeral Mountains and Lauren and I decided to make a geologic map of that area. Over a period of many years we completed the 1/48,000 scale map of the Chloride Cliff and Big Dune quadrangles. We are grateful to Mike Carr for getting it published. It was in the Funeral Mountains that we noted that west-dipping normal faults flattened at depth and joined on a common plane. We then realized that a similar fault pattern existed in the Virgin spring area.

During this time Lauren and I continued to map 7 1/2-min quadrangles independently. His work included parts of the Shoshone and Tecopa quadrangles and mine included parts of the Avawatz Pass and Leach Lake quadrangles. Later we jointly started mapping the Greenwater Range.

3. Later events and people

As interest in Death Valley geology began to grow many other people began studying various problems there. I can list only a few of them herein. The early group included Charles Hunt and Don Mabay, Don Curry, Harold Drewes, Charles Denny, Chet Wrucke, Jack Stewart, Mitchell Reynolds and Jim McAllister.

Lauren encouraged students and faculty from Pennsylvania State University to study various problems. Eugene Williams and Charles Thornton supervised student activities in Death Valley, as did Lauren. Somewhat later I began to attract students from the University of California at Davis. The Institute of Technology initiated studies in the Paramint Mountains. Faculty and students from the Massachusetts Institute of Technology and Harvard University also began working on problems in Death Valley. Laura Serpa and Terry Pavlis encouraged many students from the University of New Orleans to study a wide variety of problems in the region and deserve special mention for initiating, with Susan Sorrell, the SHEAR facility in Shoshone. Many other University faculty and students commended students in the region and I apologize for not listing them here.

Lauren has identified many students that we have worked with but I wish to identify a few others for whom I greatly appreciated having worked with. They are Roland Brady, Paul Butler, Pam Burnley, Marty Giaramita, Julie Miller, Dan Graff, Rick Kramer, Sue Hall and Mitch Casteel. There were others whose names escape me.

4. Supplemental information

I was fortunate to work in surrounding areas, usually with others and thus gain regional knowledge that was useful in understanding Death Valley geology. These areas include the Spring Mountains where I co-taught summer field geology for 2 years at the University of Nevada at Las Vegas; reconnaissance geology for the 1:250,000 Trona sheet of the California State geologic map (Cliff Gray and I mapped a large portion of this sheet); in the Fort Irwin area, we upgraded G.I. Smith's map with Roland Brady, Matt McMackin, Terry Paulis and Laura Serpa; reconnaissance mapping in the Arawatz Mountains with Dick Jahns and Lauren; field review with others of Don Kupfer's seminal work in the Silverian Hills; field trips near Las Vegas with Chester Longwell; reconnaissance geology of the Last Chance Range with J.H. Stewart; and mapping the State Range with G.I. Smith, Cliff Gray and Roland Von Huene.

5. Important events and speculations

Almost simultaneously Lauren and I recognized what we then called a basin facies of the Noonday Dolomite. My work in the Saddle Peak Hills revealed a clastic unit atop lower Noonday and beneath upper Noonday Dolomite. Lauren's work in the Ibex Hills revealed eroded lower Noonday Dolomite incorporated as basal clasts in a clastic unit overlain by upper Noonday. We later identified this as the Ibex Formation.

Our joint work in the Funeral Mountains/Virgin Spring area and individual work in nearby areas eventually led to the recognition of extension tectonics in Death Valley. Fundamental to this recognition was the documentation of the relationship of listric normal faults and an underlying planer surface now known as a detachment fault. The flattening of normal faults at depth and convergence of them were plotted by me in the Saddle Peak Hills in the late 1950s but their significance was not recognized until many years later. The map of the Saddle Peak Hills is yet to be published (2003).

An anomalous northwest-trending tertiary dike swarm in the Saddle Peak Hills and smaller ranges on each side was an enigma for many years. Our work in the Kingston Range and especially the work of Jim Calzia led to the recognition that extension in that area was to the southwest before 12.4 mya and was not overprinted by the effects of later extension to the northwest so commonly documented in most of the Death Valley region. I am now convinced that the dike swarm in the Saddle Peak Hills was emplaced during pre-12 mya extension to the southwest. Northwest-trending listric normal faults in the southern Napah Range are further documentation of extension to the southwest. A Tertiary basin on the southwest edge of the Montgomery Mountains at the north end of Pahrump Valley may record the site of a pull-away basin due to southwest extension. Likewise a similar basin may have developed on the northeast side of a northwest-trending small ridge in the northeast corner of the 15-min Avawatz Pass quadrangle. I have informally called this ridge "Fatzinger Ridge" for many years.

While mapping with Lauren in the Greenwater Range, I became intrigued with the relations of the Shoshone Volcanics and the underlying plutonic rocks of nearly the same age. Very recently I convinced Rick Haefner that we should study the area in more detail. Unfortunately, Rick died soon thereafter. Basically, the problem is one in which volcanic and lesser sedimentary rocks were deposited over a small pluton that continued to rise as it crystallized and it subsequently deformed rocks of nearly the same age and younger volcanic flows. The problem is unresolved, important, and warrants detailed study. The area is at the southern end of the Greenwater Range and is informally called "Chocolate Sunday Mountain".

Early in my career I became intrigued with the Kingston Peak Formation, named and first described by another of my heroes, Foster Hewett. Many have speculated upon its origin and most have become convinced that it is a glacial deposit. Their evidence is the presence of occasional (rare) striated stones (usually in non-layered strata) and even less common faceted stores. I do not doubt that these features are derived from a distant glacial terrain but the layers in which they are contained were deposited in submarine fans—so-called drop stones may well be carried into the area of deposition by icebergs but may have been dropped many miles from the point of origin of the icebergs.

These problems and others will continue to be a matter of debate and speculation. In addition the amount of northwest extension will also be debated. The fact that wide deficiencies of opinion exist about many facets of geology in the Death Valley region guarantees that the area will continue to be intensively studied.

I feel extremely fortunate to have become involved in the geology of such a critical area. The abundance of evidence available in such well-exposed outcrops is overwhelming. As I have said before "In Death Valley one is apt to be confused by the huge abundance of data."

In conclusion I want to list those who have been so fundamental to my accomplishments in Death Valley. They are Lauren Wright, Levi Noble, Chester Longwell, Foster Hewett, Jim Calzia, Matt McMackin, Fred Johnson, Henry Ferguson, Mitchell Reynolds, Laura Serpa, Terry Pavlis, Olaf Jenkins, Dick Jahns, Gene Williams, John Crowell, Don Carlisle, Jim Gilluly, Robert Stevenson, Brian Wernicke, Martin Miller, Roland Brown, Paul Butler, and many others. I am also grateful to the many people who have been on field trips that I have led. Their useful questions have helped improve later trips.

Lastly, I am forever grateful to my beloved wife Betty and our two children who has been extremely tolerant of my many absences from home while working in Death Valley.

Available online at www.sciencedirect.com

Earth-Science Reviews 73 (2005) 17–30

www.elsevier.com/locate/earscirev

Geological landscapes of the Death Valley region[☆]

Marli Bryant Miller

Department of Geological Sciences, University of Oregon, Eugene, OR 97403-1272, United States

Keywords: Death Valley; landforms; history; stratigraphy; photographs

1. Introduction

Death Valley's location and climate make it one of the most dramatic geological landscapes on Earth. The region lies near the western edge of cratonic North America, and so it contains a record of plate boundary effects that date back to the Proterozoic. These effects include rifting and the development of the passive margin during the Late Proterozoic, crustal shortening and Sierra Nevada magmatism largely during the Mesozoic, and crustal extension and magmatism during the Late Cenozoic. Because crustal extension continues today, the region also showcases spectacular landforms that relate to active mountain-building.

When combined with this geology, Death Valley's harsh climate makes it unique. As the hottest and

driest area in North America, both its geological record and landforms are unusually visible to geologists and non-geologists alike. It is for this reason that the national park overflows with geology field trips during the spring months, and many visitors gain a deeper understanding of Earth processes.

The authors of this volume represent several hundred years of collective experience working on the geology of the Death Valley region. Lauren Wright and Bennie Troxel alone have logged more than one hundred years. It is easy to see why so many geologists keep coming back: traveling through this landscape is like walking through a beautifully illustrated geology textbook, only better. The following photographs attempt to portray some pages of that textbook, but like all photographs, they fall well short of an actual visit.

Each photograph is keyed to a number on the accompanying geologic map (Miller and Wright, 2004) (Fig. 1). An arrow adjacent to a number indicates the direction of view. Those photographs that portray crustal extension or modern landforms appear in the first section. Those that illustrate the older geologic history appear in the second section.

[☆] All illustrations (figures and photographs) © Marli Bryant Miller.

E-mail address: millerm@uoregon.edu.

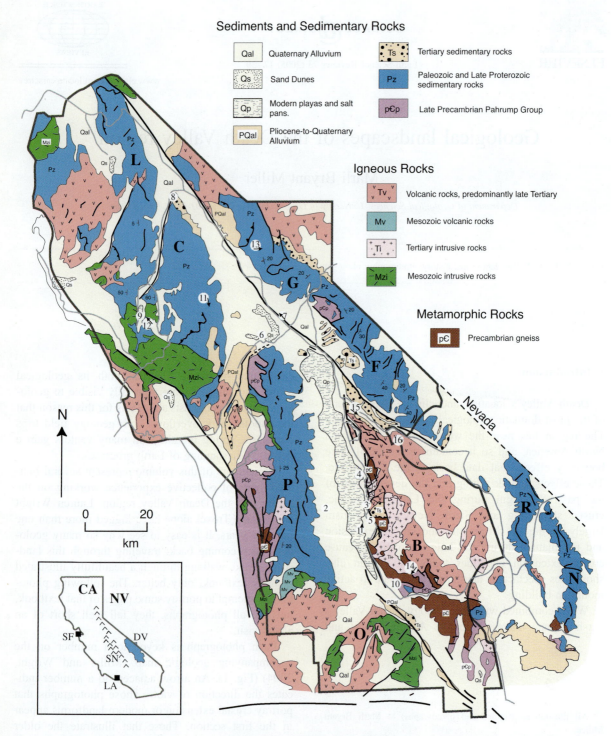

Fig. 1. Geologic map of Death Valley area, compiled by Miller and Wright (2004). Numbers and arrows depict corresponding photo location and direction of view. Abbreviations for mountain ranges are as follows: B: Black Mountains; C: Cottonwood Mountains; F: Funeral Mountains; G: Grapevine Mountains; L: Last Chance Range; N: Nopah Range; O: Owlshead Mountains; P: Panamint Mountains; R: Resting Spring Range. Abbreviations on location map of California are as follows: CA: California; DV: Death Valley National Park; LA: Los Angeles; NV: Nevada; SF: San Francisco; SN: Sierra Nevada Mountains.

2. Modern landforms and crustal extension

Photo 1. View northward along west side of Black Mountains. The Black Mountains consist predominantly of late Tertiary plutonic, volcanic, and sedimentary rock and a basement of Late Proterozoic metamorphic rock. They rise above the floor of Death Valley along the Black Mountains fault zone. The alluvial fans that spill out of each canyon tend to be relatively small because the valley floor tilts gently eastward. The Grapevine Mountains, in the far background, are approximately 80 km away.

Photo 2.

Photo 3.

Photo 2. Bajada at Hanaupah Canyon (see preceding page). In contrast to the small, well-shaped fans on the east side of Death Valley, those on the west side coalesce into a gigantic bajada at the mountain front. Jayko (2005) characterizes rates of erosion for the Panamint Mountains. A fault scarp cuts diagonally across the bajada in the bottom half of the photo. Vegetation marks locations of springs at the lowest reaches of the bajada.

Photo 3. Black Mountains fault zone (see preceding page). The Black Mountains fault zone displays all the characteristics of active faults, including an extremely abrupt and linear range front, faceted spurs, and wineglass canyons.

Photo 4. Salt pan at Badwater, view southward. Much of the valley floor of Death Valley is covered by salt that is broken into large polygons. This view to the south also shows the abrupt western edge of the Black Mountains. Messina et al. (2005) characterize polygon morphology in northern Panamint and Eureka valleys.

Photo 5. Copper Canyon turtleback. There are three turtlebacks in the Black Mountains, named because their broadly convex-upwards geometries resemble turtle shells (Curry, 1938). Each turtleback exposes a core of metamorphosed sedimentary and basement rock separated from an upper plate of sedimentary or volcanic rock by a fault zone. This photograph shows green-colored metamorphic rock faulted against red- and tan-colored sedimentary rock just south of Copper Canyon. Miller and Pavlis (2005) describe the structural evolution of the three turtlebacks and their implications for crustal extension in the region. Cemen et al. (2005) describe similar features in Turkey.

Photo 6. Mesquite Flat Sand Dunes (see next page). The Death Valley region hosts numerous sand dune fields, each of which exists in a setting that is partially protected from the wind. These dunes on Mesquite Flat lie at the northern foot of Tucki Mountain, off to the side from the main part of Death Valley. They contain both crescentic and star dunes.

Photo 7. Kit Fox Hills and Northern Death Valley fault zone (see next page). The Kit Fox Hills consist of folded and faulted Late Tertiary and Quaternary sediments. The hills end abruptly at the northern Death Valley fault zone, shown here cutting diagonally across the photo. This view to the northwest also shows much of northern Death Valley, with the Cottonwood Mountains on the left and the Grapevine Mountains on right.

Photo 6.

Photo 7.

Photo 8.

Photo 9.

Photo 8. Little Hebe Craters (see preceding page). This cluster of small phreatic explosion craters exists on the shoulder of the much larger Ubehebe Crater in northern Death Valley. A total of 13 explosions produced the entire field (Crowe and Fisher, 1973).

Photo 9. Sliding rock at Racetrack Playa (see preceding page). At an elevation of 4000 ft and a long stretch of valley to funnel the wind, Racetrack Playa hosts rocks that occasionally slide across the playa surface. No one has actually seen them move, so the actual cause has been long debated. The rock in this photo is approximately 30 cm across.

3 . Geologic history

Photo 10. Diabase sill and basement unconformity, view southward. The Crystal Spring Formation, which is the lower part of the Pahrump Group, rests depositionally on basement rocks. It is intruded by green-colored diabase sills that yielded an age of 1.08 Ga (Heaman and Grotzinger, 1992). Metamorphic reactions between the diabase and dolomite of the Crystal Spring Formation produced large deposits of talc, which have been mined throughout the Death Valley region. To the south, this photograph shows the western edge of the southern Black Mountains, and in the distant background, the Avawatz Mountains.

Photo 11. Cliffs of Paleozoic rock, Cottonwood Mountains. With the exception of the Black Mountains, every range in the Death Valley region is underlain by Late Proterozoic through Paleozoic marine sedimentary rocks. These rocks attain thicknesses of 10 km and were deposited on a long-lived passive margin, similar in many ways to the present-day eastern seaboard of North America.

Photo 12. Hunter Mountain Batholith and Paleozoic rock at Racetrack Playa (see next page). The Hunter Mountain Batholith, part of which is shown here as the dark-colored rock, is a Middle Jurassic pluton that reflects activity during early stages of the Sierran magmatic arc. Here it intrudes folded and thrust faulted Paleozoic rock (tan) to establish a pre-Middle Jurassic age for those structures. Stevens and Stone (2005) review these compressional structures. Saline Valley lies in the background; the Sierra Nevada lies in the far background, about 70 km away.

Photo 13. Folded rock in Grapevine Mountains (see next page). The Grapevine Mountains, mapped in detail by Reynolds (1969) and Niemi (2002), contain thick sequences of Paleozoic rock that was deformed into large-scale folds and faults during crustal shortening. Overlying these rocks are less deformed Tertiary sedimentary and volcanic rock.

Photo 12.

Photo 13.

Photo 14. Smith Mountain, Black Mountains. Smith Mountain, on the right side of this photograph, is underlain by the Smith Mountain pluton, one of several mid-Miocene plutons that intrude the Black Mountains. It consists largely of granite and locally contains distinctive rapakivi textures, described in detail by Calzia and Rämö (2005). Wingate Wash, a site of transtensional deformation described by Luckow et al. (2005) lies in the background.

Photo 15. Furnace Creek Formation at Zabriskie Point (see next page). The Furnace Creek Formation consists primarily of alluvial fan and playa deposits, shown here in the foreground and middle ground respectively. It was deposited from about 6–4 Ma, in the Furnace Creek Basin, an extensional basin that formed along the Furnace Creek fault zone prior to opening of modern Death Valley.

Photo 16. Angular unconformity at Ryan Mesa (see next page). This photograph shows undeformed 4 Ma basalt overlying Artist Drive Formation (left) faulted against tilted Furnace Creek Formation (right). This unconformity marks the end of active extension in the Furnace Creek Basin and the approximate onset of crustal extension in modern Death Valley. Knott et al. (2005) reconstruct the stratigraphy during this transition period.

Photo 15.

Photo 16.

References

Calzia, J.P., Rämö, O.T., 2005. Miocene rapaviki granites in the southern Death Valley region, California, USA. Earth-Science Reviews 73, 221–243. doi:10.1016/j.earscirev.2005.07.006.

Cemen, I., Tekeli, O., Seyitoglu, G., Isik, V., 2005. Are turtleback fault surfaces common structural elements of highly extended terrains? Earth-Science Reviews 73, 139–148. doi:10.1016/j.earscirev.2005.07.001.

Crowe, B.M., Fisher, R.V., 1973. Sedimentary structures in base-surge deposits with special reference to cross bedding, Ubehebe Crater, Death Valley, California. Geological Society of America Bulletin 84, 663–682.

Curry, H.D., 1938. "Turtleback" fault surfaces in Death Valley, California. Geological Society of America Bulletin 49, 1875.

Heaman, L.M., Grotzinger, J., 1992. 1.08 Ga diabase sills in the Pahrump Group, California: implications for development of the Cordilleran Miogeocline. Geology 20, 637–640.

Jayko, A.S., 2005. Late Quaternary denudation, Death and Panamint Valleys, Eastern California. Earth-Science Reviews 73, 271–289. doi:10.1016/j.earscirev.2005.04.009.

Knott, J.R., Sarna-Wojcicki, A.M., Machette, M.N., Klinger, R.E., 2005. Upper Neogene stratigraphy and tectonics of Death Valley—A review. Earth-Science Reviews 73, 245–270. doi:10.1016/j.earscirev.2005.07.004.

Luckow, H.G., Pavlis, T.L., Serpa, L.F., Guest, Bernard, Wagner, D.L., Snee, Lawrence, Hensley, T.M., Korjenkov, Andrey, 2005. Late Cenozoic sedimentation and volcanism during transtensional deformation in Wingate Wash and the Owlshead Mountains, Death Valley. Earth-Science Reviews 73, 177–219. doi:10.1016/j.earscirev.2005.07.013.

Messina, P., Soffer, P.W., Smith, W.C., 2005. Macropolygon morphology, development, and classification on North Panamint and Eureka Playas, Death Valley National Park, CA. Earth-Science Reviews 73, 309–322 . doi:10.1016/j.earscirev.2005.04.011.

Miller, M.B., Wright, L.A., 2004. Geology of Death Valley National Park, 2nd ed. Kendall-Hunt Publishing, Dubuque, Iowa (123 pp.).

Miller, M.B., Pavlis, T.L., 2005. The Black Mountains turtlebacks: Rosetta stones of the Death Valley tectonics. Earth-Science Reviews 73, 115–138. doi:10.1016/j.earscirev.2005.04.007.

Niemi, N.A., 2002. Extensional tectonics in the Basin and Range Province and the geology of the Grapevine Mountains, Death Valley Region, California and Nevada. PhD thesis, Pasadena, California Institute of Technology, 344 pp.

Reynolds, M.W., 1969. Stratigraphy and structural geology of the Titus and Titanothere Canyons area, Death Valley, California. PhD thesis, Berkeley, University of California, Berkeley, 310 pp.

Stevens, C.H., Stone, P., 2005. Structure and regional significance of the Late Permian(?) Sierra Nevada—Death Valley thrust system, East-Central California. Earth-Science Reviews 73, 103–113. doi:10.1016/j.earscirev.2005.04.006.

Available online at www.sciencedirect.com

Earth-Science Reviews 73 (2005) 31–45

www.elsevier.com/locate/earscirev

Bibliography
L.A. Wright and B.W. Troxel
1944–present

Jahns, R.H., and Wright, L.A., 1944, The Harding beryllium-tantalum-lithium pegmatites, Taos County, New Mexico: Economic Geology, v. 39, p. 96–97.

Wright, L.A., 1948, California foundry sands: California Journal of Mines and Geology, v. 44, p. 36–72.

Wright, L.A., 1948, The Globe pegmatite, Rio Arriba County, New Mexico: American Journal of Science, v. 246, p. 665–668.

Wright, L.A., 1948, White Eagle talc deposit, an example of the steatitization of granite: Geological Society of America Bulletin, v. 59, p. 1385.

Wright, L.A., 1948, Age of the basal Modelo (?) Formation in Reynier Canyon: Geological Society of America Bulletin, v. 59, p. 390.

Campbell, Ian, and Wright, L.A., 1950, Kyanite paragenesis at Ogilby, California: Geological Society of America Bulletin, v. 61, p. 1520–1521.

Wright, L.A., 1950, California talcs: American Institute of Mining and Metallurgical Engineers Transactions, v. 187, p. 122–128.

Jahns, R.H., and Wright, L.A., 1951, Gem- and lithium-bearing pegmatites of the Pala District, San Diego County, California: California Division of Mines Special Report 7A, 72 p.

Wright, L.A., 1952, Geology of the Superior talc area, Death Valley, California: California Division of Mines Special Report 20, 22 p.

Wright, L.A., Stewart, R.M., Gay Jr., T.E., and Hazenbush, G.C., 1953, Mines and mineral deposits of San Bernardino County, California: California Journal of Mines and Geology, v. 49, no. 1, p. 49–257; no. 2, p. 1–192.

doi:10.1016/j.earscirev.2005.07.011

Gay Jr., T.E., and Wright, L.A., 1954, Geology of the Talc City area, Inyo County: in Jahns, R.H., (ed.), Geology of Southern California: California Division of Mines and Geology Bulletin 170, Map Sheet 12.

Jennings, C.W., and Troxel, B.W., 1954, Ventura Basin: *in* Jahns, R.H., (ed.), Geology of Southern California: California Division Mines and Geology Bulletin 170, Geologic Guide 2, 63 p.

Noble, L.F., and Wright, L.A., 1954, Geology of the central and southern Death Valley region, California, *in* Jahns, R.H., (ed.), Geology of Southern California: California Division of Mines and Geology Bulletin 170, Chapter II, p. 59–74.

Troxel, B.W., 1954, Los Angeles Basin: *in* Jahns, R.H., (ed.), Geology of Southern California: California Division Mines and Geology Bulletin 170, Geologic Guide 3, 46 p.

Troxel, B.W., 1954, Geology of a part of the Shadow Mountains, western San Bernardino County, California: *in* Jahns, R.H., (ed.), Geology of southern California; California Division Mines and Geology Bulletin 170, Map Sheet 15, scale 1:24,000.

Wright, L.A., 1954, Geology of the Alexander Hills area, Inyo and San Bernardino Counties: *in* Jahns, R.H., (ed.), Geology of Southern California: California Division of Mines and Geology Bulletin 170, Map Sheet 17.

Wright, L.A., 1954, Geology of the Silver Lake talc deposits: California Division of Mines Special Report 38, 30 p.

Wright, L.A., 1954, Occurrence and use of nonmetallic commodities in southern California: *in* Jahns, R.H., (ed.), Geology of Southern California: California Division of Mines and Geology Bulletin 170, Chapter 8, p. 59–74.

Wright, L.A., and Troxel, B.W., 1954, Geologic guide to the western Mojave Desert and Death Valley regions: *in* Jahns, R.H., (ed.), Geology of Southern California: California Division of Mines and Geology Bulletin 170, Geologic Guide 1, 50 p.

Wright, L.A., Chesterman, C.W., and Norman, L.A., 1954, Occurrence and use of non-metallic commodities in southern California: *in* Jahns, R.H., (ed.), Geology of Southern California: California Division Mines and Geology Bulletin 170, Chapter VIII, p. 59–74.

Troxel, B.W., 1955, Geology of a part of the Shadow Mountains, west-central Mojave Desert, California: Geological Society of America Bulletin, v. 66, p. 1667.

Wright, L.A., 1955, Rainbow Mountain breccias, Amargosa Valley, California: Geological Society of America Bulletin, v. 66, p. 1670.

Wright, L.A., and Troxel, B.W., 1956, Noonday Dolomite, Johnnie Formation, Stirling Quartzite and Wood Canyon Formation in the southern Death Valley region, California: Geological Society of America Bulletin, v. 67, p. 1787.

Troxel, B.W., 1957, Abrasives: *in* Wright, L.A. (ed.), Mineral commodities of California: California Division Mines and Geology Bulletin 176, p. 23–28.

Troxel, B.W., 1957, Thorium: *in* Wright, L.A. (ed.), Mineral commodities of California: California Division Mines and Geology Bulletin 176, p. 635–640.

Troxel, B.W., 1957, Wollastonite: *in* Wright, L.A. (ed.), Mineral commodities of California: California Division Mines and Geology Bulletin 176, p. 693–698.

Troxel, B.W., Chesterman C.W., and Stinson, M.C., 1957, Uranium: *in* Wright, L.A. (ed.), Mineral commodities of California: California Division Mines and Geology Bulletin 176, p. 669–688.

Wright, L.A., 1957, (ed.), Mineral commodities of California: California Division Mines and Geology Bulletin 176, 736 p.

Troxel, B.W., 1958, Geology of the northwestern part of the northwestern part of the Shadow Mountains, western San Bernardino, California: Los Angeles, University of California MSc thesis.

Jahns, R.H., Wright, L.A., and Montgomery, A., 1959, Origin of remarkable beryl concentrations in the Harding pegmatite, New Mexico: Geological Society of America Bulletin, v. 70, p. 1725.

Wasserburg, G.J., Wetherill, G. and Wright, L.A., 1959, Ages in the Precambrian terrane of Death Valley, California: Journal of Geology, v. 67, p. 702–708.

Engel, A.E.J., and Wright, L.A., 1960, Talc and soapstone: *in* Industrial Minerals and Rocks: American Institute of Mineral and Metallurgical Engineers, p. 835–850.

Troxel, B.W., 1960, Maps used in mineral investigations: California Division Mines and Geology Mineral Information Service, v. 13, p. 1–14.

Hewett, D.F., Chesterman, C.W., and Troxel, B.W., 1961, Tephroite in California manganese deposits: Economic Geology, v. 56, p. 39–58.

Jennings, C.W., Burnett, J.L., and Troxel, B.W., 1962, Geologic map of California: Trona Sheet: California Division Mines and Geology, scale 1:250,000.

Troxel, B.W., and Morton P.K., 1962, Mines and mineral resources of Kern County, California: California Division Mines and Geology County Report 1, 370 p.

Noble, L.F., Wright, L.A., and Troxel, B.W., 1963, Fault pattern of the southern Death Valley region, eastern California: Geological Society of America Special Paper 73, p. 55.

Troxel, B.W., 1963, Mineral resources and geologic features of the Geologic map of California, Trona sheet: California Division Mines and Geology Mineral Information Service, v. 16, p. 1–7.

Chidester, A.L., Engel, A.E.J. and Wright, L.A., 1964, Talc resources of the United States: U.S. Geological Survey Bulletin 1167, 61 p.

Campbell, Ian, and Troxel, B.W., 1965, Geologic hazards: Urban mapping program of the Division of Mines and Geology: California Division Mines and Geology Mineral Information Service, v. 18, p. 161–163.

Cleveland, G.B., and Troxel, B.W., 1965, Geology related to the safety of Corral Canyon Nuclear Reactor Site, Malibu, Los Angeles County, California: California Division Mines and Geology Open File Report.

Wright, L.A., and Troxel, B.W., 1965, Talc deposits of the United States: Mining Engineering, v. 17, p. 49.

Campbell, Ian, and Troxel, B.W., 1966, Urban mapping program of the Division of Mines and Geology: *in* Landslides and subsidence: Proceedings of 2d Geologic Hazards Conference, Los Angeles, California, 1965: California Resources Agency, p. 39–42.

Hill, M.L., and Troxel, B.W., 1966, Tectonics of Death Valley region, California: Geological Society of America Bulletin, v. 77, p. 435–438.

Troxel, B.W., 1966, Wollastonite: California Division Mines and Geology Bulletin 191, p. 441–444.

Wright, L.A., and Troxel, B.W., 1966, Limitations on strike-slip displacement along the Death Valley and Furnace Creek fault zones, California: Geological Society of America Special Paper 87, p. 188–189.

Wright, L.A., and Troxel, B.W., 1966, Strata of late Precambrian–Cambrian age, Death Valley region, California-Nevada: American Association of Petroleum Geologists Bulletin, v. 50, p. 846–857.

Troxel, B.W., 1967, Sedimentary rocks of the late Precambrian and Cambrian age in the southern Salt Spring Hills, southeastern Death Valley, California: California Division Mines and Geology Special Report 92, p. 33–41.

Wright, L.A., and Troxel, B.W., 1967, Limitations on right lateral strike-slip displacement, Death Valley and urnace Creek fault zones, California: Geological Society of America Bulletin, v. 78, p. 933–949.

Smith, G.I., Troxel, B.W., Gray Jr., Cliffton, and von Huene, Roland, 1968, Geologic reconnaissance of the Slate Range, San Bernardino and Inyo Counties, California: California Division Mines and Geology Special Report 96, 33 p.

Smith, G.I., Troxel, B.W., Gray Jr., Cliffton, and von Huene, Roland, 1968, Geology of Slate Range, San Bernardino and Inyo Counties, California: Geological Society of America Special Paper 101, p. 420–421.

Troxel, B.W., 1968, Possible relationship in faults in the Mojave Desert to the San Andreas fault, *in* Geologic problems of San Andreas Fault System: Stanford University Publications in the Geological Sciences, v. 11, p. 181–182.

Troxel, B.W., 1968, Sedimentary features of the later Precambrian Kingston Peak Formation, Death Valley, California: Geological Society of America Special Paper 101, p. 341.

Troxel, B.W., and Gundersen, J.N., 1968, Geometry and origin of a Mesozoic igneous intrusive complex, Shadow Mountains, western San Bernardino County, California: Geological Society of America Special Paper 115, p. 355.

Wright, L.A., 1968, Talc deposits of the southern Death Valley-Kingston Range region, California: California Division of Mines and Geology Special Report 95, 79 p.

Cloud, P.E., Licari, G.R., Wright, L.A. and Troxel, B.W., 1969, Proterozoic eukaryotes from eastern California: Proceedings of the National Academy of Sciences, v. 62, p. 623–630.

Troxel, B.W., and Wright, L.A., 1969, Precambrian stratigraphy of the Funeral Mountains, Death Valley, California: Geological Society of America Special Paper 121, p. 574–575.

Wright, L.A., and Troxel, B.W., 1969, Evidence of northwestward crustal spreading and transform faulting in the southwestern part of the Great Basin, California and Nevada: Geological Society of America Special Paper 121, p. 580–581.

Wright, L.A., and Troxel, B.W., 1969, Chaos structure and Basin and Range normal faults—Evidence for genetic relationship: Geological Society of America Special Paper 121, p. 242.

Troxel, B.W., 1970, Anatomy of a fault zone, southern Death Valley, California: Geological Society of America Abstracts with Program, v. 2, p. 154.

Troxel, B.W., 1970, Guide to selected features of Death Valley geology, *in* Gasch, J.W., and Matthews, R.A., (eds.), Geologic guide to the Death Valley area, California: Geological Society of Sacramento, p. 40–68.

Troxel, B.W., 1970, Environmental geologists for a State geological survey: Journal Geological Education, v. 18, p. 206.

Troxel, B.W., and Gundersen, J.N., 1970, Geology of the Shadow Mountains and northeastern part of the Shadow Mountains quadrangles, western San Bernardino County, California: California Division Mines and Geology Preliminary Report 12, scale 1:24,000.

Wright, L.A., and Troxel, B.W., 1970, Summary of regional evidence for right-lateral displacement in the western Great Basin: Discussion: Geological Society of America Bulletin, v. 81, p. 2167–2174.

Wright, L.A., and Williams, E.G., 1970, Precambrian reef complex, Death Valley: Evidence for tectonic control of facies patterns and for antiquity of a major fault: Geological Society of America Abstracts with Program, p. 727.

Jahns, Richard H., Troxel, B.W., and Wright, L.A., 1971, Some structural implications of a late Precambrian–Cambrian section in the Avawatz Mountains, California; Geological Society of America Abstracts with Program, v. 3, p. 140.

Wright, L.A., and Troxel, B.W., 1971, Evidence for tectonic control of volcanism, Death Valley: Geological Society of America Abstracts with Program, v. 3, p. 221.

Wright, L.A., and Troxel, B.W., 1971, Thin-skinned megaslump model of Basin Range structure as applicable to the southwestern Great Basin: Geological Society of America Abstracts with Program, v. 3, p. 758.

Troxel, B.W., Wright, L.A., and Jahns, R.H., 1972, Evidence for differential displacement along the Garlock fault zone, California: Geological Society of America Abstracts with Program, v. 4, p. 250.

Chapman, R.H., Healey, D.L., and Troxel, B.W., 1973, Bouguer gravity map of California, Death Valley sheet: California Division Mines and Geology, scale 1:250,000.

Licari, G.R., and Troxel, B.W., 1973, Fossil algae 1.3 billion years old from eastern California rocks: California Geology, v. 26, p. 15–16.

Wright, L.A., 1973, Geology of the northeastern quarter of the Tecopa 15-minute quadrangle, San Bernardino and Inyo Counties, California: California Division of Mines and Geology Map Sheet 20.

Wright, L.A., and Troxel, B.W., 1973, Shallow-fault interpretation of Basin and Range structure, *in* DeJong, K.A. and Scholten, Robert, (ed.), Gravity and Tectonics: New York, John Wiley and Sons, p. 397–407.

Cloud, P., Wright, L.A., Williams, E.G., and Diehl, P.E., 1974, Giant stromatolites and associated vertical tubes from the Upper Proterozoic Noonday Dolomite, Death Valley region, eastern California: Geological Society of America Bulletin, v. 85, p. 1869–1882.

Spall, Henry, and Troxel, B.W., 1974, Structural and paleomagnetic studies of upper Cambrian diabase from Death Valley, California: Geological Society of America Abstracts with Program, v. 4, p. 963.

Troxel, B.W., 1974, Geologic guide to the Death Valley region, California and Nevada: *in* Guidebook, Death Valley Region, California and Nevada: Shoshone, California, Death Valley Publishing Co, p. 2–16.

Troxel, B.W., 1974, Significance of a man-made diversion of Furnace Creek wash, Death Valley, California: *in* Guidebook, Death Valley Region, California and Nevada: Shoshone, California, Death Valley Publishing Co, p. 87–90.

Williams, E.G., Wright, L.A., and Troxel, B.W., 1974, Depositional environments of the late Precambrian Noonday Dolomite, southern Death Valley region, California: Geological Society of America Abstracts with Program, v. 6, p. 276.

Williams, E.G., Wright, L.A., and Troxel, B.W., 1974, The Noonday Dolomite and equivalent stratigraphic units, southern Death Valley region, California: *in* Guidebook, Death Valley Region, California and Nevada: Shoshone, California, Death Valley Publishing Co, p. 73–77.

Wright, L.A., 1974, Geologic map of the region of central and southern Death Valley, eastern California and southwestern Nevada: *in* Guidebook, Death Valley Region, California and Nevada: Shoshone, California, Death Valley Publishing Co, p. 25.

Wright, L.A., 1974, Fault map of the region of central and southern Death Valley, eastern California and southwestern Nevada: *in* Guidebook, Death Valley Region, California and Nevada: Shoshone, California, Death Valley Publishing Co, p. 24.

Wright, L.A., Otton J.K., and Troxel, B.W., 1974, Turtleback surfaces of Death Valley viewed as phenomena of extensional tectonics: Geology, v. 2, p. 53–54.

Wright, L.A., Troxel, B.W., Williams, Eugene G., Roberts, M.T., and Diehl, P.E., 1974, Precambrian sedimentary environments of the Death Valley region, eastern California: *in* Guidebook, Death Valley Region, California and Nevada: Shoshone, California, Death Valley Publishing Co, p. 27–36.

Wright, L.A., Williams, E.G., and Troxel, B.W., 1974, Late Precambrian sedimentation, Death Valley: Evidence of prolonged and varied tectonic control of a basin and shelf environment: Geological Society of America Abstracts with Program, vol. 6, p. 278.

Troxel, B.W., 1976, Geologic features of Death Valley: California Geology, v. 29, p.182–183.

Wright, L.A., 1976, Geologic map of the region of central and southern Death Valley, eastern California and southwestern Nevada: *in* Troxel, B.W., and Wright, L.A., (eds.), Geological features, Death Valley, California: California Division of Mines Special Report 106, p. 16.

Wright, L.A., 1976, Fault map of the region of central and southern Death Valley, eastern California and southwestern Nevada: *in* Troxel, B.W., and Wright, L.A., (eds.), Geological features, Death Valley, California: California Division of Mines Special Report 106, p. 17.

Wright, L.A., 1976, Book review *Death Valley; geology, ecology, archaeology* by C.B. Hunt: American Journal of Science, v. 276, p. 923–925.

Wright, L.A., 1976, Late Cenozoic fault patterns and stress distribution in the Great Basin and westward displacement of the Sierra Nevada block: Geology, v. 4, p. 489–494.

Wright, L.A., Troxel, B.W., Williams, E.G., Roberts, M.T. and Diehl, P.E., 1976, Precambrian sedimentary environments of the Death Valley region, eastern California, *in* Troxel, B.W. and Wright, L.A., (eds.), Geologic features, Death Valley, California: California Division of Mines and Geology Special Report 106, p. 7–15.

Pierce, D., Cloud, P., and Troxel, B.W., 1977, New microfloras from the Pahrump Group, later pre-Phanerozoic, eastern California: Geological Society of America Abstracts with Program, v. 11, p. 132.

Troxel, B.W., 1977, arcuate faults and grabens in Death Valley, California: California Geology, v. 30, p. 9–11.

Troxel, B.W., Wright, L.A., and Williams, E.G., 1977, Late Precambrian history derived from the Kingston Peak Formation, Death Valley Region, California: Geological Society of America Abstracts with Program, v. 9, p. 517.

Wright, L.A., 1977, Late Cenozoic fault patterns and stress fields in the Great Basin and westward displacement of the Sierra Nevada block—Discussion: Geology, v. 5.

DeWitt, ed, Wright, L.A., and Troxel, B.W., 1979, Mesozoic metamorphic terrains in the Death Valley area, south California: Geological Society of America Abstracts with Program, v. 11, p. 413.

Troxel, B.W., and Butler, P.R., 1979, Rate of Cenozoic slip on normal faults, south-central Death Valley, California; Final report to the U.S. Geological Survey Earthquake Hazards Reduction Program, 5 p.

Troxel, B.W., and Butler, P.R., 1979, Time relations between faulting, volcanism, and fan development, south-central Death Valley, California; Final report to the U.S. Geological Survey Earthquake Hazards Reduction Program, 5 p.

Troxel, B.W., Jahns, R. H., and Butler, P.R., 1979, Quaternary and Tertiary history of offsets along the easternmost segment of the Garlock fault zone: Geological Society of America Abstracts with Program, v. 11, p. 132.

Brady III, R.H., Troxel, B.W., and Butler, P.R., 1980, Tectonic and stratigraphic elements of the northern Avawatz Mountains, San Bernardino County, California: in Fife, D.L., ed Geology and mineral wealth of the California desert: Santa Ana, CA, South Coast Geological Society, p.224–234.

Naert, K.A., Wright, L.A., and Thornton, C.P., 1980, Geology of the perlite deposits of the Northern Agua Peaks, Taos County, New Mexico: New Mexico Bureau of Mines and Mineral Resources Open File Report, 48 p.

Brady, Roland H, Troxel, B.W., 1981, Eastern termination of the Garlock Fault in the Avawatz Mountains, San Bernardino, California: Geological Society of America Abstracts with Program, vol. 13, p. 46–47.

Miller, J.M.G., Wright, L.A., and Troxel, B., 1981, The late Precambrian Kingston Peak Formation, Death Valley region, California: *in* Hambrey, P.J., and Harland, W.B. (eds.), Earth's pre-Pleistocene glacial record: Cambridge University Press, p. 745–748.

Wright, L.A., Troxel, B.W., Burchfield, B.C., Chapman, R.H., and Labotka, T.C., 1981, Geologic cross section from the Sierra Nevada to the Las Vegas Valley, eastern California to southern Nevada: Geological Society of America Map and Chart Series MC-28M, 15 p.

Cemen, I., Drake, R.E., and Wright, L.A., 1982, Stratigraphy and chronology of the Tertiary sedimentary and volcanic units at the southeastern end of the Funeral Mountains, Death Valley region, California, *in* Cooper, J.D., Troxel, B.W., and Wright, L.A., Geology of selected areas in the San Bernardino Mountains, western Mojave Desert, and southern Great Basin, California: Shoshone, CA, Death Valley Publishing Company, p. 77–88.

Troxel, B.W., 1982, Description of the uppermost part of the Kingston Peak Formation, Amargosa Rim Canyon, Death Valley region, California: *in* Cooper, J.D., Troxel, B.W., and Wright, L.A., Geology of selected areas in the San Bernardino Mountains, western Mojave Desert, and southern Great Basin, California: Shoshone, CA, Death Valley Publishing Company, p. 43–48.

Troxel, B.W., 1982, Geologic road guide, Day 2, Baker-southern Death Valley-Shoshone: *in* Cooper, J.D., Troxel, B.W., and Wright, L.A., Geology of selected areas in the San Bernardino Mountains, western Mojave Desert, and southern Great Basin, California: Shoshone, CA, Death Valley Publishing Company, p. 37–42.

Troxel, B.W., 1982, Geologic road guide, Day 3, Segment A, Shoshone to Death Valley: *in* Cooper, J.D., Troxel, B.W., and Wright, L.A., Geology of selected areas in the San Bernardino Mountains, western Mojave Desert, and southern Great Basin, California: Shoshone, CA, Death Valley Publishing Company, p. 71–76.

Troxel, B.W., 1982, Geologic road guide, Day 3, Segment C, Shoshone to Interstate 15 via Kingston Range: *in* Cooper, J.D., Troxel, B.W., and Wright, L.A., Geology of selected areas in the San Bernardino Mountains, western Mojave Desert, and southern Great Basin, California: Shoshone, CA, Death Valley Publishing Company, p 137–140.

Troxel, B.W., and Cooper, J.D., 1982, Geologic road guide, Day 3, Segment B, Shoshone to northern Nopah Range; *in* Cooper, J.D., Troxel, B.W., and Wright, L.A., Geology of selected areas in the San Bernardino Mountains, western Mojave Desert, and southern Great Basin, California: Shoshone, CA, Death Valley Publishing Company, p 89–90.

Troxel, B.W., and Heydari, Ezat, 1982, Basin and Range geology in a roadcut: *in* Cooper, J.D., Troxel, B.W., and Wright, L.A., Geology of selected areas in the San Bernardino Mountains, western Mojave Desert, and southern Great Basin, California: Shoshone, CA, Death Valley Publishing Company, p. 91–96.

Wright, L.A., 1982, The geology of Death Valley: Earth Science, v. 35, p. 11–15.

Wright, L.A., Williams, E.G., and Troxel, B.W., 1982, Precambrian and early Cambrian stratigraphic–tectonic framework of the Death Valley region, California; Geological Society of America Abstracts with Program, v. 14, p 246.

Butler, P.R., and Troxel, B.W. 1983. Geomorphic evidence for 25 km of Neogene right-lateral displacement on the Death Valley Fault Zone, California: Geological Society of America Abstracts with Program, v. 15, p. 316.

Butler, P.R., Troxel, B.W., and Verosub, R.L., 1983, Varying styles of deformation associated with strike slip on the Death Valley fault zone, southern Death Valley, California: EOS, v. 64, p. 865.

Giaramita, M., Day, H.W., and Troxel, B.W., 1983, Structural, metamorphic, and plutonic evolution of the northern Funeral Mountains, Death Valley, California: Geologic Society of America Abstracts with Program, v 15, p. 419.

Roths, P.J., Vyverberg, K.A., Troxel, B.W., McMackin, M.A., and others, 1983, Geology of the northwestern Kingston Range, San Bernardino and Inyo Counties, California: Geologic Society of America Abstracts with Program, v. 15, p 438.

Stellar, D.L., (ed.), Troxel, B.W., and Wright, L.A., 1983, Death Valley region field guide: Cypress College, California, National Association of Geology Teachers, Far West Section, 125 p.

Wright, L.A., Troxel, B.W., and Drake, R.E., 1983, Contrasting space–time patterns of extension related late Cenozoic faulting, southwestern Great Basin; Geological Society of America Abstracts with Program, v. 15, p. 287.

Miller, J.M.G., Wright, L.A., and Troxel, B.W., 1984, Correlations within the Proterozoic Kingston Peak Formation, Death Valley region, eastern California: Geologic Society of America Abstracts with Program, v 16, p. 597.

Serpa, L.T., deVoogd, B., Willemin, J., Oliver, G., Kaufman, S., Brown, L., Hauser, E., Wright, L.A., and Troxel, B. 1984, Late Cenozoic fault patterns and magma migration in Death Valley from COCORP seismic profile: EOS, v. 65, p. 985.

Wright, L.A., and Troxel, B.W., 1984, Geology of the northern half of the Confidence Hills 15-minute quadrangle, Death Valley region: the area of the Amargosa chaos: California Division Mines and Geology Map Sheet 34, scale 1:24,000.

Wright, L.A., Drake, R.E., and Troxel, B.W., 1984, Evidence for the westward migration of severe Cenozoic extension, southwestern Great Basin, California: Geological Society of America Abstracts with Program, p. 597.

Wright, L.A., Troxel, B.W., Kramer, J H., and Thornton, C.P., 1984, Type sections of two newly named volcanic units of the central Death Valley volcanic field, *in* Wright, L.A., and Troxel, B.W., Geology of the northern half of the Confidence Hills 15-minute quadrangle, Death Valley region, eastern California: California Division of Mines and Geology Map Sheet 34, p. 21–24.

Wright, L.A., Williams, E.G., and Troxel, B.W., 1984, Type section of the newly named Proterozoic Ibex Formation, the basal equivalent of the Noonday Dolomite, Death Valley region, California: *in* Wright, L.A., and Troxel, B.W., Geology of the northern half of the Confidence Hills 15-minute quadrangle, Death Valley region, California: California Division of Mines and Geology Map Sheet 34, p. 25–29.

Cemen, I., Wright, L.A., Drake, R.E., and Johnson, F.C., 1985, Cenozoic sedimentation and sequence of deformational events at the southeastern end of the Furnace Creek strike slip fault zone, Death Valley region, California: *in* Biddle, K T., and Christie-Blick, N., (eds.), Strike-slip deformation, basin formation, and sedimentation: Society of Economic Paleontologists and Mineralogists Special Publication 37, p. 127–141.

Giaramita, M.J., Burnley, P.C., Day, H.W., and Troxel, B.W., 1985, Lithologic control of deformational style in a detachment fault complex, northern Funeral Mountains, southeastern California: Geologic Society of America Abstracts with Program, v 17, p 356.

Troxel, B.W., Wright, L.A., Williams, E.G., and McMackin, M.R., 1985, Provenance of the late Precambrian Kingston Peak Formation, southeastern Death Valley region, California; Geologic Society of America Abstracts with Program, v 17, p 414.

Wright, L.A., 1985, Memorial to Richard H. Jahns: Geological Society of America Memorials, 10 p.

Butler, P.R., Troxel, B.W., and Verosub, K.L. 1986, Late Cenozoic slip rates and styles of deformation along the southern Death Valley fault zone, California: Geological Society of America Abstracts with Program, v. 18, p. 92.

deVoogd, B., Serpa, L.F., Brown, L., Hauser, E., Hoffman, S., Oliver, J., Troxel, B.W., Willemin, J., and Wright, L.A., 1986, Death Valley bright spot; a mid-crustal magma body in the southern Great Basin, California: Geology, v. 14, p. 64–67.

McMackin, M.R., Wright, L.A., Troxel, B.W., and Calzia, J.P., 1986, Cenozoic extensional faulting and plutonism in the Kingston Range and northern Mesquite Mountains, San Bernardino and Inyo Counties, CA: Geological Society of America Abstracts with Program, v. 18, p. 157.

Prave, A.R., and Wright, L.A., 1986, Isopach pattern of the Lower Cambrian Zabriskie Quartzite, Death Valley region, California: How useful in tectonic reconstructions: Geology, v. 14, p. 251–254.

Reynolds, N.W., Wright, L.A., and Troxel, B.W., 1986, Geometry and chronology of late Cenozoic detachment faulting, Funeral and Grapevine Mountains, Death Valley, California: Geological Society of America Abstracts with Program, v. 18, p 175.

Serpa, L., deVoogd, B., Wright, L.A., and Troxel, B.W., 1986, A model for the structural evolution of the central Death Valley basin from COCORP seismic data: Geological Society of America Abstracts with Program , v. 18, p. 181.

Troxel, B.W. 1986, Pleistocene and Holocene deformation on a segment of the southern Death Valley fault zone, California: *in* Troxel, B.W. (ed.), Quaternary tectonics of Southern Death Valley, California: Shoshone, CA, Field trip guide, p. 13–16.

Troxel, B.W., 1986, Significance of Quaternary fault pattern, west side of the Mormon Point Turtleback, southern Death Valley, California; a model of listric normal faults; *in* Troxel, B.W. (ed.), Quaternary tectonics of Southern Death Valley, California: Shoshone, CA, Field trip guide, p. 37–39.

Troxel, B.W., and Butler, P.R 1986, Time relations between Quaternary faulting, volcanism, and fan development, southern Death Valley, California: *in* Troxel, B.W. (ed.), Quaternary tectonics of Southern Death Valley, California: Shoshone, CA, Field trip guide, p. 31–35.

Wright, L.A., Drake, R.E., Troxel, B.W., and Thompson, R.A., 1986: Central Death Valley volcanic field, eastern California: tectonic setting, volcanic stratigraphy, and geochronology: EOS, v. 67, p 1262.

Troxel, B.W., and Wright, L.A., 1987: Tertiary extensional features, Death Valley region, eastern California: Geological Society of America Centennial Field Guide—Cordilleran Section, p. 121–132.

Troxel, B.W., McMackin, M.A., Calzia, J.P., Walker, J.D., Klepacki, D.W., and Burchfield, B.C., 1987, Late Precambrian tectonism in the Kingston Range: Discussion and reply: Geology, v. 15, p. 274–275.

Wright, L.A., Serpa, L.F., and Troxel, B.W., 1987, Tectonic–chronologic model for wrench faults related to crustal extension, Death Valley, California: Geological Society of America Abstract with Program, v. 19, p 88–89.

Butler, P.R., Troxel, B.W., and Verosub, K.L., 1988, Late Cenozoic history and styles of deformation along the southern Death Valley fault zone, California: Geological Society of America Bulletin, v. 100, p. 402–410.

Cemen, I., and Wright, L.A., 1988, Stratigraphy and chronology of the Artist Drive Formation, Death Valley region, California, *in* Gregory, J.L., and Baldwin, E J., (eds.), Geology of the Death Valley region: South Coast Geological Society Annual Field Trip Guide No. 16, p. 77–87.

DeWitt, E., Sutter, J.F., Wright, L.A., and Troxel, B.W., 1988, Ar–Ar chronology of Early Cretaceous regional metamorphism, Funeral Mts, CA: A case study of excess argon: Geological Society of America Abstracts with Program, v. 20.

Miller, J.M.G., Troxel, B.W., and Wright, L.A., 1988, Stratigraphy and paleogeology of the Kingston Peak Formation, Death Valley region, California, *in* Gregory, J.L., and Baldwin, E J., (eds.), Geology of the Death Valley region: South Coast Geological Society Annual Field Trip Guide No. 16, p. 118–142.

Serpa, L., deVoogd, B., Wright, L.A., Willemin, J., Oliver, J., Hauser, E., and Troxel, B.W., 1988, Structure of the central Death Valley pull-apart basin and vicinity from COCORP profile in the Southern Great Basin: Geological Society of America Bulletin, v. 100, p. 1437–1450.

Troxel, B.W. 1988 Multiple Quaternary tectonism, central and southern Death Valley, California: Geological Society of America Abstracts with Program, v. 20, p. 238.

Troxel, B.W., 1988, Significance of Quaternary fault pattern, west side of the Mormon Point turtleback, southern Death Valley, California; A model of listric normal faults; *in* Gregory, J.L., and Baldwin, E.J., (eds.), Geology of the Death Valley region: South Coast Geological Society Annual Field Trip Guide No. 16, p. 240–242.

Troxel, B.W., 1988, Detachment Surfaces in the southern Great Basin: a geologic traverse of the northern Funeral Mountains, Death Valley, California: *in* Weide, D.L., and Faber, M.L., (eds.), This extended land: Geological journeys in the southern Basin Range: Geological Society of America Field Trip Guidebook, p 45–49.

Wright, L.A., and Troxel, B.W., 1988, Wrench fault-related features in the Cenozoic structural framework of the Death Valley region, California-Nevada: Geological Society of America Abstract with Program, v. 20, p. 244.

Brady, R., H., Clayton, Jeff, Troxel, B.W., Verosub, R.L., Cregan, Allen, and Abrams, Michael, 1989, Thematic mapper and field investigation at the intersection of the Death Valley and Garlock Fault zones, California: New York, Elsevier, Remote Sensing of Environment, p. 207–217.

Troxel, B.W., 1989, Geologic Road Guide: Day 1, Segment 1: Baker to southern Death Valley: Pacific Section Society of Economic Paleontologists and Mineralogist Field trip guidebook, v. 61, p. 6–10.

Troxel, B.W., 1989, Geologic Road Guide: Day 2, Segment 3: Shoshone to Eastern Funeral Range: Pacific Section Society of Economic Paleontologists and Mineralogist Field trip guidebook, v. 61, p. 32–36.

Troxel, B.W., Cooper, G.D., 1989. Geologic Road Guide: Day 1, Segment 2: Wade Monument to Nopah Range: Pacific Section Society of Economic Paleontologists and Mineralogist Field trip guidebook, v. 61 p 14–20.

Troxel, B.W., and Wright, L.A., 1989, Geologic map of the central and northern Funeral Mountains and adjacent areas, Death Valley region, southern California: U.S. Geological Survey Open File Report.

Wright, L.A., 1989, Overview of the role of the strike-slip and normal faulting in the Neogene history of the region northeast of Death Valley, California–Nevada, *in* Ellis, M.A., (ed.), Late Cenozoic Evolution of the Southern Great Basin: Nevada Bureau of Mines and Geology Open File 89-1.

Wright, L.A., Troxel, B.W., and Zigler, J.L., 1989, Geologic sections to accompany geologic map of the central and northern Funeral Mountains and adjacent areas, Death Valley region, California: U.S. Geological Survey Open-file Report.

Cemen, Ibrahim, and Wright, L.A., 1990, Effect of Cenozoic extension on Mesozoic thrust surfaces in the central and southern Funeral Mountains, Death Valley, California: *in* Wernicke, B.P., (ed.), Basin and Range extensional tectonics near the latitude of Las Vegas, Nevada: Geological Society of America Memoir 176, p. 305–316.

Wright, L.A., Troxel, B.W., and DeWitt, Ed, 1990, Funeral Mountains metamorphic core complex, Death Valley Region, California-Nevada: Geological Society of America Abstracts with Program, v. 22, p. 95.

Cemen, Ibrahim, Wright, L.A., and Troxel, B.W., 1991, Miocene conglomerates of the Furnace Creek Basin, Death Valley, California: Origin and tectonic significance: Geological Society of America Abstracts with Program, v. 23, p. 82.

Wright, L.A., Thompson, R.A., Troxel, B.W., Pavlis, T.L., DeWitt, E.H., Otton, J.K., Ellis, M.A., Miller, M.G., and Serpa, L.F., 1991, Cenozoic magmatic and tectonic evolution of the east-central Death Valley region,

California, *in* Walawender, M.J., and Hannan, B.B., (eds.), Geologic excursions in southern California and Mexico: Geological Society of America Guidebook, p. 93–127.

Troxel, B.W., 1992, Neogene structural history of the Saddle Peak Hills, southern Death Valley, California: Evidence of a migrating extension direction: Geological Society of America Abstracts with Program, v. 24, p. 87.

Troxel, B.W., Calzia, J.P., and Pavlis, T.L., 1992, Southwest directed extensional features and related magmatism, southern Death Valley region and the Kingston Range: Geological Society of America Abstracts with Program, v. 24, p. 87.

Wright, L.A., and Troxel, B.W., 1992, Levi Noble, pioneer geologist of the Death Valley region: *in* Pisarowicz, James, (ed.), Proceedings, Third Death Valley Conference on history and prehistory: Death Valley Natural History Association, p. 143–157.

Wright, L.A., Troxel, B.W., and Prave, A.R., 1992, Field traverse of the Proterozoic rock units, Alexander Hills and southern Nopah Range, Death Valley region, California: Geological Society of America Penrose Conference.

Calzia, J.P., and Troxel, B.W., 1993, Absolute age constraints on the age and tectonics of the middle and late Proterozoic Pahrump group, southern Death Valley, California: Geological Society of America Abstracts with Program, v. 25, p. 17.

Wright, L.A., and Prave, A.R., 1993, Proterozoic–early Cambrian tectonostratigraphic record in the Death Valley region: *in* Reed, J.C., Jr., Bickford, M.E., Houston, R.S., Link, P.K., Rankin, D.W., Sims, P.K., and van Schmus, W.R., (eds.), Precambrian, conterminous U.S.: Boulder, Geological Society of America, The Geology of North America, v. C-2, p. 529–532.

Wright, L.A., and Troxel, B.W., 1993, Geologic map of the central and northern Funeral Mountains and adjacent areas, Death Valley: U.S. Geological Survey Miscellaneous Investigation Series, Map I-2305, scale 1:48,000.

Troxel, B.W. 1994, Right-lateral offset of ca. 28 km along a strand of the southern Death Valley fault zone, California: Geological Society of America Abstracts with Program, v. 26, p 99.

Troxel, B.W., and Calzia, J.P., 1994, Geology of the middle Miocene Ibex Pass volcanic field, southern Death Valley, California: Geological Society of America Abstracts with Program, v. 26, p 99.

Wright, L.A., and Troxel, B.W., 1994, Igneous events in the Panamint Mts.-Kingston Range Belt, Death Valley region: An overview of space–time relations: Geological Society of America Abstracts with Program, v. 26, p. 105.

Reynolds, M.W., Wright, L.A., and Troxel, B.W., 1996, Evidence for Tertiary age of recumbent folds, Grapevine and northern Funeral Mountains: Geological Society of America Abstracts with Programs, v. 28 p. 513.

Troxel, B.W., and Calzia, J.P. 1997, Geology of the southern Death Valley Region, California: Geological Society of America Abstracts with Program, v. 29, p 70–71.

Wright, L.A., and Miller, M.G., 1997, Death Valley National Park, *in* Harris, A.G., Tuttle, E., and Tuttle, S.D., (eds.), Geology of National Parks: Dubuque, Iowa, Kendall/Hunt Publishing Company, p. 610–637.

Brady III, R.H., and Troxel, B.W., 1999, The Miocene Military Canyon Formation: Depocenter evolution and constraints on lateral faulting, southern Death Valley, California: *in* Wright, L.A., and Troxel, B.W., (eds.), Cenozoic basins of the Death Valley region: Geological Society of America Special Paper 333, p. 259–275.

Cemen, I., Wright, L.A., and Prave, A.R., 1999, Stratigraphy and tectonic implications of the latest Oligocene and early Miocene sedimentary succession, southernmost Funeral Mountains, Death Valley region, California, *in* Wright, L A., and Troxel, B.W., (eds.), Cenozoic basins of the Death Valley region: Geological Society of America Special Paper 333, p. 65–86.

Knott, J.R., Sarna-Wojcicki, A.M., Klingor, R.E., Tinsley III, C., and Troxel, B.W., 1999., Late Cenozoic teprachronology of Death Valley, California: New insight into stratigraphy, paleogeography, and tectonics: *in* Slate, J L., (ed.), Proceedings of conference on status of geologic research and mapping, Death Valley National Park: U.S. Geological Survey Open-file Report 99-153, p. 115–116.

Thompson, R.A., Wright, L.A., Johnson, C.M., and Fleck, R.J., 1999, Temporal, spatial and compositional constraints on extension-related volcanism in central Death Valley, California: *in* Slate, J L., (ed.), Proceedings of conference on status of geologic research and mapping, Death Valley National Park: U.S. Geological Survey Open-file Report 99-153, p. 27–28.

Troxel, B.W., 1999, Geology of the Saddle Peak Hills 7.5 quadrangle, Death Valley, California in Slate, J.L., (ed.), Proceedings of conference on status of geologic research and mapping, Death Valley National Park: U.S. Geological Survey Open-file Report 99-153, p. 83–84.

Wright, L.A., 1999, Cenozoic basins of the central Death Valley region, eastern California, *in* Slate, J.L., (ed.), Proceedings of conference on status of geologic research and mapping, Death Valley National Park: U.S. Geological Survey Open File Report 99-153, p. 41–46.

Wright, L.A., and Troxel, B.W., 1999, Levi Noble's Death Valley, a 58-year perspective: *in* Moores, E.M., Slone, D., and Stout, D.L., (eds.), Classic concepts in Cordilleran geology: Geological Society of America Special Paper 338, p.385–411.

Wright, L.A., Green, R.C., Cemen, I., Johnson, F.C., and Prave, A.R., 1999, Tectonostratigraphic development of the Miocene–Pliocene Furnace Creek basin, Death Valley region, California: *in* Wright, L A., and Troxel, B.W., (eds.), Cenozoic basins of the Death Valley region: Geological Society of America Special Paper 333, p. 87–114.

Calzia, J.P., Troxel, B.W., Wright, L.A., Burchfiel, B.C., Davis, G.A., and McMackin, M.R., 2000, Geologic map of the Kingston Range, Southern Death Valley, California: U.S. Geological Survey Open File Report 00-412, scale, 1:31,680.

Sarna-Wojcicki, A.M, Machett, M.M., Knott, J.R., Klinger, R.E., Fleck, R.J., Tinsley, J.C. III, Troxel, B.W., Budahn, J.R., and Walker, J.P., 2001, Weaving a temporal and spatial framework for the late Neogene of Death Valley: Correlation and dating of Pliocene and Quaternary units using teprachronology, ^{40}Ar/^{39}Ar dating, and other dating methods: *in* Machette, M.N., Johnson, M.L., and Slate, J.L. (eds.), Quaternary and late

Pliocene geology of the Death Valley region: Recent observations on tectonics, stratigraphy, and lake cycles. U.S. Geological Survey Open-File Report 01-51, p. E121–135.

Calzia, J.P., Troxel, B.W., and Raumann, C.G., 2002, Geologic map of the Valjean Hills 7.5′ quadrangle, San Bernardino County, California: U.S. Geological Survey Open File Report 03-096, scale 1:24,000.

Miller, M.G., and Wright, L.A., 2002, The geology of Death Valley National Park: Dubuque, Iowa, Kendall/Hunt Publishing Company, 72 p.

Workman, J B., Menges, C.M., Page, W.R., Taylor, E.M., Elkren, E.B., Rowley, P.D., Dixon, G.L., Thompson, R.A., and Wright, L.A., 2002, Geologic map of the Death Valley ground-water model area: U.S. Geological Survey Miscellaneous Field Studies MF 2381-A.

Wright, L.A., and Troxel, B.W., 2002, Levi Noble, Geologist; His life and contributions to the geology of Death Valley, the Grand Canyon, and the San Andreas fault: U.S. Geological Survey Open File Report 02-422, 37 p.

Available online at www.sciencedirect.com

SCIENCE ⓓ DIRECT®

Earth-Science Reviews 73 (2005) 47–62

EARTH-SCIENCE
REVIEWS

www.elsevier.com/locate/earscirev

Eolian deposits in the Neoproterozoic Big Bear Group, San Bernardino Mountains, California, USA

John H. Stewart

U.S. Geological Survey, 345 Middlefield Road, Menlo Park, CA, 94025 United States

Abstract

Strata interpreted to be eolian are recognized in the Neoproterozoic Big Bear Group in the San Bernardino Mountains of southern California, USA. The strata consist of medium- to large-scale (30 cm to >6 m) cross-stratified quartzite considered to be eolian dune deposits and interstratified thinly laminated quartzite that are problematically interpreted as either eolian translatent climbing ripple laminae, or as tidal-flat deposits. High index ripples and adhesion structures considered to be eolian are associated with the thinly laminated and cross-stratified strata. The eolian strata are in a succession that is characterized by flaser bedding, aqueous ripple marks, mudcracks, and interstratified small-scale cross-strata that are suggestive of a tidal environment containing local fluvial deposits. The eolian strata may have formed in a near-shore environment inland of a tidal flat.

The Neoproterozoic Big Bear Group is unusual in the western United States and may represent a remnant of strata that were originally more widespread and part of the hypothetical Neoproterozoic supercontinent of Rodinia. The Big Bear Group perhaps is preserved only in blocks that were downdropped along Neoproterozoic extensional faults. The eolian deposits of the Big Bear Group may have been deposited during arid conditions that preceded worldwide glacial events in the late Neoproterozoic. Possibly similar pre-glacial arid events are recognized in northern Mexico, northeast Washington, Australia, and northwest Canada.
Published by Elsevier B.V.

Keywords: eolian deposits; Neoproterozoic; San Bernardino, Mountains, California, USA

1. Introduction

Proterozoic eolian deposits have been reported at only a few localities in western North America. These localities are in the Mesoproterozoic Hornby Bay Group (Ross, 1983) of the Northwest Territories,

E-mail address: stewart@usgs.gov.

0012-8252/$ - see front matter. Published by Elsevier B.V.
doi:10.1016/j.earscirev.2005.07.012

Canada; in the latest Proterozoic Browns Hole Formation in northern Utah (Christie-Blick et al., 1989); in the Mesoproterozoic Troy Quartzite in south-central Arizona (Wrucke, 1989); in the Neoproterozoic El Tápiro Formation in northwestern Mexico (Stewart et al., 2002); and in the Hazel Formation in Trans-Pecos region of westernmost Texas (Soegaard and Callahan, 1994). Worldwide, Proterozoic eolian deposits are reported elsewhere in Canada and the United States, and in Africa, Australia, Sweden, Greenland, and India (Eriksson and Simpson, 1998).

This report describes a new locality of Neoproterozoic eolian strata in the San Bernardino Mountains of southern California (Fig. 1). The deposits (1) extend the known distribution of eolian deposits, (2) contribute to

speculation that an arid desert environment existed in western North America and Australia prior to the worldwide glacial deposits of snowball earth (Hoffman and Schrag, 2000), and (3) provide information that may be useful in reconstructing of the hypothetical Neoproterozoic supercontinent of Rodinia. Paleocur-

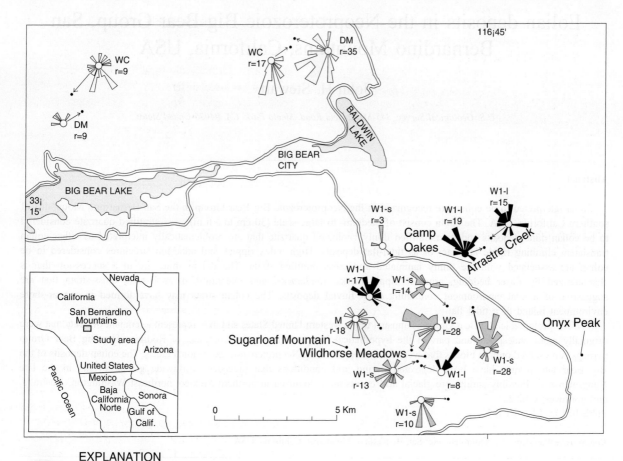

EXPLANATION

Rose diagram of paleocurrent directions

r-17   Number of readings

Symbols

WC   Wood Canyon Formation
DM   Delamar Mountain Formation
M   Moonridge Quartzite
W2   Upper member of Wildhorse Meadows Quartzite
W1-l   Medium- to large-scale cross-strata in Lower Member of Wildhorse Meadows Quartzite
W1-s   Small-scale cross-strata in Lower Member of Wildhorse Meadows Quartzite

=====   Road

Fig. 1. Index map and rose diagrams of paleocurrent directions in Neoproterozoic and Lower Cambrian strata in the San Bernardino Mountains, California. Unit in which paleocurrents were measured indicated by letter symbol or by letter and number symbol, and by different shading patterns. Rose diagrams in black are for interpreted eolian deposits.

rents in the lower part of Neoproterozoic succession in the San Bernardino Mountains have been used to indicate north and east transport of detritus from an exotic southwestern source (Barth et al., 2000a), but these paleocurrent were measured in strata here considered to be eolian. If the eolian interpretation is correct, the paleocurrent directions indicate local or regional wind patterns and may not necessarily indicate derivation of detritus from an exotic southwestern source.

In the description of eolian and associated strata in this report, small-scale, medium-scale, and large-scale cross-strata refers to cross-strata that are less than about 30 cm in length, about 30 cm to 6 m in length, and 6 m or more in length, respectively (McKee and Weir, 1953).

2. Stratigraphic setting

The eolian strata described here are in the Wildhorse Meadows Quartzite, the basal formation in the Neoproterozoic Big Bear Group (Fig. 2). The Wildhorse Meadows Quartzite rests nonconformably on the Paleoproterozoic to Mesoproterozoic Baldwin Gneiss consisting of amphibolite, muscovite biotite gneiss, schist, and minor quartzite intruded by augen and granitic gneiss intruded between 1783 and 1675 Ma (Barth et al., 2000b; A.P. Barth, written commun., 2002). Strata above the Wildhorse Meadows Quartzite (Fig. 2) consist of the remaining formations in the Big Bear Group (the Neoproterozoic Lightning Gulch Formation, Moonridge Quartzite, Greenspot Formation, and Delamar Mountains Quartzite, Fig. 2). The name Big Bear Group and the names of the formations within it were proposed by Cameron (1981, 1982). Above the Big Bear Group is the Neoproterozoic and Cambrian Wood Canyon Formation; the Cambrian Zabriskie Quartzite, Carrara Formation, Bonanza King Formation, and Nopah? Formation; the Devonian Sultan Limestone; the Mississippian Monte Cristo Limestone; and the Pennsylvanian and Permian Bird Spring Formation (Stewart and Poole, 1975; Cameron, 1981, 1982; Brown, 1991). All of these units are metamorphosed to upper greenschist to lowermost amphibolite grade (A.P. Barth, written commun., 2002).

Exposures in the San Bernardino Mountains are generally poor and the structure is complex. These factors have led to uncertainty concerning the strati-

graphy of some parts of the succession. Powell et al. (1983) mapped carbonate strata (Green Spot Formation of Cameron, 1982) on Sugarloaf Mountain as Paleozoic rather than Neoproterozoic as Cameron (1981, 1982) had done. However, the Neoproterozoic assignment of these carbonate strata seems appropriate based on the presence of stromatolites in the Green Spot Formation that resemble those found in the Precambrian rather than the Paleozoic (M.A.S. McMenamin in Stewart et al., 1984; F.A. Corsetti, 2000, oral commun.). Barth et al. (oral commun., 2002) has indicated possible differences in the Neoproterozoic stratigraphic section from area to area in the San Bernardino Mountains and, in addition, indicated the difficulty in mapping the formations proposed by Cameron.

Assignment of the Neoproterozoic and Paleozoic stratigraphic succession in the San Bernardino Mountains to major lithotectonic units is not clear. Originally, the Neoproterozoic Big Bear Group and Paleozoic strata were assigned (Cameron, 1981, 1982; Stewart et al., 1984; Brown, 1991) to the Cordilleran miogeocline—a Neoproterozoic and Paleozoic shelf deposit that formed along the western margin of Laurentia (ancestral North America) after Proterozoic rifting (Stewart, 1992). More recently, Stewart et al. (2002) indicate that the Big Bear Group, could be older than the Cordilleran miogeocline and part of the hypothetical supercontinent of Rodinia (ca. 1000 to 750 Ma) (Hoffman, 1991; Dalziel, 1992; Li et al., 1995) that existed before rifting and before development of the Cordilleran miogeocline. The Big Bear Group contains detrital zircon as young as 1.0 Ga, a characteristic of Rodinian strata in western North America (Ross et al., 1992; Rainbird et al., 1996; Stewart et al., 2001). In addition, the Neoproterozoic Big Bear Group is thick and contains abundant carbonate strata. Such a carbonate-rich succession, if it were actually part of the Cordilleran miogeocline would indicate an outer shelf position, similar to the position on the shelf of the Reed Dolomite (Stewart, 1970) in eastern California. The Lower Cambrian, and perhaps in part Neoproterozoic, Wood Canyon Formation and Lower Cambrian Zabriskie Quartzite are clearly part of the Cordilleran miogeocline and lie above the Neoproterozoic Big Bear Group. These Cambrian strata are thin and indicate an inner shelf position on the miogeocline (Stewart and Poole, 1974). The absence of Ordovician strata and the

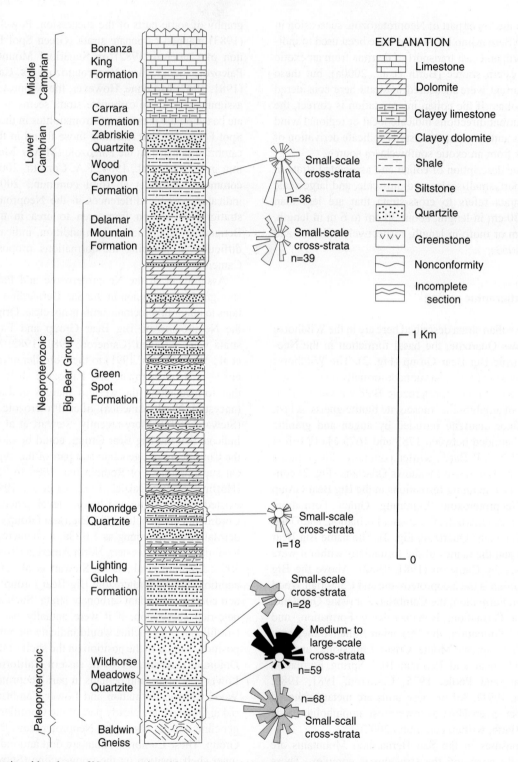

Fig. 2. Stratigraphic column of Neoproterozoic and Lower Cambrian strata in the San Bernardino Mountains, California, after Cameron (1981, 1982). Paleocurrent information from this report. Rose diagrams in black are for interpreted eolian deposits.

presence of Devonian strata unconformably on Cambrian strata in the San Bernardino Mountains are also characteristic of inner shelf deposits of the Cordilleran miogeocline (Brown, 1991). A vertical juxtaposition of Neoproterozoic outer shelf and Paleozoic inner shelf deposits would be unusual, and apparently unknown from elsewhere in eastern California and southern Nevada. Thus, the possibility that the Big Bear Group is not part of the Cordilleran miogeocline is attractive. Other features that indicate that the stratigraphic succession in the Big Bear Group is not part of the Cordilleran miogeocline are that units such as the Johnnie Formation, which is clearly part of the Cordilleran miogeocline in eastern California and southern Nevada are not recognized in the San Bernardino Mountains, although Brown (1991) suggest, I think erroneously, that a thin Johnnie Formation is present. If the Big Bear Group is truly part of the Rodinian supercontinent, it may have been preserved in a downdropped block formed in the Proterozoic, much in the same way that Proterozoic sedimentary successions are preserved in the Grand Canyon of Arizona (Timmons et al., 2001).

3. Paleocurrents

Paleocurrent directions in the Big Bear Group and the overlying Wood Canyon Formation indicate changing transport directions with time. Medium- to large-scale cross-strata interpreted as eolian in the Wildhorse Meadows Quartzite indicate paleocurrents to the north (Figs. 1 and 2). Associated small-scale tidal or fluvial cross-strata in the same unit (Figs. 1 and 2) contains variable paleocurrents directions, but with a slight tendency to be toward the west. A westerly trend is dominant in the upper part of the Wildhorse Meadows Quartzite. Paleocurrent directions are toward the south-southeast in the Moonridge Quartzite, south-southwest in the Delamar Mountain Quartzite, and toward the south in the Wood Canyon Formation (Figs. 1 and 2).

4. Eolian deposits

Strata interpreted to be eolian, or possibly eolian, are present in the Wildhorse Meadows Quartzite in two basic stratification types, medium- to large-scale

cross-strata (dune deposits) and thinly laminated strata (problematically considered to be either subcritically climbing translatent ripples or laminated tidal deposits). Structures considered to be adhesion ripples are also present. The reader should remember in reading the following descriptions that details of stratification are difficult to see because (1) in-situ exposures of strata are sparse and small, and (2) sedimentary textures and strata are commonly obscured by upper-greenschist to lowermost-amplibolite-grade metamorphism. Most strata described are on the west side of the canyon of Arrastre Creek in the NW part of sec. 26 and southwest part of sec. 23, T.2 N., R. 2 E., 0.4 to 0.8 km north-northeast of Camp Oakes (Fig. 1).

4.1. Medium- to large-scale cross-strata

4.1.1. Description

Strata containing medium- to large-scale (30 cm to >6 m) cross-strata consist of yellow-brown and light-gray, fine- to medium-grained quartzite composed of quartz, minor K-feldspar, and black opaque minerals. The latter are scattered throughout much of the quartzite and also are abundant in dark gray 1- to 2-mm-thick layers that give the strata a light and dark banded appearance. The dark opaque mineral-bearing layers are generally fine grained in contrast to the fine- to medium-grained intervening quartzite. The original texture of the sand grains is difficult to see, but obscure relic textures indicate rounded grains.

Exposures of cross-strata are generally incomplete (Figs. 3–5). Most cross-strata appear to be in trough sets, although tabular planar sets may also be present. Figs. 3 and 4 illustrates cross-strata truncated by overlying cross-strata, and Fig. 5 illustrates the wedging out of cross-strata at the bottom of a set. In most places, only a half a meter to a few meters of an individual cross-strata is exposed. In one fragmented exposure on the west side of the canyon of Arrastre Creek individual cross-strata possibly are several tens of meters long. Individual cross-stratum range in thickness from about 0.5 to 4 cm (Figs. 6 and 7), and are separated by 1- to 2-mm-thick finer-grained dark layers, the same as those mentioned above, composed of quartz, K-feldspar, and abundant black opaque minerals. Subtle inversely graded cross-strata layers are common in the upper part of cross-stratified

Fig. 3. Cross-strata in lower part of Wildhorse Meadows Quartzite. Canyon of Arrastre Creek, about 0.8 km north-northeast of Camp Oakes. Note geology pick on right for scale.

sets (Figs. 6 and 7). The original maximum dip of the cross-strata is as high as about 30°, although most readings used in the paleocurrent studies are only 10° to 20°, perhaps because readings were taken in the lower parts of downward flattening foreset cross-strata. Bed-parallel structural flattening during metamorphism may have also lowered the dip angle. Sand-filled 5- to 10-cm-wide troughs were noted rarely in exposures at right angles to the dip of the cross-strata.

Paleocurrents in the medium- to large-scale cross-strata are to the northwest, north, and northeast, and have a more consistent direction than paleocurrents in sandstone with small-scale cross strata (Figs. 1 and 2).

An unusual cross-stratified set is illustrated in Fig. 8. In this, low-angle cross-strata merge both upward and downward into horizontally laminated strata without a well-defined erosion surface at the top or bottom.

Fig. 4. Cross-strata in lower part of Wildhorse Meadows Quartzite. Canyon of Arrastre Creek about 0.4 km north-northeast of Camp Oakes. Note 14-cm-long marker pen on right for scale.

Fig. 5. Cross-strata in Wildhorse Meadows Quartzite, about 1 km west–northwest of Onyx Peak, San Bernardino Mountains. Cross-strata tangentially join lower surface of cross-stratified set.

4.1.2. Interpretation

The medium- to large-scale cross-strata are considered to be avalanche deposits on the slip-face of eolian dunes. Such an interpretation is supported by the inverse grading noted in cross-strata in the upper part of the cross-stratified set. Inverse grading in this position on eolian dunes has been noted in laboratory experiments (fig. 13 in Ahlbrandt and Fryberger, 1982), and is also characteristic of Mesozoic eolian deposits of Utah and Arizona (Loope, 1984; F. Peterson, written commun., 2000), and in modern dunes (Hunter, 1977a). In contrast to the supposed avalanche deposits described above, the cross-strata in Fig. 8 are unusual in that they do not have a slip face. This dune is very similar to an eolian dune described in the eolian Jurassic Navajo Sandstone of Utah (Hunter, 1981, fig. 9) in which the bottomset and foreset deposits are produced by grainfall and the topset deposits are subcritically climbing translatent strata.

The abundance of opaque minerals, presumably originally magnetite, is similar to the abundance of heavy minerals in some modern eolian deposits (Sharp, 1963; Inman et al., 1966; Hunter, 1977a). Two theories have been proposed to explain concentration of heavy minerals in eolian dunes: (1) heavy minerals are concentrated at the shear plain between grainfall (avalanche) deposits and underlying stable sand (Inman et al., 1966; Hunter, 1977a), and (2)

heavy minerals, because they are finer grained than quartz or feldspar in the strata, may be carried in grainfall deposits and settle onto the slip faces of dunes, in a manner similar to fine grainfall deposits on dunes described by Hunter (1977a) and Hesp and Fryberger (1988).

4.2. Thinly laminated strata

Thinly laminated strata (Fig. 10) are interstratified with the medium- to large-scale cross-strata described above. The laminated strata consist of alternating light-gray and dark-gray layers about 0.5 to 2 mm thick (Fig. 10). The light-gray layers consist of fine- to medium-grained quartz, K-feldspar, and sparse opaque minerals. The dark-gray layers consist of very fine to fine-grained quartz, feldspar, and common opaque minerals and biotite. Possible grading from the finer grained dark-gray layers to the overlying light-gray coarser layers was noted rarely, but such grading is difficult to confirm due to recrystallization of the rock during metamorphism. The top surface of thinly laminated strata locally contain ripples with ripple indexes (distance between ripple crest divided by ripple height) of 10 to 22 (Figs. 11–13). On one outcrop, the crests of the ripples were coarser than the troughs. Elsewhere, the coarseness of the crest relative to the trough could not be determined.

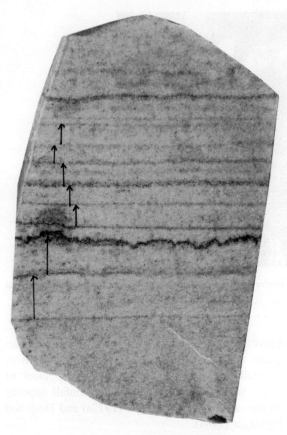

Fig. 6. Slice across dip of cross-strata in Wildhorse Meadows Quartzite. Dark layers contain abundant opaque grains. Subtle reverse grading is shown by arrows. Arrows point in stratigraphic up direction. Note stylolitic suture in lower middle part of sample. Sample from 0.7 km south of Wildhorse Meadows, San Bernardino Mountains. Sample about 13 cm from top to bottom. Interpreted to be grainflow layers on slip-face of eolian dune.

4.2.1. Interpretation

The thinly laminated strata are difficult to interpret. They may be either eolian translatent subcritically climbing ripple laminae similar to that described by Hunter (1977a,b, 1981), Fryberger et al. (1979), Kocurek (1981), Kocurek and Dott (1981), Ahlbrandt and Fryberger (1982, fig. 18A,B), Rubin and Hunter (1987), and Fryberger and Schenk (1988), Koster et al. (1993), or they may be laminated tidal deposits. In the eolian interpretation, the finer dark laminae and the coarser light laminae are produced by the passage of a single eolian ripple train. The finer laminae correspond to the trough and coarser laminae to the crest of such ripples. The scale of the lamination pair

(fine and coarse) is about 1 to 4 mm and is similar to lamination produced by eolian translatent climbing ripples in modern and ancient eolian deposits. The observation that ripples at the top of thinly laminated strata have crests that are coarser grained than the troughs supports the eolian interpretation. A coarser crest and a finer tough is characteristic of eolian ripples (Sharp, 1963; Hunter, 1977a; Fryberger and Schenk, 1988), whereas in aqueous ripples the coarse grains are concentrated in the troughs (Kocurek and Dott, 1981). The ripples at the top of laminated strata have a ripple index of 15 to 22 for ripples shown in

Fig. 7. Slice across dip of cross-strata in Wildhorse Meadows Quartzite. Dark layers contain abundant opaque grains. Subtle reverse grading is shown by arrows. Arrows point in stratigraphic up direction. Samples from 0.7 km south of Wildhorse Meadows, San Bernardino Mountains. Same locality as sample in Fig. 6. Sample about 13 cm from top to bottom. Note even laminae in lower part of sample and cross-cutting layer with disturbed stratification in upper part of sample. Even laminae are interpreted to be grainflow, and possibly grainfall layers on slip face of eolian dune. Layers range in thickness from about 4 to 25 mm. The strata with distributed stratification are interpreted to be slip-face avalanche deposits.

Fig. 8. Low angle cross-strata in Wildhorse Meadows Quartzite. Cross-strata merge with horizontally laminated strata both above and below. The cross-strata does not have a slip face. This dune is very similar to an eolian dune described in the eolian Jurassic Navajo Sandstone of Utah (Hunter, 1981, Fig. 9). At the Utah locality, foreset and bottomset deposits are of grainfall type, and topset are subcritically climbing translatent strata. Canyon of Arrastre Creek about 0.4 km north-northeast of Camp Oakes. Note pick in middle of photograph for scale.

Figs. 11 and 12 and of about 10 or higher for those in Fig. 13. These ripple indexes are more characteristic of eolian ripples than of aqueous ripples. Eolian ripples generally have ripple indexes of 10 to 40 whereas aqueous ripples have an index of 7 to 20 (Allen, 1982). The ripple marks associated with the thinly laminated strata are in a layer with the same orientation as the thinly laminated strata, and thus were originally horizontal, or nearly horizontal, in contrast to ripple marks in dunes that are oriented with their

Fig. 9. Thinly laminated strata in Wildhorse Meadows Quartzite, note pick for scale. Problematically interpreted to have been produced by translatent subcritically climbing eolian ripples (see text for alternative interpretation). X mark (above and to left of point of pick) is location of sample shown in Fig. 10. Canyon of Arrastre Creek about 0.4 km north-northeast of Camp Oakes. Note pick in lower part of photograph for scale.

Fig. 10. Slice of thinly laminated strata in Wildhorse Mountains Quartzite, problematically interpreted to have been produced by translatent subcritically climbing eolian ripples (see text for alternative interpretation). See Fig. 9 for location of sample. Canyon of Arrastre Creek about 0.4 km northeast of Camp Oakes. Sample 6 cm from top to bottom. Dark layers contain black opaque minerals (probably originally magnetite) and biotite formed by metamorphic reaction of detrital feldspar and iron-rich minerals.

long axis parallel to the down slope direction of the slip face (Inman et al., 1966; McKee, 1979; Kocurek, 1981; Ahlbrandt and Fryberger, 1982).

An alternate interpretation is that the thinly laminated sandstone is similar to laminated sandstone described in some tidal deposits (Reineck and Singh, 1975; de Klein, 1977; Weimer et al., 1982; Reading, 1996). Such a possibility is enhanced by the presence of tidal-flat deposits elsewhere in the Wildhorse Meadows Quartzite, as will be described below. A tidal-flat interpretation for the thinly laminated sandstone is supported by the somewhat irregular laminae shown in Figs. 9 and 10 which are different from the highly regular laminae of eolian translatent climbing ripples described by Hunter (1977a), Kocurek and Dott (1981) and Rubin and Hunter (1987). However, some or all of this irregularity could be due to metamorphism involving structural flattening and mineral change. If the thinly laminated sandstone is indeed tidal in origin, then the high-index ripples associated with them may be of tidal origin as well. A further possibility is that the ripples shown in Figs. 12 and 13 are indeed tidal, whereas those shown in Fig. 13 are eolian. The ripples in Fig. 13 appear to resemble eolian ripples more than those in Figs. 12 and 13, although the differences could be due in part to structural flattening and metamorphism.

The presence of biotite in the thinly laminated strata, and to a lesser extent in the cross-stratified strata, is not characteristic of modern or ancient

Fig. 11. Thinly laminated strata and ripple marks in Wildhorse Meadows Quartzite. Thinly laminated strata are on exposure below scale marker, and high index (15 to 22) ripples considered to be eolian are on top surface of thinly laminated strata. One example at this exposure showed ripples are coarser at crest than in troughs. Canyon of Arrastre Creek about 0.4 km northeast of Camp Oakes.

Fig. 12. High index (15 to 22) ripples above thinly laminated strata in Wildhorse Meadow Quartzite. Adjacent to outcrop shown in Fig. 11. Width of exposure about 4.5 m. Canyon of Arrastre Creek about 0.4 km northeast of Camp Oakes.

eolian deposits. However, the biotite does not appear to be sedimentary and is most likely the product of a metamorphic reaction of K-feldspar and iron-rich minerals (opaque minerals) during upper greenschist to lowermost amphibolite grade metamorphism. K-feldspar grains are commonly altered along their margins to muscovite, particularly in the cross-stratified strata, and to biotite in the laminated strata.

Such a relation suggests a metamorphic origin for the muscovite and biotite rather than a sedimentary origin.

4.3. Adhesion structures

Adhesion structures, formed by the adhering of dry, wind-blown sand to a wet or damp surface

Fig. 13. High index (about 10 or higher) ripples in Wildhorse Meadows Quartzite. Three parallel ripple crests shown above and to left of knife. Ripples underlain by thinly laminated strata. Canyon of Arrastre Creek about 0.4 km northeast of Camp Oakes. Knife shows scale.

(Kocurek and Fielder, 1982), are common in modern and ancient eolian deposits and on tidal flats (Kocurek, 1981; Kocurek and Fielder, 1982; Loope, 1984; Olsen et al., 1989; H.E. Clifton, 2002, personal commun.). Several examples of possible adhesion ripples were noted in the canyon of Arrastre Creek. Of these, the structures illustrated in Fig. 14 are the most like previously published examples of adhesion ripples and are similar to those illustrated in Fig. 4D of Kocurek and Fielder (1982). Structures that resemble adhesion ripples are also produced by raindrop impact (Clifton, 1977).

4.4. Associated strata

The strata interpreted to be eolian lie mostly, or perhaps entirely, within a zone from a few tens of meters to a few hundred meters above the base of the Wildhorse Meadows Quartzite. The Wildhorse Meadows Quartzite, in addition to the strata considered to be eolian, consist of very fine to fine-grained and fine- to medium-grained quartzite, and locally interstratified micaceous quartzite, quartz phyllite, laminated calc–silicate rock and quartz–plagioclase–epitote–actinolite schist (Cameron, 1981, 1982). The strata are mostly laminated to very thin bedded. Flaser bedding, aqueous ripple marks (Figs. 15 and 16), and mudcracks are common. Small-scale cross-strata in

thin trough sets are common in lenticular fine- to medium-grained quartzite layers. Two to ten meters of conglomerate is present in the basal Wildhorse Meadows Quartzite above the nonconformably underlying Paleoproterozoic to Mesoproterozoic Baldwin Gneiss.

Flaser bedding, aqueous ripple marks, mudcracks, and interstratified cross-stratified quartzite is suggestive of a tidal environment containing local fluvial deposits, in association with the eolian deposits described here.

4.5. Comparison with subaqueous medium- to large-scale cross-strata

The medium- to large-scale cross-strata described here from the Wildhorse Meadows Quartzite are in themselves not definitive of eolian deposition. Cross-strata on a similar scale are described in ancient subaqueous deposits (Kocurek and Dott, 1981) and on modern tidal and subtidal continental shelves (Belderson et al., 1982; Stride et al., 1982; Berné et al., 1988; Johnson and Baldwin, 1996). However, the presence of laminated strata problematically interpreted to have formed by eolian subcritically climbing ripples, of high-index ripples characteristic of eolian deposits, and of possible adhesion ripples are all distinctive deposits associated with eolian dunes. This associa-

Fig. 14. Possible adhesion ripples in Wildhorse Meadows Quartzite. Wind direction was from bottom to top. Canyon of Arrastre Creek about 0.4 km northeast of Camp Oakes.

Fig. 15. Low-index subaqueous ripples in lower part of Wildhorse Meadows Quartzite. Canyon of Arrastre Creek about 0.4 km northeast of Camp Oakes. Compare with high-index ripples interpreted to be eolian (Figs. 11–13).

tion gives credence to the idea that the medium- to large-scale cross-strata are eolian.

5. Regional significance

The Big Bear Group that contains the interpreted eolian deposits described here may be a unique suc-

cession which, as described above, was deposited prior to the deposition of strata of the Cordilleran miogeocline. The Big Bear Group in the San Bernardino Mountains may have been preserved in blocks bounded by Proterozoic extensional faults. Such a structural setting has been proposed for Proterozoic strata in the Grand Canyon of Arizona (Timmons et

Fig. 16. Multiple layers of low-index subaqueous ripples in lower part of Wildhorse Meadows Quartzite. Canyon of Arrastre Creek about 0.4 km northeast of Camp Oakes.

al., 2001). If this concept is true, the Big Bear Group may originally have been more widely distributed than present-day outcrops in the San Bernardino Mountains. The group apparently is older than the Cordilleran miogeocline (ca. 750 Ma) and younger than the 1100 Ma detrital zircons it contains (Vogel, 2001a,b; Stewart et al., 2001). This age span is compatible with idea that the Big Bear Group was deposited on the supercontinent of Rodinia which has a proposed age of 1200 to 750 Ma (Hoffman, 1991; Dalziel, 1992; Li et al., 1995). In this case, the Big Bear Group is similar paleogeographically to other Rodinian strata in western North America such as the Las Víboras and El Aguilla Groups in Sonora, Mexico (Stewart et al., 2002); the Crystal Springs and Beck Springs Formation in California (Wright and Prave, 1993); the upper part of the Chuar Group in Arizona (Dehler et al., 2001); the Uinta Mountains Group and Big Cottonwood Formation in Utah (Link, 1993); the Buffalo Hump Formation in Washington (Ross et al., 1992); and succession B in northwest Canada (Rainbird et al., 1996).

Although eolian dunes can develop in settings that contain tidal deposits under a humid climate, the interpreted eolian deposits of the Big Bear Group may be related to a Neoproterozoic intercontinental arid climate (Stewart et al., 2002) prior to the worldwide latest Neoproterozoic glacial events of snowball earth (Hoffman and Schrag, 2000). The time of an arid climate is characterized by late Neoproterozoic halite casts, evaporites, and eolian deposits in Australia (Warrina Supergroup) (Preiss and Forbes, 1981; Plumb, 1985; Preiss, 1987; Mc Powell et al., 1994; Walter and Veevers, 1997); by halite casts and eolian deposits in the Las Víboras Group of Sonora, Mexico (Stewart et al., 2002); by magnesite and evaporite minerals below Neoproterozoic glacial deposits in northeast Washington, USA (Miller, 2000); and by evaporite below Neoproterozoic glacial deposits in northwest Canada (Rainbird et al., 1996).

Acknowledgments

I wish to thank Andy Barth for sharing his knowledge of the geology of the San Bernardino Mountains and for his helpful criticism of the ideas presented here. However, he should not be blamed for any errors or misinterpretations presented here. D.M. Rubin and H.E. Clifton both reviewed the report. I thank them for the reviews and for help in educating me about eolian and tidal deposition. I thank D. A. John and C. Dusel-Bacon for advice on metamorphic petrography. Gordon Feely graciously provided access through the YMCA facilities at Camp Oakes to the canyon of Arrastre Creek.

References

Ahlbrandt, T.S., Fryberger, S.G., 1982. Introduction to eolian deposits. In: Scholle, P.A., Spearing, D. (Eds.), Sandstone Depositional Environments. The American Association of Petroleum Geologists, Tulsa, OK, pp. 11–47.

Allen, J.R.L., 1982. Sedimentary structures, their characteristics and physical basis. Developments in Sedimentology 30A, vol. 1. Elsevier, Amsterdam, 593 pp.

Barth, A.P., Coleman, D.S., Wooden, J.L., Stewart,, 2000a. Disassembling California: rifting and initiation of the Cordilleran miogeocline. Abstracts with Programs, vol. 32, no. 6. Geological Society of America, San Bernardino Mountains, CA, p. A-3.

Barth, A.P., Wooden, J.L., Colemen, D.S., Fanning, C.M., 2000b. Geochronology of the Proterozoic basement of southwesternmost North America, and the origin and evolution of the Mojave crustal province. Tectonics 19, 616–629.

Belderson, R.H., Johnson, M.A., Kenyon, H.H., 1982. Bedforms. In: Stride, A.H. (Ed.), Offshore Tidal Deposits. Chapman and Hall, London, England, pp. 27–55.

Berné, S., Auffret, J.-P., Walker, P., 1988. Internal structure of subtidal sandwaves revealed by high-resolution seismic reflection. Sedimentology 35, 5–20.

Brown, H.J., 1991. Stratigraphy and paleogeographic setting of Paleozoic rocks in the San Bernardino Mountains, California. In: Cooper, J.D., Stevens, C.H. (Eds.), Paleozoic Paleogeography of the Western United States II, v. 1. Pacific Section Society of Economic Paleontologists and Mineralogists, vol. 67., pp. 193–207.

Cameron, C.S., 1981. Geology of the Sugarloaf and Delamar Mountain areas, San Bernardino Mountains, California. Massachusetts Institute of Technology, Cambridge, PhD thesis, 399 pp.

Cameron, C.S., 1982. Stratigraphy and significance of the Upper Precambrian Big Bear Group. In: Cooper, J.D., Troxel, B.W., Wright, L.A. (Eds.), Geology of Selected Areas in the San Bernardino Mountains, Western Mojave Desert, and Southern Great Basin, California. Geological Society of America, Cordilleran Section Annual Meeting, Anaheim, Calif., 1982, Guidebook and Volume, pp. 5–20.

Christie-Blick, N., Mount, J.F., Levy, M., Signor, P.W., Link, P.K., 1989. Description of stops. In: Christie-Blick, N., Levy, M. (Eds.), Late Proterozoic and Cambrian Tectonics, Sedimentation, and Record of Metazoan Radiation in the Western United States. 28th International Geologic Congress, Washington, D.C. American Geophysical Union, Field Trip Guidebook T331, pp. 55–99.

Clifton, H.E., 1977. Rain impact ripples. Journal of Sedimentary Petrology 47, 678–679.

Dalziel, I.W.D., 1992. On the organization of American plates in the Neoproterozoic and the breakout of Laurentia. GSA Today, 2 (11), 237, and 240–241.

Dehler, C.M., Elrick, M., Karlstrom, K.E., Smith, G.A., Crossey, L.J., Timmons, J.M., 2001. Neoproterozoic Chuar Group (~800–742 Ma), Grand Canyon: a record of cyclic marine deposition during global cooling and supercontinental rifting. Sedimentary Geology 141–142, 465–499.

Eriksson, K.A., Simpson, E.L., 1998. Controls on spatial and temporal distribution of Precambrian eolianites. Sedimentary Geology 120, 275–294.

Fryberger, S.G., Schenk, C.J., 1988. Pin stripe laminations: a distinctive feature of modern and ancient eolian sediments. Sedimentary Geology 55, 1–15.

Fryberger, S.G., Ahlbrandt,, Andrews, S., 1979. Origin, sedimentary features, and significance of low-angle eolian "sand sheet" deposits, Great Sand Dunes National Monument, and vicinity, Colorado. Journal of Sedimentary Petrology 49 (3), 733–746.

Hesp, P., Fryberger, S.G., 1988. Eolian sediments. Sedimentary Geology 55, 1–184.

Hoffman, P.F., 1991. Did the breakout of Laurentia turn Gondwanaland inside-out? Science 252, 1409–1411.

Hoffman, P.F., Schrag, D.P., 2000. Snowball Earth. Scientific American, 68–75 (January).

Hunter, R.E., 1977a. Basic types of stratification in small eolian dunes. Sedimentology 24, 361–387.

Hunter, R.E., 1977b. Terminology of cross-stratified sedimentary layers and climbing-ripple structures. Journal of Sedimentary Petrology 47 (2), 697–706.

Hunter, R.E., 1981. Stratification styles in eolian sandstones: some Pennsylvanian to Jurassic examples from the western interior, USA. Special Publication Society of Economic Paleontologists and Mineralogists 31, 315–329.

Inman, D.L., Ewing, G.C., Corliss, J.B., 1966. Coastal sand dunes of Guerrero Negro, Baja California, Mexico. Geological Society of America Bulletin 77, 787–802.

Johnson, H.J., Baldwin, C., 1996. Shallow clastic seas. In: Reading, H.G. (Ed.), Sedimentary Environments: Process, Facies, and Stratigraphy. Blackwell Science, Oxford, England, pp. 232–280.

Klein, G. deV., 1977. Clastic Tidal Facies. CEPCO, Continuing Education Publication Company, Champaign, IL, 149 pp.

Kocurek, G., 1981. Significance of interdune deposits and bounding surfaces in aeolian dune sands. Sedimentology 28, 753–780.

Kocurek, G., Dott Jr., R.H., 1981. Distinctions and uses of stratification types in the interpretation of eolian sand. Journal of Sedimentary Petrology 51 (2), 579–595.

Kocurek, G., Fielder, G., 1982. Adhesion structures. Journal of Sedimentary Petrology 52 (4), 1229–1241.

Koster, E.A., Castel, I.I.Y., Nap, R.L., 1993. Genesis and sedimentary structures of late Holocene aeolian drift sands in northwest

Europe. In: Pye, K. (Ed.), The Dynamics and Environmental Context of Aeolian Sedimentary Systems. Geological Society London, Special Paper, vol. 72, pp. 247–267.

Li, Zheng-Xiang, Zhang, L., Christopher, M.A., 1995. South China in Rodinia: part of the missing link between Australia-Antarctica and Laurentia? Geology 23, 407–410.

Link, P.K., 1993. The Uinta Mountain Group and Big Cottonwood Formation: middle(?) and early Late Proterozoic strata of Utah. In: Reed Jr., J.C., Bickford, M.E., Houston, R.S., Link, P.K., Rankin, D.W., Sims, P.K., Van Schmus, W.R. (Eds.), Precambrian: Conterminous U.S., The Geology of North America, vol. C-2. Geological Society of America, Boulder, CO, pp. 533–536.

Loope, D.B., 1984. Eolian origin of upper Paleozoic sandstones, southeastern Utah. Journal of Sedimentary Petrology 54 (2), 563–580.

McKee, E.D., 1979. Ancient sandstones considered to be eolian. In: McKee, E.D. (Ed.), A Study of Global Sand Seas. U.S. Geological Survey Professional Paper 1052, 187–238.

McKee, E.D., Weir, G.W., 1953. Terminology for stratification and cross-stratification in sedimentary rocks. Geological Society of America Bulletin 64, 381–390.

Miller, F.K., 2000. Geologic map of the Chewelah 30′ by 60′ quadrangle, Washington and Idaho. U.S. Geological Survey Miscellaneous Field Studies Map, MF-2354.

Olsen, H., Due, P.H., Clemmensen, L.B., 1989. Morphology and genesis of asymmetric adhesion warts—a new adhesion surface structure. Sedimentary Geology 61, 277–285.

Plumb, K.A., 1985. Subdivision and correlation of late Precambrian sequences in Australia. Precambrian Research 29, 303–329.

Powell, R.E., Matti, J.C., Cox, B.F., Oliver, H.W., Wagini, A., Campbell, H.W., 1983. Mineral resource potential of the Sugarloaf roadless area, San Bernardino County, California. U.S. Geological Survey Map MF-1606-A.

Powell, C. McA., Preiss, W.V., Gatehouse, C.G., Krapez, B., Li, Z.X., 1994. South Australian record of a Rodinian epicontinental basin and its mid-Neoproterozoic breakup (~700 Ma) to form the Palaeo-Pacific Ocean. Tectonophysics 237, 113–140.

Preiss, W.V., 1987. The Adelaide geosyncline; late Proterozoic stratigraphy, sedimentation, palaeontology and tectonics. In: The Geology of South Australia. Geological Survey of South Australia Bulletin, vol. 53, 438 pp.

Preiss, W.V., Forbes, B.G., 1981. Stratigraphy, correlation, and sedimentary history of Adelaidean (Late Proterozoic) basins in Australia. Precambrian Research 15, 255–304.

Rainbird, R.H., Jefferson, C.W., Young, G.M., 1996. The early Neoproterozoic sedimentary succession B of northwestern Laurentia: correlations and paleogeographic significance. Geological Society of America Bulletin 108, 454–470.

Reading, H.G. (Ed.), 1996. Sedimentary Environments: Processes, Facies, and Stratigraphy, Third edition. Blackwell Science, Oxford, England, 704 pp.

Reineck, H.-E., Singh, I.B., 1975. Depositional Sedimentary Environments. Springer-Verlag, Berlin, 439 pp.

Ross, G.M., 1983. Proterozoic aeolian quartz arenite from the Hornby Bay Group, Northwest Territories, Canada: implications for Precambrian aeolian processes. Precambrian Research 20, 149–160.

Ross, G.M., Parrish, R.R., Winston, D., 1992. Provenance and U–Pb geochronology of the Mesoproterozoic Belt Supergroup (northwestern United States): implications for age of deposition and pre-Panthalassa plate reconstructions. Earth and Planetary Science Letters 113, 57–76.

Rubin, D.M., Hunter, R.E., 1987. Field guide to sedimentary structures in the Navajo and Entrada Sandstones in southern Utah and northern Arizona. In: Geologic Diversity of Arizona and its margins: Excursions to Choice Areas. Arizona Bureau of Geology and Mineral Technology Special Paper, vol. 5, pp. 126–138.

Sharp, R.P., 1963. Wind ripples. Journal of Geology 71, 617–636.

Soegaard, K., Callahan, D.M., 1994. Late Middle Proterozoic Hazel Formation near Van Horn, Trans-Pecos Texas: evidence for transpressive deformation in Grenville basement. Geological Society of America Bulletin 106, 413–423.

Stewart, J.H., 1970. Upper Precambrian and Lower Cambrian strata in southern Great Basin, California and Nevada. U.S. Geological Survey Professional Paper 620, 206 pp.

Stewart, J.H., 1992. Late Proterozoic and Lower Cambrian rocks. In: Burchfiel, B.C., Lipman, P.W., Zoback, M.L. (Eds.), The Cordilleran Orogen: Conterminous U.S., The Geology of North America, vol. G-3. Geological Society of America, Boulder, CO, pp. 16–20.

Stewart, J.H., Poole, F.G., 1974. Lower Paleozoic and uppermost Precambrian Cordilleran miogeocline, Great Basin, western United States. In: Dickinson, W.R. (Ed.), Tectonics and Sedimentation. Society of Economic Paleontologists and Mineralogists Special Publication, vol. 22, pp. 28–57.

Stewart, J.H., Poole, F.G., 1975. Extension of the Cordilleran miogeosynclinal belt to the San Andreas fault, southern California. Geological Society of America Bulletin 86, 205–212.

Stewart, J.H., McMenamin, M.A.S., Morales-Ramirez, J.M., 1984. Upper Proterozoic and Cambrian rocks in the Caborca region, Sonora, Mexico—physical stratigraphy, biostratigraphy, paleocurrent studies, and regional relations. U.S. Geological Survey Professional Paper 1309, 36 pp.

Stewart, J.H., Gehrels, G.E., Barth, A.P., Link, P.K., Christie-Blick, N., Wrucke, C.T., 2001. Detrital zircon provenance of Mesoproterozoic to Cambrian arenites in the western United States and northwestern Mexico. Geological Society of America Bulletin 113, 1343–1356.

Stewart, J.H., Amaya-Martínez, R., Palmer, A.R., 2002. Neoproterozoic and Cambrian strata of Sonora, Mexico: Rodinian super-continent to Laurentian Cordilleran margin. In: Barth, A. (Ed.), Contributions to Crustal Evolution of the Southwestern United States. Geological Society of America Special Paper, vol. 365, pp. 5–47.

Stride, A.H., Belderson, R.H., Kenyon, N.H., Johnson, M.A., 1982. Offshore tidal deposits: sand sheet and sand bank facies. In: Stride, A.H. (Ed.), Offshore Tidal Deposits. Chapman and Hall, London, England, pp. 95–124.

Timmons, J.M., Karlstrom, K.E., Dehler, C.M., Geissman, J.W., Heizler, M.T., 2001. Proterozoic multistage (ca. 1.1 and 0.8 Ga) extension recorded in the Grand Canyon Supergroup and establishment of northwest- and north-trending tectonic grains in the southwestern United States. Geological Society of America Bulletin 113, 163–180.

Vogel, M.B., 2001a. Detrital zircon age provenance of Late Precambrian to Early Cambrian sediments from the San Bernardino Mountains, California. Abstracts with Programs, Geological Society of America 33 (3), A45.

Vogel, M.B., 2001b. Detrital zircon age provenance of Late Precambrian to Early Cambrian sediments from the San Bernardino Mountains, California. Abstracts with Programs, Geological Society of America 33 (5), 42.

Walter, M.R., Veevers, J.J., 1997. Australian Neoproterozoic paleogeography, tectonic, and supercontinental connections: Australian Geological Society Office. Journal of Australian Geology and Geophysics 17 (1), 73–93.

Weimer, R.J., Howard, J.D., Lindsay, D.R., 1982. Tidal flats. In: Scholle, P.A., Spearing, D. (Eds.), Sandstone Depositional Environments. The American Association of Petroleum Geologists, Tulsa, OK, pp. 191–245.

Wright, L.A., Prave, A.R., 1993. Proterozoic–Early Cambrian tectonostratigraphic record in the Death Valley region, California-Nevada. In: Reed Jr., J.C., Bickford, M.E., Houston, R.S., Link, P.K., Rankin, D.W., Sims, P.K., Van Schmus, W.R. (Eds.), Precambrian: Conterminous U.S., The Geology of North America, vol. C-2. Geological Society of America, Boulder, CO, pp. 529–533.

Wrucke, C.T., 1989. The Middle Proterozoic Apache Group, Troy Quartzite, and associated diabase of Arizona. In: Jenny, J.P., Reynolds, S.J. (Eds.), Geologic Evolution of Arizona. Arizona Geological Society Digest 17, 239–258.

Available online at www.sciencedirect.com

SCIENCE d DIRECT°

Earth-Science Reviews 73 (2005) 63–78

EARTH-SCIENCE

REVIEWS

www.elsevier.com/locate/earscirev

The relationship between the Neoproterozoic Noonday Dolomite and the Ibex Formation: New observations and their bearing on 'snowball Earth'

Frank A. Corsetti [a,*], Alan J. Kaufman [b]

[a]*Department of Earth Sciences, University of Southern California, Los Angeles, CA 90089-0740, United States*
[b]*Department of Geology, University of Maryland, College Park, MD 20742-4211, United States*

Abstract

The Neoproterozoic Ibex Formation (Death Valley region, California) is commonly interpreted as a coeval basinal facies to the Noonday Dolomite carbonate platform. However, in some areas (e.g., the Black Mountains, Death Valley), the Ibex Formation is found to rest on the eroded surface of the lower Noonday Dolomite and older units. Sediment-filled grikes root from the top of the eroded lower Noonday Dolomite, followed by the subsequent deposition of the Ibex Formation. Thus, the lower Noonday Dolomite is not considered coeval with all of the Ibex Formation as they are separated by a significant unconformity.

At the type section in the Ibex Hills, the basal Ibex Formation commonly consists of polymict conglomerate and laminated mudstone; the upper surface of the mudstone is pierced by large angular clasts of underlying units, including distinctive lower Noonday Dolomite tubestone lithotypes. Here, a finely-laminated pink dolostone that records negative $\delta^{13}C$ values caps the basal Ibex conglomerate.

Several interpretations of the new observations are possible. The erosional unconformity upon which the Ibex is deposited may be glacio-eustatic in origin, the basal conglomerate would represent glaciogenic ice rafted debris, and the overlying dolostone is a classic cap carbonate (noted atop many Neoproterozoic glacial deposits worldwide). Combined with the record from underlying units, the Death Valley succession would then unambiguously record three discrete Neoproterozoic ice ages with cap carbonates in a single succession. Alternatively, the sequence boundary could represent local tectonic activity rather than glacioeustacy.
© 2005 Elsevier B.V. All rights reserved.

Keywords: Neoproterozoic; glaciation; Death Valley; Ibex Formation; Noonday Dolomite

1. Introduction

Neoproterozoic stratigraphic successions are char-acterized by glacial deposits, some of which record equatorial paleolatitudes indicative of the most extreme climate changes in Earth history (e.g., Ham-

* Corresponding author.
E-mail addresses: fcorsett@usc.edu (F.A. Corsetti),
kaufman@geol.umd.edu (A.J. Kaufman).

0012-8252/$ - see front matter © 2005 Elsevier B.V. All rights reserved.
doi:10.1016/j.earscirev.2005.07.002

brey and Harland, 1981; Crowell, 1999; Sohl et al., 1999). The environmental conditions surrounding the glaciations and their aftermath have been considered so severe that their combined effect on the biosphere is predicted to have been extreme (Hoffman et al., 1998a,b; Hoffman and Schrag, 2000). Complex animals are thought to have arisen only after the last of the snowball ice ages (Narbonne and Gehling, 2003). Most of these glacial deposits, regardless of paleolatitude, are overlain by isotopically anomalous "cap carbonates," indicating sudden increases in shallow ocean alkalinity in post-glacial oceans (e.g., Kaufman et al., 1997; Hoffman et al., 1998b; Kennedy et al., 1998). Despite the apparently global nature of the glacial deposits and associated cap carbonates, correlation has been strongly debated due to the absence of robust biostratigraphy and radiometric age constraints that commonly plague Precambrian deposits (see Knoll, 2000). Carbon and strontium isotope chemostratigraphy has been used to examine the temporal equivalence of ice ages worldwide, but these data are not unambiguous (e.g., Kaufman et al., 1997; Kennedy et al., 1998).

The stratigraphic debate revolves around the number of discrete and global ice ages that occurred in the

200 million years immediately preceding the Cambrian explosion of animal phyla (two ice ages: Kennedy et al., 1998; or more, Kaufman et al., 1997; Knoll, 2000; Hoffmann et al., 2004). Most workers agree that there were two great "cryogenic intervals" in Neoproterozoic time: an older interval, commonly termed the "Sturtian", thought to have occurred between ca. 750 and 700 Ma, and a younger interval, commonly called the Marinoan or Varanger, thought to have occurred ca. 635–600 Ma (see summary in Hoffman and Schrag, 2002 and geochronology in Hoffmann et al., 2004; Condon et al., 2005). The evidence for a third glaciation and associated cap carbonate, now dated at 585 Ma with a duration of less than 1 million years, is compelling (Myrow and Kaufman, 1999; Bowring et al., 2003). The number and magnitude of glacial pulses within these broadly defined events is hotly debated.

Since Neoproterozoic successions commonly represent more missing time than preserved strata and chronostratigraphic correlation between basins is open for debate, it is important to further document the number of discrete glacial–cap carbonate couplets in any one region. Building on the work of Miller (Miller, 1982, 1983, 1985; Miller et al., 1988), Prave

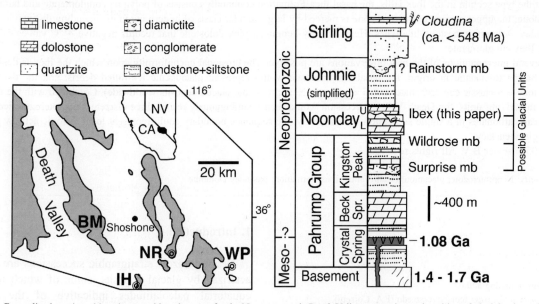

Fig. 1. Generalized stratigraphic column and location map for the Neoproterozoic Death Valley succession and position of glacial units and units possibly associated with glaciation (after Prave, 1999; Abolins et al., 2000; Corsetti and Hagadorn, 2000; Corsetti and Kaufman, 2003, this study). Location map after Stewart (1970); shaded area represents Neoproterozoic–Cambrian outcrop. BM: Black Mountains; NR: Nopah Range; IH: Ibex Hills; WP: Winters Pass Hills.

(1999) demonstrated that there were at least two Neoproterozoic glacial episodes recorded in the Death Valley succession (Prave, 1999) (Fig. 1), and correlated the lower glacial episode to the Sturtian interval and the upper to the Marinoan interval (but see Corsetti and Kaufman, 2003). Here, we continue work in the Death Valley region with a focus on the stratigraphic relations of the Neoproterozoic Noonday Dolomite and its presumed lateral equivalent, the Ibex Formation, in order to further address the number of discrete glacial–cap carbonate couplets recorded in a single stratigraphic succession.

2. Geologic background

The Noonday Dolomite is commonly interpreted as a Neoproterozoic carbonate platform notable for the presence of putative "giant stromatolitic domes" that formed large buildups along an ancient margin (Cloud et al., 1974; Williams et al., 1974a,b; Wright et al., 1974a,b, 1978). The age is poorly constrained (Fig. 1), but it has been correlated based on the chemostratigraphic record to either the post-Sturtian interval (Corsetti, 1998) or the post-Marinoan interval (Prave, 1999). It has been informally divided into a lower "algal" member (containing the domes) thought to have been deposited predominantly below wave base and an upper member composed of dolostone and sandy dolostone with abundant current related sedimentary structures, teepee structures, carbonate shrub structures and stromatolites (Stewart, 1970; Cloud et al., 1974; Williams et al., 1974a,b; Wright et al., 1974a,b, 1978; Fraiser and Corsetti, 2003). In a few localities, a middle member, interpreted to represent inter-dome deposits (Cloud et al., 1974), consists of mudstone and siltstone. Previous workers considered all members conformable, and changes in water depth during the deposition of the Noonday Dolomite were attributed to vertical tectonics (Wright et al., 1978). Prave (1999) demonstrated that the lower Noonday Dolomite represents a cap carbonate atop the glacial diamictites of the underlying Kingston Peak Formation; this unit preserves typical cap carbonate lithofacies (tubestones and sheetcrack-like cement fabrics) and a pronounced negative $\delta^{13}C$ anomaly (see also Corsetti, 1998; Abolins et al., 2000; Corsetti and Kaufman, 2003).

The Ibex Formation is commonly interpreted as a basinal facies adjacent to the contemporaneous Noonday Dolomite platform (Williams et al., 1974b; Wright et al., 1974a,b, 1984). At the type section in the Ibex Hills (Wright et al., 1984), the basal Ibex Formation rests upon diamictites and other strata correlated here to the older diamictites (the Surprise member) of the Kingston Peak Formation (e.g., Miller, 1983) based on the clast composition (abundant carbonate debris) and associated lithotypes (iron-rich, red tinted turbidites with carbonate lone-stones) (Fig. 2). These strata are distinct from the uppermost glacial deposit of the Kingston Peak Formation (the Wildrose member) that commonly contains a high percentage of basement clasts and a green muddy matrix (e.g., Miller, 1983; Prave, 1999). In addition, the contact between the underlying Kingston Peak Formation and the Ibex Formation appears mildly angular in the Ibex Hills region, suggesting a significant stratigraphic break. In the Black Mountains, the Ibex Formation rests unconformably on the Beck Spring Dolomite, Kingston Peak Formation, and lower Noonday Dolomite (Fig. 3). In this section, the lower Noonday Dolomite thins to the west while the Ibex Formation thickens; this stratigraphic relationship has been interpreted to represent thinning of the Noonday Dolomite platform as it approached the shelf edge, and the thickening of the Ibex Formation from the shelf edge into the basin (Wright et al., 1984). From these relationships, it would appear that several kilometers of strata are missing between the underlying units and the overlying Ibex Formation, represented elsewhere by glacial and interglacial siliciclastic and carbonate sedimentation, including carbonate units that record highly positive $\delta^{13}C$ values (Prave, 1999; Corsetti and Kaufman, 2003).

At the type section, the basal Ibex Formation consists of polymict conglomerate (Figs. 2A and 4) and laminated mudstone (Fig. 2B) (the Conglomerate member). The polymict conglomerate is distinct from underlying Kingston Peak Formation strata as it contains lithified clasts of the underlying diamictite, as well as the distinctive lower Noonday Dolomite tubestone. The upper surface of the mudstone is pierced by large (some greater than 5 m in diameter) angular clasts of all underlying units, including the lower Noonday tubestone (Fig. 2B, F). The Conglomerate member is overlain by a finely laminated dolo-

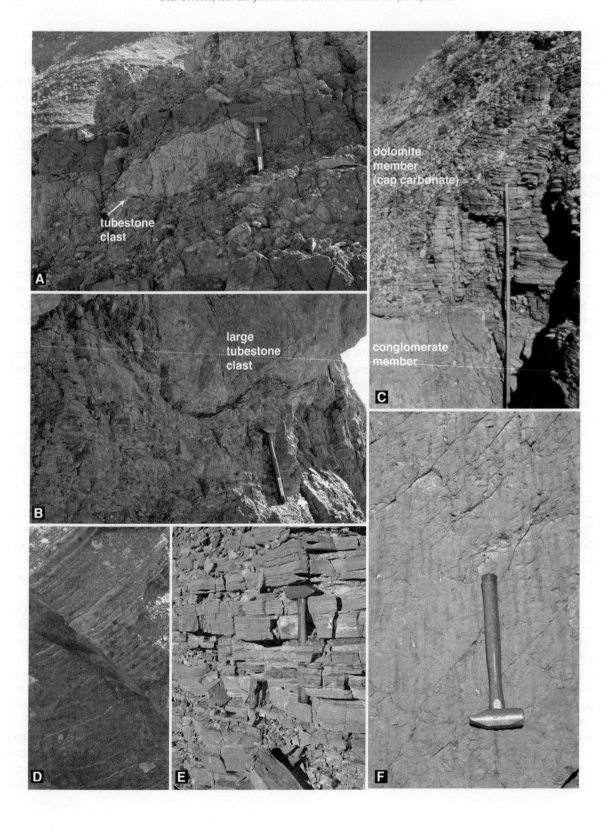

stone (the Dolomite member) (Fig. 2C), which is in turn covered by a fining then coarsening sequence of laminated purple siliciclastic shale, siltstone, and graded sands (termed the Arkose member, Fig. 2D). Two distinctive carbonate clast turbidites (~1 m thick) punctuate the primarily siliciclastic Arkose member. These beds are followed by thinly bedded limestone (the Limestone member, Fig. 2E), shaley limestone and subordinate dolostone (the Shaley Limestone member), and coarse quartz-and-dolomite clast sandstone (the Dolomite–Quartz Sandstone member).

Previous workers considered the Ibex Formation as a basinal facies to the nearby Noonday carbonate platform. The presence of lower Noonday tubestone clasts within the basal Ibex Formation conglomerate was explained as talus that fell off the platform into deeper water, and the presence of clasts from other underlying units (basement, Crystal Spring Formation, Beck Spring Dolomite, and Kingston Peak Formation) was explained as slope failure along the platform margin (e.g., Williams et al., 1974b).

3. New stratigraphic observations

New investigation has refined the interpretation of the stratigraphic relations outlined above, and allows an alternate hypothesis to be presented. It is important to note that outcrops of the Ibex Formation are discontinuous in a structurally complex area, thereby potentially complicating our view of the internal stratigraphy of this unit. We have limited our studies to the type area in the Ibex Hills and exposures in the Black Mountains, and thus our interpretations are based on extrapolation between these regions. New observations, outlined below, would suggest the lower Noonday Dolomite is separated by a significant erosional surface from the overlying units and thus temporally distinct from them.

In the Black Mountains (thought to be near the paleo-shelf edge, as discussed above; Wright et al., 1984), the contact between the lower Noonday Dolomite and the overlying Ibex Formation is not conformable and truncation of the underlying strata by the overlying unit is apparent (Fig. 5A). Sand-filled grikes, indicative of subaerial exposure, root from this surface and extend several meters into the underlying lower Noonday Dolomite (Fig. 5D); the erosional contact is overlain by pebble conglomerates of the basal Ibex Formation (Fig. 5B) followed by the Arkose member and the carbonate members. The lower member of the Noonday Dolomite is erosionally truncated from its average thickness of 200 m down to complete cutout over the course of ~2 km (rather than simply thinning and disappearing towards the shelf margin as previously suggested) (Fig. 3); the Ibex Formation is deposited upon this erosional unconformity. The magnitude of the erosional truncation is difficult to estimate, however, given uncertainties in the original thickness of the unit, especially near the shelf edge where the domes of the lower Noonday Dolomite likely stood out as major buildups.

The lower Noonday Dolomite also displays evidence for erosional truncation in platformal settings (Fig. 5C). The contact is commonly silicified and irregular; in some cases it contains a layer of carbonate breccia as well as grikes (Summa, 1993) (Fig. 5C). An erosional contact is present even in cratonal sections where dome development is absent (cf., Stewart, 1970). The contact is unremarkable, however, when compared to the incised valleys noted at the Black Mountains locality, highlighting the aerial restriction of these key features.

The Ibex Formation and the upper Noonday Dolomite form a convincing sequence deposited on the eroded top of the lower Noonday Dolomite (Fig. 6). The basal Ibex Formation is deposited on the major sequence boundary as described above, and the Dolo-

Fig. 2. The Ibex Formation. (A) Conglomerate Member, Ibex Formation, type locality of Wright et al. (1984) in the Ibex Hills. Note distinctive Lower Noonday Dolomite clast in polymict conglomerate. The conglomerate contains clasts of lithified Kingston Peak Diamictite (Surprise member), Beck Spring Dolomite, and Crystal Spring Formation. (B) Large clast of lower Noonday (>3 m wide) in finely laminated mudstone, Conglomerate member, Ibex Formation, type locality; resembles ice rafted debris. C) Top of Conglomerate member and finely laminated Dolomite member, Ibex Formation, type locality. The Dolomite member resembles a typical cap carbonate (staff is 1.5 m). (D) The siliciclastic Arkose member of the Ibex Formation (type locality). (E) The Limestone member of the Ibex Formation (type locality). (F) Detail of large tubestone clast from the lower Noonday Dolomite within the conglomerate member of the Ibex Formation. Typically, the tube fill is brown and the host rock is tan or pink. Here, the host rock is blue, suggesting that the platform and basinal depocenters record different diagenetic histories.

mite and Arkose members were emplaced during sea-level rise. The middle siltstone member of the Noonday Dolomite is tentatively equated to the shales of the lower Ibex Formation (Arkose member) and is present only where incision into the underlying platform was deep enough to preserve transgressive deposits. The bulk of the lower Noonday Dolomite platform was probably emergent during the deposition of the basal Ibex Formation and Arkose member. It is likely that the path of incised valleys may have been controlled by the presence of "giant stromatolitic domes" in the lower Noonday Dolomite, with erosion enhanced in the topographically low inter-dome areas (as suggested by Summa, 1993). The sequence architecture is more complete where incision was deeper; only the high-stand deposits of the Ibex-upper Noonday sequence are preserved in more cratonal settings. Thus, eustatic change can explain the facies patterns versus vertical basin tectonics.

4. Chemostratigraphy

Carbon and oxygen isotopic analyses were performed on stratigraphic samples from the Ibex Formation. The data and a detailed illustration of isotopic variations throughout the Ibex Formation are shown in Table 1 and plotted in Fig. 4. The finely-laminated Dolomite member atop the basal conglomerate of the Ibex Formation records a declining negative $\delta^{13}C$ trend from near -2 to $-3‰$ over 5 m of section (Fig. 4); similar trends are reproduced in equivalent sections along strike. Oxygen isotopic compositions of these fine-grained dolomicrites are remarkably constant at $\sim-8‰$ suggesting little alteration. The remaining limestone and shaley limestone members also record negative $\delta^{13}C$ values. However, $\delta^{18}O$ values of these samples are highly negative, and caution should be taken in the interpretation of isotopic trends from this interval.

Previous studies in platformal settings have shown that the lower Noonday Dolomite records negative but invariant $\delta^{13}C$ values of $\sim-3‰$ PDB (Prave, 1999; Abolins et al., 2000; Corsetti and Kaufman, 2003), rather than the declining trend noted in the Ibex Formation (Fig. 7). In the Nopah Range, the $\delta^{13}C$ values are relatively constant $\sim3‰$ through the lower Noonday Dolomite and then jump abruptly to more negative values in the basal upper Noonday, finally returning to slightly less negative values throughout the remaining upper Noonday Dolomite (Abolins et al., 2000) (Fig. 7). In more cratonal settings (Winters Pass area), $\delta^{13}C$ values record an abrupt 2‰ jump across the lower–upper Noonday contact on the platform (a pattern not inconsistent with disconformity) (Corsetti and Kaufman, 2003) (Fig. 7).

5. Discussion

Our observations from the Black Mountains area demonstrate that the lower Noonday Dolomite is separated from overlying units (the upper Noonday Dolomite on the platform and the Ibex Formation in the basin) by a significant hiatus and exposure surface (Fig. 6). The Ibex Formation and the upper Noonday Dolomite are deposited on or above this sequence boundary. In our view, it is unlikely that a coeval basinal facies would receive such volumes of siliciclastic sediments (the Arkose member) and such a dearth of carbonate sediments if a major, actively-forming carbonate platform was adjacent, supporting our interpretation that the Arkose member of the Ibex Formation is not time equivalent to the lower Noonday Dolomite.

At the type locality in the Ibex Hills, the basal Ibex Formation more closely resembles glaciomarine lithofacies than carbonate platform talus: large clasts of varying lithology, some in excess of 5 m, are noted to pierce finely laminated mud, resembling ice rafted debris. The basal Ibex conglomerate contains clasts of all the underlying units including distinctive lower Noonday Dolomite tubestone, and was not noted in our study sections to interfinger with the lower Noonday Dolomite. The finely laminated Dolomite member

Fig. 3. Top: Oblique aerial photograph of the southern Black Mts., Death Valley, looking to the southeast. Bottom: A portion of the Confidence Hills 15′ quadrangle modified from Wright and Troxel (1984), near NE corner of section 16, T21N; R3E and corresponding to much of the area shown in the aerial photograph. Note that the lower Noonday Dolomite thins to the right and the Ibex Formation thickens to the right. The map relations have alternately been interpreted to represent depositional thinning of the lower Noonday Dolomite (e.g., Wright and Troxel (1984)) or erosional cutout of the lower Noonday Dolomite (this study).

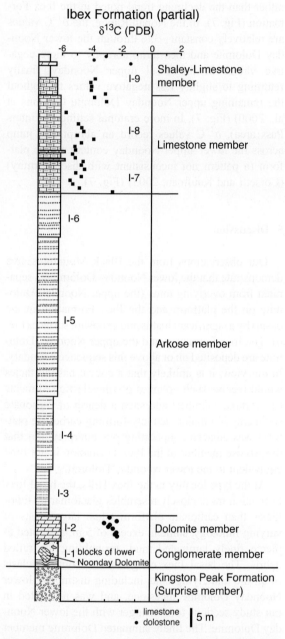

Fig. 4. Partial stratigraphic column showing important lithologic and chemostratigraphic characteristics of the Ibex Formation, type section, Saddle Peak Hills (of Wright and Troxel, 1984). Numbered letters (I-1, I-2, etc.) correspond to subunits described in detail by Wright and Troxel (1984). The Quartz–Dolomite Sandstone member (above the Shaley Limestone member) was omitted from this section but is shown in Fig. 6.

of the Ibex Formation records a declining negative $\delta^{13}C$ trend (this carbonate unit was not noted at the Black Mountains section). This trend matches those found in many other Neoproterozoic cap carbonates worldwide (Kaufman et al., 1997; Hoffman et al., 1998b; Kennedy et al., 1998). Thus, we entertain the concept that the Dolomite member represents a classic cap carbonate based upon its lithology, isotopic signature, and stratigraphic position above putative glacial strata (Figs. 2 and 4).

How does the Dolomite Member of the Ibex Formation correlate to the rest of the Noonday Dolomite? Since carbon isotope profiles simply record positive and negative excursions, it is not possible to unambiguously correlate between sections without other chronostratigraphic data. Given the data at hand, we prefer to interpret the Dolomite member of the Ibex Formation, which records a declining $\delta^{13}C$ trend, as a cap carbonate atop the Conglomerate member that is not represented in the platformal sections which may have been emergent at this time before sealevel rose enough to flood the platform, but the data are equivocal (Fig. 7). The Limestone and Shaley Limestone members of the Ibex Formation, which record a positive $\delta^{13}C$ trend, may correlate with the upper Noonday Dolomite, which also records a positive trend (Fig. 7), but these samples record anomalously low $\delta^{18}O$ values and should be interpreted with caution. The traditional correlation (A in Fig. 6) would suggest that the Dolomite Member is mostly or entirely equivalent to both the lower and upper Noonday Dolomite, but this neglects the presence of a major sequence boundary between the lower and upper Noonday Dolomite.

Building on previous work (Miller, 1983, 1985), Prave (1999) demonstrated that the Death Valley succession contains at least two glacial–cap carbonate pairs (Fig. 1). The first couplet consists of the (informal) Surprise diamictite of the Kingston Peak Formation overlain by the Sourdough limestone, and the second couplet consists of the Wildrose diamictite overlain by the lower Noonday Dolomite. Here, we demonstrate that it is permissible to interpret a third diamictite–cap carbonate pair overlying the previously defined couplets, represented by the basal Ibex Formation Conglomerate and Dolomite members. In platformal settings, the Noonday Dolomite might represent two cap carbonates stacked on top of one another. Thus, we suggest that there were at least three glacial–cap carbonate forming episodes in Neoproterozoic time.

6. The Noonday Dolomite: stacked cap carbonates?

Our interpretation of the stratigraphic relations outlined above leads us to conclude that the lower Noonday Dolomite and the upper Noonday Dolomite may represent temporally separate cap carbonates. Based on a literature survey and our own observations these enigmatic deposits would be defined based on the following criteria.

6.1. Sequence architecture

All known cap carbonates are deposited on underlying glacial strata or a correlative hiatus caused by glacial drawdown of sealevel or glacial erosion. The cap carbonates are then deposited on a transgressive surface and appear to fill some or all of the available accommodation space (Kennedy, 1996; Hoffman et al., 1998a; Hoffman and Schrag, 2002).

6.2. Negative $\delta^{13}C$ anomaly

The carbon isotopic composition of the oceans between ca. 750 and ca. 590 Ma was highly enriched (some $\delta^{13}C$ values exceed +12‰ or greater) except for short negative spikes presumed to be associated with the glaciations (e.g., Kaufman et al., 1997). All known cap carbonates display negative $\delta^{13}C$ values and are therefore anomalous with regard to the majority of the Neoproterozoic $\delta^{13}C$ record.

6.3. The presence of unique carbonate facies and fabrics

Most cap carbonates are composed of thin (meters to tens of meters), finely laminated, allodapic dolostone (much like the dolomite member of the Ibex Formation) (e.g., Hoffman and Schrag, 2002). However, other unusual carbonate textures are common, including rollup structures, sheetcrack cements, pseudo-teepee structures, tube-rock, and aragonite–

pseudomorph sea floor precipitates indicative of extraordinarily high seawater alkalinity. In addition, some caps preserve a thin layer of sedimentary barite sandwiched between basal dolostone and overlying precipitate fabrics (Hoffman and Schrag, 2002). Signs of unusually high post-glacial seawater alkalinity are most notable as discrete episodes in predominantly siliciclastic successions. The lower Noonday Dolomite has received a fair amount of attention with regard to its cap carbonate characteristics because it rests upon glacial deposits (e.g., Prave, 1999), was deposited on a flooding surface during post-glacial transgression, records negative $\delta^{13}C$ values, and contains unusual "tubestones" also found in some Namibian cap carbonates (the Maieberg Formation, Otavi Group and the Bildah Member, Witveli Group) (Hegenberger, 1987; Hoffman and Schrag, 2002; Corsetti and Grotzinger, 2005). In our view, the lower Noonday Dolomite is clearly a cap carbonate.

The upper Noonday Dolomite, now considered temporally distinct from the lower Noonday Dolomite, has received little evaluation of its potential cap carbonate status because it does not rest on a known glacial deposit. Here, we correlate the sequence boundary upon which the upper Noonday Dolomite is deposited to glacio-eustacy. It initially records negative $\delta^{13}C$ values (a declining trend, notably similar to the Dolomite Member of the Ibex Formation)(Abolins et al., 2000). In the Nopah Range, the basal upper Noonday Dolomite contains teepee structures similar to those found in other cap carbonates around the world (see Kennedy, 1996). In particular, in the Winters Pass region east of the Death Valley region, the basal upper Noonday Dolomite contains an unusual kind of sedimentary structure (alternatively interpreted as an unusual teepee structure, James et al., 2001, or an unusual cuspate megaripple, see Hoffman and Schrag, 2002). Thus, the upper Noonday Dolomite contains independent evidence that it too may warrant cap carbonate status, and as a whole, the Noonday Dolomite can be interpreted to contain two distinct cap carbonates stacked upon one another.

7. Alternate interpretations

We interpret the unconformity at the top of the lower Noonday Dolomite to represent glacially-dri-

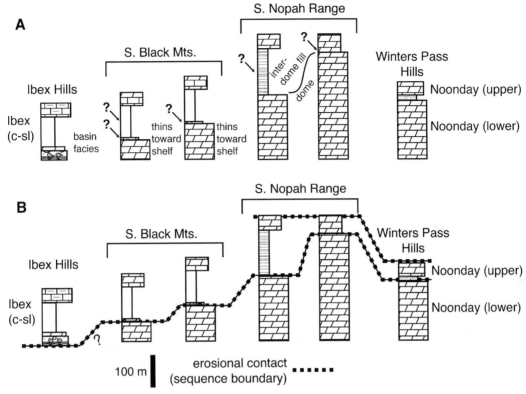

Fig. 6. Alternative correlations of the Ibex Formation and Noonday Dolomite. (A) Previously presented correlation (e.g., Williams et al., 1974a,b) where the lower and upper Noonday Dolomite are considered conformable and local thickness variations are attributed to the presence of giant stromatolitic domes between which fine sediments accumulated (e.g., the Nopah Range sections). The lower Noonday Dolomite thins and passes laterally into the Ibex Formation, a deeper basinal setting. The presence of grikes at the top of the lower Noonday Dolomite, indicative of subaerial erosion, the lack of interfingering between the lower Noonday Dolomite and the Ibex Formation, and the presence of significant amounts of siliciclastic material in the presumed equivalent Arkose member of the Ibex Formation is not consistent with this original hypothesis (see queries). (B) Alternative hypothesis (this paper), where the lower Noonday Dolomite is erosionally truncated followed by the deposition of the Ibex Formation and partially equivalent upper Noonday Dolomite (which form a sequence). This is more consistent with the new observations that indicate erosional truncation of the lower Noonday Dolomite, as discussed above, and allow for the deposition of the siliciclastic Arkose member of the Ibex Formation independent of an actively forming carbonate platform nearby.

ven sealevel drawdown. Alternatively, the unconformity may have been caused by tectonic forces. Rare faults found to crosscut the lower Noonday in the southern Nopah Range are apparently sealed by the upper Noonday (Wright et al., 1974b). The intraformational structural relations of the Noonday

Dolomite in the Panamint Range, to the west of the area presented in this report, may suggest a more complex geologic history (B. Wernicke, pers. commun., 2003), but the details remain to be examined and regional correlation of units is unclear at this time. Thus, the eustatic origin of

Fig. 5. Contact relations between the lower Noonday Dolomite and overlying units. (A) Erosional contact between the lower Noonday Dolomite and the basal Ibex Formation, Black Mts. (see Fig. 3). (B) Conglomerate at erosional contact between the Lower Noonday Dolomite and the basal Ibex Formation, Black Mts. (see Fig. 3). (C) Erosional contact between the lower Noonday Dolomite and the upper Noonday Dolomite, So. Nopah Range. (D) Grike in lower Noonday Dolomite roots from erosional surface shown in (A), Black Mts. (E) View looking north to the southern end of the Black Mountains from Highway 178 near Ashford Junction. Patches of Noonday Dolomite are visible along the Kingston Peak–Ibex Formation contact.

Table 1

Chemostratigraphic data for the Ibex Formation from the type locality of Wright and Troxel, 1984. Meters indicate height above base of member. An additional section of the Dolomite member was analyzed along strike to demonstrate the coherency of the isotopic signature

Sample Name	$\delta^{13}C$ wrt PDB	$\delta^{18}O$ wrt PDB	Meters above base of subunit
Ibex Type section, Dolomite Member			
CO2-4	−4.49	−5.20	0.5
CO2-6	−2.19	−7.99	1.5
CO2-7	−2.35	−8.33	2
CO2-8	−2.24	−7.68	2.5
CO2-9	−2.49	−7.98	3.5
CO2-11	−2.78	−8.11	4
CO2-12	−2.68	−8.02	4.5
CO2-13	−3.06	−7.94	5
CO2-14	−2.23	−7.94	5.5
CO2-16	−2.71	−6.60	6.5
Ibex type section, Limestone Member			
3-0.0	−2.63	−16.28	0
3-1.0	−4.84	−21.33	1
3-2.0	−5.27	−20.69	2
3-3.0	−5.45	−20.62	3
3-4.0	−4.72	−17.41	4
3-5.0	−5.08	−19.27	5
3-6.0	−4.49	−16.31	6
3-7.0	−4.72	−16.35	7
3-8.0	−4.60	−16.38	8
3-9.0	−5.11	−18.95	9
3-10.0	−5.29	−19.32	10
3-11.0	−5.10	−18.71	11
3-12.0	−5.13	−19.30	12
3-13.0	−4.72	−16.07	13
3-14.0	−3.89	−15.24	14
3-15.0	−4.99	−17.92	15
3-16.0	−4.56	−16.43	16
3-17.0	−4.69	−18.16	17
3-18.0	−4.92	−18.33	18
3-21.0	−5.73	−14.31	21
3-24.0	−3.04	−12.13	24
3-25.0	−3.84	−14.68	25
Ibex type section, Shaley Limestone Member			
KCO2-4-0.0	−3.12	−19.34	0
KCO2-4-1.0	−4.06	−16.79	1
KCO2-4-2.0	−4.04	−16.26	2
KCO2-4-3.0	−4.10	−16.98	3
KCO2-4-4.0	−3.80	−14.83	4
KCO2-4-8.0	−1.44	−9.32	8
KCO2-4-9.0	−3.88	−11.41	9
KCO2-4-10.0	−3.09	−10.09	10
Ibex Type section, Dolomite Member along strike from first section			
KCO2-2-0.0	−2.03	−7.91	0

Table 1 (*continued*)

Sample Name	$\delta^{13}C$ wrt PDB	$\delta^{18}O$ wrt PDB	Meters above base of subunit
Ibex Type section, Dolomite Member along strike from first section			
KCO2-2-1.0	−2.37	−8.49	1
KCO2-2-1.3	−4.41	−7.98	1.3
KCO2-2-2.2	−5.16	−8.56	2.2
Ibex Type section, Dolomite Member along strike from first section			
KCO2-2-2.7	−2.54	−7.92	2.7
KCO2-2-3.2	−3.23	−8.86	3.2
KCO2-2-3.7	−2.22	−8.91	3.7
KCO2-2-4.0	−2.29	−8.04	4

the regional relations remains the most parsimonious solution, and the striking similarity of the Conglomerate and Dolomite members of the Ibex Formation to a post-lower Noonday glacial–cap carbonate couplet is apparent.

8. Speculation and conclusions

We believe that the data presented here would argue for at least three glaciations in Neoproterozoic time with associated biogeochemical anomalies (cf., Xiao et al., 2004), although caution should be exercised here given the discontinuous outcrop pattern of the Ibex Formation. The lack of chronostratigraphic control in Neoproterozoic basins makes it difficult to determine the number of actual, discrete ice ages versus simple advance and retreat within an ice age. However, if we assume that a certain glacial magnitude is necessary to form the associated alkalinity and carbon-isotopic anomalies (that is, the severity of the ice age and the magnitude of the biogeochemical perturbations are linked), the Death Valley succession would suggest that unusually severe glaciation took place *at least* three times during Neoproterozoic time, immediately preceding the rapid evolution of megascopic animal life. If we take a simplistic approach and count the number of glacial units, cap carbonate-like units with isotopic excursions that may (or may not) indicate global glaciation, and significant sequence boundaries presumed by others to be glacio-eustatic (>100 m of erosional relief), a complex picture emerges (Fig. 9). The Death Valley succession contains three glacial

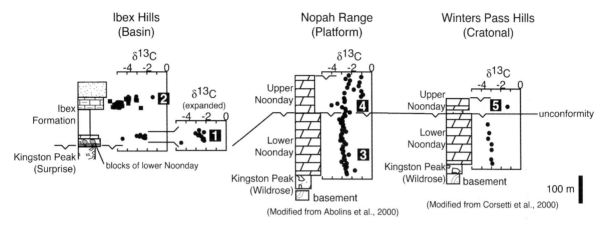

Fig. 7. Carbon isotope chemostratigraphy of the Noonday Dolomite and Ibex Formation. Since carbon isotope profiles simply record positive and negative excursions, it is not possible to unambiguously correlate between sections without other chronostratigraphic data. We prefer to interpret the Dolomite member of the Ibex Formation, which records a declining $\delta^{13}C$ trend [1], as a cap carbonate atop the Conglomerate member that is not represented in the platformal sections. The Limestone and Shaley Limestone members of the Ibex Formation, which record a positive $\delta^{13}C$ trend [2], would correlate with the upper Noonday Dolomite, which also records a positive trend, [4] and [5]. Alternatively, [1] may correlate to the basal part of [4], which both record a declining $\delta^{13}C$ trend. The traditional correlation ((A) in Fig. 6) would suggest that [1] is mostly or entirely equivalent to [3] *and* [4], but this neglects the presence of a major sequence boundary between the lower and upper Noonday Dolomite.

units with cap carbonates (discussed above), two additional cap carbonate-like units with isotopic excursions (the basal Beck Spring Dolomite and the Rainstorm carbonates in the Johnnie Formation, e.g., Corsetti and Kaufman, 2003), and one more significant sequence boundary (the incision into the Rainstorm member of the Johnnie Formation, e.g.,

Christie-Blick and Levy, 1989; Summa, 1993; Abolins et al., 2000)!

In most Neoproterozoic basins the stratigraphic record is cryptic or incomplete, which in our view is likely due to significant eustatic sea level fluctuations during the Cryogenian Era. That most successions worldwide record clear evidence for only one or two

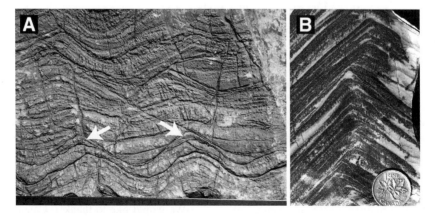

Fig. 8. (A) Teepee-like structures near the base of the upper Noonday Dolomite, Winters Pass Hills. These structures are similar to "cuspate megaripples" described by Hoffman and Schrag (2002) in the post-Icebrook cap carbonate from NW Canada (B), and reported from the Dracoisen cap carbonate, Svalbard, the Maieberg cap carbonate, Otavi Group, northern Namibia (P. Hoffman, pers. commun., 2003), and observed by Kaufman in the Bildah cap carbonate, Witveli Group, southern Namibia. The Noonday Dolomite examples are slightly smaller than the Canadian examples. Arrows point to the characteristic overlap across the apex of the structure. Photo in (B) courtesy of P. Hoffman.

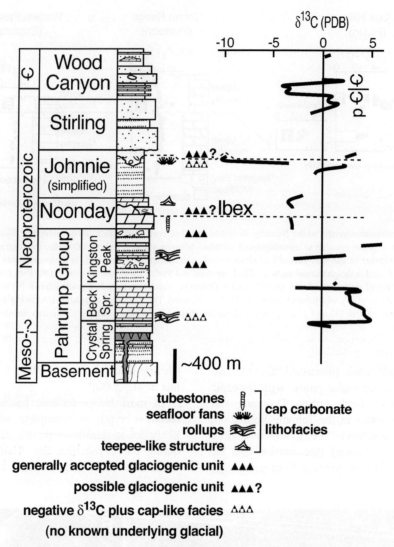

Fig. 9. Compilation of isotopic and lithologic characters of the Neoproterozoic Death Valley Succession (after Corsetti and Kaufman, 2003). The Death Valley succession presents a complex picture of Neoproterozoic Earth history, as two stratigraphic levels record accepted glacial–cap carbonate couplets with negative $\delta^{13}C$ excursions (Surprise–Sourdough members of the Kingston Peak Formation and Wildrose-lower Noonday Dolomite), and one more records a possible glacial–cap carbonate couplet with a $\delta^{13}C$ excursion (Conglomerate and Dolomite members, Ibex Formation; this paper). In addition, the major incised valley in the Rainstorm Member, Johnnie Formation, has been interpreted to represent glacial drawdown of sealevel (e.g., Abolins et al., 2000), and two other units record negative $\delta^{13}C$ excursions and cap carbonate-like facies in the absence of a known glacial deposit (basal Beck Spring Dolomite and the carbonates in the Rainstorm Member, Johnnie Formation; Corsetti and Kaufman, 2003).

ice ages highlights the dominance of base level lowering, erosion, and missing time in the stratigraphic record in a critical interval of vast climatic change and evolution of animal life. At this point, the type of clues found in the Ibex Formation, however limited in aerial extent, provide reasonable evidence for multiple ice ages and related alkalinity anomalies.

Acknowledgements

We would like to thank J. Calzia for the invitation to participate in this special publication in honor of two great geologists and mentors, Bennie Troxel and Lauren Wright. P. Marenco and N. Lorentz provided field support and valuable discussion. P. Allen, M.

Brasier, G. Halverson, and two anonymous reviews helped us greatly improve the manuscript. Peter Zeman and Vic Zawacki provided aircraft support.

References

Abolins, M., Oskin, R., Prave, T., Summa, C, Corsetti, F.A., 2000. Neoproterozoic Glacial Record in the Death Valley Region, California and Nevada. Great Basin and Sierra Nevada GSA Field Guide, volume 2. Geological Society of America, Boulder, CO, pp. 319–335.

Bowring, S.A., Myrow, P.M., Landing, E., Ramezani, J., 2003. Geochronological constraints on Neoproterozoic events and the rise of Metazoans. Astrobiology 2, 112.

Christie-Blick, N., Levy, M., 1989. Stratigraphic and tectonic framework of upper Proterozoic and Cambrian rocks in the Western United States. In: Christie-Blick, N., Levy, M., Mount, J.F., Signor, P.W., Link, P.K. (Eds.), Late Proterozoic and Cambrian Tectonic, Sedimentation, and Record of Metazoan Radiation in the Western United States. Volume Fieldtrip Guidebook T331. Am. Geophys. Union, Washington, DC. 113 pp.

Cloud, P., Wright, L.A., Williams, E.G., Diehl, P.E., Walter, M.R., 1974. Giant stromatolites and associated vertical tubes from the upper Proterozoic Noonday Dolomite, Death Valley Region, Eastern California. Geological Society of America Bulletin 85, 1869–1882.

Condon, D., Zhu, M., Bowring, S., Wang, W., Yang, A., Jin, Y., 2005. U-Pb Ages from the Neoproterozoic Doushantuo Formation, China. Science 308, 95–98.

Corsetti, F.A., 1998. Regional correlation, age constraints, and geologic history of the Neoproterozoic–Cambrian strata, southern Great Basin, USA; integrated carbon isotope stratigraphy, biostratigraphy, and lithostratigraphy [Doctoral thesis]: University of California Santa Barbara, Santa Barbara, CA, 240 pp.

Corsetti, F.A., Grotzinger, J.P., 2005. Origin of tube structures in Neoproterozoic post-glacial cap carbonates: example from Noonday Dolomite, Death Valley, United States. Palaios 20, 348–362.

Corsetti, F.A., Hagadorn, J.W., 2000. Precambrian–Cambrian transition; Death Valley, United States. Geology 28, 299–302.

Corsetti, F.A., Kaufman, A.J., 2003. Stratigraphic investigations of carbon isotope anomalies and Neoproterozoic ice ages in Death Valley, California. GSA Bulletin 115, 916–932.

Crowell, J.C., 1999. Pre-Mesozoic Ice Ages; Their Bearing on Understanding the Climate System. Geological Society of America, Boulder, CO, 106 pp.

Fraiser, M., Corsetti, F.A., 2003. Neoproterozoic carbonate shrubs: interplay of microbial activity and unusual environmental conditions in post-snowball oceans. Palaios 18, 378–387.

Hambrey, M.J., Harland, W.B., 1981. Earth's Pre-Pleistocene Glacial Record. Cambridge University Press, 1004 pp.

Hegenberger, W., 1987. Gas escape structures in Precambrian peritidal carbonate rocks. Communications of the Geological Survey of South West Africa/Namibia 3, 49–55.

Hoffman, P., Schrag, D., 2000. Snowball Earth. Scientific American 282, 68–75.

Hoffman, P.F., Schrag, D.P., 2002. The snowball Earth hypothesis: testing the limits of global change. Terra Nova 14, 129–155.

Hoffman, P.F., Kaufman, A.J., Halverson, G.P., 1998a. Comings and goings of global glaciations on a Neoproterozoic tropical platform in Namibia. GSA Today 8, 1–9.

Hoffman, P.F., Kaufman, A.J., Halverson, G.P., Schrag, D.P., 1998b. A Neoproterozoic snowball Earth. Science 281, 1342–1346.

Hoffmann, K.H., Condon, D.J., Bowring, S.A., Crowley, J.L., 2004. U–Pb zircon date from the Neoproterozoic Ghaub Formation, Namibia: constraints on Marinoan glaciation. Geology 32, 817–820.

James, N.P., Narbonne, G.M., Kyser, T.K., 2001. Late Neoproterozoic cap carbonates: Mackenzie Mountains, northwestern Canada: precipitation and global glacial meltdown. Canadian Journal of Earth Sciences 38, 1229–1262.

Kaufman, A.J., Knoll, A.H., Narbonne, G.M., 1997. Isotopes, Ice Ages, and Terminal Proterozoic Earth History. Proceedings of the National Academy of Sciences (USA), vol. 94, pp. 6600–6605.

Kennedy, M.J., 1996. Stratigraphy, sedimentology, and isotopic geochemistry of Australian Neoproterozoic postglacial cap dolostones; deglaciation, $\delta^{13}C$ excursions, and carbonate precipitation. Journal of Sedimentary Research 66, 1050–1064.

Kennedy, M.J., Runnegar, B., Prave, A.R., Hoffmann, K.H., Arthur, M.A., 1998. Two or four Neoproterozoic glaciations? Geology 26, 1059–1063.

Knoll, A.H., 2000. Learning to tell Neoproterozoic time. Precambrian Research 100, 3–20.

Miller, J.M.G., 1982. Kingston Peak Formation in the Southern Panamint Range; a Glacial Interpretation. In: Cooper, J.D., Troxel, B.W., Wright, L.A. (Eds.), Geology of Selected Areas in the San Bernardino Mountains, Western Mojave Desert, and Southern Great Basin, California. Geological Society of America Cordilleran Section Volume and Guidebook. Death Valley Publishing Company, Shoshone, CA, pp. 155–164.

Miller, J.M.G., 1983. Stratigraphy and sedimentology of the upper Proterozoic Kingston Peak Formation, Panamint Range, eastern California [Doctoral thesis]: University of California Santa Barbara, Santa Barbara, CA, 417 pp.

Miller, J.M.G., 1985. Glacial and syntectonic sedimentation; the upper Proterozoic Kingston Peak Formation, southern Panamint Range, eastern California. Geological Society of America Bulletin 96, 1537–1553.

Miller, J.M.G., Troxel, B.W., Wright, L.A., 1988. Stratigraphy and paleogeography of the Proterozoic Kingston Peak Formation, Death Valley region, Eastern California. In: Gregory, J.L., Baldwin, E.J. (Eds.), South Coast Geological Society, 1988 Field Trip: Geology of the Death Valley Region. South Coast Geological Society, Santa Ana, CA.

Myrow, P.M., Kaufman, A.J., 1999. A newly discovered cap carbonate above Varanger-age glacial deposits in Newfoundland, Canada. Journal of Sedimentary Research 69, 784–793.

Narbonne, G.M., Gehling, J.G., 2003. Life after snowball; the oldest complex Ediacaran fossils. Geology 31, 27–30.

Prave, A.R., 1999. Two diamictites, two cap carbonates, two $\delta^{13}C$ excursions, two rifts; the Neoproterozoic Kingston Peak Formation, Death Valley, California. Geology 27, 339–342.

Sohl, L.E., Christie-Blick, N., Kent, D.V., 1999. Paleomagnetic polarity reversals in Marinoan (ca. 600 Ma) glacial deposits of Australia; implications for the duration of low-latitude glaciation in Neoproterozoic time. Geological Society of America Bulletin 111, 1120–1139.

Stewart, J.H., 1970. Upper Precambrian and Lower Cambrian Strata in the Southern Great Basin, California and Nevada, U.S. Geological Survey Professional Paper 620, p. 206.

Summa, C.L., 1993. Sedimentologic, stratigraphic, and tectonic controls of a mixed carbonate–siliciclastic succession; Neoproterozoic Johnnie Formation, Southeast California [Ph.D. Dissertation thesis]. Massachusetts Institute of Technology, 616 pp.

Williams, E.G., Wright, L.A., Troxel, B.W., 1974a. Depositional environments of the late Precambrian Noonday Dolomite, southern Death Valley region, California. Geological Society of America Abstracts 6, 276.

Williams, E.G., Wright, L.A., Troxel, B.W., 1974b. The Noonday Dolomite and Equivalent Stratigraphic Units, Southern Death Valley Region, California. Guidebook; Death Valley Region, California and Nevada. Death Valley Publishing Co., Shoshone, CA, pp. 73–78.

Wright, L.A., Troxel, B.W., 1984. Geology of the northern half of the Confidence Hills 15-minute Quadrangle, Death Valley region, eastern California; the area of the Amargosa chaos, Map Sheet 34. California Division of Mines and Geology.

Wright, L.A., Troxel, B.W., Williams, E.G., Roberts, M.T., Diehl, P.E., 1974a. Precambrian Sedimentary Environments of the Death Valley Region, Eastern California. Guidebook; Death Valley Region, California and Nevada. Death Valley Publishing Co., Shoshone, CA, pp. 27–35.

Wright, L.A., Williams, E.G., Cloud, P., 1974b. Stratigraphic Cross Section of Proterozoic Noonday Dolomite, War Eagle Mine Area, Southern Nopah Range, Eastern California. Guidebook; Death Valley Region, California and Nevada. Death Valley Publishing Co., Shoshone, CA, p. 36.

Wright, L., Williams, E.G., Cloud, P., 1978. Algal and cryptalgal structures and platform environments of the late pre-Phanerozoic Noonday Dolomite, eastern California. Geological Society of America Bulletin 89, 321–333.

Wright, L.A., Williams, E.G., Troxel, B.W., 1984. Appendix II, Type section of the newly named Proterozoic Ibex Formation, the basinal equivalent to the Noonday Dolomite, Geology of the northern half of the Confidence Hills 15-minute Quadrangle, Death Valley region, eastern California; the area of the Amargosa chaos, Map Sheet 34. California Division of Mines and Geology.

Xiao, S., Bao, H., Wang, H., Kaufman, A.J., Zhou, C., Li, G., Yuan, X., Ling, H., 2004. The Neoproterozoic Quruqtagh Group in eastern Chinese Tianshan; evidence for a post-Marinoan glaciation. Precambrian Research 130, 1–26.

Available online at www.sciencedirect.com

Earth-Science Reviews 73 (2005) 79–101

EARTH-SCIENCE
REVIEWS

www.elsevier.com/locate/earscirev

Interpretation of the Last Chance thrust, Death Valley region, California, as an Early Permian décollement in a previously undeformed shale basin

Calvin H. Stevens [a,*], Paul Stone [b]

[a] *San Jose State University, San Jose, CA 95192, USA*
[b] *U.S. Geological Survey, Menlo Park, CA 94025, USA*

Accepted 1 April 2005

Abstract

The Last Chance thrust, discontinuously exposed over an area of at least 2500 km^2 near the south end of the Cordilleran foreland thrust belt in the Death Valley region of east-central California, is controversial because of its poorly constrained age and its uncertain original geometry and extent. We interpret this thrust to be Early Permian in age, to extend throughout a sedimentary basin in which deep-water Mississippian shale overlain by Pennsylvanian and earliest Permian limestone turbidites accumulated, to represent about 30 km of eastward displacement, and to be related to convergence on a northeast-trending segment of the Early Permian continental margin. Last Chance deformation occurred between the times of the Antler and Sonoma orogenies of Late Devonian–Early Mississippian and Late Permian ages, respectively, and followed Early to Middle Pennsylvanian truncation of the continental margin by transform faulting.

In the western part of the Mississippian shale basin in east-central California, the originally recognized exposures of the Last Chance thrust show Neoproterozoic and early Paleozoic strata above lower-plate Mississippian shale. Farther east, faults subparallel to bedding above, below, and within the Mississippian shale are interpreted to mark the thrust zone and to represent a regional décollement. At the eastern margin of the basin, upper-plate thrust slices of deep-water, late Paleozoic strata are interpreted to have piled up against the margin of the Mississippian carbonate shelf to form a large antiformal stack above the Lee Flat thrust, which we regard as the easternmost exposure of the Last Chance thrust. Thrust loading depressed the western part of the shelf, creating a new sedimentary basin in which about 3.5 km of younger Early Permian deep-water strata were deposited against the antiformal stack. Later, probably in the Late Permian, other thrusts, including the Inyo Crest thrust, which was subsequently overlapped by Early to Middle(?) Triassic marine strata, cut across the Last Chance thrust.

We interpret the Last Chance thrust as similar in many ways to Appalachian-type décollements in which the zone of thrusting is localized along a shale interval. The Last Chance thrust, however, has been dismembered during later geologic

---

* Corresponding author. Tel.: +1 408 924 5029; fax: +1 408 924 5053.
  E-mail address: stevens@geosun.sjsu.edu (C.H. Stevens).

events so that its original geometry has been obscured. Our model may have unrecognized analogs in other structurally complex shale basins in which the initial deformation was along a major shale unit.

Keywords: California; Death Valley; Permian; Keeler basin; décollement; thrust fault; tectonics

1. Introduction

Thrust faults and associated folds of the Cordilleran foreland thrust belt are among the most important pre-Cenozoic structures that deform Paleozoic and Neoproterozoic miogeoclinal strata in the Basin and Range province of the western United States. Many of these structures have been disrupted repeatedly during younger deformational events, some are intruded by plutons, and most are partially obscured by sedimentary or volcanic cover. This is especially true in the Death Valley region of east-central California (Figs. 1 and 2) near the south end of the thrust belt, a region of widespread Mesozoic plutonism and extreme Cenozoic extension (e.g., Dunne et al.,

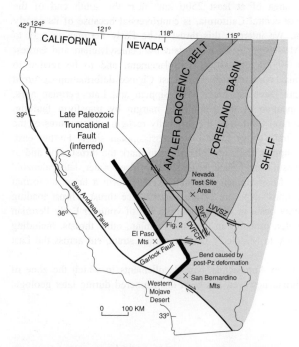

Fig. 1. Regional paleogeographic setting of east-central California, showing area (Fig. 2) in which Last Chance thrust is exposed. Modified from Stevens et al. (1991, 1992). DVFCF=Death Valley–Furnace fault; LVVSZ=Las Vegas Valley shear zone; SVF= Stewart Valley fault.

1978; Snow and Wernicke, 2000). As a result, correlation of thrusts and related folds from range to range within the region and interpretation of the original structural geometry and history of deformation are difficult.

The most significant thrust fault in the Death Valley region is the Last Chance thrust (Fig. 2), named by Stewart et al. (1966) for a widespread structure on which Neoproterozoic and lower Paleozoic rocks have been emplaced above Mississippian clastic rocks, primarily the Rest Spring Shale. Some subsequent workers (e.g., Snow, 1992) have interpreted the Last Chance thrust as part of an imbricate system of generally northeast-striking, east-vergent thrusts, although this geometry is unproven and different interpretations have been proposed (e.g., Corbett, 1990). The age of thrusting also has been controversial. The Last Chance thrust was long thought to have been active between the Middle Triassic and Middle Jurassic and broadly related to Mesozoic subduction and arc magmatism in the Sierra Nevada to the west (e.g., Dunne, 1986). More recently, however, this thrust was reinterpreted as Early Permian (Snow, 1992; Stevens and Stone, 2002) or as Middle Permian to earliest Triassic (Stevens et al., 1997), ages that carry substantially different tectonic implications. Several additional questions concerning the Last Chance thrust also remain unanswered, including the following: (1) are all exposures of the Last Chance thrust shown by Stewart et al. (1966) actually part of the same fault, or do these exposures represent two separate but related faults (e.g., Corbett, 1989; Snow, 1992)?; (2) do certain other faults in the region correlate with the Last Chance thrust as has been proposed (e.g., Stevens and Olson, 1972; Snow, 1992), or are they separate structures?; and (3) does the trace of the Last Chance thrust continue southward out of Saline Valley along the path proposed by Snow (1992) or that of Stevens et al. (1997), or is there an alternative "escape path" through the apparently continuous stratigraphy of the Inyo Mountains (see Ross, 1967a)? A number of models

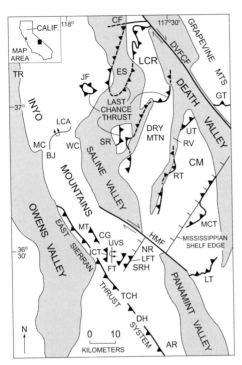

Fig. 2. Map showing trace of Last Chance thrust of Stewart et al. (1966), and other features and locations mentioned in the text. AR = Argus Range; BJ = Betty Jumbo Mine; CF = Cucomungo fault; CG = Cerro Gordo; CM = Cottonwood Mountains; DH = Darwin Hills; DVFCF = Death Valley–Furnace Creek fault; ES = Eureka salient; FT = Fishhook thrust; GT = Grapevine thrust; HMF = Hunter Mountain fault; ICT = Inyo Crest thrust; JF = Jackass Flats; LCA = Lead Canyon anticline; LCR = Last Chance Range; LFT = Lee Flat thrust; LT = Lemoigne thrust; MC = Mazourka Canyon; MCT = Marble Canyon thrust; MT = Morning Star thrust; NR = Nelson Range; RT = Racetrack thrust; RV = Racetrack Valley; SR = Saline Range; SRH = Santa Rosa Hills; TCH = Talc City Hills; TR = Tinemaha Reservoir; UT = Ubehebe thrust; UVS = Upland Valley syncline; WC = Willow Creek.

have been proposed to address these questions, but none has been fully successful.

In this paper, we critically review previous regional interpretations of the Last Chance and other thrusts in the area, point out unresolved problems and ambiguities, and propose a new structural model for resolving these issues. This model interprets the Last Chance thrust as a ramp-flat décollement similar in many ways to classic Appalachian structures such as the Pine Mountain thrust (Rich, 1934; Mitra, 1988), but now so dismembered that the original geometry is largely obscured. We propose that the Last Chance thrust cut upward through the thick miogeoclinal sec-

tion to the previously undeformed Rest Spring Shale, which it followed eastward before ramping to the surface at the original margin of the shale basin, where it locked and formed a complex antiformal stack. As interpreted here, the Last Chance thrust is of Early Permian age and distinctly older than other major thrusts in the region. At least two of these thrusts, one of which is dated as Middle Permian to earliest Triassic, apparently cut across the Last Chance thrust. These interpretations, although based on inferred structural links and therefore partly speculative, are consistent with the exposed geologic relations and are in better accord with the regional Permian paleogeography than previous models of the Last Chance thrust.

2. Regional paleogeographic setting

The area of this study in east-central California straddles the boundary between the foreland basin of the latest Devonian–Early Mississippian Antler orogenic belt and the Mississippian carbonate shelf to the southeast (Fig. 1). Deposits of the foreland basin in this area are represented primarily by the Late Mississippian Rest Spring Shale, which we interpret to have exerted substantial control on the location and geometry of the Last Chance thrust. The study area also is located close to an inferred Pennsylvanian truncational fault (e.g., Stevens et al., 1992) that has been interpreted to cut across and displace older tectonic and stratigraphic belts of the Paleozoic continental shelf, including the Antler orogenic belt. As interpreted in this report, the Last Chance thrust represents a period of deformation that took place after the Antler orogeny and continental truncation but prior to the Late Permian Sonoma orogeny, which is recognized along the trend of the older Antler belt.

3. Previous interpretations

Stewart et al. (1966) interpreted the Last Chance thrust to have originated as a single, regionally extensive, low-angle fault above which the upper plate, extending over an area of at least 2500 km², was displaced a minimum of 32 km eastward with respect to rocks below the fault. This thrust was recognized in

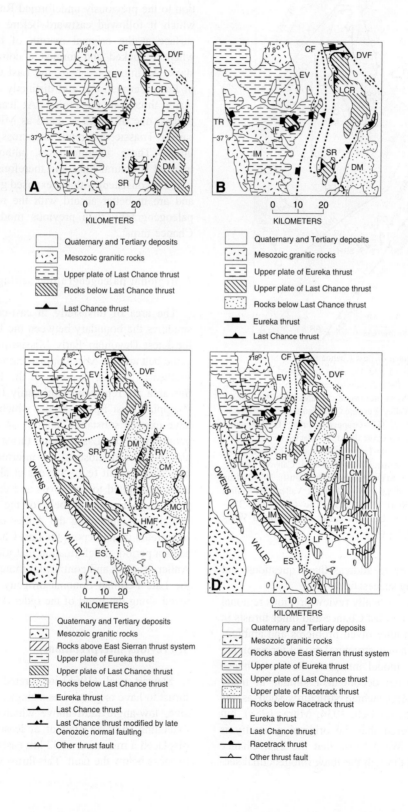

the Last Chance Range, Saline Range, Dry Mountain, and eastern Inyo Mountains (Fig. 3A) where Neoproterozoic and early Paleozoic rocks were observed to structurally overlie primarily Mississippian rocks. The age of thrusting was interpreted as post-Permian and probably Mesozoic. Later, Stevens and Olson (1972) extended the Last Chance thrust beneath the Inyo Mountains to connect with a low-angle fault on the west side of the Inyo Mountains near Tinemaha Reservoir (Fig. 2) which they referred to as the Inyo thrust.

Corbett et al. (1988) proposed that rocks above the Last Chance thrust comprise two structural plates separated by a fault later named the Eureka thrust by Corbett (1989). In this interpretation, the Eureka thrust juxtaposed two lithologically distinct sequences of Neoproterozoic and lower Paleozoic strata that Nelson (1981) had previously referred to as the Inyo (western) facies and the Death Valley (eastern) facies (Fig. 3B). Corbett (1989, Plate 17) connected the Eureka thrust with the Inyo thrust of Stevens and Olson (1972) and interpreted minimum eastward displacements of 30 and 20 km on the Eureka and Last Chance thrusts, respectively, adjusted to account for Cenozoic extension. Corbett et al. (1988) and Corbett (1989, 1990) interpreted both faults of the Last Chance thrust system as part of a continental backarc thrust belt related to the Sonoma orogeny.

Stevens and Stone (1988) inferred the existence of another thrust fault, the Lee Flat thrust, in the southeastern Inyo Mountains (Fig. 2), along which they interpreted Mississippian to Pennsylvanian basinal rocks to have been juxtaposed against coeval shelf rocks in Early Permian time. They showed that the Lee Flat thrust plate formed a large submarine ridge onlapped on both sides by younger Permian strata and suggested that the Lee Flat thrust formed as a transpressional structure related to wrench tectonism at the continental margin.

In 1992, Snow interpreted the Last Chance thrust as part of a thrust system (termed by him the Death Valley thrust belt) that also included the structurally lower Racetrack, Marble Canyon and Lemoigne thrusts. He proposed that the Last Chance thrust connected southward with the Lee Flat thrust, thereby indicating an Early Permian age for this composite fault. He considered both a one-plate and a two-plate model for the Last Chance thrust, favoring the latter (Fig. 3C) based on several lines of stratigraphic and structural evidence. In this two-plate model, Snow (1992) connected the Eureka thrust with the recumbent Lead Canyon anticline in the Inyo Mountains, similar to earlier suggestions by Stevens and Olson (1972) and Dunne (1986). These interpretations were further discussed by Stone and Stevens (1993) and Snow and Wernicke (1993).

Stevens et al. (1997) accepted the existence of a Permian thrust belt similar to that of Snow (1992) but proposed a different set of fault correlations. The Eureka thrust was connected southward to the Morning Star thrust of Elayer (1974), the Last Chance thrust was correlated with the Inyo Crest thrust of Swanson (1996), and the Racetrack thrust was considered to be a continuation of the Lee Flat thrust (Fig. 3D). These interpretations indicated a Middle Permian to earliest Triassic age for the Last Chance thrust, the independently determined age limits on the Inyo Crest thrust.

4. Problems with existing models

The studies cited above suggest that the Last Chance thrust is part of a major belt of Permian to earliest Triassic deformation, although this age interpretation depends largely on inferred structural connections. The interpretations of Corbett (1989), Snow (1992), and Stevens et al. (1997) differ in detail but all favor the existence of a two-plate model for the Last Chance thrust in which the Eureka plate overlies the Last Chance plate. In addition, the interpretations of Snow (1992) and Stevens et al. (1997) offer explanations for the southward continuation of the Last Chance thrust from the areas where it originally was defined.

Fig. 3. Generalized outcrop maps showing different previous interpretations of the Last Chance thrust. (A) Interpretation of Stewart et al. (1966), showing a single Last Chance upper plate. (B) Interpretation of Corbett et al. (1988) and Corbett (1989), in which Last Chance and Eureka plates carry Death Valley (eastern) facies and Inyo (western) facies of Neoproterozoic to Paleozoic strata, respectively. (C) Two-plate model favored by Snow (1992, Fig. 11B). (D) Interpretation of Stevens et al. (1997), which incorporates a two-plate model different from that of Snow (1992). DM=Dry Mountain block; DVF=Death Valley–Furnace Creek fault; EV=Eureka Valley; IM=Inyo Mountains; LF=Lee Flat. For other abbreviations, see caption for Fig. 2. [Note: The Cucomungo fault (CF) is shown as part of the Last Chance thrust system in (A) and part of the Eureka fault system in (B–D), although it probably is a high-angle tear fault.]

Further consideration of the regional geology, however, based in part on new field work, leads us to question some aspects of these interpretations of the Last Chance and related thrusts. Our questions revolve mainly around two issues: first, the existence of the Eureka thrust and the validity of the two-plate model, and second, the relationship between previously proposed traces of the Last Chance thrust and the regional Pennsylvanian to Early Permian paleogeography. These issues bear directly on the geometry and structural continuity of the Last Chance thrust, which in turn are central to the structural connections on which the age of thrusting has been based.

4.1. Inconclusive evidence for the two-plate model and existence of the Eureka thrust

The two-plate model of the Last Chance thrust and the concept of the Eureka thrust are primarily based on the argument that a second thrust is needed to explain the apparent juxtaposition of Neoproterozoic to early Paleozoic rocks of the Inyo and Death Valley facies in the upper plate of the Last Chance thrust as originally defined (Corbett et al., 1988; Corbett, 1989). In this model, the Eureka plate carried the Inyo facies and the underlying Last Chance plate carried the Death Valley facies. Significantly, however, only Neoproterozoic and Cambrian rocks of the Inyo facies occur near the leading edge of the Eureka plate, whereas only Cambrian and younger rocks of the Death Valley facies are exposed in the Last Chance plate of Corbett (1989). Therefore, only the Cambrian rocks can be utilized for comparison. Among these, there are well-known lithologic and stratigraphic similarities, including the presence of rocks much like the Poleta Formation (Inyo facies) in the Wood Canyon Formation (Death Valley facies), and the lithologic similarity of the Mule Spring Limestone (Inyo facies) to the middle part of the Carrara Formation (Death Valley facies) (e.g., Stewart, 1970). These similarities suggest that the Cambrian rocks change gradationally from one facies assemblage to the other, as exemplified in the Saline Range where both Inyo and Death Valley stratigraphic nomenclature has been applied to rocks of this age (Ross, 1967b; Burchfiel, 1969). In addition, nowhere do rocks of the Inyo facies lie east of correlative rocks of the Death Valley facies as one might expect if the former had been thrust over the latter.

In the northern Last Chance Range (Fig. 2), the Cucomungo fault of McKee (1985) separates areas underlain by Cambrian rocks of the Death Valley facies from Neoproterozoic rocks of the Inyo facies. This was cited by Corbett et al. (1988) and Corbett (1989) as key evidence for the existence of the Eureka thrust. The Cambrian rocks south of the fault form the basal part of the upper plate of the Last Chance thrust, as first pointed out by Stewart et al. (1966). Corbett et al. (1988) and Corbett (1989) interpreted the presence of Inyo facies rocks north of the steeply north-dipping Cucomungo fault as due to relatively recent normal displacement that down-dropped the Inyo facies rocks of the Eureka plate. At least equally plausible, however, is the original suggestion by Stewart et al. (1966) that the Cucomungo fault originated as a right-lateral strike–slip fault; this also would explain the observed stratigraphic juxtaposition. Thus, although the rocks north of the Cucomungo fault are out of place relative to the upper-plate rocks to the south, this does not necessarily suggest the existence of the Eureka thrust.

Finally, Corbett et al. (1988) and Corbett (1989) argued for the existence of two thrust faults based on structural considerations. Those authors noted that quartzite and dolomite were ductilely deformed in the Eureka plate in the western part of the region and brittlely deformed in the Last Chance plate in the eastern part of the region, and they measured greater strain intensities below the Eureka plate than below the Last Chance plate. These observations and measurements are consistent with the two-plate model, but they also would be compatible with a one-thrust model in which the upper plate thickens to the west, thereby increasing the confining pressure. Such thickening is in fact suggested by the presence of increasingly older rocks to the west in the upper plate. Other structural evidence for two thrust faults (Corbett et al., 1988; Corbett, 1989) included the observation of a 5° difference in direction of transport of the two plates, which seems insignificant, and differences in the orientations of fold axes and related structures between the two plates, which we propose are due instead to two separate deformations (see Section 11).

In summary, evidence in favor of a two-plate model of the Last Chance thrust is not conclusive and cannot be used to rule out a simpler, single-plate model similar to the original interpretation of Stewart et al. (1966).

4.2. Possible incompatibility between previous models of the Last Chance thrust and Pennsylvanian to Early Permian paleogeography

Pennsylvanian to earliest Permian rocks exposed in the Inyo Mountains and Dry Mountain block comprise a distinctive sequence of silty to sandy calcareous turbidites assigned to the Keeler Canyon Formation (Stone, 1984; Stone and Stevens, 1988). These turbidites were deposited in the Keeler basin (of Stevens et al., 2001), a successor of the southern part of the Antler foreland basin in which the Late Mississippian Rest Spring Shale had been deposited. The Keeler basin lay west of a carbonate shelf represented by coeval shallow-water limestone exposed in the eastern Cottonwood Mountains (Fig. 4). In Early Permian time, rocks of the Keeler basin were deformed and carried eastward on the Lee Flat thrust; later these rocks were overlapped by younger Early Permian strata informally called the sedimentary rocks of Santa Rosa Flat (Stevens and Stone, 1988; Stone et al., 1989). These relations tightly constrain the age of displacement on the Lee Flat thrust.

As shown in Figs. 3C, D, and 4, both Snow's (1992) connection between the Last Chance and Lee Flat thrusts and the alternative connection between the Last Chance and Inyo Crest thrusts (Stevens et al., 1997) require the Last Chance thrust to cut through the Keeler basin. In both interpretations, the Last Chance thrust separates Keeler basin rocks of very similar thickness and lithology in the southern Inyo Mountains and Dry Mountain block (Stone, 1984). This similarity, which suggests that these areas were close together when the Keeler Canyon Formation was deposited, does not preclude either previous interpretation of the Last Chance thrust but does leave both of them open to question.

5. New structural model

In light of the above and with new observations made in the southern Inyo Mountains, we have developed an alternative model in which the Last Chance thrust has a single upper plate that carried, but did not cut through, the Pennsylvanian to earliest Permian rocks of the Keeler basin. This model, which depends strongly on inferred structural connections between fault exposures that are now widely separated, is supported by field relations discussed more fully in Section 6.

The key to the new model is the Late Mississippian Rest Spring Shale, which underlies the Keeler Canyon Formation in the regional stratigraphic succession. The Rest Spring Shale is the youngest unit exposed beneath the Last Chance thrust of Stewart et al. (1966) in the northeastern Inyo Mountains, Last Chance Range, and Saline Range. East and southeast of these thrust exposures, in the southern Inyo Mountains and Dry Mountain area, the Rest Spring Shale underlies the Keeler Canyon Formation in normal stratigraphic position, but is almost invariably sheared or faulted subparallel to bedding, and its upper and lower contacts commonly are bedding-plane faults (e.g., McAllister, 1956; Burchfiel, 1969; Elayer, 1974). Thus, the Rest Spring Shale in these areas apparently defines a diffuse shear zone beneath the Keeler Canyon Formation. We propose that all of these bedding-plane faults are cor-

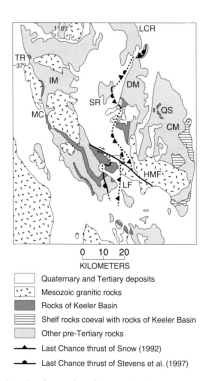

0   10   20
KILOMETERS

☐ Quaternary and Tertiary deposits

☐ Mesozoic granitic rocks

■ Rocks of Keeler Basin

☰ Shelf rocks coeval with rocks of Keeler Basin

☐ Other pre-Tertiary rocks

▲— Last Chance thrust of Snow (1992)

▲— Last Chance thrust of Stevens et al. (1997)

Fig. 4. Map showing rocks of Pennsylvanian to earliest Permian Keeler basin (from Stone, 1984) in relation to the Lee Flat thrust (=Last Chance thrust of Snow, 1992) and Inyo Crest thrust (=Last Chance thrust of Stevens et al., 1997). QS=Quartz Spring; for other abbreviations, see captions for Figs. 2 and 3. For simplicity, only selected parts of the fault traces are shown.

relative and that they connect with the Last Chance thrust exposed to the west and northwest. Eastward, we connect the shear zone in the Rest Spring Shale with the Lee Flat thrust of Stevens and Stone (1988), which coincides with the eastern edge of the Rest Spring depositional basin. Based on these inferred structural connections, we interpret the Last Chance thrust as a regional décollement that once extended continuously along the Rest Spring Shale in the eastern part of the Keeler basin.

Diagrammatic cross sections (Fig. 5) illustrate the original geometry we infer for the Last Chance thrust.

To the west, the thrust is shown as rooting in Neoproterozoic rocks beneath the Inyo Mountains as originally suggested by Stewart et al. (1966). Eastward, the thrust is shown cutting upsection as a footwall ramp to the level of the Rest Spring Shale, where it becomes a footwall flat. A large fault-bend anticline, consistent with the generally anticlinal structure of the northern Inyo Mountains (Ross, 1967a), is shown in the upper plate above the footwall ramp and flat. Neoproterozoic to lower Paleozoic upper-plate strata that dip generally eastward into the exposed fault surface (Stewart et al., 1966) are inferred to represent

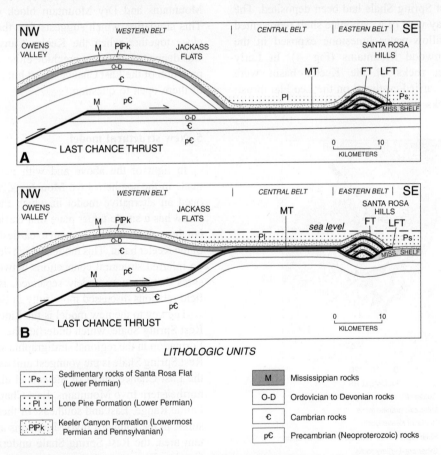

Fig. 5. Cross-sectional models of the Last Chance thrust as interpreted in this paper. (A) Section showing idealized geometry of the western ramp and fault-bend anticline, central hanging-wall flat, and eastern antiformal stack discussed in text. (B) Section modified to show a more probable actual geometry, with subsidence caused by loading beneath the fault-bend anticline and antiformal stack. FT=Fishhook thrust; LFT= Lee Flat thrust; MT=Morning Star thrust. Thrust-related structures involving rocks of Precambrian to earliest Permian age are overlapped by Lower Permian strata (Lone Pine Formation and sedimentary rocks of Santa Rosa Flat). The fault-bend anticline is based on structure of the northern Inyo Mountains, whereas the hanging-wall flat and the antiformal stack are based on geologic relations in the southeastern Inyo Mountains. In the central Inyo Mountains, about 15 km of displacement may be taken up on the Lead Canyon anticline, not shown here for the sake of simplicity. About 15 km of late Cenozoic extension east of Jackass Flats has been removed. Younger Permian to earliest Triassic thrust faults interpreted to deform the Last Chance thrust are not shown. No vertical exaggeration.

the eastern limb of this anticline. East of the anticline, the footwall flat is shown as a bedding-plane thrust that follows the Rest Spring Shale overlain by a thin hanging-wall flat composed primarily of Keeler Canyon Formation. These parts of the model are consistent with the regional geology, as is further discussed in the following section, and strongly resemble Appalachian ramp-flat thrusts such as the Pine Mountain thrust (Rich, 1934; Mitra, 1988).

At its eastern (frontal) margin, we show the Last Chance thrust as splaying and ramping to the paleosurface as the Lee Flat and Fishhook thrusts above Mississippian carbonate shelf rocks that bounded the Rest

Spring Shale depositional basin (Fig. 5). Associated with these thrusts is an Early Permian submarine ridge (Stevens and Stone, 1988), which we infer as having an internal structure similar to that of an antiformal stack of thrust sheets as defined and illustrated by various writers (e.g., Boyer and Elliott, 1982, Fig. 12). Early Permian rocks only slightly younger than the Keeler Canyon Formation depositionally onlap the antiformal stack on both its western and eastern sides, placing an upper limit on the age of thrusting.

The map trace of the Last Chance thrust, as interpreted here, is shown in Fig. 6A and B. This trace is complicated by younger folds and faults, including the

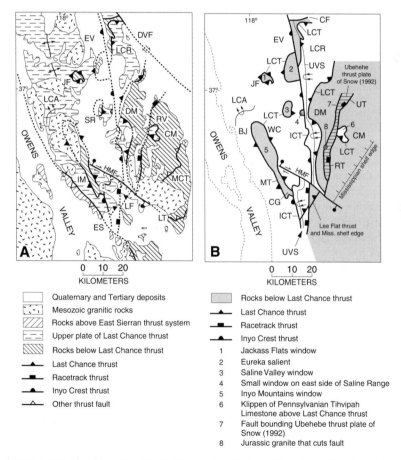

Fig. 6. (A) Generalized outcrop map showing traces of Last Chance, Racetrack, and Inyo Crest thrusts as interpreted in this report. (B) Simplified map showing traces of thrust faults, trace of Upland Valley syncline (UVS), Mississippian shelf edge, Ubehebe thrust plate of Snow (1992), and other significant features. Mississippian shelf edge in Cottonwood Mountains marks projected location of Lee Flat thrust, which is not exposed because of uplift and erosion on north side of Hunter Mountain fault. Following McAllister (1956), late Cenozoic Hunter Mountain fault (HMF) is shown as a scissors fault having opposite senses of vertical displacement at either end in addition to about 9 km of dextral offset. Large klippe of upper plate of Last Chance thrust shown near Racetrack Valley (loc. 6) is generalized to encompass all rocks of Keeler Canyon Formation in that area. For abbreviations, see captions for Figs. 2 and 3.

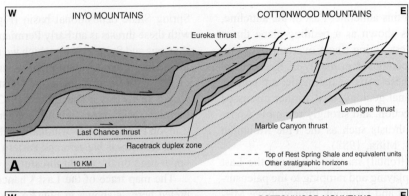

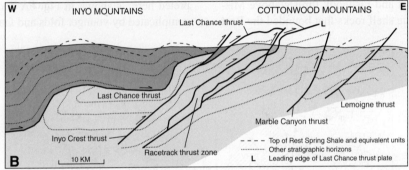

Fig. 7. (A) Cross section showing two-plate model of Last Chance and related thrusts of Snow (1992, Fig. 4). (B) The same cross section modified to conform to the one-plate model proposed in this report in which the Last Chance thrust is interpreted as a single low-angle thrust cut by the younger, steeper Inyo Crest and Racetrack thrusts.

Inyo Crest and Racetrack thrusts which are shown as cutting across the Last Chance thrust. The traces of these faults and their observed and inferred relations are discussed in Section 6.

Fig. 7 illustrates how Snow's (1992) two-plate model of the Last Chance thrust, as viewed in cross section, can be modified to conform to our proposed one-plate model. The required modifications in fault geometry, although significant, are simple and are compatible with the surface geology as shown by Snow (1992).

6. Field relations

Because of disruption by later deformation and lack of continuous exposure, the Last Chance thrust as modeled here cannot be observed in its entirety. Three major elements of the structural model, however, are exposed: a western belt, representing the western part of the footwall flat, where Neoproterozoic and lower Paleozoic rocks overlie Mississippian

rocks as young as the Rest Spring Shale; a central belt, representing the eastern part of the footwall flat, where the Keeler Canyon Formation overlies the Rest Spring Shale; and the eastern antiformal stack composed primarily of the Keeler Canyon Formation (Figs. 5 and 8).

6.1. Western belt

This belt includes exposures of the Last Chance thrust where Neoproterozoic to lower Paleozoic strata rest structurally on the Mississippian Rest Spring Shale and associated units (Stewart et al., 1966). As shown by Stewart et al. (1966), progressively younger strata of the upper plate overlie the Rest Spring Shale in a southeastward direction. In lower Mazourka Canyon in the northwestern Inyo Mountains (Fig. 2), the only area where the upper-plate stratigraphic sequence is completely preserved, the Keeler Canyon Formation depositionally overlies apparently undeformed Rest Spring Shale. As these rocks are in the upper plate of the Last Chance

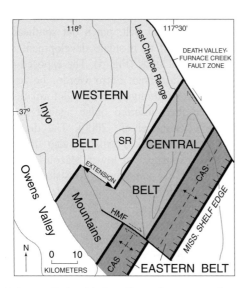

Fig. 8. Structural belts of the Last Chance thrust system. See text for explanation. Boundary between western and central belts is offset in the Saline Valley by an estimated 15 km of late Cenozoic extension. CAS=crest of antiformal stack; HMF=Hunter Mountain fault; SR=Saline Range. Western, central, and eastern belts correspond to structural belts similarly identified in Fig. 5.

of the Last Chance thrust due to later intrusion of the Jurassic Paiute Monument pluton.

6.2. Central belt

This belt is characterized by exposures of Rest Spring Shale and overlying Keeler Canyon Formation in which the Rest Spring Shale is cut by faults or shear zones subparallel to bedding at its base, internally, or at its top, and in which the lowermost part of the Keeler Canyon Formation commonly is missing due to faulting. Such relations are particularly well displayed in the southern Inyo Mountains near Cerro Gordo (Fig. 2), where Elayer (1974) mapped the Rest Spring Shale as separated from the Keeler Canyon Formation by a bedding-plane fault he called the Morning Star thrust. Both formations are strongly sheared along this fault contact. At a structurally lower level, lenticular rafts of older Mississippian strata (Tin Mountain Limestone and Mexican Spring Formation) and Keeler Canyon Formation have been inserted into the Rest Spring Shale along a complex system of anastomosing faults (Stevens et al., 1997; Stone et al., 2004), greatly increasing the structural thickness of the Rest Spring Shale interval. This thickened and deformed shale is structurally similar to ductile shale duplexes ("mushwads") described by Thomas (2001) in the Appalachian thrust belt of Alabama. In addition, about 7 km northwest of Cerro Gordo, our mapping (Stone et al., 2004) shows that the Keeler Canyon Formation is folded into a large, east-vergent, asymmetric anticline above the fault contact with the Rest Spring Shale. The Rest Spring Shale thus marks the location of a regionally extensive fault zone above which the Keeler Canyon Formation could have been displaced substantially eastward. This is consistent with our structural model (Fig. 5), which links this fault zone with exposures of the Last Chance thrust of the western belt.

The structural continuity between the western and central belts is nowhere exposed. In the Inyo Mountains, we interpret the transition to take place between the Lead Canyon anticline and the Morning Star thrust zone (Figs. 2, 6B), an area obscured by Jurassic intrusive rocks (Fig. 6A). In this area, the Last Chance thrust is inferred to cut upsection from Cambrian rocks into the Rest Spring Shale, which it

thrust, this relationship is consistent with our model (Fig. 5). The broadly antiformal upper-plate rocks in the northern Inyo Mountains (Fig. 6A) are interpreted to represent a large, originally north- to northeast-trending, fault-bend anticline highly modified by younger northwest-trending folds, and possibly steepened on its western flank by Mesozoic deformation. The form of the fault-bend anticline is similar to that in models of Salvini et al. (2001) above thrusts with ramps inclined about 35°.

Southward the anticline is inferred to extend into the west-trending, recumbent Lead Canyon anticline (Fig. 6A), which farther west apparently becomes upright, bends to the south (see Ross, 1965) and is overprinted by more recent structures. In this area, the trace of the Last Chance thrust is obscured by Jurassic plutons, but is inferred to lie between the Lead Canyon anticline and north-trending Paleozoic rocks exposed near Willow Creek and to then swing southeastward to pass east of the Cambrian rocks exposed at the Betty Jumbo Mine (Ross, 1967a) (Fig. 6A and B). The Lead Canyon anticline may mark a place where the thrust locked causing the front of the upper-plate anticline to steepen and overturn, and/or it may represent extreme distortion of the upper plate followed eastward. This interpretation is similar to that of Stevens et al. (1997) who connected the Morning Star thrust with the Eureka thrust of Corbett (1989) but considered this thrust separate from the Last Chance thrust. We here interpret the Morning Star (=Last Chance) thrust to follow the Rest Spring Shale throughout the Inyo Mountains. Later folding and faulting have complexly deformed the thrust as thus delineated, resulting in an irregular, incompletely understood fault trace. The dominant later structure is the Cerro Gordo anticline as mapped by Conrad and McKee (1985), which we interpret to fold the thrust into a northwest-trending antiform that encircles the crest and east side of the Inyo Mountains (Fig. 6A, B). Mississippian and older rocks of the lower plate are exposed in this large structural window, the Inyo Mountains window, beneath the Rest Spring Shale as shown in Fig. 6B.

The transition between the western and central belts, covered by Quaternary alluvium and disrupted by younger structures, is also inferred to take place between the Saline Range and an area south of Dry Mountain (Figs. 6 and 8). Here we place the Last Chance thrust (Fig. 6A, B) along a major fault first mapped by Burchfiel (1969) at the base of the Rest Spring Shale and later interpreted by Snow (1992) as the base of the Racetrack duplex zone. Southward from this area, the narrow outcrop of Rest Spring Shale extending to Saline Valley north of the late Cenozoic Hunter Mountain fault (Fig. 6A, B) is faulted against the Keeler Canyon Formation on the east (McAllister, 1956); we consider this fault to be the continuation of the Last Chance thrust.

In the northern Cottonwood Mountains (Fig. 6A, B), we place the trace of the Last Chance thrust at the contact between the Rest Spring Shale and Pennsylvanian Tihvipah Limestone (=lower Keeler Canyon Formation) as mapped by McAllister (1952). This contact is everywhere faulted and structurally resembles the Morning Star thrust near Cerro Gordo.

Permian strata (Lone Pine Formation) that depositionally overlie the Keeler Canyon Formation in the central belt consist primarily of thin-bedded siltstone and mudstone representing a deep-water marine environment (Stone et al., 2000). These rocks, which reach a thickness of more than 1 km in the middle of the central belt, pinch out southeastward and are interpreted to postdate development of the eastern antiformal stack (Fig. 5). The exposed pinch-out of these strata near Cerro Gordo, however, is largely a result of younger Permian deformation that has obscured their inferred onlap against the antiformal stack (Stone et al., 2000). The deep-water rocks of the Lone Pine Formation also thin north-westward to exposures in lower Mazourka Canyon (Ross, 1965). This thinning is consistent with the relations shown in our structural model (Fig. 5) in which these rocks pinch out against the upper-plate anticline of the western belt. Coarse-grained sandstone and conglomerate (Stone et al., 2000), which could have been derived from erosion of the western anticline, occur in the upper few hundred meters of the Lone Pine Formation near Mazourka Canyon (Fig. 2).

6.3. Eastern belt

This belt is characterized by rocks and structures that we interpret to represent an antiformal stack of complexly folded and thrust-faulted basinal Mississippian to Early Permian rocks, mostly Rest Spring Shale and Keeler Canyon Formation, which formed above a large basement irregularity (the Late Mississippian carbonate shelf margin) at the tip of the Last Chance thrust (=Lee Flat thrust) (Figs. 5 and 8). This stack, which formed a large, antiformal submarine ridge in Early Permian time (Stevens and Stone, 1988; Stone et al., 1989), is exposed in the southern Inyo Mountains about 9 km southeast of Cerro Gordo (Fig. 2). We infer this ridge to be underlain by the Lee Flat thrust, now covered by Cenozoic alluvium and basalt, which juxtaposed the basinal Rest Spring Shale and Keeler Canyon Formation against the Mississippian carbonate shelf margin to the southeast. Exposed within the upper plate of the Lee Flat thrust is the Fishhook thrust, which repeats the Keeler Canyon Formation and part of the underlying Mississippian section (see Fig. 4 of Stevens and Stone, 1988). We interpret the ridge to have attained a topographic relief of about 3.5 km above the adjacent shelf, the approximate thickness of the younger Permian sedimentary rocks of Santa Rosa Flat subsequently deposited against its southeast side. This amount of relief suggests the existence of at least one thrust, in addition to the Fish-hook thrust, which would have to lie buried in the subsurface. The Lee Flat, Fishhook, and inferred lower thrusts represent the primary structures of the antiformal stack modeled in Fig. 5.

The structural connection between the antiformal stack of the eastern belt and the Morning Star (=Last Chance) thrust of the central belt (Figs. 5 and 8) cannot be seen in outcrop. Rocks representing the upper plate of the Fishhook thrust, however, can be followed westward to a point about 3 km southeast of Cerro Gordo (Fig. 6B). There, these rocks are juxta-posed against the upper plate of the Morning Star thrust by a fault we interpret as the northward exten-sion of the Middle Permian to earliest Triassic Inyo Crest thrust of Swanson (1996). We interpret this later faulting to have obscured the western part of the antiformal stack and to have buried its original struc-tural connection with the Morning Star thrust.

The presence of the thick, younger Permian sedi-mentary rocks of Santa Rosa Flat that onlapped the antiformal stack from the southeast suggests that the stack depressed the older shelf, creating a new sedi-mentary basin (see also Snow and Wernicke, 1993). To the northeast, related rocks (Darwin Canyon For-mation) representing this same basin are recognized in the east-central Cottonwood Mountains, whereas the Lone Pine Formation, inferred to onlap the opposite side of the antiformal stack, is recognized in the Dry Mountain area (Stevens and Stone, 1988; see Figs. 2 and 8). Neither the stack nor the Lee Flat thrust is exposed between those two areas, however, presum-ably due to later uplift and erosion. We interpret the Mississippian shelf edge, recognized on the basis of facies relations (Stevens et al., 1991), to mark the projected position of the tip of the Lee Flat thrust in the Cottonwood Mountains (Fig. 6B) as it does in the southern Inyo Mountains. A dextral offset of about 9 km separates this projected trace from the inferred trace of the Lee Flat thrust south of the Hunter Mountain fault (Burchfiel et al., 1987).

The inferred connection between the Lee Flat thrust at the base of the antiformal stack with the Morning Star and Last Chance thrusts brings the trace of this composite structure to the southeastern side of the Inyo Mountains and thus provides the necessary "escape path" for the Last Chance thrust south of Saline Valley (Fig. 6A).

7. Age of thrusting

The age of the Last Chance thrust is not tightly constrained in the area where it was originally recog-nized. If this thrust connects with the Fishhook and Lee Flat thrusts, as suggested by Snow (1992) and here, however, its age is Early Permian (middle Wolf-campian) based on detailed stratigraphic relations between deformed Keeler Canyon Formation in the antiformal stack and the overlapping sedimentary rocks of Santa Rosa Flat (Stevens and Stone, 1988; Stone et al., 1989). Northwest of the antiformal stack, the sharp contact between the Keeler Canyon and Lone Pine Formations probably reflects the initiation of thrusting, which may have continued during deposition of the lowermost part (member A) of the Lone Pine Formation. Thrusting clearly predated deposition of the stratigraphically higher member B of the Lone Pine Formation, which contains limestone turbidites coeval with carbonate bank deposits that overlap the antiformal stack southeast of Cerro Gordo (Stone et al., 2000).

8. Direction and amount of displacement

Evidence first discussed by Stewart et al. (1966) established that the upper plate of the Last Chance thrust moved generally eastward relative to the lower plate. Later, based on a study of asymmetric shear indicators, Corbett (1989) interpreted a N80°E trans-port direction for his Last Chance plate and N75°E for his Eureka plate. As discussed in Section 11, however, we interpret the structures upon which these determi-nations were based to have formed during a Late Permian deformation rather than during displacement on the Last Chance thrust. Here, we infer the direction of displacement of the Last Chance upper plate to be perpendicular to the boundary between our western and central structural belts (Fig. 8), which marks the eastern margin of the fault-bend anticline modeled in Fig. 5. This direction of displacement, about S65°E, is compatible with the orientations of F1 folds in the Last Chance Range, about N23°E, and upper-plate folds near Jackass Flats, about N13°E (Corbett et al., 1988; Corbett, 1989), both of which we interpret to have formed concurrently with displacement on the Last Chance thrust.

The amount of displacement on the Last Chance thrust is uncertain. Stewart et al. (1966) estimated a minimum displacement of about 32 km based on the distance between outcrops of the thrust at Jackass Flats and the southeastern Last Chance Range. That estimate, however, did not account for the possibility of as much as 15 km of Cenozoic extension across Saline and Eureka Valleys (Snow, 1992). A more conservative minimum displacement of 20–25 km is measured between outcrops of the thrust in an east-southeast direction across the Last Chance Range (Fig. 6A, B). In addition to the displacement on the Last Chance thrust, Corbett (1989) suggested a minimum displacement of about 30 km on the Eureka thrust, based on the distance between exposure of that thrust at Jackass Flats and the Inyo thrust of Stevens and Olson (1972) on the west side of the Inyo Mountains, interpreted as equivalent to the Eureka thrust. As discussed in Section 10.1, however, we no longer consider the Inyo thrust to exist as originally conceived; therefore, it is not a factor in estimating displacement on the Last Chance thrust.

In the southern Inyo Mountains (Fig. 2), displacement on the Last Chance thrust can be estimated on the basis of several interrelated arguments. There are no localities where basinal Keeler Canyon Formation is known to have been thrust eastward across more than just the margin of the Mississippian carbonate shelf. Therefore, most or all of the displacement on the Last Chance thrust must be accommodated in the antiformal stack at the tip of the thrust (Fig. 5). This stack apparently attained a height of about 3.5 km above the shelf, the thickness of Early Permian sediment deposited against the stack. Based on an estimated additional 1 km or so of original relief between the shelf and the Keeler Canyon basin floor, the total amount of structural thickening in the stack probably is about 5 km. Assuming an original stratigraphic thickness of 1.2 to 1.5 km for each plate (Keeler Canyon Formation plus some Mississippian section) and width of the stack of about 12 km, based on the distance (about 6 km) from the crest to the eastern margin of the stack, the amount of shortening necessary to build the stack would be about 30 km. In the central Inyo Mountains, as much as 15 km of the total shortening may be taken up by the Lead Canyon anticline (Snow, 1992).

9. Further discussion of Last Chance thrust model

Three aspects of our model of the Last Chance thrust require more detailed discussion, in part because of significant differences with previous interpretations. These concern evidence for only limited subaerial exposure of the western fault-bend anticline, transmission of stress below a very thin upper plate, and development of the antiformal stack.

9.1. Limited subaerial exposure of the western fault-bend anticline

Lower Permian coarse-grained sedimentary rocks that could have been derived from erosion of the fault-bend anticline modeled in the western part of the thrust system (Fig. 5) are not abundant, suggesting that the anticline was not deeply eroded during or immediately after thrusting. The only notable sandstones and conglomerates of an appropriate age are in the upper part of the Lone Pine Formation in the western Inyo Mountains, and these may have been deposited in response to a later (late Wolfcampian to Leonardian) tectonic event (Stone et al., 2000) rather than having been specifically related to the Last Chance thrust. For these reasons, we suggest that the anticline probably did not form a massive subaerial uplift such as that illustrated in Fig. 5A but instead may have formed mostly below sea level, as suggested by Snow (1992) and shown schematically in Fig. 5B.

9.2. Transmission of stress below a thin upper plate

As shown in Fig. 5, our model requires a thin sheet (1.2–1.5 km thick), consisting primarily of the Keeler Canyon Formation, to be transported eastward along a décollement above and within the Rest Spring Shale for a distance of 30 km. Although such thin-skinned deformation may seem unlikely, Corbett (1989, p. 200) showed that the greatest principle strain in the basal-plate strata of the Last Chance thrust is equal to or greater than that in the upper-plate strata, suggesting that "the basal plate (primarily the Rest Spring Shale) was a deformable wedge responsible for much of the upper plate lateral transport." Thus, a very thin upper plate, a rigid beam in the sense of Jacobeen and Kanes (1975), could have ridden relatively passively on highly strained rocks of the lower plate. The structural features of the Last Chance thrust, especially the thickness and lack of deformation of the upper plate, are very similar to those shown by Mitra (1988) for the Pine Mountain thrust in the southern Appalachians. Another regional décollement (Manning Canyon detachment), similar to the Last Chance thrust in involving tens of kilometers of displacement of Pennsylvanian and Permian rocks over Mississippian shale, has been proposed in eastern Nevada and Utah (Allmendinger and Jordan, 1981). That detachment is of inferred Jurassic age and thus is not directly related to the Last Chance thrust; apparently, however, it does reflect a similar style of tectonic transport.

As pointed out by Wiltschko and Eastman (1983) and shown in cross sections by others (e.g., Thomas, 2001), basement warps and faults tend to lead to the development of thrust fault ramps. The apparent lack of such ramps or other major deformational features in the thin upper plate of the Last Chance thrust thus suggests that the Keeler basin was essentially undeformed prior to development of the décollement.

9.3. Development of the antiformal stack

The antiformal stack modeled at the tip of the Last Chance thrust system (Fig. 5) is an unusual feature that we interpret to have formed where the Rest Spring Shale pinched out against the margin of the Mississippian carbonate shelf and the thrust ramped to the surface. We suggest that there the upper plate, no longer underlain by Rest Spring Shale, moved only a short distance across the shelf before becoming locked, causing the thrust to splay. This process led to structural thickening of the upper plate and development of the antiformal stack. The inferred stacking of thrust sheets was accompanied by complex folding as has been documented for the Fishhook thrust (Stevens and Stone, 1988). Rocks as old as the Early Mississippian Tin Mountain Limestone were locally carried in the upper and lower plates of the Fishhook thrust, possibly as the result of decapitation of minor anticlines formed in the evolving thrust zone.

10. Relationship to other thrust faults

In our model, the Last Chance thrust is a single thrust that formed prior to any other significant deformation in the area. As already discussed, our model incorporates the Morning Star and Lee Flat thrusts, originally described as unconnected local features, as integral parts of the Last Chance thrust. Below, other thrust faults that previously have been related to the Last Chance thrust are reinterpreted in the context of our model.

10.1. Inyo thrust

Recent work in the Tinemaha Reservoir area (Stevens and Stone, 2002; Fig. 2) has resulted in significant stratigraphic and structural revisions to the mapping of Stevens and Olson (1972). This work shows that most of the main fault surface regarded as the Inyo thrust by Stevens and Olson (1972) probably is a late Cenozoic low-angle normal fault rather than an older thrust. Thus, we here abandon the original concept of the Inyo thrust and no longer regard any structures in the Tinemaha Reservoir area to be directly connected eastward with the Last Chance thrust.

10.2. Inyo Crest thrust

The Inyo Crest thrust (Fig. 2) was first mapped by Swanson (1996) in the southern Inyo Mountains and later was described by Stevens et al. (1997). It is associated with a footwall syncline (Upland Valley syncline) cored by Early Permian strata correlative with those that overlap the Fishhook thrust (part of the Last Chance thrust system) to the east; the syncline is overlapped in turn by the Early Triassic Union Wash Formation. Thus, the Inyo Crest thrust and Upland Valley syncline are Middle Permian to earliest Triassic (probably Late Permian) in age, distinctly younger than the Fishhook and Lee Flat (=Last Chance) thrusts. As noted earlier, we here interpret the Inyo Crest thrust to cut the Morning Star (=Last Chance) thrust southeast of Cerro Gordo (Fig. 6A, B). Previously, we interpreted the Inyo Crest thrust as the southern extension of the Last Chance thrust, which we did not consider the same as either the Morning Star or the Lee Flat thrust (Stevens et al., 1997).

We now interpret the Inyo Crest thrust and Upland Valley syncline to be structures of regional significance. As shown in Fig. 6A and B, we interpret these structures to extend northward from the southern Inyo Mountains, where the thrust has brought up rocks of the lower plate of the Last Chance thrust, through the Nelson Range (Werner, 1979) and western part of Dry Mountain (Burchfiel, 1969) to the northern Last Chance Range where the Last Chance thrust is locally folded and overturned to the east (Wrucke and Corbett, 1990). Near Dry Mountain, exposures of Ordovician through Devonian strata, which structurally overlie late Paleozoic rocks and originally were considered part of the upper plate of the Last Chance thrust (Stewart et al., 1966; Burchfiel, 1969), are here interpreted as lower plate rocks of that thrust fault brought up on the Inyo Crest thrust.

The Upland Valley syncline and its proposed correlatives trend approximately N10°W, approximately parallel with the trace of the Inyo Crest thrust, suggesting ENE displacement on this thrust.

10.3. Marble Canyon and Lemoigne thrusts

The Marble Canyon thrust in the central Cottonwood Mountains (Fig. 2) involves strata as young as Early Permian (middle Wolfcampian) and is intruded by plutonic rocks no younger than about 230 Ma (earliest Late Triassic) (Snow, 1992; Stevens et al., 1997). Permian rocks in the lower plate of the Marble Canyon thrust are coeval with rocks that onlap the upper plate of the Lee Flat thrust in the southeastern Inyo Mountains, showing that this thrust postdates the Last Chance thrust as modeled herein. We therefore consider it likely that the Marble Canyon thrust is coeval with the Inyo Crest thrust. This interpretation is consistent with the geometric similarity between the large footwall syncline beneath the Marble Canyon thrust and the Upland Valley syncline beneath the Inyo Crest thrust.

The structurally lower Lemoigne thrust, located in the southern Cottonwood Mountains (Fig. 2), involves strata as young as Early Permian (early middle Wolfcampian) and predates the Jurassic Hunter Mountain batholith (Snow, 1992; Stevens et al., 1997). Despite the lack of better age control, we agree with Snow (1992) that this fault likely is approximately coeval

with the nearby, structurally similar Marble Canyon thrust.

The Marble Canyon thrust is located near the Mississippian carbonate shelf edge (Stevens et al., 1991) (Fig. 2) and probably cuts across or overprints the leading edge of the Last Chance thrust, although this relation is nowhere exposed. Uplift and erosion of the eastern part of the Marble Canyon thrust plate north of the dextral Hunter Mountain scissors fault probably explains why the antiformal stack and the leading edge of the Last Chance thrust are not exposed in their expected positions in the central Cottonwood Mountains. The Lemoigne thrust lies east of the Mississippian shelf margin and therefore has no direct relationship with the Last Chance thrust.

10.4. Racetrack and Ubehebe thrusts

The Racetrack thrust of McAllister (1952) and the Ubehebe thrust of Snow (1992) both are located in the Racetrack Valley area (Fig. 2), east of the originally recognized outcrops of the Last Chance thrust. The Racetrack thrust is part of a complex fault zone that places Cambrian and Ordovician strata on Mississippian rocks along the east side of Racetrack Valley; the Ubehebe thrust, reactivated by Cenozoic normal faulting (Snow, 1992), places Cambrian on Mississippian and older rocks along the margins of northern Racetrack Valley. In early studies, the Racetrack thrust was interpreted to root beneath Racetrack Valley and to carry rocks of the Dry Mountain block over those of the Cottonwood Mountains (McAllister, 1952); the Last Chance thrust, in turn, was interpreted to overlie the Dry Mountain block (e.g., Stewart et al., 1966). Later, however, Snow (1992) proposed that the Dry Mountain block and Cottonwood Mountains are structurally continuous and that both originally underlay the Last Chance thrust, the upper plate of which was later dropped down into Racetrack Valley by Cenozoic normal faults (Fig. 3C). Snow inferred a connection between the Ubehebe and Last Chance thrusts and reinterpreted the Racetrack thrust as part of a duplex zone at the base of the Last Chance allochthon, rather than as a separate, rooted thrust (Figs. 3C, 7A).

One of the keys to Snow's (1992) analysis is the interpretation that both sides of Racetrack Valley are bounded by major normal faults of Cenozoic age. We agree that evidence for Cenozoic normal faulting along the east side of Racetrack Valley is strong. The interpretation by Snow and White (1990) and Snow (1992) that the prominent east-dipping fault on the west side of the valley is a Cenozoic normal fault along which the Ubehebe (=Last Chance) thrust plate was displaced downward relative to the Dry Mountain block, however, is doubtful. As mapped by Burchfiel (1969), this fault is intruded by granitic rocks (loc. 8 on Fig. 6B) assigned to the Jurassic Hunter Mountain batholith and appears to be a reverse rather than a normal fault; our own reconnaissance observations in the area support this interpretation. These relations show that the fault in question is unlikely to be the Cenozoic normal fault required by Snow's (1992) analysis.

Snow (1992) also pointed out that the Ordovician Eureka Quartzite is thicker in Racetrack Valley (185 m) than in either the Dry Mountain block to the west or the Cottonwood Mountains to the east (120–135 m). On this basis, Snow suggested that the Eureka Quartzite in this area is part of an allochthonous thrust plate derived from west of the Dry Mountain block. The Eureka Quartzite does thin eastward into the Cottonwood Mountains, but there is no evidence that it thickens westward from the Dry Mountain area as would be consistent with Snow's interpretation. Farther south, in fact, the Eureka Quartzite thins appreciably northwestward from the Darwin area into the west-central Inyo Mountains (Stevens, 1986). Thus, the thickness of the Eureka Quartzite in Racetrack Valley cannot be used as evidence of a thrust plate derived from the west; perhaps the Racetrack Valley locality simply represents the maximum thickness of a unit that thins both east and west.

In summary, we do not accept the proposition that the rocks in Racetrack Valley represent a downdropped part of the Last Chance allochthon that originally overlay the Dry Mountain block as proposed by Snow (1992). Instead, we follow McAllister (1952) in interpreting these rocks as part of the upper plate of the Racetrack thrust. The fault on the west side of Racetrack Valley could be a backthrust coeval with the Racetrack thrust, along which these rocks were uplifted relative to the Dry Mountain block.

As noted previously, we interpret the faulted contact between the Rest Spring Shale and Tihvipah Limestone (=lower Keeler Canyon Formation), exposed a short distance east of the Racetrack thrust complex, to represent part of the Last Chance thrust. We infer that

the Racetrack thrust complex cuts this faulted contact, although this relation is not exposed, and thus postdates the Last Chance thrust (Fig. 6A, B). The simplest interpretation of the Ubehebe thrust is that it was originally continuous with the Racetrack thrust from which it is now separated by alluvium. The age of the composite Racetrack–Ubehebe thrust is not tightly constrained but it could be coeval with the Inyo Crest, Marble Canyon, and Lemoigne thrusts.

If the Racetrack thrust extends south of the Hunter Mountain fault, which seems probable, its position is uncertain. Considering about 9 km of right-lateral displacement on the Hunter Mountain fault, the trace of the Racetrack thrust should extend through the Lee Flat area where only granitic rocks and alluvium are exposed. In this area, we infer the Racetrack thrust to cut the Lee Flat (=Last Chance) thrust as shown in Fig. 6A and B.

11. Regional relations of Permian deformation

Convergence along the continental margin probably was responsible for development of both the Early Permian Last Chance thrust and the younger Permian to earliest Triassic thrust belt that includes the Inyo Crest thrust. During that time span, it is now thought that the continental margin consisted of two segments (Figs. 1 and 9). The older, northeast-trending segment, developed by rifting in Precambrian time, subsided slowly as a passive margin into the Devonian, when rocks along that margin were deformed during the Late Devonian to Early Mississippian Antler orogeny. The younger, southeast-trending segment formed by truncation of the older margin, probably in the Early to Middle Pennsylvanian, coeval with the origin of the Keeler basin (Stevens et al., 1992, 2001). The basis for the truncation hypothesis is the abrupt termination of lower and middle Paleozoic facies belts against the present Sierra Nevada. Several lines of evidence, especially the apparent southeastward displacement of deformed eugeoclinal rocks now exposed in the El Paso Mountains, suggest that the truncating structure was a sinistral transform fault zone (Stone and Stevens, 1988; Walker, 1988; Stevens et al., 1992; Dickinson, 2000).

In east-central California, the pre-truncation continental margin (approximately coincident with the

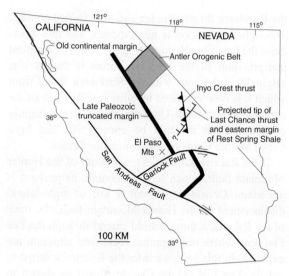

Fig. 9. Schematic map illustrating approximate spatial relationship between Last Chance thrust, Inyo Crest thrust, and continental margin segments. Last Chance thrust is approximately parallel to older (pre-Pennsylvanian), northeast-trending margin segment; cross-cutting Inyo Crest thrust is more closely parallel to younger (Pennsylvanian–Permian), northwest-trending truncated margin segment.

western edge of the Antler orogenic belt) and pre-Pennsylvanian facies belts presently trend about N20°E (e.g., Stevens and Greene, 1999), whereas the late Paleozoic truncated margin and Pennsylvanian and Permian facies belts trend about N30°W (Stevens et al., 1992). (Note that these trends and others mentioned in this paper have not been altered to conform to reconstructions of Cenozoic extension proposed by Snow and Wernicke, 2000.) Because we interpret the change in orientation of the Pennsylvanian and Permian facies belts to be a product of the truncational event, all of the Permian contractional events discussed here are considered to have taken place after the continental margin had changed from dominantly transform in the Pennsylvanian to at least partially convergent.

11.1. Tectonic significance of transport directions on the Last Chance thrust and Late Permian to earliest Triassic thrusts

The Last Chance thrust and the younger thrusts of Inyo Crest age show significant differences in direction of displacement. As discussed earlier, the inferred

orientation of the fault bend anticline and the other structural belts above the Last Chance thrust (Fig. 8) suggests that the direction of displacement was about S65°E, whereas the orientation of the Upland Valley syncline and its correlative structures below the Inyo Crest thrust indicates the displacement direction on that thrust to have been about N80°E.

Our inferred direction of displacement on the Last Chance thrust differs by about 35° from that proposed by Corbett (1989), who interpreted the directions of displacement on the Last Chance and Eureka thrusts as N80°E and N75°E, respectively, based primarily on asymmetric shear indicators and orientation of principle strain axes. Our analysis of Corbett's (1989) structural data indicates that a different interpretation is probable. Especially significant are data on map- and outcrop-scale folds. Plots of fold axes in the Last Chance Range (Corbett et al., 1988; Corbett, 1989) show F1 folds with a trend of about N23°E and F2 folds with a trend of N18°W. In the Jackass Flats area (Fig. 2), Corbett et al. (1988) and Corbett (1989) reported folds similar in orientation to F1 of the Last Chance Range in the upper plate of the Eureka thrust and folds similar in orientation to F2 in the lower plate. (Both fold sets in the Last Chance Range trend about 10° more eastward than those at Jackass Flats, probably because of bending of all rocks and structures in the former area which is much closer to the dextral Death Valley–Furnace Creek fault zone than the latter.) Corbett et al. (1988) and Corbett (1989) interpreted the north-northwest-trending folds in these areas as perpendicular to and coeval with displacement on the Last Chance and Eureka thrusts. In contrast, we interpret the north-northeast-trending folds (F1) as approximately perpendicular to and coeval with displacement on the Last Chance thrust, and the north-northwest-trending folds (F2) as approximately perpendicular to and coeval with displacement on the younger Inyo Crest and related thrusts.

The above interpretation is most convincing in the northern Last Chance Range (Fig. 6B), where the Last Chance thrust is deformed by F2 folds of Corbett (1989), the orientations of which are compatible with structures associated with the Inyo Crest thrust and Upland Valley syncline. In addition, the east-northeast-striking Cucomungo fault in this area is approximately parallel to our inferred direction of displacement on the Inyo Crest thrust and therefore could have probably originated as a right-lateral tear fault related to that thrust. In contrast, directly south of the Cucomungo fault, an overturned syncline trending about N35°E, which was mapped by Corbett (1989) in lower plate Mississippian rocks, is compatible with our inferred east-southeasterly direction of displacement on the Last Chance thrust.

Based on the direction of displacement on the Last Chance thrust and orientation of continental margin segments, we conclude that the Last Chance thrust followed an original zone of weakness that approximately paralleled the older, north-northeast-trending continental margin and controlled the direction of displacement of the upper plate (Fig. 9). In contrast, the younger Inyo Crest and related thrusts are more closely parallel to and may have been related to compression along the late Paleozoic truncated continental margin, which was at an angle to original facies trends as emphasized by Snow and Wernicke (2000).

11.2. Northeastward extent of the Last Chance thrust

Several thrust faults east of Death Valley (now offset on the right-lateral Death Valley–Furnace Creek fault zone), located in a structural position similar to that of the Last Chance thrust with regard to the Paleozoic continental margin, have been correlated with the Last Chance thrust by previous workers. The most significant of these are the Grapevine thrust in the Grapevine Mountains in California (Fig. 2) and the Belted Range thrust at the Nevada Test Site (e.g., Reynolds, 1976; Snow, 1992; Cole and Cashman, 1999; see Fig. 1 for general location). Like classic exposures of the Last Chance thrust, these thrust faults place lower Paleozoic rocks on Mississippian clastic strata broadly correlative with the Rest Spring Shale. Unlike the Last Chance thrust in most of east-central California, however, both the Grapevine and Belted Range thrusts locally override Pennsylvanian limestone in addition to Mississippian clastic rocks (Reynolds, 1976; Cole and Cashman, 1999). It is possible either that the character of the Last Chance thrust changed northeastward or that the Grapevine and Belted Range thrusts are not correlative with the Last Chance thrust.

11.3. Southward extent of the Last Chance thrust

The Last Chance thrust presumably extended southward from the southeastern Inyo Mountains but cannot be identified with confidence because younger structures, probably associated with the Jurassic East Sierra thrust system (Dunne et al., 1983; Dunne, 1986) have overprinted all earlier structures in these areas. In both the Talc City and western Darwin Hills (Fig. 2), however, the presence of Rest Spring Shale and Keeler Canyon Formation with Mississippian carbonate-shelf rocks exposed immediately to the east (Stone et al., 1989) suggests proximity to the Last Chance thrust. South of the Darwin Hills, the Last Chance thrust presumably lies west of the Argus Range in which all of the exposed Mississippian rocks are of carbonate-shelf facies.

Farther south, any correlation of faults with the Last Chance thrust is highly speculative. Permo-Triassic folding, metamorphism, and plutonism have been documented in the western Mojave Desert and San Bernardino Mountains (e.g., Miller and Cameron, 1982; Miller et al., 1995; see Fig. 1), but there is no evidence that these events had any direct relationship with the Last Chance thrust. Snow (1992) proposed that the Permian Death Valley thrust belt, including the Last Chance thrust, continued southward into the El Paso Mountains (Fig. 1), where complex upper Paleozoic stratigraphic relations and unconformities (Carr et al., 1997) conceivably are related to Early Permian compressive deformation. As part of that interpretation, Snow (1992) suggested that Early Permian thrusting could be responsible for juxtaposition of the oceanic El Paso Mountains terrane and the continental margin, in contrast to the alternative interpretation that the El Paso Mountains terrane was emplaced by left-lateral faulting along the late Paleozoic truncated margin (e.g., Stevens et al., 1992). The former interpretation seems unlikely because it would require the Death Valley thrust belt of Snow (1992) to swing from well within the miogeocline in the Inyo Mountains far westward into the eugeocline, and data of Miller et al. (1995) suggest that emplacement of the El Paso terrane onto continental crust occurred much later.

If continental truncation did occur as suggested by Stevens et al. (1992), the portion of the Antler foreland basin in which the Rest Spring Shale was deposited probably was truncated north of the El Paso Mountains (Figs. 1 and 9). If the Last Chance thrust extended south of this point along the truncated margin, it would be expected to differ geometrically from that to the north because the Rest Spring Shale would not have been involved and no décollement would have developed.

12. Structural development

Initial contraction during the middle Wolfcampian resulted in development of the Last Chance fault ramp and the beginning of formation of the fault ramp anticline. The fault propagated across the previously undeformed Keeler basin along the weak Mississippian Rest Spring Shale to form a décollement. At the basin margin, where the fault encountered the older carbonate shelf, the thrust was deflected upward. The upper plate moved only a short distance beyond the older carbonate shelf margin because the growing structure, above which the upper plate was required to rise, initiated additional imbricate thrusts and development of an antiformal stack. Probably late in the deformational event, an incipient duplex structure ("mushwad") developed in the ductile shale a short distance behind the anticlinal stack. This structure did not initiate additional thrusts probably because deformation ceased before the mushwad became thick enough to deflect the thrust upward through the upper plate.

Emplacement of the Last Chance thrust was followed by renewed marine deposition during middle to late Early Permian time, producing a sedimentary sequence as thick as 3.5 km in front of the antiformal stack (Stone et al., 1989, 2000). In Middle Permian to earliest Triassic time, all of the older rocks and the Last Chance thrust itself were deformed by a second episode of thrusting represented by the Inyo Crest and related thrusts. Depositional overlap by Lower Triassic marine strata of the Union Wash Formation marked the end of this deformational episode.

13. Conclusions

The Last Chance thrust formed in the Early Permian and represents one of several contractional

events that affected east-central California in late Paleozoic and early Mesozoic time. The magnitude of displacement and regional extent of the Last Chance thrust show that it was a major event occurring in the early stages of development of a convergent plate margin that led to widespread subduction and arc magmatism beginning in the Late Triassic in this part of the Cordilleran region. This event was distinct from the better known Antler (Late Devonian to Early Mississippian) and Sonoma (Late Permian) orogenies and followed truncation of the continental margin by transform faulting that began in the Early to Middle Pennsylvanian.

The structural model of the Last Chance thrust presented here differs from all previous models in linking the older-on-younger exposures of the originally recognized Last Chance thrust with a bedding-parallel shear zone associated with the Rest Spring Shale and Keeler Canyon Formation to the east, which we interpret to represent a younger-on-older thrust flat. These faults in turn are linked eastward to the Lee Flat thrust and the overlying antiformal stack. These inferred structural links result in an integrated structural model based on established concepts of ramp-flat thrusting and on the regional late Paleozoic paleogeography. Although the critical links are unproven, our model serves as an alternative to existing interpretations of the Last Chance thrust, all of which are revealed as questionable when critically evaluated.

This study supports many of Snow's (1992) conclusions about the geometry, timing, and regional significance of Permian deformation in the Death Valley thrust belt. In contrast to Snow's ideas, however, we (1) interpret the Last Chance allochthon as a single structural plate and do not recognize the existence of a structurally higher Eureka thrust, (2) show that the Last Chance thrust is distinctly older than the other thrusts of the Death Valley thrust belt of Snow (1992), and (3) consider the Last Chance thrust to be largely a décollement controlled by the Mississippian Rest Spring Shale, whereas the other thrusts of the Death Valley thrust belt originated as steeper, cross-cutting faults not confined to a specific stratigraphic horizon.

As interpreted here, the Last Chance thrust is important as an example of décollement-style thrust faulting in a previously undeformed shale basin that was highly deformed and dismembered during subsequent events. It is possible that similar structures exist in other complexly deformed shale basins but have not been recognized because of problems similar to those encountered in reconstruction of the Last Chance thrust. The approach we have taken here, based in part on deductive reasoning from established thrust-fault models and regional paleogeographic considerations, could prove useful in understanding the structural relations in other such basins where field relations alone are inconclusive.

Acknowledgments

We are grateful to Jonathan Miller, Robert Miller, Jack Stewart, and Chet Wrucke, who read various earlier versions of the manuscript, and the critical evaluations of Andy Barth, George Dunne, Keith Howard, Sven Morgan, Walt Snyder, Doug Walker, and Brian Wernicke. Discussions with Robert Miller and Richard Sedlock have been especially helpful.

References

Allmendinger, R.W., Jordan, T.E., 1981. Mesozoic evolution, hinterland of the Sevier orogenic belt. Geology 9, 308–313.

Boyer, S.E., Elliott, D., 1982. Thrust systems. American Association of Petroleum Geologists Bulletin 66, 1196–1230.

Burchfiel, B.C., 1969. Geology of the Dry Mountain quadrangle, Inyo County, California. Special Report-California Division of Mines and Geology 99, 19 p., 1:62,500.

Burchfiel, B.C., Hodges, K.V., Royden, L.H., 1987. Geology of Panamint Valley-Saline Valley pull-apart system, California: palinspastic evidence for low-angle geometry of a Neogene range-bounding fault. Journal of Geophysical Research 92 (B10), 10,422–10,426.

Carr, M.D., Christiansen, R.L., Poole, F.G., Goodge, J.W., 1997. Bedrock geologic map of the El Paso Mountains in the Garlock and El Paso Peaks 7-1/2' quadrangles, Kern County, California. U.S. Geological Survey Miscellaneous Investigations Series Map I-2389, 9 p., 1:24,000.

Cole, J.C., Cashman, P.H., 1999. Structural relationships of pre-Tertiary rocks in the Nevada Test Site region, southern Nevada. U.S. Geological Survey Professional Paper 1607 (39 pp.).

Conrad, J.E., McKee, E.H., 1985. Geologic map of the Inyo Mountains Wilderness Study Area, Inyo County, California. U.S. Geological Survey Miscellaneous Field Studies Map MF-1733-A, 1:62,500.

Corbett, K.P., 1989. Structural geology of the Last Chance thrust system, east-central California. Los Angeles, University of California, PhD thesis. 245 pp.

Corbett, K.P., 1990. Basin and Range extensional tectonics at the latitude of Las Vegas, Nevada: Discussion. Geological Society of America Bulletin 102, 267–268.

Corbett, K.P., Wrucke, C.T., Nelson, C.A., 1988. Structure and tectonic history of the Last Chance thrust system, Inyo Mountains and Last Chance Range, California. In: Weide, D.L., Faber, M.L. (Eds.), This Extended Land. Field Trip Guidebook, Geological Society of America, Cordilleran Section, and University of Nevada, Las Vegas, pp. 269–292.

Dickinson, W.R., 2000. Geodynamic interpretation of Paleozoic tectonic trends oriented oblique to the Mesozoic Klamath–Sierran continental margin in California. In: Gehrels, G.E., Soreghan, M.J. (Eds.), Paleozoic and Triassic Paleogeography and Tectonics of Western Nevada and Northern California, Special Paper-Geological Society of America, vol. 347, pp. 209–245.

Dunne, G.C., 1986. Geologic evolution of the southern Inyo Range, Darwin Plateau, and Argus and Slate Ranges, east-central California—an overview. In: Dunne, G.C. (Ed.), Mesozoic and Cenozoic Structural Evolution of Selected Areas, East-central California. Guidebook and Volume, Field Trips 2 and 14. Geological Society of America, Cordilleran Section, Los Angeles, CA, pp. 3–21.

Dunne, G.C., Gulliver, R.M., Sylvester, A.G., 1978. Mesozoic evolution of rocks of the White, Inyo, Argus, and Slate Ranges, eastern California. In: Howell, D.G., McDougall, K.A. (Eds.), Mesozoic Paleogeography of the Western United States, Pacific Coast Paleogeography Symposium, vol. 2. Pacific Section, Society of Economic Paleontologists and Mineralogists, pp. 189–207.

Dunne, G.C., Moore, S.C., Gulliver, R.M., Fowler, J., 1983. East Sierran thrust system, eastern California. Abstracts with Programs-Geological Society of America 15 (5), 322.

Elayer, R.W., 1974. Stratigraphy and structure of the southern Inyo Mountains, Inyo County, California. San Jose, CA, San Jose State University, MS thesis. 121 pp.

Jacobeen Jr., F., Kanes, W.H., 1975. Structure of Broadtop synclinorium, Wills Mountain anticlinorium, and Allegheny frontal zone. American Association of Petroleum Geologists Bulletin 59, 1136–1150.

McAllister, J.F., 1952. Rocks and structure of the Quartz Spring area, northern Panamint Range, California. Special Report-California Division of Mines 25 (38 pp.).

McAllister, J.F., 1956. Geology of the Ubehebe Peak quadrangle, California. U.S. Geological Survey Geologic Quadrangle Map GQ-95, 1:62,500.

McKee, E.H., 1985. Geologic map of the Magruder Mountain quadrangle, Esmeralda County, Nevada, and Inyo County, California. U.S. Geological Survey Geologic Quadrangle Map GQ-1587, 1:62,500.

Miller, E.L., Cameron, C.S., 1982. Late Precambrian to Late Cretaceous evolution of the southwestern Mojave Desert, California. In: Cooper, J.D., Troxel, B.W., Wright, L.A. (Eds.), Geology of Selected Areas in the San Bernardino Mountains, Western Mojave Desert, and Southern Great Basin, California, Geological Society of America Field Trip Guidebook, pp. 21–34.

Miller, J.S., Glazner, A.F., Walker, J.D., Martin, M.W., 1995. Geochronologic and isotopic evidence for Triassic–Jurassic emplacement of the eugeoclinal allochthon in the Mojave Desert region, California. Geological Society of America Bulletin 107, 1441–1457.

Mitra, S., 1988. Three-dimensional geometry and kinematic evolution of the Pine Mountain thrust system, southern Appalachians. Geological Society of America Bulletin 100, 72–95.

Nelson, C.A., 1981. Basin and Range province. In: Ernst, W.G. (Ed.), The Geotectonic Development of California, Englewood Cliffs, New Jersey. Prentice-Hall, pp. 203–216.

Reynolds, M.W., 1976. Geology of the Grapevine Mountains, Death Valley, California: a summary. In: Troxel, B.W., Wright, L.A. (Eds.), Geologic Features, Death Valley, California. Special Report-California Division of Mines and Geology, vol. 106, pp. 19–25.

Rich, J.L., 1934. Mechanics of low-angle overthrust faulting as illustrated by Cumberland thrust block, Virginia, Kentucky, and Tennessee. American Association of Petroleum Geologists Bulletin 18, 1584–1596.

Ross, D.C., 1965. Geology of the Independence quadrangle, Inyo County, California. U.S. Geological Survey Bulletin 1181-O, 64 p., 1:62,500.

Ross, D.C., 1967a. Generalized geologic map of the Inyo Mountains region, California. U.S. Geological Survey Miscellaneous Geological Investigations Map I-506, 1:125,000.

Ross, D.C., 1967b. Geologic map of the Waucoba Wash quadrangle, Inyo County, California. U.S. Geological Survey Geologic Quadrangle Map GQ-612, 1:62,500.

Salvini, F., Storti, F., McClay, K., 2001. Self-determining numerical modeling of compressional fault-bend folding. Geology 29, 839–842.

Snow, J.K., 1992. Large-magnitude Permian shortening and continental-margin tectonics in the southern Cordillera. Geological Society of America Bulletin 104, 80–105.

Snow, J.K., Wernicke, B., 1993. Large-magnitude Permian shortening and continental-margin tectonics in the southern Cordillera: reply. Geological Society of America Bulletin 105, 280–283.

Snow, J.K., Wernicke, B., 2000. Cenozoic tectonism in the central Basin and Range: magnitude, rate, and distribution of upper crustal strain. American Journal of Science 300, 659–719.

Snow, J.K., White, C., 1990. Listric normal faulting and synorogenic sedimentation, northern Cottonwood Mountains, Death Valley region, California. In: Wernicke, B.P. (Ed.), Basin and Range Extensional Tectonics near the Latitude of Las Vegas, Nevada, Memoir-Geological Society of America, vol. 176, pp. 413–445.

Stevens, C.H., 1986. Evolution of the Ordovician through Middle Pennsylvanian carbonate shelf in east-central California. Geological Society of America Bulletin 97, 11–25.

Stevens, C.H., Greene, D.C., 1999. Stratigraphy, depositional history, and tectonic evolution of Paleozoic continental-margin rocks in roof pendants of the eastern Sierra Nevada, California. Geological Society of America Bulletin 111, 919–933.

Stevens, C.H., Olson, R.C., 1972. Nature and significance of the Inyo thrust fault, eastern California. Geological Society of America Bulletin 83, 3761–3768.

Stevens, C.H., Stone, P., 1988. Early Permian thrust faults in east-central California. Geological Society of America Bulletin 100, 552–562.

Stevens, C.H., Stone, P., 2002. Correlation of Permian and Triassic deformations in the western Great Basin and eastern Sierra Nevada: evidence from the northern Inyo Mountains near Tinemaha Resevoir, east-central California. Geological Society of America Bulletin 114, 1210–1221.

Stevens, C.H., Stone, P., Belasky, P., 1991. Paleogeographic and structural significance of an Upper Mississippian facies boundary in southern Nevada and east-central California. Geological Society of America Bulletin 103, 876–885.

Stevens, C.H., Stone, P., Kistler, R.W., 1992. A speculative reconstruction of the middle Paleozoic continental margin of southwestern North America. Tectonics 11, 405–419.

Stevens, C.H., Stone, P., Dunne, G.C., Greene, D.C., Walker, J.D., Swanson, B.J., 1997. Paleozoic and Mesozoic evolution of east-central California. International Geology Review 39, 788–829.

Stevens, C.H., Stone, P., Ritter, S.M., 2001. Conodont and fusulinid biostratigraphy and history of the Pennsylvanian to Lower Permian Keeler basin, east-central California. Brigham Young University Geology Studies 46, 99–142.

Stewart, J.H., 1970. Upper Precambrian and Lower Cambrian strata in the southern Great Basin, California and Nevada. U.S. Geological Survey Professional Paper 620 (206 pp.).

Stewart, J.H., Ross, D.C., Nelson, C.A., Burchfiel, B.C., 1966. Last Chance thrust—a major fault in the eastern part of Inyo County, California. U.S. Geological Survey Professional Paper 550-D, D23–D34.

Stone, P., 1984. Stratigraphy, depositional history, and paleogeographic significance of Pennsylvanian and Permian rocks in the Owens Valley–Death Valley region, California. Stanford, CA, Stanford University, PhD thesis. 399 pp.

Stone, P., Stevens, C.H., 1988. Pennsylvanian and Early Permian paleogeography of east-central California: implications for the shape of the continental margin and the timing of continental truncation. Geology 16, 330–333.

Stone, P., Stevens, C.H., 1993. Large-magnitude Permian shortening and continental-margin tectonics in the southern Cordillera: discussion. Geological Society of America Bulletin 105, 279–280.

Stone, P., Dunne, G.C., Stevens, C.H., Gulliver, R.M., 1989. Geologic map of Paleozoic and Mesozoic rocks in parts of the Darwin and adjacent quadrangles, Inyo County, California. U.S. Geological Survey Miscellaneous Investigations Series Map I-1932, 1:31,250.

Stone, P., Stevens, C.H., Spinosa, C., Furnish, W.M., Glenister, B.F., Wardlaw, B.R., 2000. Stratigraphic relations and tectonic significance of rocks near the Permian-Triassic boundary, southern Inyo Mountains, California. Geological Society of America Map and Chart Series MCH086, 32 p., 1:12,000.

Stone, P., Dunne, G.C., Conrad, J.E., Swanson, B.J., Stevens, C.H., Valin, Z.C., 2004. Geologic map of the Cerro Gordo Peak 7.5′ quadrangle, Inyo County, California. U.S. Geological Survey Scientific Investigations Map 2851, 1:24,000.

Swanson, B.J., 1996. Structural geology and deformational history of the southern Inyo Mountains east of Keeler, Inyo County,

California. Northridge, California State University, MS thesis. 125 pp.

Thomas, W.A., 2001. Mushwad: Ductile duplex in the Appalachian thrust belt in Alabama. American Association of Petroleum Geologists Bulletin 85, 1847–1869.

Walker, J.D., 1988. Permian and Triassic rocks of the Mojave Desert and their implications for timing and mechanisms of continental truncation. Tectonics 7, 685–709.

Werner, M.R., 1979. Superposed Mesozoic deformations, southeastern Inyo Mountains, California. Northridge, California State University, MS thesis. 69 pp.

Wiltschko, D., Eastman, D., 1983. Role of basement warps and faults in localizing thrust fault ramps. In: Hatcher Jr., R.D., Williams, H., Zietz, I. (Eds.), Contributions to the Tectonics and Geophysics of Mountain Chains, Memoir-Geological Society of America, vol. 158, pp. 177–190.

Wrucke, C.T., Corbett, K.P., 1990. Geologic map of the Last Chance quadrangle, California and Nevada. U.S. Geological Survey Open-file Report 90-647-A, 1:62,500.

Available online at www.sciencedirect.com

Earth-Science Reviews 73 (2005) 103–113

www.elsevier.com/locate/earscirev

Structure and regional significance of the Late Permian(?) Sierra Nevada–Death Valley thrust system, east-central California

Calvin H. Stevens [a,*], Paul Stone [b]

[a] *San Jose State University, San Jose, CA 95192, USA*
[b] *U.S. Geological Survey, Menlo Park CA 94025, USA*

Accepted 1 April 2005

Abstract

An imbricate system of north-trending, east-directed thrust faults of late Early Permian to middle Early Triassic (most likely Late Permian) age forms a belt in east-central California extending from the Mount Morrison roof pendant in the eastern Sierra Nevada to Death Valley. Six major thrust faults typically with a spacing of 15–20 km, original dips probably of 25–35°, and stratigraphic throws of 2–5 km compose this structural belt, which we call the Sierra Nevada–Death Valley thrust system. These thrusts presumably merge into a décollement at depth, perhaps at the contact with crystalline basement, the position of which is unknown. We interpret the deformation that produced these thrusts to have been related to the initiation of convergent plate motion along a southeast-trending continental margin segment probably formed by Pennsylvanian transform truncation. This deformation apparently represents a period of tectonic transition to full-scale convergence and arc magmatism along the continental margin beginning in the Late Triassic in central California.
© 2005 Elsevier B.V. All rights reserved.

Keywords: California; Death Valley; Sierra Nevada; thrust fault; Permian; Sonoma orogeny; Morrison orogeny; tectonics

1. Introduction

Multiple tectonic events have affected the east-central California region (Fig. 1), resulting in a complex geologic puzzle. The work of many geologists has revealed at least eight major deformational events in this region: (1) Antler-age (earliest Mississippian)

* Corresponding author. Tel.: +1 408 924 5029; fax: +1 408 924 5053.
E-mail address: stevens@geosun.sjsu.edu (C.H. Stevens).

0012-8252/$ - see front matter © 2005 Elsevier B.V. All rights reserved.
doi:10.1016/j.earscirev.2005.04.006

contraction (Schweickert and Lahren, 1987, 1993a; Greene et al., 1997); (2) development of the Pennsylvanian–earliest Permian Keeler Basin that cut across earlier facies belts (Stevens et al., 2001), interpreted to be the result of truncation of the pre-Pennsylvanian continental margin (Walker, 1988; Stone and Stevens, 1988; Barth et al., 1997); (3) Early Permian emplacement of the Last Chance thrust (Snow, 1992; Stevens and Stone, 2005-this volume); (4) a Late Permian(?) thrusting event (the subject of this report); (5) emplacement of the Golconda allochthon in the Mid-

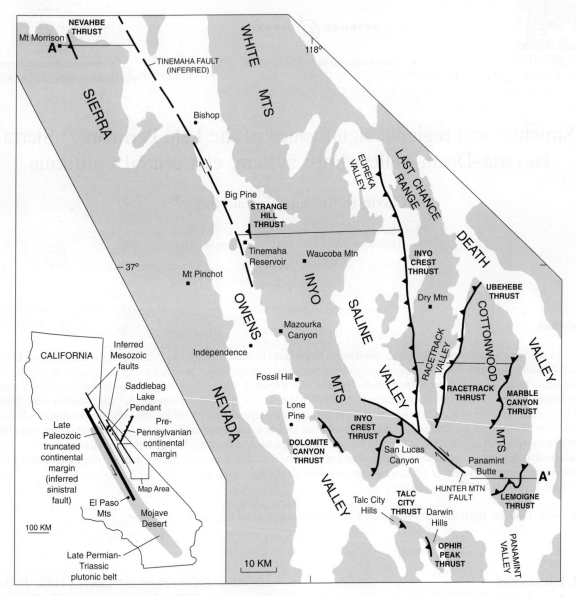

Fig. 1. Map of east-central California showing faults of Sierra Nevada–Death Valley thrust system and other features referred to in the text. Inset shows location of study area in relation to inferred late Paleozoic and pre-Pennsylvanian segments of the continental margin.

dle Triassic (Speed, 1984; Schweickert and Lahren, 1987, 1993a); (6) initiation of the Sierran magmatic arc in the Late Triassic (Bateman, 1992; Schweickert and Lahren, 1993a,b); (7) development of the East Sierran thrust system mostly, if not entirely, in the Jurassic (Dunne, 1986); and (8) extensional and dextral faulting in the late Cenozoic (Snow and Wernicke, 2000). A synthesis of the Paleozoic and Mesozoic events was presented by Stevens et al.

(1997), but work since then shows that some of the interpretations on Permian thrusting made in that report require modification.

The purpose of this paper is to bring together and elucidate the evidence for a major Late Permian(?) deformational event, the extent of which had not been recognized previously. The model constructed to explain the origin of this thrust belt is speculative but conforms to all known data.

2. Recognition of Late Permian(?) thrusting: the Inyo Crest thrust

Existence of a Late Permian(?) thrust faulting event in east-central California first became apparent through the work of Swanson (1996), who mapped and named the Inyo Crest thrust in the southern Inyo Mountains (Fig. 1). This thrust, where exposed, cuts only rocks of the Pennsylvanian–earliest Permian Keeler Canyon Formation, but a large footwall syncline (Upland Valley syncline) folds younger Permian (middle Wolfcampian to Leonardian) units correlative with rocks that overlap the Fishhook and Lee Flat thrusts of Stone et al. (1989) in the easternmost Inyo Mountains. If the Fishhook and Lee Flat thrusts are part of the Last Chance thrust zone, as interpreted by Snow (1992) and Stevens and Stone (2005-this volume), it is clear that the Inyo Crest thrust and Upland Valley syncline are younger than the Last Chance thrust. Smithian (middle Early Triassic) rocks (Stone et al., 1991) overlap the syncline and place a cap on the age of deformation (Stevens et al., 1997; Stevens and Stone, 2002). Thus, the Inyo Crest thrust was active between the late Early Permian and middle Early Triassic.

In a companion study focused primarily on the Last Chance thrust (Stevens and Stone, 2005–this volume), we postulate that the Inyo Crest thrust and Upland Valley syncline were originally continuous with a fault and large overturned footwall syncline in San Lucas Canyon (Fig. 1). From there, these structures extend northward to exposures mapped by Burchfiel (1969) on the west side of Dry Mountain. We further interpret the same structures to extend northward under the alluvium of Eureka Valley to the Last Chance Range, where the Last Chance thrust and associated strata are folded into an overturned syncline (Wrucke and Corbett, 1990) thought to be a continuation of the Upland Valley syncline. If these correlations are correct, the Inyo Crest thrust and footwall syncline have a present regional trend of slightly west of north (Fig. 1).

The original dip of the Inyo Crest thrust in the Inyo Mountains, corrected for subsequent gentle southeastward tilting of the overlapping Triassic strata, was about 35°W (Swanson, 1996), but the amount of displacement there is unknown. On the west side of Dry Mountain, rocks as old as Early Ordovician have been emplaced over the Mississippian Rest Spring Shale (Burchfiel, 1969) on a complex set of faults that dip 15–35°W. Here, the stratigraphic throw is about 2 km. Most previous workers (e.g., Stewart et al., 1966) have considered these faults part of the Last Chance thrust zone, but we consider them related to the Inyo Crest thrust for reasons given elsewhere (Stevens and Stone, 2005-this volume).

3. Other Late Permian(?) thrusts and definition of the Sierra Nevada–Death Valley thrust system

We propose that several additional thrust faults in east-central California, most of which have been recognized for many years, are coeval with the Inyo Crest thrust and represent a single period of deformation. These are, from east to west (Fig. 1), the Lemoigne thrust (Hall, 1971), the Marble Canyon thrust (Stadler, 1968), the Racetrack thrust (McAllister, 1952), the Strange Hill thrust (Stevens and Stone, 2002), and several thrust faults in the Mount Morrison roof pendant in the eastern Sierra Nevada, the most important being the Nevahbe thrust (Stevens and Greene, 2000). All of these thrusts have field relations that suggest or permit an age similar to that of the Inyo Crest thrust, and they all possess various physical characteristics similar to those of the Inyo Crest thrust. These include a northerly trend (uncorrected for possible vertical-axis rotation of crustal blocks such as those discussed by Snow and Wernicke, 2000), relatively low westward original dip (mostly 25–35°), eastward sense of displacement of the upper plate, stratigraphic throw generally in the range of 2–5 km, and the common presence of a large, east-vergent footwall syncline. These characteristics link these faults together into a group for which we here propose the name Sierra Nevada–Death Valley thrust system.

Many of the faults that we include in the Sierra Nevada–Death Valley thrust system were previously considered part of the Death Valley thrust belt of Snow (1992), which was interpreted also to include the Last Chance thrust. The new name is needed because this late Early Permian to middle Early Triassic thrust system, as defined herein, excludes the Early Permian Last Chance thrust, which is older and is seen to be cut by two of the thrusts of this system, and includes several additional faults that extend the belt westward into the eastern Sierra Nevada.

Below we summarize the structural characteristics and age relations of each fault here included in the Sierra Nevada–Death Valley thrust system.

3.1. Lemoigne thrust

This thrust in the Panamint Butte area of the southern Cottonwood Mountains emplaces Cambrian rocks (primarily Bonanza King Dolomite) over Pennsylvanian and Lower Permian strata assigned to the Keeler Canyon and Owens Valley formations by Hall (1971), but reassigned to the Bird Spring Formation and Owens Valley Group, respectively, by Stone (1984) and Stone et al. (1987). This relationship indicates a stratigraphic throw of about 3 km. Rocks in the lower plate of the thrust are folded into a large, east-vergent footwall syncline that can be seen on the west side of the range from U.S. Highway 190 in Panamint Valley.

The youngest rocks in the lower plate of the Lemoigne thrust are of Middle Wolfcampian age (Stone, 1984) and approximately coeval with rocks that overlap faults correlated with the Last Chance thrust in the southeastern Inyo Mountains (Stevens and Stone, 2005-this volume). This relationship indicates that the Lemoigne thrust is younger than the Last Chance thrust and could be coeval with the Inyo Crest thrust. Rocks of both the upper and lower plates of the Lemoigne thrust are intruded by the Hunter Mountain batholith of Middle Jurassic age (Hall, 1971; Dunne, 1986), which places an upper limit on the age of faulting.

The present subhorizontal orientation of the Lemoigne thrust is not original. Tertiary rocks in depositional contact with Paleozoic rocks in the lower plate of the thrust strike approximately N–S and dip about 35°E (Hall, 1971, cross-section A–A′), approximately parallel to the underlying Paleozoic rocks, indicating that all rocks and pre-Tertiary structures have been tilted. Restoration of this tilt indicates that the thrust had a pre-Tertiary strike of approximately N–S and a dip of about 35°W, an orientation similar to that of the Inyo Crest thrust.

3.2. Marble Canyon thrust

This thrust in the central Cottonwood Mountains emplaces lower Paleozoic rocks on strata as young as the Darwin Canyon Formation of Early Permian (mid-

dle to late Wolfcampian) age (Stone, 1984; Stone et al., 1987). The stratigraphic throw on the thrust, which is everywhere excised by Cenozoic normal faults (Snow, 1990, 1992), is about 3 km. Similar to the Lemoigne thrust, rocks in the lower plate of the thrust are folded into a large, east-vergent footwall syncline (Dry Bone syncline of Snow, 1990). The thrust itself predates intrusion of the Hunter Mountain batholith (Stadler, 1968), and the Dry Bone syncline is cut by apophyses of a stock no younger than 230 Ma (early Late Triassic) and possibly older than 245 Ma (Middle Triassic) according to Snow et al. (1991). These relations suggest a temporal correlation between the Lemoigne, Marble Canyon and Inyo Crest thrusts.

Exposures of the overturned Dry Bone footwall syncline show that the axial plane of this fold is generally subhorizontal (Stone, 1984; Snow, 1990, 1992). Eastward Tertiary tilting in this part of the Cottonwood Mountains suggested to Snow (1990), however, that the axial plane of the Dry Bone syncline originally dipped about 40°W, and that the Marble Canyon thrust probably dipped between 20° and 40°W. This interpretation is consistent with mapping by Stone (1984) that shows Pennsylvanian rocks folded by the Dry Bone syncline in the Cottonwood Canyon area in depositional contact with Tertiary sandstone dipping about 40°E. The pre-Tertiary orientation of the Marble Canyon thrust was thus similar to that of the Inyo Crest and Lemoigne thrusts.

3.3. Racetrack thrust

The Racetrack thrust of McAllister (1952) and related faults place Lower Cambrian strata as old as the Zabriskie Quartzite on the Mississippian Rest Spring Shale and older rocks along the margins of Racetrack Valley (Burchfiel, 1969). If the Ubehebe thrust of Snow (1992) is continuous with the Racetrack thrust, as we interpret it to be (Stevens and Stone, 2005-this volume), rocks as old as the earliest Cambrian Wood Canyon Formation are in the upper plate and the maximum stratigraphic throw is about 5 km (Snow, 1990). Faults of this zone strike approximately N–S and dip 5–30°W (McAllister, 1952; Snow, 1992).

Two quite different interpretations of the nature of the Racetrack and related thrusts have been advanced. Snow (1992) considered the Racetrack and other nearby thrusts to represent an Early Permian duplex zone associated with the Last Chance thrust, and the Ubehebe thrust at the northeast end of Racetrack Valley to be part of the Last Chance thrust. In contrast, we (Stevens and Stone, 2005-this volume) interpret the composite Racetrack–Ubehebe thrust to displace the Last Chance thrust, which in this area we place between the Rest Spring Shale and the overlying Pennsylvanian Tihvipah Limestone (lower Keeler Canyon Formation). If our interpretation is correct, the Racetrack–Ubehebe thrust is younger than the Last Chance thrust and probably coeval with the other thrusts of the Sierra Nevada–Death Valley thrust system, with which it shares several common geologic characteristics.

3.4. Strange Hill thrust

Near Tinemaha Reservoir in the northwestern Inyo Mountains, the Strange Hill thrust of Stevens and Stone (2002) places Devonian rocks assigned to the Squares Tunnel Formation and Mount Morrison Sandstone on Mississippian to Pennsylvanian rocks of the Kearsarge Formation, Rest Spring Shale, and Keeler Canyon Formation. The lower plate rocks are folded into a footwall syncline (Mule Spring syncline) that is depositionally overlapped by the Early to Middle(?) Triassic Union Wash Formation. The Strange Hill thrust, the overturned limb of the Mule Spring syncline, and the overlapping Triassic rocks all have been deformed by later folds and faults and generally dip steeply to the west; reconstruction of the Triassic rocks to horizontal (Stevens and Stone, 2002) suggests that the thrust and axial plane of the syncline originally dipped westward, perhaps as steeply as 25°. The amount of displacement on the Strange Hill thrust is uncertain, but the relatively small stratigraphic throw suggests a displacement probably less than 2 km. The pre-middle Early Triassic age of this thrust and the structural characteristics suggest a tectonic relationship to the Inyo Crest thrust and other faults of the Sierra Nevada–Death Valley thrust system.

3.5. Thrusts in the Mount Morrison pendant

Six thrust faults in the Mount Morrison roof pendant northeast of the Laurel–Convict fault described by Stevens and Greene (2000) and Greene and Ste-

vens (2002) have characteristics similar to those of the Inyo Crest thrust. These faults are, from west to east (structurally higher to lower), the Laurel Mountain, Convict Creek, Mount Morrison, McGee Creek, Esha Canyon, and Nevahbe thrusts (see Fig. 5 of Stevens and Greene, 2000). All of these faults strike northerly to northwesterly and presently dip eastward; they are inferred to have originally dipped westward and later to have been rotated and overturned (Stevens and Greene, 2000). These faults were active between Middle Permian and Middle or Late Triassic time (Stevens and Greene, 2000), and Stevens and Stone (2002) have argued that they formed at the same time as the Strange Hill thrust in the Tinemaha Reservoir area.

The stratigraphic throw on each of the five western faults in the Mount Morrison pendant is small and displacement probably is small as well. Three of these faults, the Laurel Canyon, Mount Morrison, and McGee Creek, are associated with large footwall synclines. The structurally lowest thrust, the Nevahbe thrust, apparently is the most significant. On this fault, Cambrian and Ordovician rocks of the Mount Aggie and Convict Lake formations have been emplaced over the Mississippian Bright Dot Formation and Pennsylvanian–earliest Permian Mount Baldwin Marble representing a stratigraphic throw of at least 2 km. The steeply dipping Bright Dot Formation lies west of the Mount Baldwin Marble. Therefore, if these rocks were originally west-dipping and have been rotated to their present steep eastward dip, similar to all other rocks and structures northeast of the Laurel–Convict fault (Stevens and Greene, 2000), they may represent the overturned limb of a large footwall syncline similar to those associated with most of the other faults in the Sierra Nevada–Death Valley system.

4. Regional cross-section of the Sierra Nevada–Death Valley thrust system

The cross-section shown in Fig. 2 illustrates our concept of the regional structure of the Sierra Nevada–Death Valley thrust system. This cross-section, although generally based on the surface geology on and near line A–A′ shown in Fig. 1, is an interpretive model that attempts to depict the original configuration of the thrust system. The section thus ignores Mesozoic plutons and shows the thrusts in

C.H. Stevens, P. Stone / Earth-Science Reviews 73 (2005) 103–113

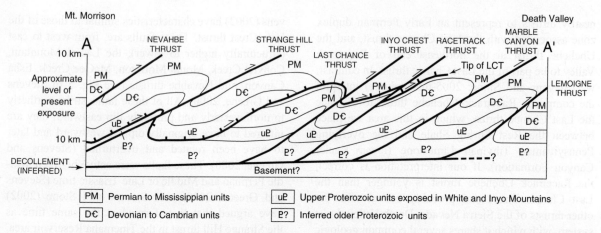

Fig. 2. Interpretive cross-section of the Sierra Nevada–Death Valley thrust system and Last Chance thrust from Mount Morrison roof pendant in the eastern Sierra Nevada to Death Valley. No vertical exaggeration. See text for discussion. Line of section (A–A′) is shown on Fig. 1.

what we interpret to be their most likely original orientations.

In preparation of this cross-section, the Mount Morrison pendant first was restored 65 km to the south along the cryptic Tinemaha fault (Stevens and Greene, 1999). The westernmost part of the section is based on the geology in this pendant and in the westernmost Inyo Mountains near Tinemaha Reservoir (Stevens and Greene, 2000; Greene and Stevens, 2002; Stevens and Stone, 2002). The section across the northern Inyo Mountains is based on the mapping and cross-sections of Nelson (1966, 1971) in the Waucoba Mountain and Waucoba Spring 15′ quadrangles. In Saline Valley, the line of section was translated 35 km southward along the inferred trace of the Inyo Crest thrust to the Dry Mountain 15′ quadrangle, where the section is based on the mapping and cross-sections of Burchfiel (1969). Nine kilometers of extension in a northwest direction, parallel with and approximately equal to late Cenozoic dextral displacement on the Hunter Mountain fault (Burchfiel et al., 1987), was removed from the section in Saline Valley. In the central Cottonwood Mountains, the section is based on the mapping and cross-sections of McAllister (1952) and Snow (1990). Finally, the line of section was translated about 30 km southward on the projected trace of the Marble Canyon thrust into the Panamint Butte 15′ quadrangle, where the mapping and cross-sections of Hall (1971) were used to continue the section across the southern Cottonwood Mountains to Death Valley.

In the cross-section, formations are grouped by age. From younger to older, the units shown are Mississippian–Permian, Cambrian–Devonian, a younger Proterozoic unit that represents all Proterozoic rocks actually exposed at the surface in the White and Inyo Mountains, and a queried older Proterozoic unit of unknown lithology and thickness. Dips of the Lemoigne, Marble Canyon, Racetrack, Inyo Crest, and Strange Hill thrusts are based on field relations noted previously; the original dip of the Nevahbe thrust cannot be determined from field relations and is arbitrarily shown as 35°. The Racetrack, Inyo Crest, and Strange Hill thrusts are shown as cutting the older Last Chance thrust, which is modeled as a ramp-flat thrust consistent with the interpretations of Stevens and Stone (2005-this volume). Because section A–A′ cuts obliquely across the northeast-trending Last Chance thrust, and because of complications arising from the translation of the line of section in Saline Valley, the amount of displacement on the Last Chance thrust and the width of the upper plate ramp anticline appear smaller in this section than that indicated by Stevens and Stone (2005-this volume).

The cross-section shows our inference that thrusts of the Sierra Nevada–Death Valley system merge at depth into a regional décollement, a common feature of thrust belts (e.g., Price and Hatcher, 1983). The depth to the inferred décollement is unconstrained; however, it must be deeper than the oldest Proterozoic units exposed at the surface and it probably is deeper than any strata cut by the Last Chance thrust (Fig. 2).

The inferred décollement is shown dipping gently to the west, which is typical for the upper surface of the basement in a thrust belt where the sedimentary cover has been stripped off (Woodward et al., 1985).

5. Other possible faults of the Sierra Nevada–Death Valley thrust system

Other faults that may be part of the Sierra Nevada–Death Valley thrust system crop out on the west side and south of the southern Inyo Mountains (Fig. 1). These include the Ophir Peak thrust and the Talc City thrust zone (Hall and MacKevett, 1962; Stone et al., 1989), and the Dolomite Canyon thrust zone (Merriam, 1963; Kelley, 1973). These thrusts share some characteristics with those of the Sierra Nevada–Death Valley system, but their relations are not clear because of Jurassic overprinting by structures of the East Sierran thrust system.

5.1. Ophir Peak thrust

The Ophir Peak thrust, exposed near the crest of the Darwin Hills (Fig. 1), is structurally quite similar to several of the thrusts in the Sierra Nevada–Death Valley system. This thrust emplaces rocks as old as Devonian over the Early Permian Osborne Canyon Formation, indicating a stratigraphic throw of about 1.5 km. The Osborne Canyon Formation, stratigraphic equivalents of which are interpreted to overlap the Lee Flat (=Last Chance) thrust in the southeastern Inyo Mountains (Stone et al., 1989), is folded into an east-vergent footwall syncline similar to other synclines characteristic of the Sierra Nevada–Death Valley thrust system. The Ophir Peak thrust is intruded by a 174-Ma (Jurassic) pluton that is deformed by the East Sierran thrust system. These age relations indicate that the Ophir Peak thrust could be part of the Sierra Nevada–Death Valley thrust system.

5.2. Talc City thrust zone

Thrust faults exposed in the Talc City Hills differ somewhat from those of the Sierra Nevada–Death Valley system in their apparent low dips and more westerly trends, but they could be of similar age. Stone et al. (1989) correlated the lowest of three major thrust faults recognized in this area with the Ophir Peak thrust, which is pre-174 Ma and post-early Early Permian (see above). If this correlation is correct and the three thrusts recognized by Stone et al. (1989) are all the same age, they also may be related to the Sierra Nevada–Death Valley thrust system.

5.3. Dolomite Canyon thrust zone

The Dolomite Canyon thrust zone, exposed on the west flank of the southern Inyo Mountains, is a complex of at least four faults that emplace rocks as old as Silurian on rocks as young as Pennsylvanian (Kelley, 1973). These faults differ from those of the Sierra Nevada–Death Valley thrust system in having a more westerly trend and lacking footwall synclines, but their age may be similar. The Dolomite Canyon thrusts may be kinematically related to an anticline in the upper plate which is intruded by the 174-Ma Long John Canyon pluton (Dunne et al., 1978), and one of the thrusts near the south end of this fault belt is cut (Kelley, 1973) by a pluton of probable Middle Jurassic age (George Dunne, oral communication, 2002). These relations, although not definitive, permit a temporal correlation with the Sierra Nevada–Death Valley thrust system.

6. Age of deformation

Stratigraphic and structural relations in the areas discussed above show that deformation resulting in development of the Sierra Nevada–Death Valley thrust system took place between late Early Permian and middle Early Triassic time. A more precise age, however, can be interpreted from other geologic relations in the region. First, it is probable that the deformation discussed here is represented by an angular unconformity that separates rocks as young as Wuchiapingian (early Late Permian) from middle Early Triassic strata in the southern Inyo Mountains (Stone et al., 2000). Deformation in the region also probably predated deposition of the Early Triassic Candelaria Formation, which has been identified in both west-central Nevada and the Saddlebag Lake pendant, the latter only about 50 km north of the Mount Morrison pendant. This unit apparently represents continuous deposition and was unaffected by any important syndepositional contractile deformation, from earliest Triassic (Griesbachian) to late Early Triassic time (Speed, 1984). Therefore, we suggest that the most likely age of the Sierra Nevada–Death Valley thrust system is Late Permian.

Structures at Tinemaha Reservoir and Mount Morrison pendant, here considered part of the Sierra Nevada–Death Valley thrust system, represent a tectonic event that Stevens and Greene (2000) referred to as the Morrison orogeny. Stevens and Greene (2000) considered this event as older than the Sonoma orogeny following Speed (1984) and Schweickert and Lahren (1987), who interpreted the latter as coeval with Middle Triassic emplacement of the Golconda allochthon in west-central Nevada and the Saddlebag Lake pendant. However, deformation of the Permian and older Havallah Sequence during the Sonoma orogeny as originally defined in north-central Nevada predated deposition of the Koipato Formation, the upper part of which contains late Early Triassic ammonoids (Silberling, 1973). Thus, the age of the Sonoma orogeny (sensu stricto) is constrained to be older than the time of emplacement of the Golconda allochthon in west-central Nevada and the Saddlebag Lake pendant. It follows that the age of deformation resulting in development of the Sierra Nevada–Death Valley thrust system could be the same as that of the Sonoma orogeny. The location of the Havallah Sequence at the time of the Sonoma orogeny, however, is unknown, and this deformation could have taken place at some distance from the North American continental margin. Because the Morrison and Sonoma orogenies involve rocks of such different character that were deformed in distant locales, we here recognize them as separate events and consider the Sierra Nevada–Death Valley thrust belt as a product of the Morrison orogeny.

7. Comparison with other thrust systems

The overall structure of the Sierra Nevada–Death Valley thrust system appears to be rather simple (Fig. 2) and compares closely with the distal parts of well known thrust systems such as those in the Cordilleran and Appalachian thrust belts (e.g., Price and Hatcher, 1983) and the Alps east of Lake Geneva (e.g., Boyer and Elliott, 1982). In these systems, thrusts commonly

are spaced at 7- to 10-km intervals along with some more closely spaced thrusts, although spacing may be as much as 15–30 km in the Cordilleran thrust belt (e.g., Kulik and Schmidt, 1988), similar to that in the Sierra Nevada–Death Valley system.

8. Origin of the Sierra Nevada–Death Valley thrust system

The western North American continental margin apparently consisted of two segments during the Permian (Fig. 1). The northern segment, which was formed by rifting during the Neoproterozoic and was passive until the Late Devonian–Early Mississippian Antler orogeny, extends about N 20°E from the latitude of Independence, California. The orientation of this segment is based largely on lower Paleozoic facies trends and the frontal margin of the Antler orogenic belt (Stevens and Greene, 1999). The southern segment is interpreted to have formed by transform truncation of the original margin initiated in the Pennsylvanian (Stone and Stevens, 1988; Walker, 1988) and extends about S 30°E from the latitude of Independence.

Late Paleozoic and Mesozoic deformational and plutonic belts generally trend subparallel to one or the other of these margin segments, as shown diagrammatically in Fig. 3. The oldest late Paleozoic deformational belt, represented by the Last Chance thrust (Stevens and Stone, 2005-this volume), approximately parallels the Neoproterozoic to middle Paleozoic continental margin segment. We therefore interpret this deformation to have been related to tectonic activity along this northeast-trending margin. In contrast, the Sierra Nevada–Death Valley thrust system is more closely parallel to the southeast-trending continental margin segment and many later features, including the Tinemaha strike-slip fault, the East Sierran thrust system, and the Sierra Nevada batholith (Fig. 3). Thus, the Sierra Nevada–Death Valley thrust system probably formed in response to events occurring along that margin.

In a belt extending from the southern Sierra Nevada through the El Paso Mountains and into the Mojave Desert (Fig. 1), Dunne and Saleeby (1993), Miller et al. (1995), and Barth et al. (1997) have identified numerous pre-Jurassic plutons that appar-

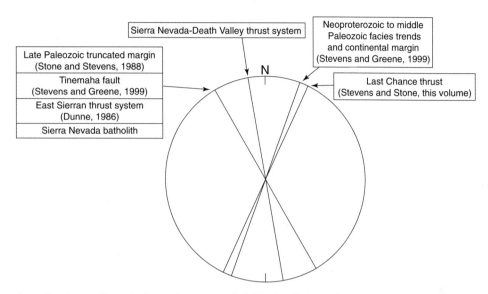

Fig. 3. Orientations of major pre-Cenozoic features in east-central California relative to the Neoproterozoic–middle Paleozoic continental margin and late Paleozoic truncated-margin segments.

ently represent an early episode of subduction and arc magmatism along this southeast-trending continental margin. These plutons are mostly Triassic in age, but some are Late Permian (Andy Barth, written communication, 2002). Thus, deformation associated with the Sierra Nevada–Death Valley thrust system may have been linked tectonically to the initiation of convergent plate motion and the earliest phase of arc development all along the southeast-trending continental margin during the Late Permian as inferred by numerous previous workers.

9. Summary and conclusions

The Sierra Nevada–Death Valley thrust system forms a belt more than 100 km wide, stretching from pendants within the Sierra Nevada eastward to Death Valley. The thrust faults that form this belt are characterized by original dips generally of about 25–35°W, stratigraphic throws of 2–5 km, and large footwall synclines. The faults are typically spaced on the order of 15–20 km, except for the closer distribution of small faults in the Mount Morrison pendant.

In all areas, age constraints are compatible with a late Early Permian to middle Early Triassic age of deformation, with Late Permian being preferred on the basis of regional relationships. We view all the faults

of the Sierra Nevada–Death Valley thrust system as products of the Morrison orogeny, which is essentially coeval with the Sonoma orogeny recognized in north-central Nevada.

Deformation represented by the Sierra Nevada–Death Valley thrust system, as well as all subsequent contractional deformations in the region, are here considered to have been related to compressional tectonics along a southeast-trending segment of the continental margin. This Late Permian(?) contractional deformation apparently reflects the initiation of plate convergence and subduction that later led to voluminous Sierran arc magmatism.

Acknowledgments

We are grateful to George Dunne and Norm Silberling for their careful reviews of the manuscript. Robert B. Miller helped greatly, especially in the construction of the cross-section.

References

Barth, A.P., Tosdal, R.M., Wooden, J.L., Howard, K.A., 1997. Triassic plutonism in southern California: southward younging of arc initiation along a truncated continental margin. Tectonics 16, 290–304.

Bateman, P.C., 1992. Plutonism in the central part of the Sierra Nevada batholith, California. U.S. Geological Survey Professional Paper 1483, 186 pp.

Boyer, S.E., Elliott, D., 1982. Thrust systems. American Association of Petroleum Geologists Bulletin 66, 1196–1230.

Burchfiel, B.C., 1969. Geology of the Dry Mountain quadrangle, Inyo County, California. Special Report-California Division of Mines and Geology 99, 19 pp.

Burchfiel, B.C., Hodges, K.V., Royden, L.H., 1987. Geology of Panamint Valley-Saline Valley pull-apart system, California: palinspastic evidence for low-angle geometry of a Neogene range-bounding fault. Journal of Geophysical Research 92 (B10), 10422–10426.

Dunne, G.C., 1986. Geologic evolution of the southern Inyo Range, Darwin Plateau, and Argus and Slate ranges, east-central California—an overview. In: Dunne, G.C. (Ed.), Mesozoic and Cenozoic Structural Evolution of Selected Areas, East-central California. Geological Society of America, Cordilleran Section, Guidebook and Volume, Trips 2 and 14, pp. 3–21.

Dunne, G.C., Saleeby, J.B., 1993. Kern Plateau shear zone, southern Sierra Nevada—new data concerning age and northward continuation. Abstracts with Programs. Geological Society of America 25 (5), 33.

Dunne, G.C., Gulliver, R.M., Sylvester, A.G., 1978. Mesozoic evolution of rocks of the White, Inyo, Argus, and Slate Ranges, eastern California. In: Howell, D.G., McDougall, K.A. (Eds.), Mesozoic Paleogeography of the Western United States. Society of Economic Paleontologists and Mineralogists, Pacific Section, Book 8, pp. 189–207.

Greene, D.C., Stevens, C.H., 2002. Geologic map of Paleozoic rocks in the Mount Morrison pendant, eastern Sierra Nevada, California. California Division of Mines and Geology Map Sheet 53, scale 1:24,000.

Greene, D.C., Schweickert, R.A., Stevens, C.H., 1997. Roberts Mountains allochthon and the western margin of the Cordilleran miogeocline in the Northern Ritter Range pendant, eastern Sierra Nevada, California. Geological Society of America Bulletin 109, 1294–1305.

Hall, W.E., 1971. Geology of the Panamint Butte quadrangle, Inyo County, California. U.S. Geological Survey Bulletin 1299, 67 pp.

Hall, W.E., MacKevett, E.M., 1962. Geology and ore deposits of the Darwin quadrangle, Inyo County, California. U.S. Geological Survey Professional Paper 368, 87 pp.

Kelley, J.S., 1973. Structural geology of a portion of the southwest quarter of the New York Butte quadrangle, Inyo County, California [MS thesis]. San Jose State University, San Jose, Calif., 93 pp.

Kulik, D.M., Schmidt, C.J., 1988. Region of overlap and styles of interaction of Cordilleran thrust belt and Rocky Mountain foreland. In: Schmidt, C.J., Perry Jr., W.J. (Eds.), Interaction of the Rocky Mountain Foreland and the Cordilleran Thrust Belt. Geological Society of America, Memoir 171, 75–98.

McAllister, J.F., 1952. Rocks and structure of the Quartz Spring area, northern Panamint Range, California. Special Report. California Division of Mines 25, 38 pp.

Merriam, C.W., 1963. Geology of the Cerro Gordo mining district, Inyo County, California. U.S. Geological Survey Professional Paper 408, 83 pp.

Miller, J.S., Glazner, A.F., Walker, J.D., Martin, M.W., 1995. Geochronologic and isotopic evidence for Triassic–Jurassic emplacement of the eugeoclinal allochthon in the Mojave Desert region, California. Geological Society of America Bulletin 107, 1441–1457.

Nelson, C.A., 1966. Geologic map of the Waucoba Mountain quadrangle, Inyo County, California. U.S. Geological Survey Map GQ-528, scale 1:62,500.

Nelson, C.A., 1971. Geologic map of the Waucoba Spring quadrangle, Inyo County, California. U.S. Geological Survey Map GQ-921, scale 1:62,500.

Price, R.A., Hatcher Jr., R.D., 1983. Tectonic significance of similarities in the evolution of the Alabama-Pennsylvania Appalchians and the Alberta–British Columbia Canadian Cordillera. In: Hatcher Jr., R.D., Williams, H., Zietz, I. (Eds.), Contributions to the Tectonics and Geophysics of Mountain Chains. Geological Society of America, Memoir 158, 149–160.

Schweickert, R.A., Lahren, M.M., 1987. Continuation of Antler and Sonoma orogenic belts to the eastern Sierra Nevada, California, and Late Triassic thrusting in a compressional arc. Geology 15, 270–273.

Schweickert, R.A., Lahren, M.M., 1993a. Tectonics of the east-central Sierra Nevada-Saddlebag Lake and northern Ritter Range pendants. In: Lahren, M.M., Trexler Jr., J.H., Spinosa, C. (Eds.), Crustal Evolution of the Great Basin and Sierra Nevada. Geological Society of America, Cordilleran/Rocky Mountain Section, Guidebook. Department of Geological Sciences, University of Nevada, Reno, pp. 313–351.

Schweickert, R.A., Lahren, M.M., 1993b. Triassic–Jurassic magmatic arc in eastern California and western Nevada: arc evolution, cryptic tectonic breaks, and significance of the Mojave-Snow Lake fault. In: Dunne, G.C., McDougall, K.A. (Eds.), Mesozoic Paleogeography of the Western United States - II. Pacific Section Society of Economic Paleontologists and Mineralogists, pp. 227–261.

Silberling, N.J., 1973. Geologic events during Permo-Triassic time along the Pacific margin of the United States. In: Logan, A., Hills, L.V. (Eds.), The Permian and Triassic Systems and Their Mutual Boundary. Canadian Society of Petroleum Geologists, Memoir, vol. 2, pp. 345–362.

Snow, J.K., 1990. Cordilleran orogenesis, extensional tectonics, and geology of the Cottonwood Mountains area, Death Valley region, California and Nevada [PhD thesis]. Harvard University, Cambridge, Mass., 533 pp.

Snow, J.K., 1992. Large-magnitude Permian shortening and continental-margin tectonics in the southern Cordillera. Geological Society of America Bulletin 104, 80–105.

Snow, J.K., Wernicke, B.P., 2000. Cenozoic tectonism in the central Basin and Range: magnitude, rate, and distribution of upper crustal strain. American Journal of Science 300, 659–719.

Snow, J.K., Asmeron, Y., Lux, D.R., 1991. Permian–Triassic plutonism and tectonics, Death Valley region, California and Nevada. Geology 19, 629–632.

Speed, R.C., 1984. Paleozoic and Mesozoic continental margin collision zone features. In: Lintz Jr., J. (Ed.), Western Geological Excursions. Geological Society of America Annual Meeting, Reno, Nevada, 1984 Field Trip Guidebook, vol. 4, pp. 66–80.

Stadler, C.A., 1968. The geology of the Goldbelt Spring area, northern Panamint Range, Inyo County, California [MS thesis]. University of Oregon, Eugene, 78 pp.

Stevens, C.H., Greene, D.C., 1999. Stratigraphy, depositional history, and tectonic evolution of Paleozoic continental-margin rocks in roof pendants of the eastern Sierra Nevada, California. Geological Society of America Bulletin 111, 919–933.

Stevens, C.H., Greene, D.C., 2000. Geology of Paleozoic rocks in eastern Sierra Nevada roof pendants, California. In: Lageson, D.R., Peters, S.G., Lahren, M.M. (Eds.), Great Basin and Sierra Nevada. Geological Society of America Field Guide, vol. 2, pp. 237–254.

Stevens, C.H., Stone, P., 2002. Correlation of Permian and Triassic deformations in the western Great Basin and eastern Sierra Nevada: evidence from the northern Inyo Mountains near Tinemaha Reservoir, east-central California. Geological Society of America Bulletin 114, 1210–1221.

Stevens, C.H., Stone, P., 2005. Interpretation of the Last Chance thrust, east-central California, as an Early Permian dûcollement in a previously undeformed shale basin. Earth-Science Reviews 73, 79–101. doi:10.1016/j.earscirev.2005.04. 005.

Stevens, C.H., Stone, P., Dunne, G.C., Greene, D.C., Walker, J.D., Swanson, B.J., 1997. Paleozoic and Mesozoic evolution of east-central California. International Geology Review 39, 788–829.

Stevens, C.H., Stone, P., Ritter, S.M., 2001. Conodont and fusulinid biostratigraphy and history of the Pennsylvanian to Lower Permian Keeler basin, east-central California. Brigham Young University Geology Studies 46, 99–142.

Stewart, J.H., Ross, D.C., Nelson, C.A., Burchfiel, B.C., 1966. Last Chance thrust—a major fault in the eastern part of Inyo County, California. U.S. Geological Survey Professional Paper 550-D, D23–D34.

Stone, P., 1984. Stratigraphy, depositional history, and paleogeographic significance of Pennsylvanian and Permian rocks in the Owens Valley–Death Valley region, California [PhD thesis]. Stanford University, Stanford, Calif., 399 pp.

Stone, P., Stevens, C.H., 1988. Pennsylvanian and Early Permian paleogeography of east-central California: implications for the shape of the continental margin and timing of continental truncation. Geology 16, 330–333.

Stone, P., Stevens, C.H., Magginetti, R.T., 1987. Pennsylvanian and Permian stratigraphy of the northern Argus Range and Darwin Canyon area, California. U.S. Geological Survey Bulletin 1691, 30 pp.

Stone, P., Dunne, G.C., Stevens, C.H., Gulliver, R.M., 1989. Geologic map of Paleozoic and Mesozoic rocks in parts of the Darwin and adjacent quadrangles, Inyo County, California. U.S. Geological Survey Miscellaneous Investigations Series Map I-1932, scale 1:31,250.

Stone, P., Stevens, C.H., Orchard, M.J., 1991. Stratigraphy of the Lower and Middle(?) Triassic Union Wash Formation, east-central California. U.S. Geological Survey Bulletin 1928, 26 pp.

Stone, P., Stevens, C.H., Spinosa, C., Furnish, W.M., Glenister, B.F., Wardlaw, B.R., 2000. Stratigraphic relations and tectonic significance of rocks near the Permian–Triassic boundary, southern Inyo Mountains, California. Geological Society of America Map and Chart Series MCH086, 32 pp.

Swanson, B.J., 1996. Structural geology and deformational history of the southern Inyo Mountains east of Keeler, Inyo County, California [MS thesis]. California State University, Northridge, 125 pp.

Walker, J.D., 1988. Permian and Triassic rocks of the Mojave Desert and their implications for timing and mechanisms of continental truncation. Tectonics 7, 685–709.

Woodward, N.B., Boyer, S.E., Suppe, J., 1985. An outline of balanced cross-sections, 2nd ed. Studies in Geology, vol. 11. University of Tennessee Dept. of Geological Sciences, 172 pp.

Wrucke, C.T., Corbett, K.P., 1990. Geologic map of the Last Chance quadrangle, California and Nevada. U.S. Geological Survey Open-file Report 90-649-B, scale 1:62,500.

Available online at www.sciencedirect.com

Earth-Science Reviews 73 (2005) 115–138

www.elsevier.com/locate/earscirev

The Black Mountains turtlebacks: Rosetta stones of Death Valley tectonics

Marli B. Miller [a,*], Terry L. Pavlis [b]

[a] Department of Geological Sciences, University of Oregon, Eugene, OR 97403-1272, United States
[b] Department of Geology and Geophysics, University of New Orleans, New Orleans, LA 70148, United States

Accepted 1 April 2005

Abstract

The Black Mountains turtlebacks expose mid-crustal rock along the western front of the Black Mountains. As such, they provide keys to understanding the Tertiary structural evolution of Death Valley, and because of the outstanding rock exposure, they also provide valuable natural laboratories for observing structural processes. There are three turtlebacks: the Badwater turtleback in the north, the Copper Canyon turtleback, and the Mormon Point turtleback in the south. Although important differences exist among them, each turtleback displays a doubly plunging antiformal core of metamorphic and igneous rock and a brittle fault contact to the northwest that is structurally overlain by Miocene–Pleistocene volcanic and/or sedimentary rock.

The turtleback cores contain mylonitic rocks that record an early period of top-southeastward directed shear followed by top-northwestward directed shear. The earlier formed mylonites are cut by, and locally appear concurrent with, 55–61 Ma pegmatite. We interpret these fabrics as related to large-scale, basement-involved thrust faults at the turtlebacks, now preserved as areally-extensive, metamorphosed, basement over younger-cover contacts.

The younger, and far more pervasive, mylonites record late Tertiary extensional unroofing of the turtleback footwalls from mid-crustal depths. Available geochronology suggests that they cooled through 300 °C at different times: 13 Ma at Badwater; 6 Ma at Copper Canyon; 8 Ma at Mormon Point. At Mormon Point and Copper Canyon turtlebacks these dates record cooling of the metamorphic assemblages from beneath the floor of an ~11 Ma Tertiary plutonic complex. Collectively these relationships suggest that the turtlebacks record initiation of ductile extension before ~14 Ma followed by injection of a large plutonic complex along the ductile shear zone. Ductile deformation continued during extensional uplift until the rocks cooled below temperatures for crystal plastic deformation by 6–8 Ma. Subsequent low-angle brittle fault slip led to final exposure of the igneous and metamorphic complex.

The turtleback shear zones can constrain models for crustal extension from map-view as well as cross-sectional perspectives. In map view, the presence of basement-involved thrust faults in the turtlebacks suggest the Black Mountains were a basement

---

 * Corresponding author.
   E-mail addresses: millerm@uoregon.edu (M.B. Miller), tpavlis@uno.edu (T.L. Pavlis).

high prior to late Tertiary extension. In cross-section, the turtleback geometries and histories are most compatible with models that call on multiple faults rather than a single detachment to drive post-11 Ma extension.

Keywords: turtlebacks; Death Valley; extension; ductile shear zones; basin; range

1. Introduction

The three Black Mountains turtlebacks have long been considered important but enigmatic features of Death Valley geology. They were named "turtlebacks" by Curry (1938) because their convex-upwards morphologies resemble turtleshells. In terms of bedrock geology, however, the turtlebacks stand out because they offer the only exposures of crystalline basement rock along the Black Mountains front (Curry, 1938, 1954; Fig. 1). They are probably the single most important features in deciphering Death Valley's complex tectonic story.

Each of the turtlebacks displays the essential features of Cordilleran Metamorphic Core Complexes (Davis, 1980; Lister and Davis, 1989): a ductilely deformed metamorphic core, an overlying body of highly faulted upper crustal rock, and at their northwest margins, a brittle extensional fault zone between them. Additionally, mylonitic rocks of the core reflect a sense-of-shear that is similar to, but older than, the fault zone. Post- to syn-mylonitization folding about axes subparallel to shear zone transport produced up to 4 km of structural relief both at high angles and parallel to transport. Thus, each turtleback individually represents a "core complex" that occupies an area of only about 6 km^2, but collectively, the turtlebacks represent different parts of a large-scale extensional system. This complex 3D geometry provides outstanding opportunities to observe both shallow- and moderately deep-level processes of shear zone evolution.

The Death Valley region is an especially fertile area for testing models for crustal extension. There, tectonic interpretations generally fall into one of two categories (Fig. 2). One category, the "Rolling Hinge" model (Stewart, 1983; Hamilton, 1988; Wernicke et al., 1988; Snow and Wernicke, 2000), calls for ~80 km of horizontal translation and unroofing of the Black Mountains on a detachment fault of regional extent. Because the Black Mountains contain ~11 Ma plutons, this faulting must postdate 11

Ma. This model views strike-slip faults, such as the en echelon Furnace Creek and Sheephead faults, as upper crustal edges of the detachment system. The other category calls for extension by slip on numerous, distinct fault zones (Wright and Troxel, 1984; Wright et al., 1991; Serpa and Pavlis, 1996; Miller and Prave, 2002). This category, the "Pull-Apart" model, calls on the strike-slip faults to penetrate deeply into the crust and drive extension between their terminations, similar to the model proposed by Burchfiel and Stewart (1966) for modern Death Valley. Both interpretations rely heavily on findings from the turtlebacks because they each describe the turtlebacks as the principal shear zones. In the Rolling Hinge model, the turtlebacks are different exposures of the same detachment fault; in the Pull-Apart model, the turtlebacks define three of the largest, and distinct, faults.

Both tectonic interpretations also rely on differing views of the original configuration of the Mesozoic fold-thrust belt. The Rolling Hinge model requires an originally narrow thrust belt in which the Panamint and Chicago Pass thrusts, now separated by about 80 km, were originally the same structure. Detachment faulting during late Tertiary extension cut this fault and displaced the two parts to their present locations (Figs. 1 and 2A). By contrast, the Pull-Apart model requires an originally wider thrust belt in which the Panamint and Chicago Pass thrusts originated as separate features (Fig. 2B).

The turtlebacks are here described as "Rosetta Stones" because they provide keys to decipher both the original fold-thrust belt geometry as well as the geometry and mechanisms of late Tertiary extension. Their locations at the western edge of the Black Mountains place them in the middle of the fold and thrust belt. Most recently, Miller and Friedman (1999) and Miller (2003) have found evidence in their footwalls for mid-crustal, pre-55 Ma, basement-involved thrust faulting. The addition of the Black Mountains to the fold-thrust belt helps constrain its original

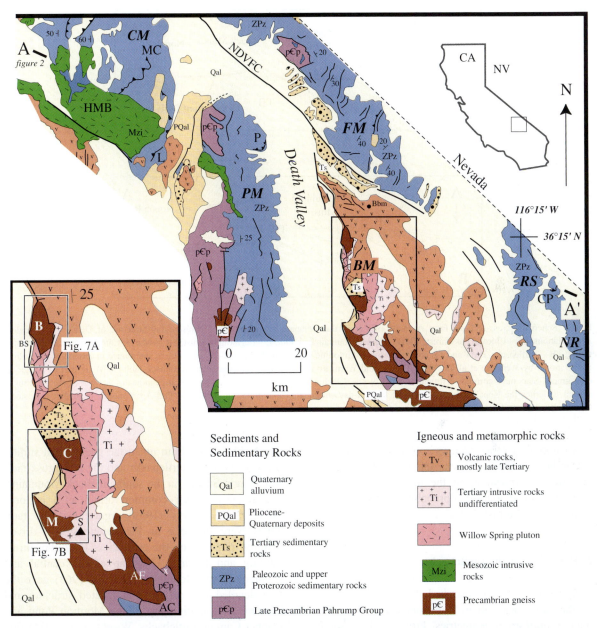

Fig. 1. Map of the central Death Valley region; inset shows close-up view of Black Mountains. Abbreviations: AC—Amargosa Chaos; AF—Amargosa Fault; B—Badwater turtleback; Bbm—Billie Borate Mine; BM—Black Mountains; BS—Badwater Spring; C—Copper Canyon turtleback; CM—Cottonwood Mountains; CP—Chicago Pass thrust; FM—Funeral Mountains; HMB—Hunter Mountain batholith; L—Lemoigne thrust; M—Mormon Point turtleback; MC—Marble Canyon thrust; NR—Nopah Range; P—Panamint thrust; PM—Panamint Mountains; RS—Resting Spring Range; S—Smith Mountain.

geometry to one that favors the Pull-Apart model and severely limits the Rolling Hinge model.

The turtlebacks therefore offer world-class field laboratories to study the processes of structural geology and tectonics—the former through their superb exposures of mid- and upper-crustal fault zones and associated fault rocks and structures, and the latter through the insights they can give to models of crustal

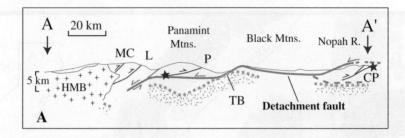

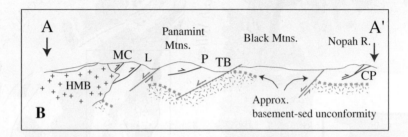

Fig. 2. Schematic cross-sections from A to A′ of Fig. 1 that compare interpretations of present-day structure of Death Valley region. Abbreviations: CP—Chicago Pass thrust; HMB—Hunter Mountain batholith; L—Legmoigne thrust; MC—Marble Canyon thrust; P—Panamint thrust; TB—Turtlebacks. BMT and TB are both projected into line of section. A: Present-day structure as interpreted in context of the Rolling Hinge model by Wernicke et al. (1988) and Snow and Wernicke (2000). Note that extensional turtleback faults (TB) are part of a regional detachment. Stars mark correlative structures offset by the detachment. Dashed line beneath Nopah range indicates that part of the detachment system underlies the Nopah Range as well. B: Present-day structure as interpreted in context of Pull-Apart model. Note that extensional turtleback faults are deep-seated faults.

extension. This paper reviews contributions made by researchers on the turtlebacks and concludes with our preferred interpretation of the turtlebacks as being long-lived shear zones that probably reactivated basement-involved thrust faults.

2. Research on the turtlebacks

Curry (1938) first described the crystalline cores of the turtlebacks, their fault zones, and their distinctive antiformal geometries. From north to south, Curry (1938, 1954) described three turtlebacks in the Black Mountains, the Badwater, Copper Canyon, and Mormon Point turtlebacks (Figs. 1 and 3A,B,C). He

and Noble (1941) both noted the geometrical similarities between the turtlebacks and the Amargosa fault near Virgin Spring, in the southern Black Mountains, and suggested their common origin as a regional thrust fault. Drewes (1959) argued that the fault zones were Pliocene or Pleistocene gravity-driven features, imposed on the turtlebacks long after thrust faulting and folding. His map of the central and southern Black Mountains (Drewes, 1963) gave the first detailed look of the complicated structure and hanging wall stratigraphy of the Mormon Point and Copper Canyon turtlebacks. Wright et al. (1974) first recognized that the turtleback fault zones were rooted normal faults that played an important role in Death Valley's late Cenozoic extensional story.

Fig. 3. Photographs of the Black Mountains turtlebacks. In each, view is to north. Abbreviations: B—Badwater turtleback; C—Copper Canyon turtleback; M—Mormon Point turtleback. A: View of Badwater turtleback. Fold symbol and letter "c" corresponds to fold axes as depicted on Fig. 7A. Arrow marks location of Fault #2 of the Badwater turtleback fault system. B: Copper Canyon turtleback. Footwall is green; Hanging wall is red and tan. Arrow marks location of Copper Canyon turtleback fault. C: Mormon Point turtleback. Tan color marks footwall marble; greenish rock marks basement gneiss.

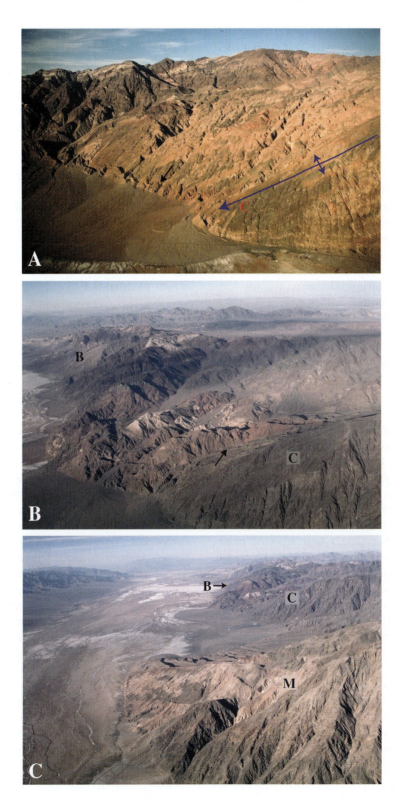

Otton (1976, 1977) mapped the Mormon Point and Copper Canyon turtlebacks at 1:24,000 and described most of the relations that are key to interpretations today.

A common theme of this early research is the importance of pre-Cenozoic deformation in shaping the turtlebacks. Wright et al. (1974) and Otton (1976), for example, recognized the relevance of the turtlebacks to crustal extension, but called on Mesozoic shortening to form their antiformal geometries. However, with the recognition of the extensional nature of metamorphic core complexes and their associated ductilely deformed rocks elsewhere (e.g. Davis et al., 1982, 1986; Miller et al., 1983) as well as the recognition that the Willow Spring pluton in the Black Mountains was only 11 Ma (Asmerom et al., 1990), most later workers emphasized the extensional features in the turtlebacks (e.g. Miller in Wright et al., 1991; Pavlis in Wright et al., 1991; Miller, 1992a,b; Holm et al., 1992). By the mid-1990s, most workers accepted that Tertiary extension, not Mesozoic shortening, was the primary cause of the turtleback geometries (Mancktelow and Pavlis, 1994; Holm et al., 1994a,b).

Otton (1976), however, mapped older-over-younger contacts at the Mormon Point and Copper Canyon turtlebacks, which Holm (1992) interpreted as minor thrust faults related to the Mesozoic contraction. Using U–Pb dating of zircons, Miller and Friedman (1999) found pegmatite at the Badwater turtleback that locally cross-cut mylonitic fabrics to be 55 Ma; they interpreted the older fabrics as related to the Mesozoic Sevier Orogeny. Miller (2003) later reported more structural data to propose that a significant basement thrust existed at the Badwater turtleback in addition to the thrust faults at the Mormon Point and Copper Canyon turtlebacks, and concluded that the thrusts were probably early Tertiary in age. This paper further develops that argument to show the relevance of these structures to the pre-extension geometry of the fold-thrust belt.

Several recent studies of the turtleback fault zones and their hanging walls have clarified important aspects of the northern Black Mountains stratigraphy, turtleback fault geometries and timing of slip. Miller (1991) and Keener et al. (1993) found evidence for successive generations of turtleback fault zones, and interpreted that they were cut by

present-day faults in front of the range. This interpretation is questioned by Hayman et al. (2003). Pavlis et al. (1993) and Burchfiel et al. (1995) interpreted the kinematic framework of the embayed mountain front at Mormon Point. Greene (1997) mapped stratigraphy and structures of the hanging wall of the Badwater turtleback north of Natural Bridge canyon. Knott (1998) documented Pleistocene slip on the Badwater and Mormon Point turtlebacks and described fault segmentation along the length of the Black Mountains front. Nemser (2001) and Hayman et al. (2003) modeled faulting in the upper plates of the Badwater and Mormon Point turtlebacks. Most recently, Dee et al. (2004) found evidence for two stages of ductile deformation at the Mormon Point turtleback, the most significant of which predated 9.5 Ma.

3. Features of the turtlebacks

The lower plate, brittle fault zone, and upper plate define the three principal components of each turtleback (Figs. 3 and 4). As suggested by studies of metamorphic core complexes elsewhere (e.g. Lister and Davis, 1989) and documented by Miller (1992a,b), Holm (1992), Pavlis et al. (1993), Keener et al. (1993), Nemser (2001), and Hayman et al. (2003), brittle deformation of the hanging walls fit into the same extensional kinematic picture as that of the fault zones, which in turn have approximately the same sense of shear as the footwall mylonites. Each component also exhibits to varying degrees the antiformal geometries characteristic of the turtlebacks. The footwalls, however, include Proterozoic and both Neogene and Paleogene intrusive rock and so have geologic histories that predate Neogene extension. Here, each component of the turtlebacks, as well as their antiformal shapes, will be reviewed separately.

3.1. Turtleback footwalls

Each turtleback footwall consists of quartz-feldspar gneiss and marble with minor amounts of pelitic schist. They are intruded by a host of igneous rocks that date from the early to late Tertiary. Most footwall rocks, except for the youngest intrusions, exhibit a range of tectonite fabrics. Consequently, the igneous

Fig. 4. Photograph of three principal elements of the turtlebacks as seen at Badwater turtleback. Note listric normal fault in hanging wall directly above geologist.

rocks have been critical in sorting out episodes of deformation.

3.1.1. Footwall stratigraphy

The gneiss, with a U–Pb zircon age of 1.7 Ga (Wasserburg et al., 1959; DeWitt et al., 1984), belongs to the regional crystalline basement complex of the Mojave Province (Condie, 1992). The marble and schist are metasedimentary assemblages that are arguably derived from the overlying miogeoclinal section. Most workers have traditionally assigned these metasedimentary rocks to the Noonday Dolomite and Johnnie Formations (Wright et al., 1974, 1991; Otton, 1976; Holm, 1992). Most still assign part of the section to the Noonday Dolomite, but some now question the interpretation of the associated, non-dolomitic rock. In the lowest exposed levels of the Badwater turtleback, the exposed section of pelitic rock contains about 50 m of feldspathic quartz mylonite, suggestive of an arkosic protolith (Fig. 5A). This observation, plus the relative abundance of calcite marble with silty laminae, led Miller (1992a,b) to suggest that the clastic rocks were Crystal Spring Formation, possibly overlain by Noonday Dolomite. At Mormon Point, Holm (1992) described diamictite-like gneiss structurally beneath the Noonday(?) marble that he argued

was Kingston Peak Formation. We dispute this interpretation, noting that the "clasts" are actually deformed granitic sills.

Intrusive rocks of the footwalls consist of early Tertiary pegmatite and a variety of mafic to felsic late Tertiary plutons and dikes. Based on U–Pb zircon ages, the pegmatite is 55 ± 3 Ma at the Badwater turtleback (Miller and Friedman, 1999), and 61 Ma at the Copper Canyon turtleback (R. Friedman, 2000, personal communication). The pegmatite consists almost entirely of feldspar and quartz, with minor muscovite and/or biotite and few accessory minerals. It is most voluminous at the Badwater turtleback, where it constitutes approximately 30% of the footwall, but becomes less voluminous southward, where it constitutes less than 10% of either the Copper Canyon or Mormon Point turtlebacks. Typically, the pegmatite forms lens-shaped bodies from 1 m to about 10 m in width that are concordant to foliation in adjacent rock, but it locally forms dikes. Miller and Friedman (1999) interpreted the lens-shaped bodies as boudins and the dikes as parts of locally preserved "rafts" within ductilely deformed, late Tertiary extensional shear zones. Lee (2003) used geochemistry to argue that the pegmatites were derived directly from their metamorphic host rocks.

Fig. 5. Photograph of south wall of deep canyon in Badwater turtleback and stratigraphy of metasedimentary rock.

The late Tertiary intrusions consist of the Willow Spring pluton, Smith Mountain Granite and associated felsic bodies, and latite and diabase dikes. These rocks are described in detail by Holm et al. (1994a,b), Wright et al. (1991), Otton (1976) and Drewes (1963). The Willow Spring pluton and the Smith Mountain Granite are a comagmatic Late Miocene mafic (hornblende gabbro to quartz diorite) and felsic (biotite rapikivi granite) assemblage respectively that show magma mingling textures along a mutually intrusive contact (Meurer, 1992; Pavlis in Holm et al., 1994a,b). In addition to this field evidence, geochronological data from the plutonic complex are broadly supportive of this conclusion with two U–Pb zircon dates from the Willow Spring pluton of 11.6 ± 0.2 Ma by Asmerom et al., 1990 and 10.93 ± 0.75 Ma by DeWitt (reported in Wright et al., 1991). These ages are only slightly older than a 10.8 Ma Ar/Ar cooling date on orthoclase from the structurally highest exposed level of the Smith Mountain Granite (Pavlis in Holm et al., 1994a,b) and a U–Pb zircon age of 10.44 ± 0.22 Ma (Miller et al., 2004).

An important feature of the Late Miocene intrusive complex is that the intrusives form a sill-like body intruded along the structural top of each turtleback (Pavlis in Holm et al., 1994a,b; Pavlis, 1996). This sill is itself stratified with the mafic Willow Springs assemblage structurally beneath the felsic Smith Mountain granite. This geometry is undoubtedly not coincidental given the evidence that the Smith Mountain and Willow Springs assemblages are phases of the same plutonic body and so probably reflects density stratification within the original magma chamber of the sill. This geometry also led to dramatic thermal manifestations in the Copper Canyon and Mormon Point turtlebacks as the hot plutonic sheet above them cooled (Pavlis, 1996). Effects included: (1) partial to complete destruction of older fabrics by recrystallization; (2) growth of new mineral assemblages suggestive of low-P, upper amphibolite facies conditions (olivine in marble and muscovite breakdown to form K-spar+ sillimanite in pelites); and (3) cooling to biotite closure temperatures at 6–8 Ma (Holm and Dokka, 1993; Pavlis and Snee, unpublished data), which is 3–5 m.y. after plutonism, but consistent with the slower cooling of rocks below a large intrusive sheet (e.g. Pavlis, 1996).

3.1.2. Footwall fabrics

Footwall rocks that pre-date the Smith Mountain granite exhibit tectonite fabrics at all exposed structural levels in the turtlebacks. These fabrics fall into two categories: those that involve, and those that are cut by, the pegmatite. Those that involve the pegmatite dominate the footwalls. They show top-to-the-NW sense-of-shear, consistent with late Tertiary extension (Fig. 6A). This shear sense determination is unambiguous, as it is based on numerous outcrop and microscopic criteria, including shear bands, asymmetric porphyroclasts and mica fish, and preferred grain-shape orientations (e.g. Miller, 1992a,b; Holm, 1992).

A late Tertiary age for these fabrics is further indicated by three other structural and geochronologic findings. First, the fabrics affect the base of the 11.6 Ma Willow Spring pluton (Holm, 1992; Pavlis in Wright et al., 1991; Pavlis, 1996). Second, Holm et al. (1992) obtained Ar–Ar biotite ages of 13 Ma, 6 Ma, and 8 Ma for the Badwater, Copper Canyon and Mormon Point turtlebacks, respectively, to indicate when each turtleback cooled through ~300 °C. As 300 °C is the approximate temperature required for crystal-plasticity in quartz (Mitra, 1978; Passchier and Trouw, 1996), these ages help define when mylonitization ceased in gneissic rocks of each footwall; calcite-dominated marbles, which require temperatures of about 250 °C (Schmid, 1982), could continue to deform ductilely. Third, Miller (1992a,b) noted a continuous evolution of the mylonite zone at the Badwater turtleback across the ductile–brittle transition

into brittle normal faulting with a clear connection to Tertiary extension. Miller (1999a,b) then applied the findings by Holm et al. (1992) to suggest that most of the deformation at the Badwater turtleback was between about 16–13 Ma.

At the Mormon Point turtleback, Miller and Friedman (2003) obtained a U–Pb zircon age of 9.53 ± 0.04 Ma from silicic dikes that cut most of the shear zone. Dee et al. (2004) found that these dikes were deflected into the shear zone and internally deformed at the highest structural levels. Based on an approximate 45° angle of deflection, they argued for a post 9.5 Ma angular shear strain of only about one. Pre-9.5 Ma shear strains, however, were likely much greater.

In contrast, early fabrics that are cut by the pegmatite appear as locally preserved "rafts" within the late Tertiary zone. Foliations and lineations of these rocks are parallel to those of the late Tertiary fabrics and so can be identified only where pegmatite is present. They typically show a top-to-the-east sense of shear (Holm, 1992; Turner and Miller, 1999; Miller, 2003) to suggest a pre-late Cenozoic shortening origin. Miller (2003) suggested they were at least partly coeval with intrusion of the pegmatite to make them of early Tertiary age. We argue later in this paper that the

Fig. 6. Large asymmetric pegmatite clast, deformed with top-northwest (right) shear sense at the Badwater turtleback. Plane of photograph is parallel to stretching lineations.

parallelism of both sets of fabrics suggests the turtle-back shear zones are in fact reactivated thrust faults.

3.1.3. Metamorphism

Each turtleback footwall experienced amphibolite-to upper amphibolite facies metamorphism prior to

unroofing by late Tertiary extension. This grade is evident by the presence of crystal-plastically strained feldspar, indicative of temperatures greater than about 450 °C (Tullis and Yund, 1980; Pryer, 1993), as well as the metamorphic mineral assemblages in pelitic and carbonate rocks. At the Badwater Turtle-

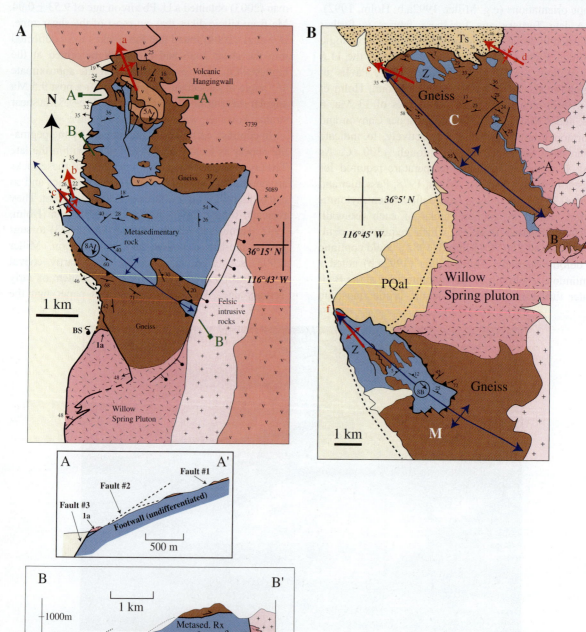

back, kyanite-bearing schists, in association with garnet, rutile, ilmenite, and quartz (GRAIL) yielded a pressure of 700 ± 100 MPa for an approximate depth of 20 km and temperature between 500 °C and 675 °C (Whitney et al., 1993). At Mormon Point, Pavlis (1996) described k-spar and sillima-nite-bearing schists to indicate much higher tempera-ture second-sillimanite conditions. These high grade fabrics at Mormon Point are closely associated with the Miocene plutonic assemblage, which suggests that they formed during the Late Miocene (Pavlis, 1996). Furthermore, amphibolite samples collected from both the Copper Canyon and Mormon Point turtlebacks contain embayed garnets overgrown by a rim of plagioclase, a texture that suggests a higher pressure garnet amphibolite assemblage was over-printed by a lower pressure, hornblende-plagioclase assemblage.

To date no samples have been collected at either Mormon Point or Copper Canyon that contain low-variance mineral assemblages useful for thermobaro-metry. However, hornblende geobarometry of the Wil-low Spring pluton (Holm and Dokka, 1993) and Smith Mountain Granite (Meurer, 1992) indicates pressures no greater than 300 MPa in the Copper Canyon and Mormon Point turtlebacks during the Late Miocene. These pressures correspond to intru-sion depths of 10–11 km in the turtleback footwalls (Holm and Wernicke, 1990; Holm and Dokka, 1993). Furthermore, Meurer (1992) found that intrusion depths increased northward in the Willow Spring pluton, consistent with exhumation along an initially more steeply dipping fault.

3.1.4. Footwall folds

Each turtleback footwall is the structural culmina-tion of a northwest–southeast trending, doubly plun-ging antiform. These folds are second-generation (F2) structures relative to the main metamorphic foliation (S1) in the footwalls, and are composite features representing different styles of overprinting at differ-ent positions within large-scale ductile shear zones. The age of the S1 foliation is unclear, but undoubtedly includes Mesozoic and/or early Tertiary metamorph-ism. However, it was clearly re-used and transposed by S2 during the early stages of the extensional deformation.

Because the turtleback folds are largely defined by this foliation, they must have formed during or after the extension. At Badwater, Miller (1999a,b) applied a slip-line analysis (Hansen, 1971) to asymmetric minor folds on the northeast limb to suggest that they formed within a ductile shear zone during top-to-N 63°W transport. Each turtleback antiform is locally subpar-allel to the overlying brittle fault zone, the "turtleback faults" discussed in the following section. It is this feature that gives rise to the geomorphic expression (Fig. 3) first described by Curry (1938) and later attributed to extensional fault-reactivations of the foliation by Wright et al. (1974). This generalization, however, is an oversimplification when each turtle-back is examined in detail.

Specifically, the principal footwall folds vary con-siderably from one turtleback to another. At the Bad-water turtleback, the hingeline is poorly defined, but the northeastern "limb" displays abundant outcrop-scale non-cylindrical folds of the basement gneiss with metasedimentary rock. These folds typically trend northwest and range from upright to recumbent. Mormon Point exhibits a high degree of non-cylind-rical folding of both the principal antiform and sub-parallel minor folds. The axial surface of the principal antiform is nonplanar, as it ranges from nearly upright at the southeast end of the turtleback to gently inclined

---

Fig. 7. Maps and cross-sections of Black Mountains turtlebacks. Dark blue fold symbols indicate approximate location of footwall antiformal hinge, whereas red fold symbols indicate locations of apparent fold hinges in fault surface. Locations of photographs in Figs. 5A and 8A,B are shown in circles with arrows that designate view direction. A: Map and cross sections of Badwater turtleback. Abbreviations: BS—Badwater Spring; N—Natural Bridge Canyon. Badwater Turtleback normal fault shown as line separating gneiss and metasedimentary rock from Quaternary alluvium, "volcanic hanging wall," or Willow Spring Pluton. Badwater Turtleback thrust fault shown by heavy line with teeth that separates metasedimentary rock from structurally overlying basement gneiss. As bottom of metasedimentary rock is not exposed, its contact is queried in cross section B-B'. Dip-slip separation on high-angle normal fault on east side of cross-section B-B' is only approximate. Composite cross-section A-A' shows geometry of brittle fault system projected into line A-A' of map. Fault 1a on map corresponds to fault 1a on cross-section. C: Map of Copper Canyon (C) and Mormon Point (M) turtlebacks, modified from Otton (1976) and Holm (1992). Exposures of basement gneiss above Noonday Dolomite (Z) exist at location A. Large enclave of gneiss that structurally overlies dolomite is at location B. Exposed thrust fault at Mormon Point is marked by heavy line with teeth.

towards the northwest. By contrast, the Copper Canyon turtleback only locally displays complex infolding of basement and metasedimentary rock, and overall exhibits a macroscopic upright to steeply inclined geometry with a gentle northeast limb and a steep southwest limb.

To varying degrees, the footwall folds of each turtleback also display doubly-plunging geometries (e.g. Miller, 1992b; Mancktelow and Pavlis, 1994; Pavlis in Holm et al., 1994a,b). As a result, the relationship between the folds and structurally overlying faults varies across each turtleback. To the southeast, fold hinges plunge southeastward beneath the structurally overlying Miocene plutonic complex and/or Precambrian basement (Fig. 7A,B). As reported by Mancktelow and Pavlis (1994), this contact is itself deformed, with ductile fabrics in the floor of the plutons that are parallel to the those in the turtleback footwalls. Northwestward, fold hinges become horizontal and then plunge moderately northwestward, typically steeper than the structurally overlying brittle fault. The structurally overlying brittle faults, in turn, display less structural relief than the folds. Therefore, at both Copper Canyon and Mormon Point, the brittle faults cut across structural section on both limbs of the antiforms to expose the structurally overlying Miocene plutonic complex in the core of the synclinoria between them (Fig. 7B). At the Badwater turtleback, the structurally deepest part of the turtleback is exposed near its center.

This general form of the turtleback anticlinoria suggested to Mancktelow and Pavlis (1994) and Pavlis (1996) that the folds formed as part of a continuous deformational sequence related to distributed transtensional shear, after intrusion of the plutonic complex. In this scenario, the transtension created localized shortening that buckled low- to moderate-angle extensional shear zones (S2), superimposed on the turtleback main-phase foliation (S1). This buckling generated large-scale structural relief in the shear zones, which ultimately were transected by brittle faults. Serpa and Pavlis (1996) extended this concept to the broader regional scale. They noted that the F2 folds in the turtleback footwalls trend more northerly than younger folds and so probably reflect greater amounts of finite clockwise vertical axis rotation. Similarly, Miller (1999a,b) found that lineations in mylonitic

gneiss trended more northerly than those in calcite marble at the Badwater turtleback. As the gneiss reflected earlier, higher temperatures than the marble, these data may also reflect clockwise rotation during transtension.

3.1.5. Basement thrust faults

Each turtleback exhibits areally extensive non-planar contacts where basement gneiss overlies younger, metasedimentary rock. Miller (2003) interpreted these contacts as SE-directed, basement-involved thrust faults that were re-deformed during late Tertiary extension. At the Badwater turtleback, several hundred meters to approximately 1 km of gneiss overlies the carbonate at its northern and southern edges respectively (Figs. 7A and 8A). Near the center of the Badwater turtleback, thin bodies of gneiss structurally overlie, and are infolded with, carbonate rock. As the exposed trace of the gneiss-over-carbonate contact is approximately 4.5 km long in the direction of transport, Miller (2003) concluded that the thrust fault had at least that much displacement. At the Copper Canyon turtleback, small bodies of gneiss structurally overlie carbonate rock at the contact with the Willow Spring pluton, and as an enclave within the pluton (Fig. 7B; Otton, 1976; Holm, 1992). At Mormon Point, ~300 m of gneiss structurally overlies the metasedimentary rock northwest of Smith Mountain (Figs. 7B and 8B; Otton, 1976; Holm, 1992).

Miller (2003) further noted exposures of pre- and syn-pegmatite mylonitic rocks that display top-to-the-east and -southeast fabrics. These exposures, although relatively rare, are most abundant at deep structural levels within each turtleback. As the pegmatite is early Tertiary in age (Miller and Friedman, 1999) and some of the mylonitic fabrics appeared to be concurrent with its intrusion, Miller (2003) argued that both mylonitization and basement thrusting were probably early Tertiary.

Except for the presence of cross-cutting pegmatite and the sense-of-shear, the early mylonitic foliation is indistinguishable from the late Tertiary, extension-related foliation. Both sets of foliations are concordant and both display similarly oriented stretching lineations. In some localities where both foliatons are preserved together, the later foliation appears to merge with the older one (Fig. 9). We therefore sug-

Fig. 8. Photographs of exposed older-over-younger contacts at Badwater and Mormon Point turtlebacks. Photograph locations are shown as "8A" and "48B" in Fig. 7A and B, respectively. A: At south side of Badwater turtleback; view toward south. B: On Mormon Point turtleback; view toward east.

gest that the extensional turtleback shear zones originated as basement thrust faults, possibly as late as early Tertiary, but were reactivated and almost completely transposed during late Tertiary extension.

3.2. Turtleback hanging walls

The delineation of the "hanging wall" of each turtleback is somewhat subjective, as the footwalls

Fig. 9. Photograph of concordant foliation in pegmatite and previously deformed gneiss. Note that pegmatite clearly cuts across foliation in the gneiss on the left edge of the photograph and immediately behind the clump of grass. The gneiss also exhibits a much greater intensity of foliation, but that the foliation continues into the pegmatite. Kinematic indicators from the gneiss yield top-southeastward shear directions; those from the pegmatite yield top-northwestward shear directions.

of each turtleback plunge northwestward and may continue for some distance beneath Death Valley. They also plunge southeastward beneath the Miocene plutonic complex. The Panamint Range could therefore be considered as part of the hanging wall of either turtleback, whereas the Badwater turtleback could be considered as part of the Copper Canyon turtleback's hanging wall. Alternatively, when viewed to the south, the turtleback hanging walls consist entirely of plutonic rock and structurally overlying Proterozoic basement. This discussion is therefore limited to the *immediate* hanging walls: rocks that directly overlie the visible portions of the brittle turtleback fault zones.

Numerous studies of the immediate hanging walls of each turtleback have clarified the ages and types of rock present, as well as aspects of the structural geology of each one. In general, each turtleback has a hanging wall that consists almost entirely of Miocene to Pleistocene sedimentary and/or volcanic rock. However, significant differences exist between each one. Similarly, the structures within each hanging wall reflect aspects of late Tertiary extension, but vary considerably in specific details.

Rocks of the Badwater turtleback hanging wall consist of the Miocene to Pliocene Artist Drive Formation, Greenwater Volcanics (Greene, 1997), megabreccia deposits of these same rocks, and Pleistocene fanglomerate deposits. The volcanic flows and interbedded sedimentary rocks lie in the hanging wall immediately north of the turtleback, whereas the megabreccia and fanglomerate deposits lie in the hanging wall immediately west and above the central part of the turtleback. To the south, brecciated rock of the Willow Spring Pluton lies in the hanging wall.

Miller (1999b) interpreted the megabreccia deposits as having originated in landslides, and suggested that the brittle turtleback fault surface acted as a gravitational sliding surface in its later history. Fanglomerate deposits along the west side of the Badwater Turtleback that stratigraphically overlie the megabreccia deposits provide evidence for Pleistocene unroofing of the Badwater turtleback (Cichanski, 1990). Deposits of 0.67 Ma Lava Creek B tephra, as reported by Knott et al. (2001a,b) and Hayman et al. (2003), establishes their Pleistocene age, while an increase in the proportion of metamorphic clasts upwards documents exposure of the footwall to erosion.

Hanging wall rocks of the Copper Canyon turtle-back consist, in upwards succession, of Late Miocene Willow Springs diorite, Tertiary volcanic rocks and Copper Canyon Formation, and fanglomerate and megabreccia deposits (Drewes, 1963). The Copper Canyon Formation dominates the hanging wall. It consists of approximately 3000 m of alluvial fan deposits, in fault contact with the turtleback, that grade laterally (northward) into playa and freshwater lake deposits, and basalt flows. Holm et al. (1994a,b), using Ar/Ar, obtained an age of ~6 Ma to ~3 Ma for its deposition. The Copper Canyon Formation is significant because it records local basin development in a transtensional setting, pre-sumably an early phase of the formation of modern Death Valley (Ellis and Trexler, 1991). Its fine-grained deposits also display abundant well-pre-served Neogene mammal and bird tracks (Curry, 1938; Scrivner and Bottjer, 1986).

At the Mormon Point turtleback, hanging wall rocks consist of two distinct assemblages (Otton, 1977; Knott et al., 2001a,b): a Pleistocene section and older Pliocene (?) section. The older Pliocene (?) rocks dip much more steeply than the Pleistocene rocks and are limited to a prism of rock bounded by the turtleback fault and the Willow Wash fault, a major hanging-wall fault above the turtleback fault. Pleistocene rocks occur locally as nearly flat lying deposits atop an angular unconformity on the older rocks but are primarily exposed as variably tilted, faulted and exhumed rocks in the hanging wall of the Willow Wash fault. Burchfiel et al. (1995) made a detailed structural/stratigraphic map of the area in which they defined 11 separate units, marked by rapid facies changes and much interfingering. They argued that the sediments were deposited along the margin of modern Death Valley and later uplifted as the range-bounding normal fault stepped out into the playa. Knott et al. (2001a,b) and Hayman et al. (2003) reported the presence of the 0.76 Ma Bishop Ash and 0.67 Ma Lava Creek B Ash within the younger deposits providing the first clear evidence of a Pleis-tocene depositional age.

Structures in the turtleback hanging walls consist predominantly of high-angle brittle faults that termi-nate downward at the turtleback detachments in both listric and planar geometries (Fig. 4). This relation is visible in the many small canyons that cut into the footwalls of the Badwater and Mormon Point turtle-backs and in the cliffs of Copper Canyon Formation immediately north of the Copper Canyon turtleback. Knott et al. (1997) argued that the multiple stran-dlines at Mormon Point actually consisted only of one strandline that had been offset by multiple nor-mal faults. Hayman et al. (2003) found that, at Bad-water and Mormon Point, these faults defined a conjugate set with an approximately vertical maxi-mum compressive stress. They estimated that these faults accounted for approximately 600 m of Pleis-tocene and younger horizontal extension at Mormon Point. This interpretation contrasts with Keener et al. (1993) who inferred that the Mormon Point turtle-back fault was not active and was cut by a high-angle fault at depth.

Each turtleback also displays some degree of folding in its hanging wall. The most intensely folded hanging wall exists at Copper Canyon, where Drewes (1963) mapped a series of southeast-plunging asymmetric folds over an area of approxi-mately 10 km^2. The most prominent anticline has a mapped length of more than 3 km and verges north-northeastward. At Mormon Point, Hayman et al. (2003) reported a roll-over anticline adjacent to the detachment fault. Immediately north of Badwater, several kilometer-scale folds are visible in the Artist Drive Formation.

3.3. Turtleback fault zones

The turtleback fault zones are the discrete brittle surfaces that separate the ductilely deformed mid-crustal rock of the footwalls from the upper crustal rock of the hanging walls. Superb exposures of these fault zones exist along the west front of each turtle-back (Fig. 3A,B,C). In general, the turtleback faults dip westward or northwestward at shallow angles (<35°), show evidence for a latest episode of predo-minantly normal slip, and display a wide variety of cataclastic fault rocks. Moreover, they each appear to have been active as recently as the late Pleistocene (Knott, 1999; Hayman et al., 2003). Outstanding questions persist regarding their specific geometries and their relations to the rest of the Black Mountains frontal fault zone.

3.3.1. Fault geometries and relation to frontal fault zone

The classic interpretation of the turtleback fault geometries holds that they are antiformal and broadly coincident with the antiforms in the footwalls (Curry, 1954; Wright et al., 1974; Otton, 1976; Holm and Dokka, 1993). The turtlebacks do display geometries suggestive of both antiforms and synforms, the hinges of which are shown on Fig. 7A,B, but it is not yet clear if these features actually represent single, curved fault surfaces. Instead, several of the larger "folds" are probably the poorly exposed intersections of two or more differently oriented, subplanar fault surfaces. Miller (1991) argued this case at Badwater (hinge a on Fig. 7A), as did Pavlis et al. (1993) for Mormon Point (hinge f on Fig. 7B). Additionally, the apparent folds at Copper Canyon (hinges d and e on Fig. 7B) remain ambiguous, as they could be interpreted as either intersecting faults or kink-like folds in a single fault. Two minor folds at Badwater (hinges b and c on Fig. 7A) do appear to reflect true undulations in the fault, as the fault trace is continuous across the fold hinges. The antiformal hinge (hinge c) coincides with the antiformal hinge of the footwall.

The turtleback fault geometries are also important to understanding the structural evolution of the fault systems, as well as their genetic relation to the Black Mountains frontal fault zone. At Badwater, the turtleback fault consists of at least three distinct fault surfaces, which decrease in age but increase in dip towards the west. Fig. 7A shows these surfaces as faults #1, #2, #3. Miller (1991) attributed these relations to progressive rotation and abandonment of fault surfaces through time, with fault #3, the range-bounding fault, cutting fault #2. While the Miocene–Pliocene volcanic rocks of the hanging wall dip 20–30° eastward to support this interpretation, Knott (1998) and Hayman et al. (2003) noted that the Pleistocene material of the hanging wall has not rotated. Therefore, significant rotation of the turtleback fault and footwall must have taken place prior to deposition of the Pleistocene material, possibly before initiation of slip on fault #2.

At Mormon Point, a gravity and magnetic study by Keener et al. (1993) suggested a similar geometry to Badwater in that the principal turtleback fault appeared to be cut by a later, range-bounding fault.

Hayman et al. (2003), however, argued an alternative geometry, where the exposed detachment fault (fault #2 at Badwater) is the active, controlling structure, and the present, high-angle, range-frontal fault terminates downward at the detachment. To support their interpretation, they noted that, at Mormon Point, the hanging wall has been extended by approximately 600 m by minor normal faults that terminate downward onto the detachment. We note, however, that 600 m of slip is only about one third of the total likely slip predicted across the Black Mountains frontal faults during the time interval considered by Hayman et al. (2003). Geodetic data and Quaternary offsets in northern Death Valley (Williams et al., 1999) indicate the Death Valley fault system should be extending the region ~2.9 km/m.y.; i.e. since deposition of the Bishop Ash (~600 ka) approximately 1.8 km of horizontal motion should have occurred across the frontal faults. Therefore, more slip has presumably accumulated across buried faults, including the frontal faults, than the exposed fault array at Mormon Point. Alternatively, the cumulative slip on hanging wall faults reflects only a portion of the slip on the main detachment fault (D. Cowan, personal communication, 2003).

We consider the relation between the turtleback faults and the frontal fault zone of the Black Mountains to be an open question. Hayman et al. (2003) document features that require the turtleback faults to be recently active. This activity makes it difficult to envision the turtleback faults as cut by the frontal fault zones, unless there is cycling back and forth between gravitationally driven slip on the turtleback faults and rooted slip on the frontal fault zones as suggested by Miller (1999b) for Badwater. Miller (1992a,b, 1999a,b) also showed that fault 2 of the Badwater turtleback fault system coincides with the Black Mountains range front about 2 km south of Badwater, and cuts structurally upwards into volcanic rocks immediately north of Natural Bridge Canyon (Fig. 7A). In this way, the fault appears to be more of an abandoned range front fault than a true detachment. The Badwater turtleback detachment, if there is one, is most likely fault #1 on Fig. 7A.

3.3.2. Kinematics and fault rocks

As inferred from slickenline orientations, the most recent episode of slip on the turtleback faults has been normal (Pavlis et al., 1993; Miller, 1999b; Hayman et al., 2003). However, the faults likely had an earlier history of right-lateral oblique motion, consistent with the slip direction on the rest of the Black Mountains fault zone (Struthers, 1990; Brogan et al., 1991; Slemmons and Brogan, 1999). The most recent, normal slip therefore appears to be a departure from a more long-standing norm.

At Badwater, evidence for earlier oblique motion comes from the fault geometry, folded fault rocks, slickenlines, and high-angle faults in the footwall (Miller, 1999b). Hinge c in Figs. 3A and 7A, marks a true fold in the fault plane that plunges about N 60°W. Miller (1999b) argued that for the hanging wall to remain intact, slip had to parallel the hinge. Additionally, a Hansen slip line analysis of mesoscopic folds in the fault zone yielded a northwest-directed, oblique vector, and slickenlines on crystalline rocks south of hinge C indicated oblique motion. Miller (1999b) suggested that because these slickenlines were on crystalline rock, they were more durable and therefore reflected an earlier episode of slip than the dip-parallel ones that are preserved solely on the underside of the relatively soft hanging wall. Finally, the footwall is cut by numerous, demonstrably oblique high-angle faults.

At Mormon Point and Copper Canyon, Pavlis et al. (1993) used large-scale fault geometry and the distribution of fault rocks to infer oblique motion. There, calcite-rich implosion breccias exist within right-stepping planar sections of the turtleback faults. These breccias form through episodes of sudden dilatancy (Sibson, 1986), which would most likely occur through seismic right-lateral oblique motions on the fault zone. Additionally, Pavlis et al. (1993) and Keener et al. (1993) documented oblique slip in several localities along southwest dipping segments of both the Mormon Point and Copper Canyon turtleback fault based on complex slickenline arrays in fault-rocks below the fault surface. As at Badwater, the footwalls in both Mormon Point and Copper Canyon turtleback are cut by more steeply dipping, oblique-slip faults that do not appear to cut into the hanging walls (Pavlis et al., 1993).

Besides the implosion breccias near Mormon Point, each turtleback fault displays spectacular exposures of fault gouge, foliated and non-foliated cataclasite, and microbreccia. These fault rocks have been the subject of several studies, including those of fault behavior

(Miller, 1996), kinematics (Cladouhos, 1999a,b), and fault rock evolution (Hayman, 2002). Fault rocks are best developed below the discrete sliding surface of the turtleback fault. They typically show a downward structural progression from clay-rich gouge into foliated cataclasite and then into a variable damage zone of cataclasis and hydrothermal alteration that extends up to several tens of meters into the footwall. Cowan et al. (1997, 2003) showed that these zones correlated with decreasing shear strains away from the fault zone. Miller (1996) documented mesoscopically ductile flow in the fault gouge, and Cowan and Miller (1999) showed that episodes of ductile flow in gouge appear to alternate with episodes of discrete sliding. Alternating localized and non-localized events conflict with other studies of fault rock development that argue for continuous localization of slip on developing fault zones (e.g. Chester and Logan, 1986; Logan et al., 1992).

3.3.3. Estimates of slip

Because the brittle turtleback faults separate such disparate rock types, slip estimates tend to be based on indirect evidence. Several observations, however, suggest that the amount of slip for each brittle fault is limited to only a few kilometers at most. To a first approximation, the most relevant feature is the Willow Springs pluton, as visible portions of it exist in the hanging walls and footwalls of the both the Badwater and Copper Canyon turtlebacks (Figs. 1 and 7A).

At the Badwater turtleback, a small fault-bounded exposure of mylonitic basement rock exists in the hanging wall of Fault #2 near Badwater spring (Fig. 7A). Based on this exposure, Miller (1991) inferred approximately 2 km of slip on Fault #2. As indirect support of this estimate, we note that north of Natural Bridge canyon, Fault #2 cuts upwards into the volcanic hanging wall, with rocks of the Artist Drive Formation in both its hanging wall and footwall (Fig. 7A). At Mormon Point, Knott et al. (2001a,b) estimated 438 m of Pleistocene heave on the turtleback fault based on an offset bed of Bishop Ash.

4. Discussion and conclusions

Among the structural features of Death Valley, the turtlebacks contain some of the most critical information relevant to testing models of crustal extension. In particular, models are constrained by two types of information from within the turtlebacks: (1) observations of the turtleback shear zones themselves, and (2) implications of the basement-involved thrust faults of the turtlebacks for the pre-extensional geometry of the fold-thrust belt in Death Valley. This information has implications that extend far beyond Death Valley, as the region is frequently cited as a type example for continental extensional tectonics.

Tectonic models for the Death Valley region include both regional map-view reconstructions that attempt to constrain the regional strains (e.g. Serpa and Pavlis, 1996; Snow and Wernicke, 2000) and models of the extensional process in cross-sectional views (e.g. Holm et al., 1992; Miller, 1999a). Although these approaches provide important illustrations of the extensional process, they are limited to two dimensions. In fact, the turtlebacks are complex three-dimensional entities that formed over ~14 m.y. of extensional overprinting on an older Mesozoic framework. We submit that many of the controversies regarding the Death Valley region in general, and the turtleback systems in particular, stem largely from the inability to communicate observations of these multiple dimensions. Even the coauthors of this paper are not in complete agreement on several details, but our approach here is to clarify our view of the four-dimensional history. Following previous presentations, we consider the problem in the context of two-dimensional map and cross-sectional views, but we acknowledge that these views are oversimplifications of the geometric evolution.

4.1. Map-view reconstructions of regional deformation

Several tectonic models for the Death Valley region emphasize large-scale, map-view reconstructions based on the correlation of regional structures and stratigraphic assemblages within the Death Valley region. These controversies were initiated by the correlation of the Panamint thrust in the northern Panamint Mountains with the Chicago Pass thrust system in the Nopah and Resting Spring ranges (Wernicke et al., 1988). Other workers, however, have challenged this correlation on different lines of evidence (e.g. Wright et al., 1991; Serpa and Pavlis, 1996; Miller and Prave, 2002). The resolution of this question is critical because

extensional magnitudes vary radically in different versions of these reconstructions.

As described by Miller (2003), the presence of basement-involved thrust faults in the turtlebacks help define the pre-extension configuration of the fold-thrust belt in Death Valley, and therefore provide keys to interpreting the crustal extension. As argued below, their presence suggests that the Black Mountains were a structural high prior to Tertiary extension. It further suggests that restoration of the Cottonwood Mountains above the Black Mountains is unlikely, yet this configuration is required if the Panamint and Chicago Pass thrusts were the same thrust displaced by a master detachment system.

Evidence for a pre-extensional structural high in the Black Mountains is supported by the general absence of Paleozoic sedimentary rocks and the deep stratigraphic level of the pre-Tertiary unconformity in the Black Mountains. In the Amargosa Chaos, ~12 Ma sedimentary and volcanic rocks lie depositionally on late Proterozoic to early Cambrian sedimentary rocks (location "x" on Fig. 1 inset; Wright and Troxel, 1984; Topping, 1993). At the Billie Mine in the northern Black Mountains, ~14 Ma sedimentary rocks of the early Furnace Creek basin depositionally overlie early Cambrian rock (Fig. 1; Cemen et al., 1985; Wright et al., 1999; Miller and Prave, 2002). These contacts contrast markedly with the northern Panamint Mountains where late Paleozoic rocks lie directly beneath the Tertiary unconformity. It is therefore likely that the thrust systems now exposed in the turtlebacks produced significant structural relief above what is now the Black Mountains and that the late Paleozoic cover was largely stripped from the Black Mountains prior to Tertiary extension. Correlation of the Panamint thrust with the Chicago Pass thrust across this structural high is suspect because structural relief created by the thrust structures would disturb the simple thrust belt geometries required by the correlation. The correlation also places Late Paleozoic rocks and thrust belt structures of the Cottonwood Mountains structurally above and east of the turtleback basement thrusts. This restoration conflicts with the observed rocks below the Tertiary unconformity or anywhere within the Black Mountains.

An alternative explanation, that the exposed Black Mountains cover rocks are entirely allochthonous and so can not reliably indicate a basement high, is unlikely. In this scenario, the cover rocks should contain slices that sample all parts of the hanging wall assemblage, including late Paleozoic rocks (Fig. 10). Furthermore, those rocks should be preserved beneath the Tertiary unconformity. In fact, the youngest Paleozoic sedimentary rocks recognized in the Black Mountains are lower Paleozoic.

Moreover, the thrust structures among the ranges are not age correlative. In the Cottonwood Mountains, thrust structures are Permian to Triassic in age (Snow et al., 1991), which is much older than the apparent late Mesozoic–early Cenozoic age for the Black Mountains thrust faults. Finally, the Paleozoic rock of the Cottonwood Mountains is intruded by the Jurassic Hunter Mountain batholith while the Proterozoic and early Paleozoic rock of the Panamint Moun-

tains are intruded by the Cretaceous Harrisburg Pluton (Fig. 1). If either of these ranges originated above the Black Mountains, then the plutons must have intruded through them. However, no equivalent Mesozoic intrusive rock have been recognized in the Black Mountains (Fig. 10C).

4.2. Cross-sectional views of the turtlebacks

In cross-sectional view, the principal controversy surrounding the turtlebacks centers on the theme of a master detachment system with a rolling hinge (e.g. Holm et al., 1992) vs. models that call on a polygenetic origin for the turtlebacks (e.g. Wright et al., 1991; Mancktelow and Pavlis, 1994; Serpa and Pavlis, 1996; Pavlis, 1996). A related controversy concerns the inferred connection, in space and time,

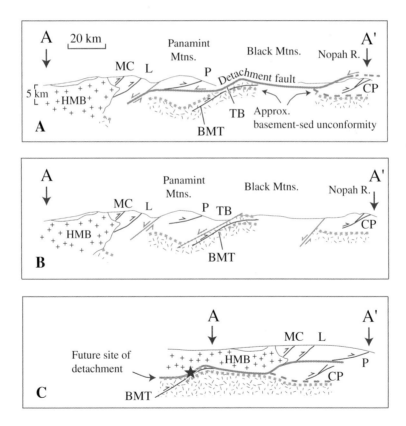

Fig. 10. Schematic cross-sections from A to A′ on Fig. 1 that account for presence of basement thrust faults (BMT) in the Black Mountains. A: depiction of present-day structure according to Rolling Hinge Model. B. Depiction of present-day structure according to Pull-Apart Model. C: Correlation of Chicago Pass and Panamint thrusts requires transport over basement high in Black Mountains and also places Hunter Mountain Batholith on top basement rock of the Black Mountains.

between the turtleback systems and structurally higher-level structures in the Black Mountains. These structures include those in the Amargosa Chaos area to the south (Wright and Troxel, 1984; Topping, 1993; Topping in Holm et al., 1994a,b) and low-angle normal fault systems that frame the Black Mountains to the east. The structural relationships between the Badwater and the other turtlebacks also represent an important detail because the Badwater turtleback is geochronologically distinct and lies in the hanging wall of the Copper Canyon turtleback.

Existing Ar–Ar geochronology for the turtlebacks can be interpreted to satisfy both models. Holm et al. (1992) found that, from south to north, the Mormon Point, Copper Canyon, and Badwater turtlebacks cooled through 300 °C at about 8 Ma, 6 Ma and 13 Ma, respectively. They interpreted the cooling ages at Mormon Point and Copper Canyon as evidence of the hinge rolling northwestward from Mormon Point to Copper Canyon and the much older age at Badwater as an earlier, deep-seated

event. We agree that the old age at Badwater reflects an earlier event, but the 8 Ma and 6 Ma ages at Mormon Point and Copper Canyon do not require a rolling hinge. Instead, they probably reflect diachronous cooling related to two simultaneously operating processes: (1) cooling from the emplacement of the ~11 Ma Black Mountains plutonic suite and (2) unroofing of the turtlebacks by extension, presumably at different times. In this context we suggest that two features of the turtlebacks are particularly relevant: the geometries of the turtleback folds, and the location of much of the Badwater turtleback beyond the northern terminus of the Willow Springs pluton (Fig. 7A).

The latter feature suggests that intrusion of the Willow Spring pluton dominated the cooling history of the southern turtlebacks (Holm and Dokka, 1993; Pavlis, 1996), but likely had little effect on the Badwater turtleback. As a result, the older cooling ages in the Badwater turtleback reflect a minimum age for the initial ductile deformation related to extension in the

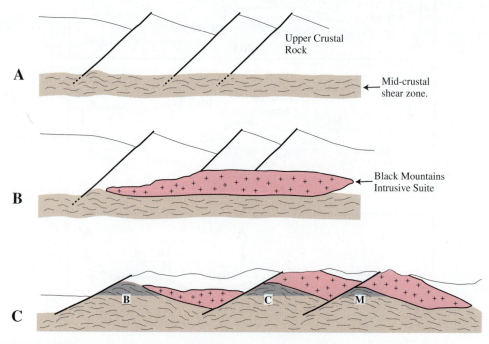

Fig. 11. Schematic cross-sections to illustrate principal extensional events at the turtlebacks. A: Prior to ~11 Ma, upper crustal faults terminated at mid-crustal shear zone. B: Intrusion of Black Mountains intrusive suite at ~11 Ma. These sill-like plutons intruded along the top of the shear zone and recrystallized fabrics at Copper Canyon and Mormon Point turtlebacks, but had little effect at Badwater Turtleback. C: Continued extensional faulting and ductile shear exhumed the intrusive suite and underlying mid-crustal shear zone. Dark-shaded areas represent approximate present-day exposure of metamorphic portion of each turtleback footwall.

region. This early history was also present in the other turtlebacks, but was largely obliterated at ~11 Ma when the Black Mountains intrusive suite was emplaced above them, completely recrystallizing pre-existing microstructures and resetting all low-T geochronometers. This conclusion is consistent with onset of volcanism and incipient extensional basin development prior to ~14 Ma (Wright et al., 1999). For example, Miller and Prave (2002) found that activity on the pre-13 Ma Badwater Turtleback fault likely influenced sedimentation in the early Furnace Creek Basin.

Similarly, the variability of fold geometries of the turtlebacks, especially between the highly non-cylindrical Mormon Point turtleback and the nearly upright Copper Canyon turtleback, are difficult to reconcile with the concept of a single, master rolling hinge. Moreover, their doubly plunging geometries are also inconsistent, because a migrating hinge should leave behind a single low-angle shear zone.

Our conceptual view of the turtleback evolution is that they all share a common, early history as parts of a mid-crustal shear zone at ~12–16 Ma that at least in part reactivated a pre-existing basement thrust fault (Fig. 11A). This shear zone locally reached upwards to connect with supracrustal fault zones as described by Miller (1992a,b, 1999a,b) at the Badwater turtleback. When the Black Mountains intrusive suite was emplaced at ~11 Ma, however, it cut across many of these faults and recrystallized fabrics in the subjacent ductile shear zone (Fig. 11B). Consequently, the thermally unmodified older shear zone is only preserved at the Badwater turtleback. At the Mormon Point and Copper Canyon turtlebacks, the older shear zone is now represented only by the thermally overprinted tectonite fabrics of the southeastward plunging segments of the turtleback antiforms. Continued crustal extension caused the exhumation of the intrusive suite and eventual formation of the brittle turtleback faults, shown in Fig. 11C as dismembering the intrusive suite. This view of the turtlebacks, in which much of the ductile deformation occurred before emplacement of the Miocene intrusives, and most brittle deformation occurred after their emplacement, differs significantly from the rolling hinge model which requires that most of the brittle and ductile deformation post-dated intrusion.

Acknowledgements

Our views of the turtlebacks have benefited from discussions with Darrel Cowan, Richard Friedman, Nick Hayman, Dan Holm, Jeff Knott, Laura Serpa, Bennie Troxel, Brian Wernicke, and Lauren Wright. This manuscript was greatly improved through detailed reviews by Darrel Cowan and Greg Davis.

References

Asmerom, Y., Snow, J.K., Holm, D.K., Jacobsen, S.B., Wernicke, B.P., Lux, D.R., 1990. Rapid uplift and crustal growth in extensional environments: An isotopic study from the Death Valley region, California. Geology 18, 227–230.

Brogan, G.E., Kellogg, K.S., Slemmons, D.B., Terhune, C.L., 1991. Late Quaternary faulting along the Death Valley-Furnace Creek fault system, California and Nevada. U.S. Geological Survey Bulletin, 23 pp.

Burchfiel, B.C., Molnar, P., Zhang, P., Deng, Q., Zhang, W., Wang, Y., 1995. Example of a supradetachment basin within a pull-apart tectonic setting: Mormon Point, Death Valley, California. Basin Research 7, 199–214.

Cemen, I., Wright, L.A., Drake, R.E., Johnson, F.C., 1995. Cenozoic sedimentation and sequence of deformational events at the south-eastern end of the Furnace Creek strike-slip fault zone, Death Valley, California. In: Biddle, K.T., Christie-Blick, N. (Eds.), Strike-slip Deformation, Basin Formation, and Sedimentation. Society of Economic Paleontologists and Mineralogists Special Publication, vol. 37, pp. 127–141.

Chester, F.M., Logan, J.M., 1986. Implications for mechanical properties of brittle faults from observations of the Punchbowl fault zone, California. Pure and Applied Geophyics 124, 79–106.

Cichanski, M.A., 1990. Stratigraphy and structure of the upper plate of the Badwater Turtleback, Death Valey, California. B.S. thesis, University of Washington, Seattle, WA, 23 pp.

Cladouhos, T.T., 1999a. Shape preferred orientations of survivor grains in fault gouge. Journal of Structural Geology 21, 419–436.

Cladouhos, T.T., 1999b. A kinematic model for deformation within brittle shear zones. Journal of Structural Geology 21, 437–449.

Condie, K.C., 1992. Proterozoic terranes and continental accretion in southwestern North America. In: Condie, K. (Ed.), Proterozoic Crustal Evolution. Developments in Precambrian Geology, vol. 10, pp. 447–480.

Cowan, D.S., Miller, M., 1999. Distributed flow vs. localized slip in Late Cenozoic low-angle fault zones, Death Valley, California. Seismological Research Letters 70, 247.

Cowan, D.S., Cladouhos, T.T., Morgan, J.K., 1997. Kinematic evolution of fault rocks in brittle shear zones, Death Valley, California. Abstracts with Programs, Geological Society of America 29 (6), 200.

Cowan, D.S., Cladouhos, T.T., Morgan, J.K., 2003. Structural geology and kinematic history of rocks formed along low-angle normal faults, Death Valley, California. Geological Society of America Bulletin 115, 1230–1248.

Curry, H.D., 1938. Turtleback fault surfaces in Death Valley, California. Geological Society of America Bulletin 49, 1875.

Curry, H.D., 1954. Turtlebacks in the central Black Mountains, Death Valley, California. Bulletin, California Division of Mines and Geology 170, 53–59.

Davis, G.H., 1980. Structural characteristics of metamorphic core complexes, southern Arizona. In: Crittenden Jr., M.D., Coney, P.J., Davis, G.H. (Eds.), Cordilleran Metamorphic Core Complexes. Geological Society of America Memoir, 153, 35–78.

Davis, G.A., Anderson, J.L., Martin, D.L., Krummenacher, E.G., Armstrong, R.L., 1982. Geologic and geochronologic relations in the lower plate of the Whipple detachment fault, Whipple Mountains, southeast California: a progress report. In: Frost, E., Martin, D. (Eds.), Mesozoic–Cenozoic Tectonic Evolution of the Colorado River Region, California, Arizona and Nevada. Cordilleran Publishers, San Diego, pp. 408–432.

Davis, G.A., Lister, G.S., Reynolds, S.J., 1986. Structural evolution of the Whipple and South Mountains shear zones, southwestern United States. Geology 14, 7–10.

Dee, S., Miller, M., Friedman, R.M., 2004. Pre- and post 9.5 Ma shear zone activity at the Mormon Point turtleback, Death Valley, California. Abstracts with Programs, Geological Society of America 36 (5).

DeWitt, E.H., Armstrong, R.L., Sutter, J.F., Zartman, R.E., 1984. U–Th–Pb, Rb–Sr, and Ar–Ar mineral and whole-rock isotopic systematics in a metamorphosed granitic terrane, southeastern California. Geological Society of America Bulletin 95, 723–739.

Drewes, H., 1959. Turtleback faults of Death Valley, California, a reinterpretation. Geological Society of America Bulletin 70, 1497–1508.

Drewes, H., 1963. Geology of the Funeral Peak Quadrangle, California, on the eastern flank of Death Valley. U.S. Geological Survey Professional Paper 413, 78 pp.

Ellis, M.A., Trexler, J.H., 1991. Basin-margin development in pull-apart settings; an example from Death Valley, California. Abstracts with Programs, Geological Society of America 23 (5), 82.

Greene, R.C., 1997. Geology of the northern Black Mountains, Death Valley, California. U.S. Geological Survey Open-File Report, OFR 97-79, 110 pp.

Hamilton, W.D., 1988. Detachment faulting in the Death Valley region, California and Nevada. U.S. Geological Survey Bulletin 1790, 51–85.

Hansen, E., 1971. Strain Facies. Springer-Verlag, New York, 207 pp.

Hayman, N.W., 2002. A textural progression of brittle fault rocks along the strikes of the Death Valley detachments. Abstracts with Programs, Geological Society of America 34 (6), 249.

Hayman, N.W., Knott, J.R., Cowan, D.S., Nemser, E., Sarna-Wojcicki, A., 2003. Quaternary low-angle slip on detachment faults in Death Valley, California. Geology 31, 343–346.

Holm, D.K., 1992. Structural, thermal and paleomagnetic constraints on the tectonic evolution of the Black Mountains crys-

talline terrain, Death Valley region, California, and implications for extensional tectonism. PhD thesis, Harvard University, Cambridge, Massachusetts, 237 pp.

Holm, D.K., Dokka, R.K., 1993. Interpretation and tectonic implications of cooling histories: An example from the Black Mountains, Death Valley extended terrane, California. Earth and Planetary Science Letters 116, 63–80.

Holm, D.K., Wernicke, B.P., 1990. Black Mountains crustal sections, Death Valley extended terrain, California. Geology 18, 520–523.

Holm, D.K., Snow, J.K., Lux, D.R., 1992. Thermal and barometric constraints on the intrusion and unroofing history of the Black Mountains, Death Valley, CA. Tectonics 11, 507–522.

Holm, D.K., Fleck, R.J., Lux, D.R., 1994a. The Death Valley turtlebacks reinterpreted as Miocene–Pliocene folds of a major detachment surface. Journal of Geology 102, 718–727.

Holm, D.K., Pavlis, T.L., Topping, D.J., 1994b. Black Mountains crustal section, Death Valley region, California: Geological investigations of an active margin. In: McGill, S.F., Ross, T.M. (Eds.), Geological Society of America, Cordilleran Section, Annual Meeting: Guidebook, pp. 31–54.

Keener, C., Serpa, L.F., Pavlis, T.L., 1993. Faulting at Mormon Point, Death Valley, California: a low-angle fault cut by high-angle faults. Geology 21, 327–330.

Knott, J.R., 1998. Late Cenozoic Terphrochronology, Stratigraphy, geomorphology, and neotectonics of the western Black Mountains piedmont, Death Valley, California: implications for the spatial and temporal evolution of the Death Valley fault zone. PhD thesis, University of California, Riverside, Riverside, California, 407 pp.

Knott, J.R., 1999. Quaternary displacement on the Badwater and Mormon Point turtleback (low-angle normal) faults, Death Valley, California. Abstracts with Programs, Geological Society of America 31 (7), 301.

Knott, J.R., Tinsley, J.C., Wells, S.G., 1997. Quaternary faulting across the 180 Ka abrasion platform at Mormon Point, Death Valley, California: scarps vs. strandlines. Abstracts with Programs, Geological Society of America 29 (6), 437.

Knott, J.R., Sarna-Wojcicki, A.M., Meyer, C.E., Tinsley III, J.C., Wells, S.G., Wan, E., 2001a. Late Cenozoic stratigraphy and tephrochronology of the western Black Mountains piedmont, Death Valley, California: Implications for the tectonic development of Death Valley. In: Wright, L.A., Troxel, B.W. (Eds.), Tertiary basins in the Death Valley Region, California. Special Paper, Geological Society of America, 333, 345–366.

Knott, J.R., Sarna-Wojcicki, A.M., Tinsley III, J.C., Wells, S.G., 2001b. Late Quaternary tectonic-geomorphic development and pluvial lakes at Mormon Point. In: Machette, M.N., Johnson, M.L., Slate, J.L. (Eds.), Quaternary and late Pliocene geology of the Death Valley region: Recent observations on tectonics, stratigraphy and lake cycles (Guidebook for the 2001 Pacific Cell - Friends of the Pleistocene fieldtrip). Denver, CO, U.S. Geological Survey Open File Report 01-51: 92–103.

Lee, K.F., 2003. Geology of granitic pegmatites in the Badwater, Copper Canyon, and Mormon Point turtlebacks, Black Mountains, Death Valley, California. M.S. thesis, University of Oregon, Eugene, Oregon, 68 pp.

Lister, G.S., Davis, G.A., 1989. The origin of metamorphic core complexes and detachment faults formed during Tertiary continental extension in the northern Colorado River region, USA. Journal of Structural Geology 11, 65–94.

Logan, J.M., Dengo, C.A., Higgs, N.G., Wang, Z.Z., 1992. Fabrics of experimental fault zones: their development and relationship to mechanical behavior. Fault Mechanics and Transport Properties of Rocks. Academic Press, pp. 33–67.

Mancktelow, N.S., Pavlis, T.L., 1994. Fold-fault relationships in low-angle detachment systems. Tectonics 13, 668–685.

Meurer, K.J., 1992. Tectonic evolution of Smith Mountain—Gold Valley region, Death Valley, California. M.S. thesis, University of New Orleans, New Orleans, LA.

Miller, M., 1991. High-angle origin of the currently low-angle Badwater turtleback fault, Death Valley, California. Geology 19, 372–375.

Miller, M., 1992a. Brittle faulting induced by ductile deformation of a rheologically stratified rock sequence, Badwater Turtleback, Death Valley, California. Geological Society of America Bulletin 104, 1376–1385.

Miller, M., 1992b. Structural and kinematic evolution of the Badwater Turtleback, Death Valley, California. PhD thesis, University of Washington, Seattle, Washington, 155 pp.

Miller, M., 1996. Ductility in fault gouge from a normal fault system, Death Valley, California: a mechanism for fault-zone strengthening and relevance to paleoseismicity. Geology 24, 603–606.

Miller, M., 1999a. Implications of ductile strain on the Badwater Turtleback for pre-14 Ma extension in the Death Valley region, California. In: Wright, L.A., Troxel, B.W. (Eds.), Tertiary Basins in the Death Valley Region, California. Special Paper Geological Society of America, vol. 333, pp. 115–126.

Miller, M., 1999b. Gravitational reactivation of an extensional fault system, Badwater turtleback, Death Valley, California. In: Wright, L.A., Troxel, B.W. (Eds.), Tertiary Basins in the Death Valley Region, California. Special Paper Geological Society of America, vol. 333, pp. 367–376.

Miller, M., 2003. Basement-involved thrust faulting in a thin-skinned fold and thrust belt, Death Valley, California, USA. Geology 31, 31–34.

Miller, M., Friedman, R.M., 1999. Early Tertiary magmatism and probable Mesozoic fabrics in the Black Mountains, Death Valley, California. Geology 27, 19–22.

Miller, M., Friedman, R.M., 2003. New U–Pb zircon ages indicate major extension in Death Valley, CA predated 10 Ma: implications for models of crustal extension. Abstracts with Programs, Geological Society of America 35 (6), 26.

Miller, M., Prave, A.R., 2002. Rolling hinge or fixed basin?: A test of continental extensional models in Death Valley, California, United States. Geology 30, 847–850.

Miller, E.L., Gans, P.B., Garing, J., 1983. The Snake Range decollement: an exhumed mid-Tertiary brittle–ductile transition. Tectonics 2, 239–263.

Miller, M., Friedman, R.M., Dee, S., 2004. Pre-10 Ma TIMS U–Pb zircon age for the Smith Mountain Granite, Death Valley, California: implications for timing of major extension. Abstracts with Programs, Geological Society of America 36 (5).

Mitra, G., 1978. Microscopic deformation mechanisms and flow laws in quartzites within the South Mountain Anticline. Journal of Geology 86, 129–152.

Nemser, E.S., 2001. Kinematic development of upper plate faults above low-angle normal faults in Death Valley, CA. M.S. thesis, University of Washington, Seattle, Washington, 102 pp.

Noble, L.F., 1941. Structural features of the Virgin Spring area, Death Valley, California. Geological Society of America Bulletin 52, 941–1000.

Otton, J.K., 1976. Geologic features of the Central Black Mountains, Death Valley, California. In: Troxel, B.W., Wright, L.A. (Eds.), Geologic Features of Death Valley. California Division of Mines and Geology Special Report, vol. 106, pp. 27–33.

Otton, J.K., 1977. Geology of the central Black Mountains, Death Valley, California: the turtleback terrane. PhD thesis, Pennsylvania State University, University Park, Pennsylvania. 155 pp.

Passchier, C.W., Trouw, R.A.A., 1996. Microtectonics. Springer, Berlin, 289 pp.

Pavlis, T.L., 1996. Fabric development in syn-tectonic sheets as a consequence of melt- dominated flow and thermal softening of the crust. Tectonophysics 253, 1–31.

Pavlis, T.L., Serpa, L.F., Keener, C., 1993. Role of seismogenic processes in fault-rock development: an example from Death Valley, California. Geology 21, 267–270.

Pryer, L.L., 1993. Microstructures in feldspars from a major crustal thrust zone: the Grenville Front, Ontario, Canada. Journal of Structural Geology 15, 21–36.

Schmid, S.M., 1982. Microfabric studies as indicators of deformation mechanisms and flow laws operative in mountain building. In: Hsu, K.J. (Ed.), Mountain Building Processes. Academic Press, London, pp. 95–110.

Scrivner, P.J., Bottjer, D.J., 1986. Neogene Avian and Mammalian tracks from Death Valley National Monument, California: their context, classification, and preservation. Palaeogeography, Palaeoclimatology, Palaeoecology 57, 285–331.

Serpa, L.F., Pavlis, T.L., 1996. Three-dimensional model of the Cenozoic history of the Death Valley region, southeastern California. Tectonics 15, 1113–1128.

Sibson, R.H., 1986. Brecciation processes in fault zones: inferences from earthquake rupturing. Pure and Applied Geophysics 124, 159–175.

Slemmons, D.B., Brogan, G.E., 1999. Quaternary strike-slip components of the Death Valley fault between the Furnace Creek and southern Death Valley fault zones. In: Slate, J.L., (Ed.), Proceedings of conference on status of geologic research and mapping in Death Valley National Park. U.S. Geologic Survey Open File Report 99-153, pp. 152–153.

Snow, J.K., Wernicke, B.W., 2000. Cenozoic tectonism in the central Basin and Range: Magnitude, rate, and distribution of upper crustal strain. American Journal of Science 300, 659–719.

Snow, J.K., Asmerom, Y., Lux, D.R., 1991. Permian–Triassic plutonism and tectonics, Death Valley region, California and Nevada. Geology 19, 629–632.

Stewart, J.H., 1983. Extensional tectonics in the Death Valley area, California: transport of the Panamint Range structural block 80 km northwestward. Geology 11, 153–157.

Struthers, J., 1990. Evidence for right-lateral oblique slip on the Black Mountains range front: Desolation Canyon, Death Valley, California. B.S. thesis, University of Washington, Seattle, Washington, 24 pp.

Topping, D.J., 1993. Paleogeographic reconstruction of the Death Valley extended region: evidence from Miocene large rock-avalanche deposits in the Amargosa Chaos Basin, California. Geological Society of America Bulletin 105, 1190–1213.

Tullis, J., Yund, R.A., 1980. Hydrolytic weakening of experimentally deformed Westerly granite and Hale albite rock. Journal of Structural Geology 2, 439–451.

Turner, H., Miller, M., 1999. Kinematic indicators in a pre-55 Ma ductile shear zone, Death Valley, California. Abstracts with Programs, Geological Society of America 31 (7), 369.

Wasserburg, G.J.F., Wetherill, G.W., Wright, L.A., 1959. Ages in the Precambrian terrane of Death Valley, California. Journal of Geology 67, 702–708.

Wernicke, B.P., Axen, G.J., Snow, J.K., 1988. Basin and Range extensional tectonics at the latitude of Las Vegas, Nevada. Geological Society of America Bulletin 100, 1738–1757.

Whitney, D.L., Hirschmann, M., Miller, M.G., 1993. Zincian ilmenite-ecandrewsite from a pelitic schist, Death Valley, California, and the paragenesis of $(Zn,Fe)TiO_3$ solid solution in metamorphic rocks. Canadian Mineralogist 31, 425–436.

Williams, T.B., Johnson, D.J., Miller, M.M., Dixon, T.H., 1999. GPS-determined constraints on interseismic deformation along active

fault zones within the Death Valley region, southeastern California. In: Slate, J.L., (Ed.), Proceedings of conference on status of geologic research and mapping in Death Valley National Park. United States Geologic Survey Open File Report 99-153, p. 149.

Wright, L.A., Troxel, B.W., 1984. Geology of the north 1/2 Confidence Hills 15° quadrangle, Inyo County, California. California Division of Mines and Geology Map Sheet 34, scale 1:24,000.

Wright, L.A., Otton, J.K., Troxel, B.W., 1974. Turtleback surfaces of Death Valley viewed as phenomena of extensional tectonics. Geology 2, 53–54.

Wright, L.A., Thompson, R.A., Troxel, B.W., Pavlis, T.L., DeWitt, E.H., Otton, J.K., Ellis, M.A., Miller, M., Serpa, L.F., 1991. Cenozoic magmatic and tectonic evolution of the east-central Death Valley region, California. In: Walawender, M.J., Hanan, B.B. (Eds.), Geologic Excursions in Southern California and Mexico. Department of Geological Sciences, San Diego State University, San Diego, pp. 93–127.

Wright, L.A., Green, R., Cemen, I., Johnson, R., Prave, A., 1999. Tectonostratigraphic development of the Miocene–Pliocene Furnace Creek Basin, Death Valley, California. In: Wright, L.A., Troxel (Eds.), Tertiary basins in the Death Valley Region, California. Special Paper Geological Society of America, 333, 87–114.

Available online at www.sciencedirect.com

Earth-Science Reviews 73 (2005) 139–148

www.elsevier.com/locate/earscirev

Are turtleback fault surfaces common structural elements of highly extended terranes?

Ibrahim Çemen [a,*], Okan Tekeli [b], Gűrol Seyitoğlu [b], Veysel Isik [b]

[a] School of Geology, Oklahoma State University, Stillwater, OK, USA
[b] Department of Geological Engineering, Ankara University, Ankara, Turkey

Abstract

The Death Valley region of the U.S.A. contains three topographic surfaces resembling the carapace of a turtle. These three surfaces are well exposed along the Black Mountain front and are named the Badwater, Copper Canyon, and Mormon Point Turtlebacks. It is widely accepted that the turtlebacks are also detachment surfaces that separate brittlely deformed Cenozoic volcanic and sedimentary rocks of the hanging wall from the strongly mylonitic, ductilely deformed pre-Cenozoic rocks of the footwall.

We have found a turtleback-like detachment surface along the southern margin of the Alasehir (Gediz) Graben in western Anatolia, Turkey. This surface qualifies as a turtleback fault surface because it (a) is overall convex-upward and (b) separates brittlely deformed hanging wall Cenozoic sedimentary rocks from the ductilely to brittlely deformed, strongly mylonitic pre-Cenozoic footwall rocks. The surface, named here Horzum Turtleback, contains striations that overprint mylonitic stretching lineations indicating top to the NE sense of shear. This suggests that the northeasterly directed Cenozoic extension in the region resulted in a ductile deformation at depth and as the crust isostatically adjusted to the removal of the rocks in the hanging wall of the detachment fault, the ductilely deformed mylonitic rocks of the footwall were brought to shallower depths where they were brittlely deformed.

The turtleback surfaces have been considered unique to the Death Valley region, although detachment surfaces, rollover folds, and other extensional structures have been well observed in other extended terranes of the world. The presence of a turtleback fault surface in western Anatolia, Turkey, suggests that the turtleback faults may be common structural features of highly extended terranes.
© 2005 Elsevier B.V. All rights reserved.

Keywords: turtleback surfaces; Death Valley; western Turkey; Alasehir (Gediz) graben; detachment surfaces

1. Introduction

One of the most spectacular features of Death Valley National Park in eastern California, is the high relief of the Black Mountains front. This topo-

* Corresponding author.
  E-mail address: icemen@okstate.edu (I. Çemen).

0012-8252/$ - see front matter © 2005 Elsevier B.V. All rights reserved.
doi:10.1016/j.earscirev.2005.07.001

graphic feature contains three topographic surfaces strikingly similar to the carapace of a turtle because of their overall convex-upward shapes (Fig. 1). Curry (1938, 1954) named them the Badwater, Copper Canyon, and Mormon Point Turtlebacks (Fig. 1). The American Geological Institute, *Glossary of Geology* (fourth edition) defines the term turtleback as "An extensive, smooth, and curved topographic surface, apparently unique to the Death Valley (Calif.) region, that resembles the carapace of a turtle or the nose of a large, elongate dome with an amplitude up to a few thousand meters."

During the 1990s, the term Turtleback fault emerged as a special type of tectono-morphologic surface distinguished by its unique topographic expression and structural setting (Miller, 1991, 1992; Davis and Reynolds, 1992). Another turtle-

back structure was mapped by Cichanski (1993) in the southwestern Panamint Mountains of the Death Valley region. However, turtleback structures were still considered unique to the Death Valley region. We question, however, that such features are unique to the Death Valley region because we have mapped a turtleback-like fault surface along the southern margin of the Alasehir (Gediz) graben in western Anatolia. This surface, which we name Horzum Turtleback, qualifies as turtleback fault surface because it resembles to the carapace of a turtle with its overall convex-upward geometry. It also has a metamorphic core separated from an overlying body of highly faulted upper crustal rocks by a brittle fault zone.

In this paper, we describe the similarities between the turtleback surfaces in the Death Valley region,

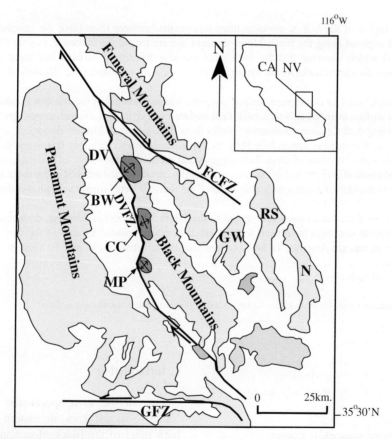

Fig. 1. Map of the Death Valley region showing location of the three turtlebacks. Abbreviations: BW=Badwater Turtleback; CC=Copper Canyon Turtleback and MP=Mormon Point Turtleback; FCFZ=Furnace Creek fault zone; DVFZ=Death Valley fault zone; GFZ=Garlock fault zone; RS=Resting Spring Range; N=Nopah Range. Insert shows the location of the Death Valley region.

and the Horzum Turtleback surface. We will first briefly describe geologic setting of the turtleback surfaces in the Death Valley region with a special emphasis on the Badwater Turtleback because it has been well studied by Miller (1991, 1992, and 1999). We will then provide a detailed description of the geology of the Horzum Turtleback including its structural setting, footwall and hanging wall lithologies, and fault rocks.

Death Valley of eastern California and the Alasehir (Gediz) Graben of western Turkey are located in regions where continental crust has been highly extended during the Cenozoic times. The first is located in the Basin and Range province of North America and the second is located in the Aegean region of Eurasia. Therefore, we will also point out the possibility that the turtleback faults are common structural features of highly extended terranes such as detachment surfaces, accommodation zones, and rollover folds.

2. Geologic overview of the Death Valley turtlebacks

The Death Valley region is characterized by innumerable normal to oblique-slip faults produced by pervasive Cenozoic Basin and Range extension which was initiated in mid-Miocene time (Çemen et al., 1985, 1999; Wernicke et al., 1989; Wright et al., 1991, 1999). Lauren Wright, James Otton, and Bennie Troxel conducted the first structural study of the turtleback surfaces in the early 1970s. Wright et al. (1974) identified the turtleback surfaces as low angle normal faults that were formed during the Basin and Range extension. A COCORP seismic reflection data also indicate the presence of low angle normal faults across the Death Valley region and suggest that upper crustal faults terminate at a mid-crustal, subhorizontal detachment zone that locally includes molten material (Serpa et al., 1988). Holm et al. (1992) interpreted the turtlebacks as folded Miocene detachment surfaces. The estimate of total magnitude of normal slip on the detachment zone ranges from over 50 km (Wernicke et al., 1989) to under 5 km (Wright and Troxel, 1984; and Miller, 1991). Recently, Miller (2003) interpreted the Badwater turtleback as a reactivated thrust fault surface.

The turtleback faults (Figs. 1 and 2) are parts of the detachment surfaces that separate brittlely deformed Cenozoic volcanic and sedimentary rocks of the hanging wall from the strongly mylonitic metamorphosed rocks of the footwall (Wernicke, 1985; Miller, 1991; and Holm et al., 1992). They contain mylonitic rocks and overprinting brittle faults that reflect deformation at different depths along progressively denuding shear zones (Miller, 1992). Stretching lineations are common features of the footwall. The brittle deformation is also indicated by the presence of about 1-m thick breccia and gouge and a 10–30-cm-thick zone of flow-banded or foliated gouge adjacent to the hanging wall rocks (Cladouhos, 1999).

3. Badwater Turtleback

The Badwater Turtleback separates the Cenozoic Artist Drive Formation on its hanging wall from the metamorphic basement in its footwall (Fig. 2). It is a broadly antiformal topographic surface resembling the carapace of a turtle similar to the Copper Canyon and Mormon Point Turtlebacks (Figs. 1 and 2). It consists of a strongly metamorphosed footwall that is separated from a brittlely faulted hanging wall by the west dipping brittle turtleback fault zone (Fig. 2). Miller (1991) reported that the sense of hanging wall transport recorded in the footwall mylonites and in the brittle features on the turtleback surface are similar, both are towards the west and northwest. Miller (1992) found that the Badwater Turtleback fault surface contains fault rocks that underwent mylonitization during early stages of deformation and overprinted brittle deformation at later stages. He also reported that the turtleback contains two sets of brittle faults; early stage faults that record the change from ductile deformation to brittle deformation and later stage faulting.

4. Geologic overview of the West Anatolia Turtleback

The Aegean extended region (Fig. 3) experienced a series of continental collisions from Late Cretaceous to Eocene, which led to the formation of the Izmir–Ankara Neo-Tethyan suture zone (Fig. 3A) (e.g., Sengor and Yilmaz, 1981; Stamplie, 2000). Post-colli-

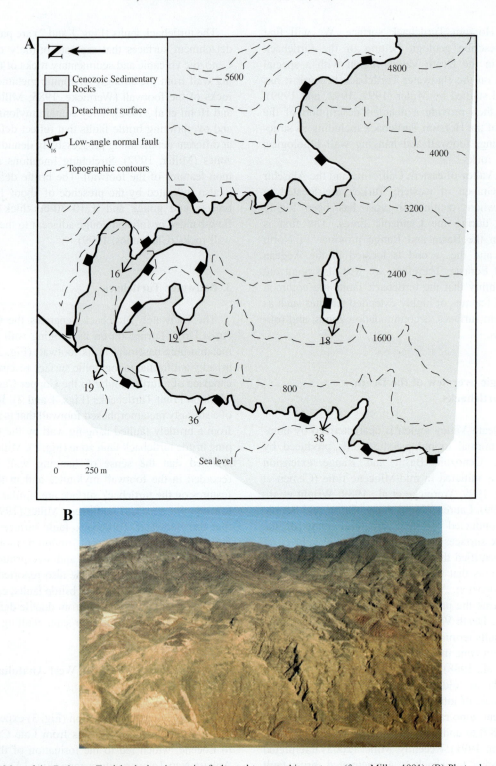

Fig. 2. (A) Map of the Badwater Turtleback showing major faults and topographic contours (from Miller, 1991). (B) Photo showing general view of the Badwater Turtleback, looking northeastward (Courtesy of Martin Miller).

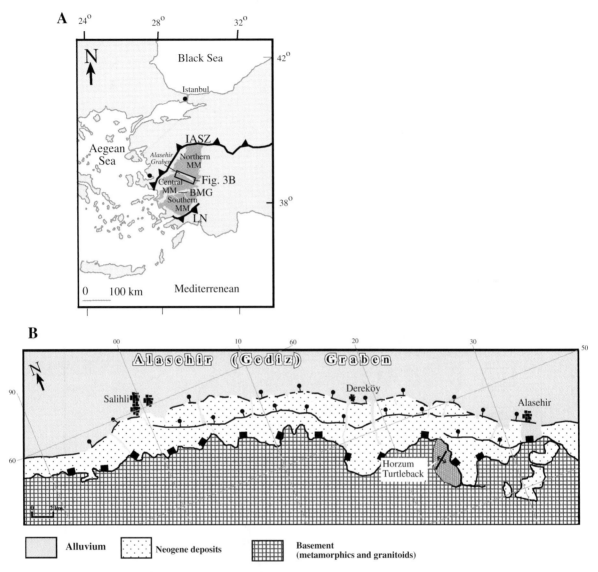

Fig. 3. (A) Map of western Anatolia showing the location of the Menderes Massif and the Alasehir graben. Abbreviations: IASZ = Izmir–Ankara suture zone; MM = Menderes Massif; BMG = Buyuk Menderes Massif. (B) Generalized geologic map of the southern margin of the Alasehir graben showing the location of the Horzum Turtleback. (Modified from Isik et al., 2003).

sional extension in the region, although controversial in its timing and nature, caused the exhumation of several Alpine metamorphic belts. The largest of these, the Menderes Massif, accommodated a significant amount of extension, covering more than 40,000 km^2 area between the Izmir–Ankara Neo-Tethyan suture in the north and the Lycian Nappes to the south (Fig. 3A). The Massif is divided into northern, central, and southern sections based on the presence

of E–W-trending grabens. The central Menderes Massif is located between the Alasehir (Gediz) Graben to the north and the Buyuk Menderes Graben to the south. (Fig. 3A).

Three major faults cross the Alasehir (Gediz) Graben roughly in an E–W direction (Fig. 3B). The main fault (Fig. 3B) is located on the southern side of the graben and is a low-angle detachment surface (Hetzel et al., 1995; Emre, 1996; Seyitoglu et al., 2000). The

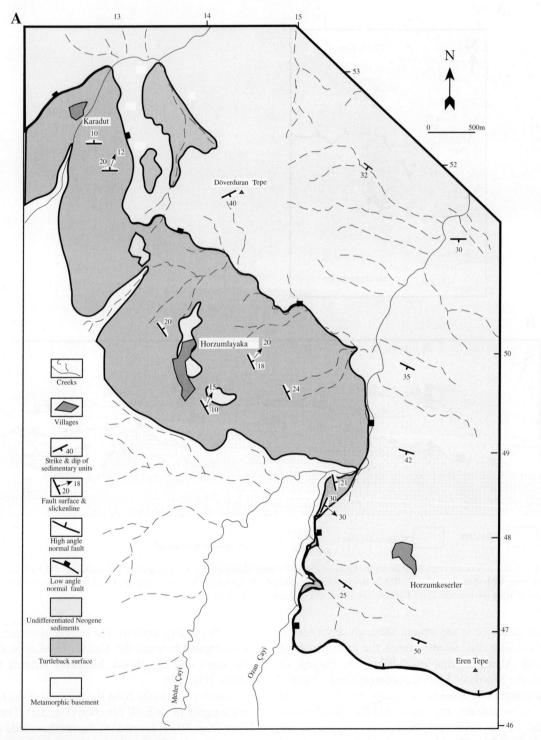

Fig. 4. (A) Detailed geologic map of the Horzum Turtleback surface and surrounding area; (B) Photo showing the Horzum Turtleback surface looking westward.

B

Fig. 4 (*continued*).

detachment surface has been referred to as the Kuzey Detachment by Gessner et al. (2001) and the Alasehir Detachment by Isik et al. (2003).

A

B

Fig. 5. (A) Photo showing the general appearance of the fault rocks exposed along the Horzum Turtleback surface. (B) Slickenlines on the Horzum Turtleback indicating brittle deformation.

5. Horzum Turtleback

The Alasehir (Kuzey) Detachment makes a southward bend to the SE of the town of Alasehir where its surface resembles the turtleback surfaces of the Death Valley region (Çemen et al., 2000), because of its overall convex-upward shape (Fig. 4). We mapped this area at a scale of 1:25,000 (Fig. 4A), measured fault-slip directions in many localities along this surface, and examined thin sections of oriented fault rock samples under a petrographic microscope. We name this surface as the Horzum Turtleback because it is close to the village of Horzumlayaka (Fig. 4A).

The Horzum Turtleback surface in western Anatolia shows all the geologic and geomorphologic characteristics of the turtleback surfaces of the Death Valley region. First, it is a smooth, broadly antiformal

Fig. 6. Photomicrograph showing cataclasite from the footwall of the Horzum Turtleback. Subrounded to rounded clasts lie in a fine-grained quartz–serisite–epidote–iron-oxide matrix.

fault surface, resembling the carapace of a turtle and separating strongly metamorphosed footwall rocks from the brittlely faulted hanging wall rocks. Second, it experienced mylonitization during early stages of deformation that has been later overprinted by a brittle deformation. Third, there are some patches of the Cenozoic sedimentary units on top of the Horzum Turtleback surface that are the relics of the detachment process (Fig. 5A). Following is the summary of our field and laboratory evidence for these three major characteristics that are found along the Horzum Turtleback surface.

The smooth, broadly antiformal topographic fault surface that forms the footwall of the Horzum Turtleback fault is composed of brittlely deformed cohesive fault rocks that are fine grained, dark gray to black, and massive in texture. The surface contains well-developed slickenlines (Fig. 5), indicative of a brittle deformation.

The protoliths of the fault rocks are low- to medium-grade metasedimentary rocks which are composed of schist, marble, and quartzite. The lower part of the metasedimentary rocks contains psammitic schists with minor calcite marble intercalations. Its upper part is marked by pelitic schist with interbeds of quartzite and marble. The schist has a mineral assemblage including garnet, indicative of greenschist–amphibolite facies metamorphism.

Cohesive fault rocks with more matrix and fine-grained clasts and massive texture are observed near the fault surface. However, the metamorphic host rock that maintains pre-existing texture contains anastomosing shear zones a few meters below the surface (Fig. 5A). It is clearly observed in the field that sheared metamorphic rocks are transitional into cohesive fault rocks towards the turtleback surface. The sense of shear indicators in the footwall mylonites and brittle features such as slickenlines along the Horzum Turtleback surface indicate that the hanging wall rocks transported northeastward. The slickenlines overprint the stretching lineations that were formed during the ductile deformation phase of the turtleback formation.

Thin section analysis of the oriented fault rock samples suggests that most of the fault rocks can be classified as cohesive cataclasite, while some can be classified as cohesive breccia (Passchier and Trouw, 1996). The cohesive cataclasites reveal the presence of microscopic grains imbedded within a microscopic/submicroscopic cataclasite matrix of chlorite, sericite, epidote, calcite, and iron oxide (Fig. 6). Most grains are subrounded to rounded clasts of quartz and plagioclase feldspar. Microstructures within the quartz clasts (including undulose extinction, subgrains, and recrystallized grains) suggest that clasts were affected by a previous mylonitic deformation. There are also rock fragments with mylonitic deformation. The cohesive cataclasites, in general, do not contain a well-developed foliation. However, there is recognizable cataclastic foliation (cf. Chester et al., 1985) where an abundance of phyllosilicate minerals occur and thin shear zones develop. This foliation is observed in hand samples as shear bands varying in length (a few to tens of cm) and width (a few mm to a few cm).

The cohesive breccias are more clast-supported than the cohesive cataclasites and include more rock fragments than mineral clasts (quartz, feldspar). These clasts are mostly subrounded but there are also angular to rounded clasts. The rock fragments show a well-developed mylonitic foliation that was developed in general as shape-preferred orientation (SPO).

The presence of rock fragments and mineral grains having the feature of mylonitic deformation in the cohesive cataclasites and cohesive breccias indicates that the fault rocks had experienced a previous ductile (mylonitic) deformation before the start of the brittle deformation. The mylonitic rocks must have been produced during an earlier stage, when footwall rocks were experiencing ductile deformation at depth along the Alasehir (Kuzey) Detachment surface. In other words, the mylonitic deformation occurred at first in the middle crust. It was later overprinted by a brittle deformation in the upper crust as the detachment surface reached to the shallower levels of the crust during the exhumation of the central Menderes Massif, due to the Cenozoic extension in the region.

The sedimentary rocks in the hanging wall rocks of the Horzum Turtleback are, in general, red colored, thick to medium bedded conglomerates and sandstones that were deposited as a lateral alluvial fan (Cohen et al., 1995). They get coarser grained toward the upper part of the section and contain dominantly boulder conglomerates around Eren Tepe (Fig. 4B). All sedimentary layers are tilted (up to 50°) to the south towards the Horzum Turtleback fault surface. There are some patches of the sedimentary units on top of the Horzum Turtleback surface that are the relics of detach process (Fig. 4B).

6. Discussion and conclusions

The turtleback surfaces were considered unique to the Death Valley region, where the crust has been highly extended during the Cenozoic Basin and Range extension. Our detailed mapping in the western Anatolia extended terrane suggests that the southern margin of the Alasehir graben contains a turtleback fault surface, which is very similar in its geometry and tectonic setting to the turtleback faults of the Death Valley region. We suggest that the turtleback surfaces are probably much more common than previously thought. They may, in fact, be typical structural features of highly extended terranes along with detachment surfaces, accommodation zones, and rollover folds. They may be present in other highly extended terranes such as the Aegean Islands located to the west of western Anatolia (Fig 3A).

An important aspect of the turtleback surfaces in Death Valley, California, and in western Anatolia, Turkey is that the features indicative of brittle deformation such as slickenlines overprint the earlier stretching lineations, indicative of ductile deformation. This common observation suggests that turtleback faults were formed as ductile shear zones at mid-crustal levels. As crustal extension continued, they were brought to shallower levels of the crust where they experienced brittle deformation. Therefore, the turtlebacks can be considered as a window to the processes of deformation in the middle and upper crusts.

Acknowledgements

This research was supported by an NSF-TÜBITAK Cooperative Research Grant (9810811): Extensional tectonics in West Anatolia, Turkey, and its comparison to the Death Valley extended terrane in the U.S. Discussions at various times with Yucel Yilmaz, Terry Pavlis, Laura Serpa, Brian Wernicke, Bennie Troxel, and Metin Yazman were very helpful. We sincerely thank Lauren A. Wright and Martin G. Miller for their critical reviews.

References

Çemen, I., Wright, L.A., Drake, R., Johnson, F.C., 1985. Cenozoic sedimentation and sequence of deformational events at southern end of Furnace Creek strike-slip fault zone, Death Valley region, California. In: Biddle, K.T., Christie-Blick, N. (Eds.), Strike-Slip Deformation and Basin Formation. Society of Economic Paleontologists and Mineralogists Special Publication, vol. 37, pp. 127–141.

Çemen, I., Wright, L.A., Prave, T., 1999. Stratigraphy and tectonic implications of the latest Oligocene and Early Miocene sedimentary succession, southernmost Funeral Mountains, Death Valley region, California. In: Wright, L.A., Troxel, B.W. (Eds.), Cenozoic Basins of the Death Valley region. Geological Society of America Special Paper 333, 65–86.

Çemen, I., Tekeli, O., Seyitoğlu, G., 2000. A turtleback surface along the southern margin of the Alasehir Graben, West Anatolia, Turkey. Abstracts with Programs, International Scientific Congress of the Aegean Region, p. 21.

Chester, F.M., Friedman, M., Logan, J.M., 1985. Foliated cataclasites. Tectonophysics 111, 139–146.

Cichanski, M.A., 1993. "Turtleback" structure in the southwestern Panamint Mountains, Death Valley region, California. Abstracts with Programs, Geological Society of America, Cordilleran Section Meeting, vol. 25, p. 21.

Cladouhos, T.T., 1999. Shape preferred orientations of survivor grains in fault gouge. Journal of Structural Geology 21, 419–436.

Cohen, H.A., Dart, C.J., Akyüz, H.S., Barka, A., 1995. Syn-rift sedimentation and structural development of the Gediz and Büyük Menderes graben, western Turkey. Journal of the Geological Society (London) 152, 629–638.

Curry, D., 1938. "Turtleback" fault surfaces in Death Valley, California (abs.). Geological Society of America Bulletin 49, 1875.

Curry, D., 1954. Turtlebacks in the central Black Mountains, Death Valley, California. In: Jahns, R.H. (Ed.), Geology of Southern California. California Division of Mines Bulletin 170, 53–59.

Davis, G., Reynolds, J., 1992. Structural Geology of Rocks and Regions, second edition. John Wiley and Sons, New York, 775 pp.

Emre, T., 1996. Gediz grabeni'nin jeolojisi ve tektonigi (Geology and tectonics of the Gediz graben). Turkish Journal of Earth Sciences 5, 171–185.

Gessner, K., Ring, U., Passchier, C.W., Johnson, C., Hetzel, R., Gungor, T., 2001. An active bivergent rolling-hinge detachment system: Central Menderes metamorphic core complex in western Turkey. Geology 29, 611–614.

Hetzel, R., Ring, U., Akal, C., Troesch, M., 1995. Miocene NNE-directed extensional unroofing in the Menderes massif, southwestern Turkey. Journal of the Geological Society (London) 152, 639–654.

Holm, D.K., Snow, J.K., Lux, D.R., 1992. Thermal and barometric constraints on the intrusive and unroofing history of the Black Mountains: implications for timing, initial dip, and kinematics of detachment faulting in the Death Valley region, California. Tectonics 11, 507–522.

Isik, V., Seyitoglu, G., Çemen, I., 2003. Ductile–brittle transition along the Alasehir shear zone and its structural relationship with the Simav Detachment, Menderes Massif, western Turkey. Tectonophysics 374, 1–18.

Miller, M.G., 1991. High angle origin of the currently low-angle Badwater Turtleback fault, Death Valley, California. Geology 9, 372–375.

Miller, M.G., 1992. Brittle faulting induced by ductile deformation of a rheologically stratified rock sequence, Badwater Turtleback, Death Valley California. Geological Society of America Bulletin 104, 1376–1385.

Miller, M.G., 1999. Implications of ductile strain on the Badwater Turtleback, for pre-14 Ma extension in the Death Valley region, California. In: Wright, L.A., Troxel, B.W. (Eds.), Cenozoic Basins of the Death Valley Region. Geological Society of America Special Paper 333, 115–126.

Miller, M.G., 2003. Basement-involved thrust faulting in a thin-skinned fold-and-thrust belt, Death Valley, California, USA. Geology 31, 31–34.

Passchier, C.W., Trouw, R.A.J., 1996. Microtectonics. Springer, Berlin, 775 pp.

Sengor, A.M.C., Yilmaz, Y., 1981. Tethyan evolution of Turkey: a plate tectonic approach. Tectonophysics 75, 181–241.

Serpa, L., deVoogd, B., Wright, L.A., Willemin, J., Oliver, J., Hauser, E., Troxel, B.W., 1988. Structure of the central Death Valley pull-apart basin and vicinity from COCORP profiles in the southern Great Basin. Geological Society of America Bulletin 100, 1437–1450.

Seyitoglu, G., Çemen, I., Tekeli, O., 2000. Extensional folding in the Alasehir Graben, West Anatolia, Turkey. Journal of the Geological Society of London 157, 1097–1100.

Stamplie, G.M., 2000. Tethyan oceans. In: Bozkurt, E., Winchester, J.A., Piper, J.D.A. (Eds.), Tectonics and Magmatism in Turkey

and Surrounding Area. Geological Society of London, Special Publications 173, 1–23.

Wernicke, 1985. Uniform-sense normal simple shear of the continental lithosphere. Canadian Journal of Earth Sciences 22, 108–125.

Wernicke, B., Snow, J.K., Axen, G.J., Burchfiel, B.C., Hodges, K.V., Walker, J.D., Guth, J., 1989. Extensional Tectonics in the Basin and Range province between the southern Nevada and Colorado Plateau: 28th International Geological Congress Field Trip Guidebook, T138, 80 pp.

Wright, L.A., Troxel, B.W., 1984. Geology of the northern half of the confidence Hills quadrangle. Death Valley region, California: the area of the Amargosa chaos: California Division of Mines and Geology Map Sheet, 34, scale 1:24,000, 31 pp.

Wright, L.A., Otten, J., Troxel, B.W., 1974. Turtleback fault surfaces of Death valley viewed as phenomena of extension. Geology 2, 53–54.

Wright, L.A., Thompson, R.A., Troxel, B.W., Pavlis, T.L., DeWitt, E.H., Otton, J.K., Ellis, M.A., Miller, M.G., Serpa, L.F., 1991. Cenozoic magmatic and tectonic evolution of the east-central Death valley region, California. In: Walawender, M.J., Hanan, B.B. (Eds.), Geologic excursions in southern California and Mexico: San Diego, California. Geological Society of America 1991 Annual Meeting Guidebook, pp. 93–127.

Wright, L.A., Greene, R.C., Cemen, I., Johnson, F.C., Prave, A.R., Drake, R.E., 1999. Tectonostratigraphic development of the Miocene–Pliocene Furnace Creek Basin and related features, Death Valley region, California. In: Wright, L.A., Troxel, B.W. (Eds.), Cenozoic Basins of the Death Valley Region. Geological Society of America Special Paper 333, 115–126.

Available online at www.sciencedirect.com

Earth-Science Reviews 73 (2005) 149–176

EARTH-SCIENCE
REVIEWS

www.elsevier.com/locate/earscirev

Large-scale gravity sliding in the Miocene Shadow Valley Supradetachment Basin, Eastern Mojave Desert, California

G.A. Davis [a,*], S.J. Friedmann [b]

[a] *Department of Earth Sciences, University of Southern California, Los Angeles, CA 90089-0740, USA*
[b] *Energy and Environmental Directorate, MC L-640, Lawrence Livermore National Laboratory, Livermore, CA 94550, USA*

Accepted 1 April 2005

Abstract

The Miocene Shadow Valley basin in the eastern Mojave Desert of California developed above the active west-dipping Kingston Range-Halloran Hills extensional detachment fault system between 13.5 and ca. 7 mybp. Although mass-wasting processes are common phenomena in supradetachment basins, the Shadow Valley basin is an exceptional locale for the study of such processes, especially rock-avalanches and gravity sliding. A score of megabreccias, interpreted as rock-avalanche deposits, and half that number of very large (>1 km², up to 200 m thick), internally intact gravity-driven slide sheets are interbedded with various sedimentary facies. The slide sheets, variably composed of Proterozoic crystalline rocks and Proterozoic, Paleozoic, and Tertiary sedimentary strata, moved across both depositional and erosional surfaces in the basin. Although the majority consist of Paleozoic carbonate rocks, the largest slide sheet, the Eastern Star crystalline allochthon, contains Proterozoic gneisses and their sedimentary cover and is now preserved as klippen atop Miocene lacustrine and alluvial fan deposits over an area >40 km². Estimates of slide sheet runouts into the basin from higher eastern and northern source terranes range from approximately a few km to >10 km; in most cases the exact provenances of the slide blocks are not known.

The basal contacts of Shadow Valley slide sheets are characteristically knife sharp, show few signs of lithologic mixing of upper- and lower-plate rocks, and locally exhibit slickensided and striated, planar fault-like bases. Pronounced folding of overridden Miocene lacustrine and fan deposits beneath the Eastern Star allochthon extends to depths up to 40 m at widely scattered localities. We conclude that this slow moving slide sheet encountered isolated topographic asperities (hills) and that stress transfer across the basal slide surface produced folding of footwall strata. Synkinematic gypsum veins in footwall playa sediments, with fibers up to 12 cm long, have trends and shear senses compatible with the direction and sense of displacement of the overriding crystalline allochthon. The undisturbed veins, which closely parallel the base of the slide sheet, attest to high fluid presence and pressure in the playa sediments—factors facilitating allochthon movement across them. The long length of the fibers, indicative of a protracted dilational process, is incompatible with a catastrophic rate of emplacement. We believe that the only explanation for slow displacement of this allochthon and other gravity driven slide sheets across the landscape is that they formed as slumps on high, steep bedrock slopes and that their elevated heads drove their toes across lower fan and playa deposits. Initial detachments from bedrock sources were facilitated by pre-existing structural and stratigraphic anisotropies.

---

* Corresponding author. Tel.: +1 213 740 6726; fax: +1 213 740 8801.
  E-mail address: gdavis@usc.edu (G.A. Davis).

0012-8252/$ - see front matter © 2005 Elsevier B.V. All rights reserved.
doi:10.1016/j.earscirev.2005.04.008

Detachment of the Eastern Star allochthon from the bedrock of Shadow Mountain likely occurred by inversion along the playa-ward dip of a preexisting Mesozoic thrust fault within Proterozoic rock units.

Keywords: slide blocks; gravity tectonics; Shadow Valley; Kingston Wash; Mojave Desert

1. Introduction

The Miocene Shadow Valley basin lies within the southern portion of the Mesozoic foreland fold and thrust belt in the easternmost Mojave Desert, California (Fig. 1; Burchfiel and Davis, 1971, 1988). The basin formed above the active Kingston Range-Halloran Hills extensional detachment fault between 13.5 and ca. 7 mybp, during which time it was being displaced and internally extended above the shallow-dipping fault (Fig. 2; Davis et al., 1993; Friedmann et al., 1994, 1996; Fowler et al., 1995; Friedmann, 1999). Sedimentation patterns in the basin were influenced by a number of tectonic factors, among them

the corrugated geometry of the detachment fault and its breakaway zone, the emplacement of the shallow-seated Kingston Range pluton across the detachment in northern parts of the early Shadow Valley basin (Davis et al., 1993), contemporaneous normal faulting of the allochthonous basin fill, and late stage folding associated with sinistral strike-slip faulting beneath both Kingston Wash (Davis and Burchfiel, 1993) and the Kingston Spring area.

Four unconformity-bounded informal members referred to as Member I, II, etc., by Friedmann (1999) constitute its 2.5–3.5 km-thick fill (Figs. 2 and 3). Volcanic activity, primarily andesitic to rhyolitic, was concentrated in Members I and II from ca.

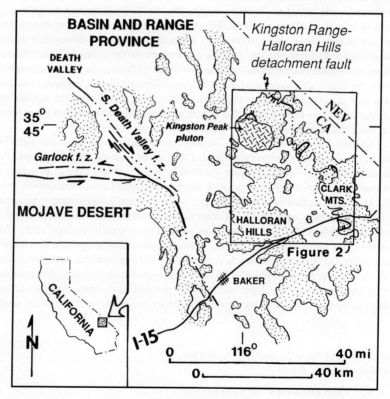

Fig. 1. Location map of Kingston Range-Halloran Hills detachment fault, eastern San Bernardino County, California.

13 to 11 Ma (Friedmann, 1999) Megabreccias occur in Members I, II, and IV, and slide blocks occur in all four members, but most importantly in Members II and III. Collectively, megabreccias and slide blocks constitute 10–15% of the Shadow Valley basin fill, but mass-wasting processes account for 30–50% of the total fill when lahars and debris flow conglomerates are included. Most of the mass-wasting events in Shadow Valley probably occurred during semi-arid or arid conditions. The presence of relatively deep perennial lakes between 12 and 11 Ma suggest a semi-arid to moderate climate, although an arid climate has not been ruled out. Gypsiferous playa deposits between 11 and 10 Ma indicate that Shadow Valley was arid or semi-arid during that time (Friedmann, 1999).

Although mass-wasting processes are characteristic phenomena in supradetachment basins (e.g. Yarnold, 1993; Yin and Dunn, 1992; Forshee and Yin, 1995; Friedmann et al., 1994; Friedmann and Burbank, 1995; Friedmann, 1997, 1999), the Shadow Valley basin is an exceptional locale for the study of such processes, especially rock-avalanches and gravity sliding. A score of megabreccias, interpreted as rock-avalanche deposits, and half that number of very large sheets of structurally intact rock (>1 km^2, up to 200 m thick), interpreted as gravity-driven slide (=glide) blocks, crop out in transverse and longitudinal cross-sections intercalated with various sedimentary facies. The Shadow Valley basin megabreccias have been described elsewhere (Friedmann, 1997), but the spectacular gravity-driven blocks of the basin (Fig. 4) have received little previous detailed attention and some published misinterpretation. Their impressive scale and the excellence of their exposure warrant this account. The Shadow Valley basin provides an unparalleled opportunity to examine Neogene examples of grand-scale, well-preserved gravity slide blocks and to learn about their transport history and mechanisms.

2. General attributes of Shadow Valley basin slide blocks

Large allochthonous sheets within the Shadow Valley basin are composed of pre-Tertiary units, with the partial exception of the Valjean allochthon as discussed below. The Shadow Valley basin allochthons range in present areal extent from approximately 0.5 to >40 km^2 and include Proterozoic crystalline rocks and Proterozoic, Paleozoic, and Tertiary strata. Some Shadow Valley slide blocks (e.g. Francis, 1109, central and southern Eastern Star, Valley Wells) moved into active depositional environments, but others (e.g. northern Eastern Star, Valjean, Carrara-Bonanza King) were emplaced on erosional surfaces developed across previously tilted and faulted basin units. Estimates of minimum block run-out from their source terranes range from a few kilometers to >15 km, although in most cases the exact provenances of the slide blocks are not known.

Throughout the western United States, many internally intact slide blocks were originally mapped as thin thrust sheets. Lasky and Webber (1949), for example, described the Artillery Mountains thrust plate in western Arizona as a Laramide structure. In reality it is a slide block of Proterozoic gneiss encased in a Miocene fanglomerate sequence above the Whipple-Rawhide detachment fault (Eric Frost, personal communication, 1985). Hewett (1955, 1956) first described the slide sheets of Precambrian rocks in the Shadow Mountains area, but interpreted them as thrust-fault klippen of Tertiary age, an interpretation challenged by Burchfiel and Davis (1971). There are several reasons for such structural misinterpretations. First, many workers originally thought that thrust plates were gravitationally driven, making the distinction between slide blocks and thrusts sheets essentially semantic and one of scale (e.g. Hubbert and Rubey, 1959; Kehle, 1970). Second, gravity-gliding of rock units into active depositional sites always places older rocks atop younger strata. Third, the basal contacts of the allochthonous blocks strongly resemble shallow low-angle faults, complete with gouge zones, brittle microfabrics, slickensided and striated surfaces, and breccia (friction) carpets as described below for the Shadow Valley basin examples. Fourth, the very large areal extent of some of supradetachment basin slide blocks (see below) makes an origin for them by gravitational sliding difficult to understand.

The structurally and stratigraphically intact nature of the largest Shadow Valley slide sheets is the major criterion in setting them apart from rock-avalanche breccias (Fig. 4). The orientation of strata within the slide sheets is variable, both because of earlier defor-

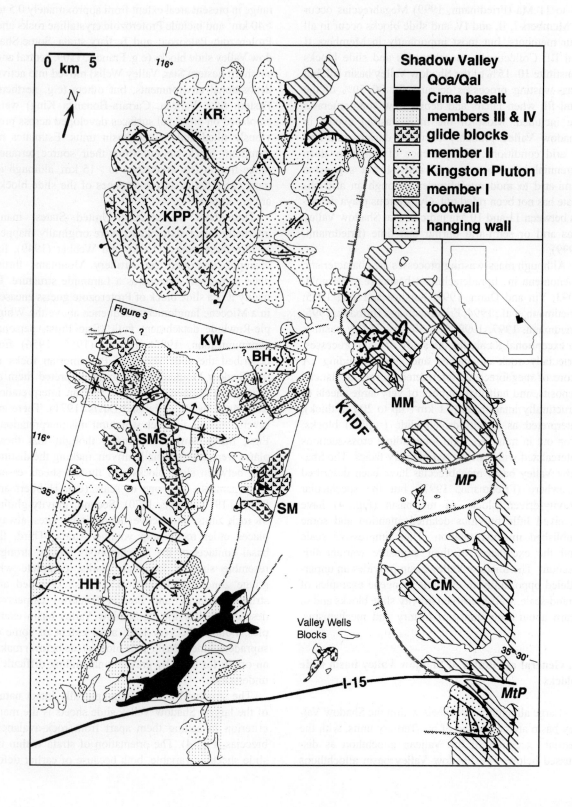

Shadow Valley

- Qoa/Qal
- Cima basalt
- members III & IV
- glide blocks
- member II
- Kingston Pluton
- member I
- footwall
- hanging wall

0   km   5

116°

N

KR

KPP

Figure 3

KW

BH

SMS

SM

HH

MM

KHDF

MP

CM

MtP

Valley Wells
Blocks

I-15

116°

35° 30'

35° 30'

mation and the geometry of the initial detachment surface with respect to bedding. Mappable folds and thrust faults within several of the larger sheets are products of the Mesozoic contractional history of the surrounding Cordilleran foreland fold and thrust belt. In some allochthons bedding and stratigraphic contacts dip into the basal detachment at moderate to high angles (e.g. Fig. 4D).

The bases of gravity avalanche deposits are typically irregular, show variable degrees and styles of admixture of footwall and avalanche material, and characteristically resemble depositional contacts more than tectonic ones. In contrast, the basal contacts of Shadow Valley slide blocks are characteristically knife sharp, and show few signs of lithologic mixing of upper- and lower-plate rocks (Fig. 4A). Locally, they exhibit slickensided and striated, planar fault-like bases (Fig. 4B). Basal contacts are not, however, always planar. Considerable variation in basal dip, perhaps related to preexisting surficial topography, is typical of the larger allochthons.

Basal contacts are typically overlain by a breccia (friction) carpet of variable thickness. In the Francis allochthon of Cambrian Bonanza King carbonates, for example (see below), the basal breccias are as thick as 25 m. Parke (1994), however, reports that basal breccias are lacking above approximately 1.5 km of the 4.1-km-long exposed base of the sheet (see also Fig. 7B for another slide sheet base lacking basal breccias). Friction carpet breccias at the thin feathered edges of slide blocks, where intact upper portions of the allochthon are not present, can be misinterpreted as landslide deposits (Parke, 1994). Sharp-walled, generally rather planar clastic dikes containing footwall sediments intrude upwards into the breccia carpets of several allochthons, but have been noted at the base of one large rock-avalanche deposit as well.

Structural features acquired during transport are varied. Normal faults in several blocks either flatten into or terminate downwards at the basal detachment. It seems likely that their formation was related to slide sheet emplacement and internal extension, given the listric geometry of some faults and their characteristic high strike angles to the direction of allochthon transport. Boyer and Hossack (1992) have recognized similar syn-sliding faults within Tertiary slide blocks in western Wyoming. The Eastern Star allochthon of Proterozoic basement and cover units contains steep strike-slip faults that Parke (1994) and Davis interpret as tear faults active during allochthon transport. A variety of other structural features can be used to assess local directions of transport. These include striations and other slickenlines on the basal slide surface, clastic dikes in the lower portions of the slide blocks that represent tensional phenomena developed at high angles to block movement, and inclined foliation in gouge at the base of the plate. A thin (0–20 cm) layer of gouge consisting chiefly of clay-sized particles separates portions of some slide blocks from underlying strata. The gouge can be massive, but commonly contains moderately inclined foliation surfaces (P-foliation, Cladouhos, 1999) that dip in the direction opposite to transport, and thus can be used as kinematic indicators.

Valuable information on transport direction can also be supplied by less direct means. For example, the stratigraphy of Proterozoic sedimentary units in the Eastern Star slide sheet requires, as amplified below, an eastern provenance for the allochthon. With some caution, the provenance of clasts in coarse-grained sediments interbedded with the slide blocks can also provide kinematic constraints on block movement. Two major trends can be determined from the diverse kinematic indicators. The predominant sedimentary transport direction throughout most of the Shadow Mountains area is to the WSW (Friedmann et al., 1996), which is consistent with geologic evidence for an easterly provenance for the majority of Shadow Valley basin slide blocks. A subordinate direction of allochthon and sediment transport, confined to the northern Shadow Mountains area directly south of Kingston Wash, is to the south from the pluton-uplifted Kingston Range (Fig. 2; Davis et al., 1993; Fowler et al., 1995). This trend

Fig. 2. Simplified geological map of the Shadow Valley basin and surrounding region (modified from Friedmann, 1997, Fig. 1). Thick solid (exposed) and gray (buried) line trace of the Miocene Kingston Range-Halloran Hills (KHDF) detachment fault, which cuts across Mesozoic thrust faults of the foreland fold-thrust belt (barbed lines). BH=Blacksmith Hills; CM=Clark Mountain; HH=Halloran Hills; I-15=Interstate highway 15; KPP=Kingston Peak pluton; KR=Kingston Range (includes KPP); KW=Kingston wash; MM=Mesquite Mountains; MP=Mesquite Pass; MtP=Mountain Pass; SM=Shadow Mountain; SMS=Shadow Mountains.

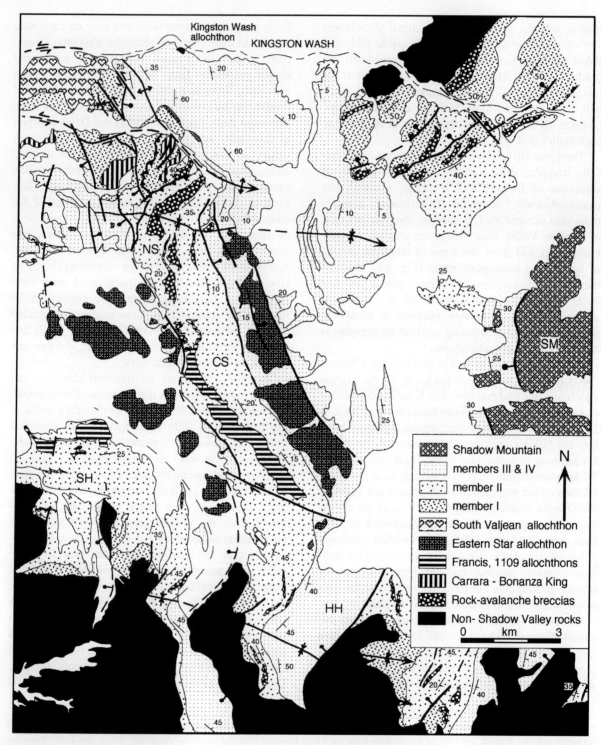

Fig. 3. Simplified geological map of the central Shadow Valley basin (modified from Friedmann, 1997, Fig. 2). Geology primarily by G. Davis (1990–1997, unpublished) with additions from Bishop (1994), Fowler (1992), Friedmann (1995), and Parke (1994). NS=northern Shadow Mountains; CS=central Shadow Mountains; SH=Silurian Hills; HH=Halloran Hills; SM=Shadow Mountain.

Fig. 4. (A) Slide sheet 1109 m containing Cambrian Bonanza King carbonate resting on Miocene fanglomerate deposits. View to north across Evening Star wash in south-central Shadow Mountains. Persons in far lower right provide scale. Footwall strata are essentially undeformed. (B) Striated, planar basal contact of sheet 1109 where it crosses Evening Star wash. (C) View of west side of hill 988 m, ca 3 km ESE of the Eastern Star mine; relief is ca 110 m. Hill is capped by Proterozoic gneiss (light color) and intrusive diabase (dark color) in the Eastern Star slide sheet. The sheet rests on subparallel, interlayered Miocene fluviatile and playa deposits along a knife-sharp contact (Fig. 7A). (D) View to northeast of Cambrian Bonanza King carbonate slide block resting on poorly exposed alluvial fan deposits, northeast corner of Silurian Hills (Fig. 3). The block, which is about 75 m high, is correlated with the Francis slide sheet (see text).

is consistent with northern slide block geometries, and is supported by associated sediments containing pluton and metamorphic aureole detritus from a Kingston Range provenance.

3. Major slide blocks in the Shadow Valley Basin

3.1. The Eastern Star crystalline allochthon

The most widespread and impressive gravity slide sheet (or sheet complex) in the Shadow Mountains basin is the Eastern Star allochthon (the "crystalline

allochthon" of Davis et al., 1991, 1998; Friedmann, 1997), so named for the fact that it caps hills surrounding the Eastern Star mine in the west-central Shadow Valley basin (Fig. 5; U.S.G.S. Kingston Spring 1:24,000 topographic sheet). It consists largely of Proterozoic quartzo-feldspathic gneiss, overlying Late Proterozoic sedimentary and metasedimentary rocks of the Crystal Springs Fm. of the Pahrump Group, and ca. 1.07 Ga diabasic rocks (Heaman and Grotzinger, 1992) that intrude both basement and cover. Locally, carbonate rocks of the Noonday Dolomite, the basal unit of the Cordilleran miogeosyncline in this region, overlie Crystal Spring and diabase units

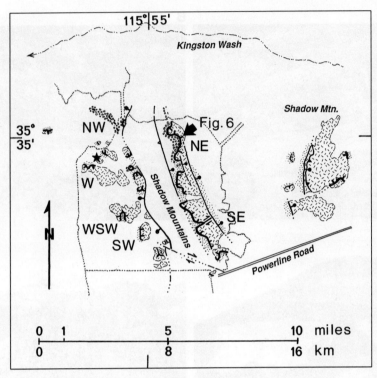

Fig. 5. Location map of klippen of the Eastern Star crystalline allochthon (irregular dash pattern), Shadow Mountains area, shown against an outline of the city of San Francisco to emphasize its large areal extent. Location of the Eastern Star mine is shown by the small star. Locations of klippen discussed in the text are given by NE, SE, etc. Small circular pattern beneath klippen denotes overridden fluviatile sedimentary strata; stippled pattern denotes overridden lacustrine deposits, widely gypsiferous.

unconformably. Isoclinally folded Noonday dolomites and Crystal Spring quartzites, the latter with a well-defined subvertical axial plane foliation, are present in the northeasternmost klippe (NE, Figs. 5 and 6).

The present outcrop pattern of the Eastern Star allochthon(s) has been strongly influenced by post-emplacement sedimentation and erosion, N–S-striking normal faulting, and broad folding with E–W trends. Numerous large and small klippen are found scattered throughout an area in excess of 40 km² (Fig. 5), although it is not clear if the Eastern Star allochthon moved as a single coherent slab or is a composite of various klippen that moved independently of each other (see below). The two largest and easternmost klippen (NE and SE, Fig. 5) form a nearly continuous north–south belt of exposures ca. 8 km long, with a combined area of 8–9 km² and preserved structural thicknesses of up to 200 m; the original thickness and areal extent of the eroded slide sheet are unknown.

It is possible to reconstruct the basin's geologic terrane prior to WSW-movement of the Eastern Star allochthon across it. Although most of the allochthon moved over upper Member III playa sediments containing evaporitic gypsum, its northern portion moved across previously faulted, east-tilted (ca. 10–15°), and eroded volcaniclastic fan/sand flat deposits with a northern provenance (Fig. 6). Erosion is indicated by the remnants of channels cut into the volcaniclastic section that were less than 10 m deep and were filled with a highly distinctive boulder- to cobble-sized assemblage of miogeoclinal Proterozoic quartzite (probably Sterling Quartzite), Paleozoic carbonate clasts, and granitic rocks resembling those of the Cretaceous Teutonia quartz monzonite of the eastern Mojave region. This clast assemblage is preserved widely, but discontinuously beneath the allochthon (stars, Fig. 6), most typically as a non-bedded (presumably sheared and disrupted), distinctive gray to orange-weathering "gouge-like" diamictite. Fan de-

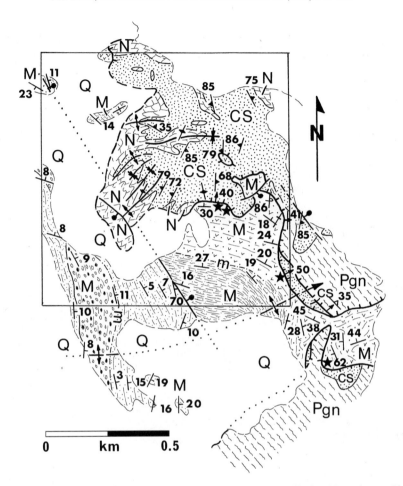

Fig. 6. Geologic map of northern end of the NE klippe of the Eastern Star allochthon, eastern Shadow Mountains; see Fig. 5 for location. Square outline is a 1 km² area from the USGS 1:24 000 Kingston Spring Quadrangle. The base of the Eastern Star slide block is hatchured; the heavy dashed line against Quaternary alluvium (Q) is the approximate buried trace of the allochthon's base. Geologic units: Pgn=undifferentiated Proterozoic gneiss and diabase; CS=Proterozoic Crystal Spring Fm. quartzites; N=Neoproterozoic Noonday Dolomite; M=undifferentiated Miocene fluviatile and lacustrine units (partial "v" pattern respresents volcaniclastic rocks); m=marker bed; stars indicate footwall outcrops of sheared quartzite-and granite-bearing conglomerate lying unconformably on the tilted, eroded Miocene section.

posits with the same lithology, but much greater thickness than that of the footwall channels, bury small northwestern klippen of the allochthon (NW, Fig. 5), and indicate that these klippen were emplaced earlier than larger klippen to the east.

The basal contact of the Eastern Star allochthon is generally marked by an extremely thin (10–50 cm) reddish, "clay"-rich foliated gouge. Sheared and brecciated upper-plate rocks constitute a brittle friction carpet at the base of the plate that varies in thickness between about 5 and 25 m (Fig. 7A). However, at one southwestern locality (WSW, west of 1089 T), there is no friction carpet beneath a klippe of Proterozoic

gneisses more than 120 m thick that overlies shallow-dipping lacustrine siltstones (Fig. 7B). The brittle deformational style of Crystal Spring quartzites above the basal contact in the northeasternmost klippe (NE, 0.3 km east of VABM 1125.3; Figs. 6, 7C and 8) is worthy of mention. The basal contact at this locality is very sharply defined with no mixing of upper-and lower-plate lithologies along it. A north-striking, subvertical foliation in the quartzites provides a reference surface that can be followed downwards towards the basal contact (Fig. 8). At distances of 40–50 m above the contact, the foliation is essentially pristine. A faint in situ fracturing of the quartzites becomes discernible

Fig. 7. (A) Base of Eastern Star slide sheet, west side of hill 988T (Fig. 4C), showing sharp contact with underlying Miocene sediments and overlying 15–20 m-thick friction carpet of brecciated diabase. The gradational upper contact (see arrows) of the friction carpet is overlain by more intact diabase. (B) Base of Eastern Star slide sheet on southwest side of klippe hill 1089 (SW, Fig. 5). Proterozoic gneisses and granitic sills (light color) are truncated sharply by the basal slide block contact. No friction carpet is present at this locality above the concordant footwall lacustrine sediments. (C) View to northwest of Eastern Star crystalline allochthon in NE map area of Fig. 6 (in area of 40° northward dip of the basal contact). Bedding in lower-plate clastic units, the basal contact of the slide sheet, and shear surfaces in Crystal Spring quartzitic rocks in the base of the sheet are all parallel. A listric normal fault that dips to the left (WSW) and flattens into the shear zone at the base of the plate is indicated by the black arrow. (D) Southward view of disharmonically folded lower-plate volcaniclastic strata, 1 km south of the southwestern corner of the km² outlined area, Fig. 6. Arrows define base of the slide sheet below Proterozoic gneiss.

at somewhat closer distances. However, as the base of the slide sheet is approached, the fracturing increases in intensity producing an in situ shattering that progressively obliterates the foliated rocks *without* rotating the foliation away from the vertical. Shattering is accompanied by the formation of a very fine-grained red quarzitic breccia that encloses the larger, unrotated fragments. Within 5–7 m of the fault, the intensely shattered rocks and the remnant foliation are abruptly transposed by multiple shear surfaces that lie subparallel to the basal fault (Fig. 8). One of these shear surfaces is the lower flattened part of a listric normal

fault that feeds into the basal shear zone (Fig. 8). Boyer and Hossack (1992, their Fig. 7) describe and illustrate a similar "strike-parallel segmentation fault" in a western Wyoming slide block of Cambrian carbonate rocks lying above Neogene volcaniclastic units.

3.1.1. Footwall deformation

Physical disruption and mixing of footwall strata are present at a number of localities below the allochthon, usually within 5–10 m of the basal contact and especially when lacustrine silts are major footwall lithologies. Disharmonically folded member III sedi-

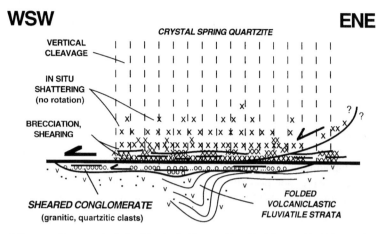

Fig. 8. Composite idealized diagram of structural relationships in the Eastern Star allochthon and its Miocene footwall in the area of Fig. 6. Upper-plate structures and sheared conglomerate depict base of slide block in the vicinity of its 40° dip (Figs. 6 and 7C). Lower-plate structures depict folded footwall strata relationships as illustrated in Fig. 7D.

ments commonly underlie the allochthon for distances up to 40 m (Figs. 7D, 8 and 9). Such deformation is not, surprisingly, confined to the poorly consolidated lacustrine sediments of low strength. Volcaniclastic strata beneath the northeastern and northwesternmost klippen (NE and NW, Fig. 5) typically exhibit pronounced folding. Folds are strongly disharmonic and have variable hinge orientations. Asymmetric anticlinal folds have geometries suggestive of dragging of lower-plate strata by the allochthon to the northwest or west. Such folds typically exhibit shallow-dipping bedding near the contact; bedding then steepens abruptly westward and downward into steep to over-turned attitudes with predominantly northerly strikes (Figs. 7D and 8). Based on the presence of such footwall deformation, the allochthon once extended at least 1 km north of its present northernmost exposures in the NE klippe (Figs. 5 and 6).

Whereas footwall strata are profoundly deformed at some localities, at others such deformation is

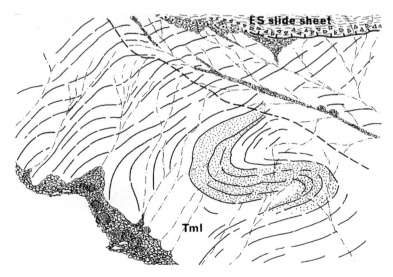

Fig. 9. Traces of bedding in tilted and folded Miocene lacustrine siltstones (Tml) and interbedded volcaniclastic sandstones on slopes beneath Proterozoic gneiss (Pgn) in the Evening Star allochthon (Pgn), eastern Shadow Mountains; line drawing from photograph. Footwall deformation at this site, WSW of hill 1243 m, extends ca. 40–45 m below the slide sheet. Stippled sandstone beds in the S-shaped anticline-syncline pair span about 10 m of elevation. View to north.

absent. For example, on the northern limb of the anticline pictured in Fig. 6, volcaniclastic footwall strata dip uniformly in the same direction as the base of the slide sheet although at a somewhat shallower angle (Fig. 7C). We explain such disparity in deformational styles at nearby localities as varying responses between surface topography and the sliding allochthon. Footwall deformation can, thus, be interpreted as an interference effect between the moving slide sheet and positive topographic asperities, i.e. hills. Where such asperities were not present, the allochthon moved across footwall strata with a mini-

mum of stress transmission. Where the allochthon, more than 200 m thick, encountered hills of low relief (several tens of meters or less), stresses were transmitted into footwall strata that became folded with variable, but predominantly westward vergence.

3.1.2. Kinematics of allochthon displacement

Kinematic indicators for the Eastern Star allochthon are varied in nature, but are quite consistent in indicating its transport to the W or WSW, a direction also required by the distribution of provenance lithologies in this part of the foreland fold and thrust belt.

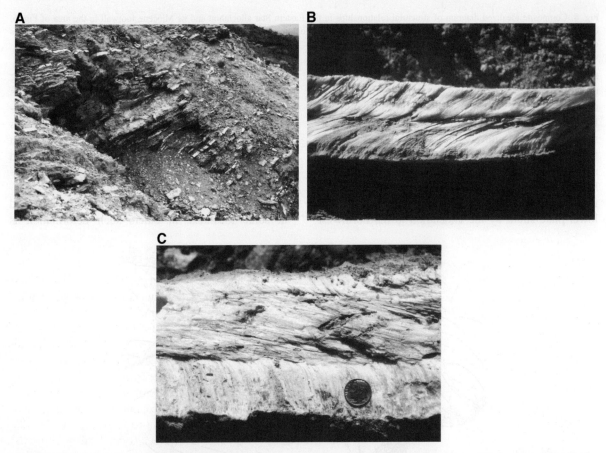

Fig. 10. (A) View to the northwest of southwest-dipping gypsiferous playa siltstones and parallel gypsum veins on the east side of hill 988 (Fig. 4C). The basal contact of the Evening Star crystalline allochthon lies ca. 3 m above the highest vein in this picture. Most of the veins pictured here are ca. 2–6 cm thick, lie parallel to the base of the slide sheet, and are not physically disrupted except by minor slumping of the soft enclosing sediments. Their unsheared internal geometries are interpreted as indicating vein dilation and fiber growth during southwestward displacement of the overlying sheet. (B) Close up view of part of a gypsum vein from within the area of (A). Fiber growth outward from the vein median line shows an increasingly oblique dilation (top to the right) compatible with a SW-sense of displacement of the overriding allochthon. Vein thickness is 7 cm. (C) Composite gypsum vein illustrating two generations of fiber development, both with geometries compatible with southwestward dilation (top to the left). The strongly oblique fibers in the youngest central part of the vein are up to 12 cm long.

Slickenside striae measured at two northern exposures of the basal contact 3 km apart, together with the inclination of gouge foliation and the asymmetry of footwall folds indicate transport to S 70°W. An upper-plate listric normal fault that flattens westward into the basal contact of the northeasternmost klippe (NE, Figs. 7C and 8) has a strike of N 15°E. Subhorizontal footwall lake sediments that lie between the two southwestern klippen (WSW and SW) are broken by conjugate, moderately to steeply dipping normal faults defining a N 59°E–S 59°W direction of stratal extension—a direction compatible with the passage of the Eastern Star allochthon over the lacustrine beds. Fiber growth of gypsum in dilational veins beneath the allochthon, which we believe formed synchonously with slide sheet emplacement, also indicate its southwestward displacement (see below; Figs. 10–12).

Emplacement of the allochthon was accompanied by its internal segmentation along tear faults. In the southeasternmost klippe (SE, Fig. 5) an intrusive contact between diabase and a steep NE-dipping section of Crystal Spring strata is dextrally offset ca. 1 km by a subvertical N 60°E strike-slip fault that is confined to the allochthon. It is interpreted by Parke (1994) as a tear fault that accommodated differential velocities of transport within the allochthon

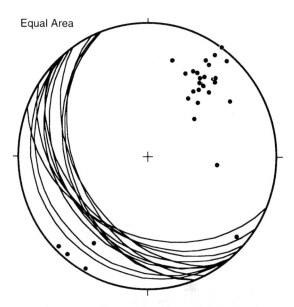

Fig. 12. Equal area projection of gypsum veins (great circles) and gypsum fibers (dots) that lie below the Eastern Star allochthon on the east side of hill 988T, ca. 3 km ESE of the Eastern Star mine. The southwest dip of the veins (Fig. 10) and the basal contact of the overlying slide sheet are due to late folding.

during its emplacement. Four other probable tear faults in the western klippen (W, Fig. 5) in the vicinity of the Eastern Star Mine have subvertical dips and strikes that range from N 65°E through EW to N 80°W.

3.1.3. Allochthon provenance

The Proterozoic units in the Eastern Star allochthon were derived from the Winters Pass thrust plate in the Clark Mountains fold and thrust belt to the east (Burchfiel and Davis, 1971, 1988; Davis et al., 1993). This NE-directed plate, the highest in the Jura-Cretaceous thrust stack of the eastern Mojave region, carries Proterozoic gneisses and a supracrustal cover of younger Proterozoic and Paleozoic units. A likely source for the southeasternmost klippen of the Eastern Star allochthon appears to be the Winters Pass thrust plate as exposed at Shadow Mountain (Fig. 5), 6 km N 60°E of the southeasternmost klippe (SE, Fig. 5). Recall that the transport direction of this part of the allochthon, based on a major tear fault within it, was S 60°W.

Shadow Mountain is an inlier of Precambrian and Paleozoic units rising some 220 m above the Quaternary alluvium of present day Shadow Valley. Pro-

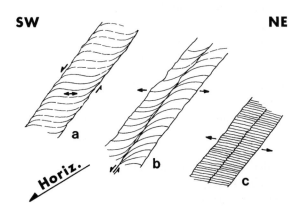

Fig. 11. Varying geometries of gypsum fibers in dilational veins within lacustrine sediments below the Eastern Star crystalline slide sheet (Figs. 10 and 11); only representative fibers are shown from actual examples. All exhibit oblique dilation compatible with a ca S 35°W sense of displacement of the overriding allochthon. Fiber growth in vein a is outward from its dilating walls; veins b and c grow outward from their median lines (cf. Ramsay and Huber, 1983).

terozoic granitic gneiss, ca. 1.6–1.8 Ga (DeWitt, 1980), and a steep, east-dipping section of overlying Crystal Springs strata are intruded by at least five diabase sills and dikes (Fowler, 1992). These rocks lie above Cambrian Carrara marbles and phyllites along a gently east-dipping thrust fault correlated by Burchfiel and Davis (1971, 1988) with the Winters Pass thrust. Although once considered by Burchfiel and Davis (1971) as being in situ, Fowler (1992) offers circumstantial evidence that Shadow Mountain is itself a slide block in the Miocene Shadow Valley basin.

The geology of the southeasternmost Eastern Star klippe (SE, Fig. 5) is so similar to that of Shadow Mountain that we confidently interpret Shadow Mountain as a source area for that klippe. In both areas, the steep, east-dipping nonconformity between the Crystal Springs Fm. and basement gneisses is largely intruded by Late Proterozoic diabase. There are multiple sheetlike diabase intrusions in the basement rocks of both areas. A shallow ($<10°$), west-dipping thrust fault in the Winters Pass plate at Shadow Mountain, with a northeasterly displacement of approximately 1 km, is almost certainly an imbricate splay off the thrust. A shallow-dipping thrust fault involving Proterozoic strata is also present in the southeasternmost klippe.

If Shadow Mountain was indeed the source terrane for the Eastern Star allochthon, then its bedrock units must once have been exposed over a much larger area than at the present time. As stated above, we are not certain if the widespread Eastern Star klippen are erosional remnants of a once continuous sheet, or moved independently from a common provenance area. However, the consistency of slide block transport to the west and west-southwest and the widespread occurrence of similar gypsiferous sediments beneath most of the klippen argues for emplacement of the allochthon (or allochthons) at essentially the same time and under similar circumstances. Whether the Eastern Star allochthon was once continuous or is an aggregation of independently moving blocks, a minimum runout distance of 9–10 km over a presumably horizontal playa is required for the NE and westernmost blocks (Fig. 5). If the SE allochthon was derived from Shadow Mountain, as we believe, its runout distance from that source is approximately 6 km. Finally, the Shadow Mountain block itself may

have been transported several kilometers west from an original position in the Winters Pass plate.

3.2. The Francis and 1109T allochthons (South carbonate block of Friedmann, 1997)

The second most extensive allochthon in the Shadow Valley basin is apparently represented by a number of isolated slide blocks, of which the largest underlies "Francis" peak (1232.9 m) in the southern Shadow Mountains (Fig. 3). Paleozoic carbonate rocks at Francis peak form an imposing strike ridge and are completely enclosed in an east-dipping (ca. $25° ± 5°$) section of Miocene fluviatile redbed gravels of basin Member II (Parke, 1994; Friedmann, 1999). The Francis allochthon consists primarily of recrystallized Middle Cambrian Bonanza King carbonate rocks, although higher stratigraphic units, now metamorphosed and bleached, are locally present. Other slide blocks of Bonanza King carbonates lie within Member II redbeds in areas up to 8 km to the west on the northeastern margin of the Silurian Hills (SH, Fig. 3; Bishop, 1994; Fig. 4D). If these blocks are erosional remnants of the Francis allochthon, as we believe likely, then its original areal extent was more than 30 km^2.

The Francis block is 4.1 km in length along its strike, has a maximum map width of 720 m, and a maximum structural thickness of about 145 m (Parke, 1994). Carbonate strata within it typically dip discordantly into the basal contact. Within the northern part of the Francis peak allochthon older Mesozoic folds, faults, and stratigraphic continuity have been preserved within an intact carbonate cap above the block's 0–25 m-thick friction carpet. The boundary between friction carpet and intact cap is variably gradational, or a fault, or a series of faults.

Parke (1994) has mapped the 1109 allochthon, a smaller, thinner (< 45 m) glide block that overlaps and lies above the northern end of the Francis allochthon and continues another 3 km to the north as an increasingly thinner breccia sheet (Fig. 3). This sheet is spectacularly exposed on the steep northern wall of Evening Star wash 2 km NW of Francis peak, where its planar, knife-sharp base sits concordantly on undisturbed, well stratified alluvial fan deposits (Fig. 4A,B). Bishop (1997, his Fig. 3) interprets this slide sheet as a rock-avalanche deposit, perhaps because it contains extensively brecciated Cambrian Bonanza King carbonates and a capping unit of polymict carbonate breccia derived from Middle Paleozoic units. We believe that the characteristics of its basal contact support a slide block origin and that its internal brecciation is a function of its relative thinness.

Member II redbeds lie conformably both below and above the 1109 sheet and carry distinctive clasts of Cambrian Tapeats Sandstone that can only have been derived from areally restricted cratonal outcrops several tens of kilometers to the east. A west-inclined paleoslope indicated by the Tapeats clasts in enclosing fan deposits favors westward displacement of the slide sheet as well. However, striae on the planar base of this slide sheet (Fig. 4B) trend between N 8–13°W and plunge gently northward (Parke, 1994). Nearby clastic dikes in the base of the sheet have strikes between N 65°E and N 73°E. These indicators suggest an alternative kinematic explanation, i.e. that the 1109 block moved down a NNW-dipping bedrock paleoslope in the Halloran Hills to the south (HH, Fig. 3) and out into the alluvial domain with its west-dipping paleoslope. Bishop (1994) has mapped Member II strata as continuing another 8 km or so to the southeast in the Halloran Hills. If the Francis and 1109 slide blocks were derived from that direction and moved north-northwestward into the Shadow Valley basin of Member II time, their minimum transport distance was 8 km. We consider the provenance of the 1109 sheet to be unresolved.

3.3. Valjean allochthon ("Tertiary" block of Friedmann, 1987)

One of the most interesting, and puzzling, slide blocks in the Shadow Valley basin is the Valjean allochthon, named after "Valjean" VABM (801.5 m) hill north of Kingston Wash (U.S.G.S. Valjean Hills 1:24,000 topographic sheet). With a presently preserved area of at least 14 km^2, half of the identified slide block lies north of the wash and half to the south (Fig. 13). The two halves of the allochthon have been displaced from one another by an oblique sinistral slip fault that strikes ca. N 70°E beneath the alluviated wash, has a lateral displacement of 4 ± 1 km, and drops its northern wall down. Davis and

Burchfiel (1993) have speculated that the fault might represent an eastern branch of the Garlock fault before that structure was offset by the younger southern Death valley fault zone (Davis and Burchfiel, 1973).

3.3.1. Stratigraphy

The Valjean allochthon contains a thick (>1500 m) south-dipping section of pre-Tertiary miogeoclinal units and an overlying sequence of Shadow Valley basin strata (Figs. 13 and 14) that is largely repeated in the north and south blocks. Miogeoclinal units include the Eocambrian Noonday Dolomite and Sterling Quartzite (the Johnnie Fm. that normally separates them may be hidden beneath Quaternary sand deposits), the Eocambrian–Cambrian Wood Canyon Fm., the Cambrian Zabriskie Quartzite, Carrara Fm., Bonanza King Fm., and the Cambrian–Ordovician Nopah Fm. Because of obscuring windblown sand deposits the slide block has been defined with certainty as extending only 3 km north of Kingston Wash. However, it very likely continues at least another 5 km farther north to include a thick south-dipping section of Kingston Peak Fm. redbed sandstones that underlie a north-trending ridge with minor summits above 800 m. Carbonate rocks tentatively assigned to the Noonday Dolomite sit more-or-less concordantly on reddish Kingston Peak clastic rocks at the southern end of this ridge (about 1 km west of the northern corner of Fig. 13).

The Valjean allochthon is atypical in the Shadow Valley basin in that it includes Tertiary strata in addition to pre-Mesozoic units. A Member I(?) Miocene section that rests unconformably on Paleozoic carbonate rocks in the slide sheet is a very distinctive, high energy sequence of interlayered coarse clastic rocks, megabreccias, and megablocks (stratigraphically more-or-less intact blocks with strike lengths in the range of 250–750 m). Clastic rocks are predominantly carbonate clast conglomerate and sedimentary breccia of Paleozoic lithologies; quartzite and siliceous volcanic clasts are greatly subordinate. Unlike other Member I sections of the Shadow Valley basin, the Valjean sequence contains relatively few volcanic units; those that are present include volumetrically minor tuffaceous siltstone, siliceous volcanic rocks, and hornblende andesite breccia/

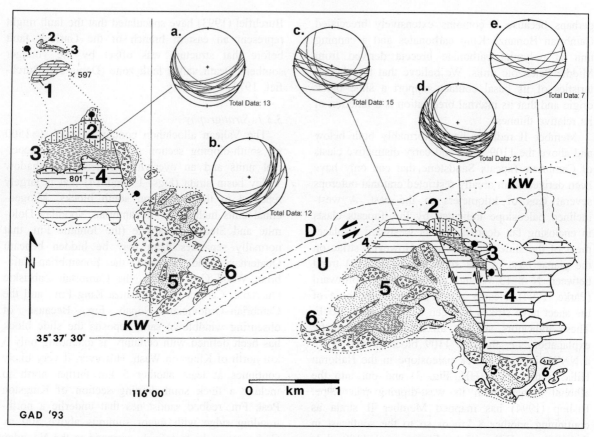

Fig. 13. Geologic map of Valjean slide sheet northwest and southeast of Kingston Wash (KW). The two hilly areas of bedrock north and south of the wash are interpreted as once continuous portions of the slide sheet subsequently offset along a sinistral strike-slip fault (Garlock?, Davis and Burchfiel, 1993) beneath the wash. The hatchured contact south of the wash is the slide sheet's basal contact; north of the wash, this contact is everywhere covered by Quaternary sediments. Units: 1=Eocambrian Noonday Dolomite; 2=Eocambrian-L. Cambrian Sterling Quartzite and Wood Canyon Fm.; L. Cambrian Zabriskie Quartzite (all tectonically thinned); 3=Cambrian Carrara Fm. (thinned); 4=Cambrian to Ordovician carbonate strata; 5=Shadow Valley basin Member I(?) carbonate clast conglomerate and sedimentary breccia; minor lacustrine siltstones and siliceous volcanic rocks; 6=undifferentiated Paleozoic carbonate megablocks and rock avalanche breccias. All stratigraphic units (1–5) have variable southerly dips, typically >40°. Equal-area stereoplots show bedding dip in each of five structural domains: a=pre-Tertiary bedrock of northern Valjean block; b=Miocene section of northern block; c=southern Valjean block west of westernmost dextral fault; d=eastern section of southern block; e=footwall strata east of d.

lahar. The most noticeable aspect of this Miocene section is the great abundance within it of thin lenses of Early to Middle Paleozoic carbonate megabreccias and more or less intact megablocks of Cambrian Carrara and Bonanza King Fms. More than two dozen megabreccia bodies were mapped in the 750-m-thick Miocene section preserved in the Valjean block south of Kingston Wash (Fig. 14). Several Bonanza King dolomite megablocks rest on thin foundations of sheared Carrara red shale, a relationship suggesting stratigraphic controls on their initial detachment. Collectively, debris flows, megabreccias

and megablocks make up between ca. 50% and 70% of that section.

3.3.2. Internal structure

Strata in the northern block, both Miocene and older, dip southwards at angles ranging between 30° and 60° (Fig. 13); dips in the largely correlative section of the southern block are generally steeper (40–80°). The subhorizontal base of the Valjean allochthon south of Kingston Wash lies at elevations of ca. 700 ± 30 m, but was not directly observed. This part of the Valjean allochthon overlies two

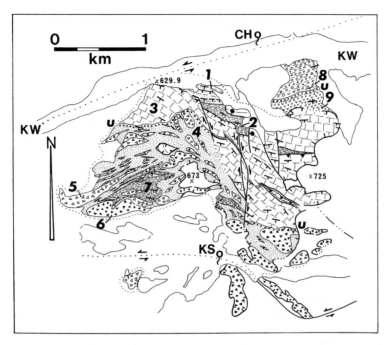

Fig. 14. Detailed geologic map of the Valjean slide sheet south of Kingston Wash. Geographic localities: CH=Coyote Holes; KS=Kingston Spring. Map units: 1=Sterling Quartzite, Wood Canyon Fm., and Zabriskie Quartzite; 2=Carrara Fm. (thinned); 3=M. Cambrian to Ordovician carbonate strata; 4=Member I(?) Miocene strata; 5=Cambrian Bonanza King carbonate rock avalanche breccias and megablocks; 6=Devonian (?)-Mississippian carbonate rock avalanche breccias and megablocks; 7=Carrara megablock (strongly sheared); 8=lower plate Neoproterozoic Kingston Peak Fm; 9=Member I basal conglomerate and overlying andesitic and basaltic volcanic rocks; u=unconformity below contrasting basal Miocene sections in the allochthon (4) and below it (9). See Fig. 13 c, d, and e for orientations of bedding in this map area.

contrasting, steeply south-dipping (>60°) sections of Member I sedimentary and volcanic rocks that are assignable to different facies and are separated by a steep E–W sinistral (?) shear zone. Neither of the two footwall Member I facies is stratigraphically similar to the Member I(?) sequence in the Valjean allochthon.

The Miocene section of the south block and its Bonanza King basement are offset dextrally by two steep faults (Fig. 14) that are confined to the slide block. Their collective horizontal component of displacement is about 1 km. The western of the two faults has an average strike of ca. N 30°W and cuts the Bonanza King-Carrara contact, but appears to end downwards at a slip horizon within the Wood Canyon Fm. The eastern fault strikes ca. N 10°W, probably truncates the western fault, and terminates downward at the slip surface separating the Carrara and Bonanza King Fms. There are two possible explanations for the two Neogene faults. One is that they were tear faults active within the allochthon during its sliding, thus

indicating its gravitationally induced transport to the NNW or SSE. This explanation, however, encounters difficulties in explaining how the hypothesized tear faults could terminate downwards into steeply dipping slip surfaces that are now truncated by the base of the slide block. A preferred explanation is that the faults developed prior to detachment of the Valjean allochthon, when the slip surfaces at which they terminate were still shallow dipping, as was the basement-Miocene stratigraphic sequence.

3.3.3. Provenance

Because the base of the Valjean allochthon has not been seen, there is no direct kinematic evidence for its sense of transport. The southward dip of strata in both the allochthon and in its overridden footwall is suggestive of transport and rotation at high angles to this orientation. If so, the slide block must have come from the region north of Kingston Wash. A southern source for the allochthon is not possible because that area was the active depocenter of the Shadow Valley basin from Member I through Member IV time (Friedmann, 1999).

The Valjean allochthon contains a generally concordant south-dipping stratigraphic section that is at least 1500–2000 m thick (Noonday Dolomite and higher units) and is truncated at a high angle (>40–60°) by the base of the slide sheet (Fig. 13). Similarly, the overridden footwall section of different stratigraphy and Miocene facies just south of Kingston Wash (Fig. 14) is also steeply south dipping. In fact, all footwall Member I through mid-Member III strata south of Kingston Wash dip southward across a map distance of about 4.5 km. These strata constitute either a south-dipping homocline or, alternatively, the south limb of a major E–W trending anticline, the crest and north flank of which lie somewhere north of Kingston Wash. [Although it is tempting to relate such folding to forceful intrusion of the shallow Kingston Peak pluton, emplacement of the ca. 12.4 Ma pluton (Calzia, 1990; Davis et al., 1993; Fowler et al., 1995) cannot account for tilting or folding of Shadow Valley Member III units as young as ca. 11 Ma; Friedmann, 1999].

The internal geometry of the Valjean slide block could have been produced by a shallow south-dipping detachment surface (ca. 20°) cutting through a steeper, previously south-tilted basement and cover section. However, this section cannot be placed in a geologic context with its disparate footwall sections, nor with the homoclinal or anticlinal limb geometry of those footwall units. What is not clear is why such a detachment surface would cut discordantly across a 1.5–2 km-thick section of quartzites, carbonates, conglomerates, megabreccias, and megablocks at such a low angle of initial dip. Finally, uncertainties regarding the extent of the slide block north of Kingston Wash hamper determination of its transport distance from a source area. For these various reasons, including its derivation from north of Kingston Wash, the Valjean allochthon is the most enigmatic of the Shadow Valley slide sheets.

The timing of emplacement of the Valjean allochthon is uncertain, but a Member III time (between ca. 11.2 and 10.8 Ma) is favored. The allochthon overrides only a south-tilted facies of Member I, but higher Member II and early Member III units, are similarly south-tilted. A carbonate slide block that appears to be the southern end of the Valjean alloch-

thon is offset by a sinistral E–W-striking fault south of Kingston Wash (Fig. 14); that fault was active after 11.2 Ma and prior to ca. 10.8 Ma in Member III time (Friedmann, 1999; Davis, unpublished mapping).

3.4. Carrara-Bonanza King allochthon (North carbonate blocks of Friedmann, 1997)

Carrara and Bonanza King miogeoclinal strata comprise one of the oldest of the Shadow Valley slide sheets and now lie unconformably above Member I strata in the north-central Shadow Mountains. The best exposures of this allochthon are found 3.5–4 km S 80°E of Kingston Spring (Figs. 3 and 14). An isolated portion of this slide sheet is also present to the west, on the north flank of the 965 m hill, ca. 2 km S 60°E of the spring (1:24,000 Kingston Spring sheet). This Carrara-Bonanza King allochthon is instructive from two standpoints. Firstly, its eastern portion was emplaced on a shallow-dipping surface eroded across three panels of Member I strata that had been displaced by east-side down normal faults (Fig. 15). The two panel-bounding faults are the earliest known record of extensional faulting within the Shadow Valley basin (>12.0 Ma). After emplacement of the Carrara-Bonanza King slide sheet across the fault-bounded panels and deposition of Member II strata above it, both faults were reactivated with an opposite sense of displacement (west-side down), thus offsetting the overlying allochthon (Fig. 15). Secondly, the now fault-separated portions of the allochthon all display a basal unit less than 10 m thick of sheared Carrara red shale beneath overlying dark gray to black Bonanza King dolomite. Carrara shales lie along nearly 2 km of the basal contact and demonstrate that the sheet's detachment from its original locale was stratigraphically controlled. The Carrara-Bonanza King contact is a common zone of detachment and decollement within the Mesozoic Clark Mountain thrust complex to the east (Burchfiel and Davis, 1971, 1988).

The provenance and runout distance of this allochthon are not known with certainty, but two lines of circumstantial evidence indicate that it came from the north. Firstly, in the central and northern Shadow Valley basin the only other slide blocks and megablocks with similar Carrara red shale-Bonanza King lithologies lie in the Member I(?) section of the Val-

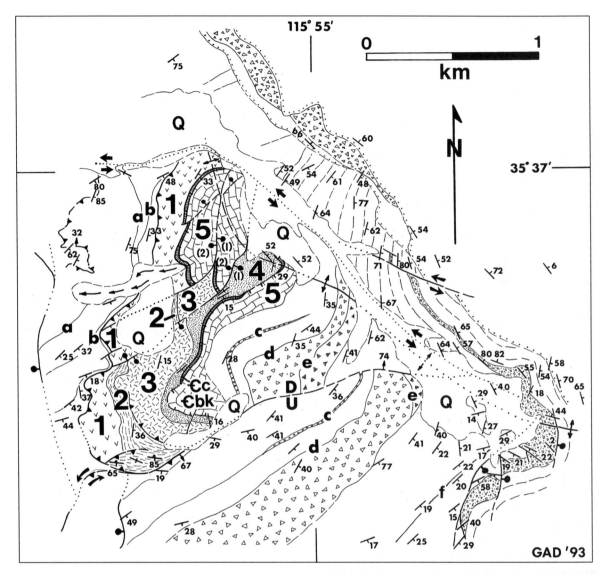

Fig. 15. Detailed geologic map of the Carrara-Bonanza King slide sheet ca 4 km east of Kingston Spring (Fig. 14). The slide sheet (5) contains Carrara (Cc) and Bonanza King (Cbk) strata and lies above an unconformity developed across different fault-separated units of Member I strata (from west to east: units 1, 3, and 4). Following slide block emplacement, the footwall faults (1) offsetting Member I units were reactivated with an opposite sense of displacement (2). Member I units: 1 = andesitic and basaltic volcanic rocks; 2 = siltstone, sandstone, altered tuff; 3 = silicic tuff breccia; 4 = sandstone, conglomerate; Q = undifferentiated Quaternary units. Selected stratigraphic contacts are designated by a, b, c ..., f; the various triangular patterns identify carbonate rock avalanche (megabreccia) deposits. A folded, major post-Member II, pre-Member III sinistral strike-slip fault lies beneath Quaternary alluvium north of the slide sheet; the same unexposed fault lies beneath Kingston Spring (Fig. 14) and may be an early expression of the Garlock fault (Davis and Burchfiel, 1993). A northwest-trending anticline folds the slide sheet and all Member I through Member IV strata in the north-central Shadow Mountains.

jean allochthon (Figs. 13 and 14), which must have had a northern provenance. Secondly, the slide sheet was emplaced across an erosional surface separating Member I (> 13.0 Ma) and Member II strata (ca. 12.0 Ma). This hiatus coincides in time with the forceful emplacement of the Kingston Peak Pluton to the north (ca. 12.4 Ma) and uplift of its wallrocks (Calzia, 1990), thus possibly leading to detachment and southward displacement of the Carrara-Bonanza King sheet.

3.5. Kingston Wash allochthon (Pahrump megablock of Friedmann, 1997)

A small slide block composed of highly brecciated brick red clastic rocks of the Proterozoic Kingston Peak Fm. (Pahrump Group) is spectacularly exposed on the south side of Kingston Wash (Fig. 3). The megablock, less than 1 km^2 in area, overlies boulder-bearing alluvial fan strata of Member IV that are composed primarily of abundant detritus shed from the Miocene Kingston Peak pluton. The basal contact truncates the fan deposits at a low (5°) angle. Locally, the block's substrate was incised by channels up to 1 m deep that are filled with angular cobble-pebble conglomerate composed predominantly (more than 85%) of sedimentary clasts of the Proterozoic Kingston Peak Fm. This channel composition is in marked contrast to the underlying fan deposits that contain less than a few percent of Kingston Peak sedimentary clasts. Field relationships indicate that the channels were filled by detritus shed from the overlying allochthon as it advanced across the landscape.

The glide block's basal contact consists of a relatively thin (5–50 cm), foliated gouge zone containing well developed sense-of-shear indicators that are definitive in indicating allochthon transport to the SW. Included among these indicators are slickenlines, asymmetric folds within the gouge and a minor, gouge-filled, south-vergent thrust that rises from the basal contact at a small angle and imbricates the basal breccias. Areally extensive exposures of Kingston Peak Fm. lie along the southeastern margin of the Kingston Peak pluton, ca. 6–8 km NE of the slide sheet's locality. This Kingston Range is a requisite source area for this young slide block, but extensive alluviation along the southern margin of the Range and the history of late (post-Member IV), large sinistral strike-slip displacement along Kingston Wash preclude definitive determination of the allochthon's provenance. Sinistral separation of the Kingston Wash slide block from the section of in situ Kingston Peak rocks described above would have the relative effect of moving the source area farther from the allochthon.

3.6. Valley Wells slide blocks

Low hills north of Valley Wells Rest Area on U.S. Interstate 15 west of Cima Road (Fig. 2) are underlain by east-dipping (ca. 15° ± 10°) Shadow Valley basin fanglomerates and intercalated carbonate megabreccias and gravity-driven slide sheets (Davis, unpublished mapping; Davis et al., 1993; Bishop, 1994). The hills have an areal extent of about 4 km^2 and a maximum relief of ca. 60 m. The most prominent hill is held up by a 1-km-long slide block that carries and cuts off discordantly the Winters Pass thrust fault and its upper-and lower-plate units. Proterozoic gneiss and its cover of Noonday Dolomite and higher miogeoclinal units sit in thrust contact atop Johnnie Formation strata and underlying Noonday rocks. Both upper- and lower-plate rocks (exclusive of the carbonate) share a common NE-plunging mylonitic stretching lineation. The fanglomerates that enclose this block and lesser, stratigraphically lower slide blocks of Noonday and Bonanza King carbonate rocks contain a distinctive clast assemblage from a suite of cratonal units that crop out only near Mountain Pass 16 km to the east (garnetiferous granitic gneiss, Mesoproterozoic syenite, Jurassic Aztec Sandstone, and Delfonte volcanic rocks; cf. Fleck et al., 1994). The slide block containing the Winters Pass thrust must also have been derived from the east, but from a much shorter distance than the alluvial fans that enclose it. Geologic relations that indicate the probable position of the alluvium-covered Winters Pass thrust along the west side of the Clark Mountains support a maximum westward transport of the slide block that now contains it of ca. 8 km.

4. Insights on gravity-induced sliding provided by the Shadow Valley basin sheets

4.1. Provenance

Gravity-driven allochthons in the various stratigraphic members of the Shadow Valley basin were derived from topographically elevated areas in the Kingston Range to the north (e.g. Valjean, Carrara-Bonanza King, and Kingston Wash slide blocks), to the east (e.g., Eastern Star, Shadow Mountain?, Valley Wells blocks), and possibly, although not likely, from the Halloran Hills to the south or southeast (e.g. Francis, 1109 [m] slide blocks). The stratigraphic interspersing of rock avalanche deposits (megabreccias) with slide blocks in all four members of the Shadow Valley basin sequence attests to steep slopes of the slide block-spawning areas during basin formation.

Steep slopes were variably formed during the evolution of the Kingston Range-Halloran Hills extensional detachment fault and the supradetachment Shadow Valley basin. Slumping of footwall rock units of the Kingston Range-Halloran Hills detachment system into the overlying Shadow Valley basin could have been facilitated by high relief generated at the system's breakaway zone and by seismic activity related to extensional faulting. Fowler et al. (1995) present evidence that at the time of its initiation the detachment fault east of the present Kingston Range dipped 30–35° westward across subhorizontal Tertiary and pre-Tertiary strata in both hanging wall and footwall positions. Downward displacement (relative) of the hanging wall could have produced a pronounced scarp.

The detachment itself was corrugated in the direction of transport, cutting up and down across the west-rooting thrust plates of the foreland fold and thrust belt (Davis et al., 1993). Corrugation amplitudes and wavelengths are up to 1.5 km and 10–15 km, respectively. This initial west-plunging, east–west trending corrugated geometry was responsible for the development of a complex footwall topography with west-facing headlands (the "antiformal" corrugations) and west-draining recessive valleys (the "synformal" corrugations; cf. Friedmann et al., 1996; Friedmann, 1999). Given steep initial westward dips of the detachment fault (Fowler et al., 1995), the west-facing headlands might well have resembled the prominent, steep turtleback fault surfaces of modern Death Valley (Wright et al., 1974; Miller and Pavlis, this volume), thus providing the gravitative potential for the transfer of bedrock units into a proximal lacustrine basin in the form of giant slumps and chaotic rockfall deposits. We visualize just such a paleotopographic setting for the Shadow Mountain block and the Eastern Star allochthon. In contrast, the recessive valleys between the antiformal headlands were the primary pathways for west-flowing river systems draining and eroding the footwall of the detachment fault complex and carrying

sediment into the supradetachment basin (Friedmann, 1995; Friedmann and Burbank, 1995; Friedmann et al., 1996; Friedmann, 1999). The potential for mixing of fluviatile, lacustrine, and gravity-driven sedimentation in the topographically complex environment of the breakaway zone of the upper plate of the detachment fault is obvious.

Evolution of northern parts of the Shadow Valley basin was dramatically affected by the intrusion at ca. 12.4 Ma of the Kingston Peak pluton (Calzia, 1990). This large (ca. 130 km^2) Miocene pluton was intruded to shallow levels (<4 km) across the then-active Kingston Range-Halloran Hills detachment fault. Movement on the northern, pluton-pinned portions of the detachment ceased and its upper plate and the overlying supradetachment basin were physically uplifted (Davis et al., 1993; Fowler et al., 1995). A million year-long hiatus, ca. 13–12 Ma, between basin Members I and II in the eastern area of Kingston Wash is reflective of the pluton's emplacement and its disruptive effects on northern basin sedimentation (Friedmann, 1995; Friedmann et al., 1994).

The relief of the Miocene Kingston Range during basin Members II, III, and IV time is not known, but must have been considerable. North-derived slide blocks occur at the base of Member II (Carrara-Bonanza King allochthon), and within Members III and IV time (Valjean allochthon and Kingston Wash megablock respectively). Even after removal of the Kingston Peak pluton's country rock roof and upper levels of the pluton itself by erosion and large-scale mass-wasting (Topping, 1993), the present relief on the south flank of the Kingston Range is 1.6 km.

4.2. Controls on initial detachment of the Shadow Valley slide blocks

Much attention is paid in the analysis of slide block emplacement to the mechanics of sliding. Much less attention seems to be paid to the other half of slide block genesis, i.e. reasons for or controls on the initial detachment of the slide blocks from their bedrock source areas. In most of the Shadow Valley slide blocks, preexisting stratigraphy and structure are cut off discordantly at the base of the blocks. The major exceptions to this observation are the Carrara/Bonanza King allochthon and megablocks within the Tertiary section of the Valjean allochthon where initial block detachments occurred near the top of Middle Cambrian Carrara red shales that underlie a several kilometer-thick section of Paleozoic carbonate rocks. This juxtaposition of competent over incompetent strata was also responsible in Mesozoic time for widespread younger over older detachments in the Clark Mountains thrust belt (Burchfiel and Davis, 1971, 1988).

In the Bearpaw Mountains, Montana, Gucwa and Kehle (1978) suggested that periods of creep led to weakening of the basal decollément of a very large ($500 \ km^2$) slide mass above the Colorado Shale, lowering the angle of internal friction from 7° to less than 3°. They presented evidence that the mudstone detachment had a rheology similar to salt, possibly through work-softening associated with creep. Slow creep above 10–25° dipping shale substrates has also been advanced as one possible mechanism for the detachment of large slide blocks off the Dakota Group sandstones along the Colorado Front Range during the past 30 Ma (Braddock and Eicher, 1962; Braddock, 1978). This phenomenon might well have facilitated initial detachment of the Shadow Valley Carrara-Bonanza King block in Carrara shales, but would on lithologic grounds be inapplicable to most others.

The problem of detachment appears especially vexing when the gravity-driven sheet consists principally of Proterozoic gneisses, the foliation of which is cut discordantly by the base of the allochthonous sheet. Geologic relations at Shadow Mountain (Fig. 5) offer an explanation for how initial detachment of an Eastern Star allochthon might have occurred. Mention has been made of a west-dipping thrust fault in Shadow Mountain that is presumably an imbricate splay of the Winters Pass thrust. The Shadow Mountain thrust cuts at a high angle, ca. 60°, to the unconformity between basement gneisses and the Proterozoic Crystal Springs Fm. Eastward ductile displacement of this contact across the fault was approximately 1 km, but drag folding of brecciated footwall Crystal Springs strata indicate a younger phase of brittle, west-directed slip along the fault (Burchfiel and Davis, 1971; Fowler, 1992). Given our interpretation that the SE Eastern Star allochthon was derived from the Shadow Mountain block, the relations just described offer an explanation for its initial detachment. We believe that detachment is best explained by Miocene reactivated downdip slip

of gneiss, diabase, and Proterozoic strata along a west-dipping imbricate splay in the Winters Pass plate. This splay would have been at a higher structural level than the imbricate thrust still exposed at Shadow Mountain. That a pre-existing Mesozoic thrust fault can undergo Cenozoic opposite sense brittle reactivation is documentable at Shadow Mountain and from numerous studies of inversion tectonics in the Cordilleran foreland fold and thrust belt and elsewhere (e.g. Boyer and Hossack, 1992; Constenius, 1996). Ivins et al. (1990) discuss the mechanics of extensional reactivation of abandoned thrusts when the frictional strength of the preexisting structure is significantly less than that of the surrounding rocks. They conclude (op. cit., p. 303) that "Frictional strength ratios of 3 or greater could account for extremely shallow normal faults (dips 10–20°) without consideration of pore pressures in excess of the least principal stress or of principal stress systems rotated away from the gravity vector."

If Shadow Mountain is also a slide block, as most evidence indicates, its initial detachment may have been controlled by mylonitic fabrics in the lower plate of the Winters Pass thrust itself (cf. Fowler, 1992, Pl. 2). The anisotropy of thrust-related and thrust-parallel mylonitic foliation above and below the base of the Winters Pass plate (Burchfiel and Davis, 1971, 1988) might have controlled gravity-induced detachment of those rocks from their location in the Clark Mountains fold-thrust belt.

We conclude that the combination of steep slopes and pre-existing inclined anisotropies, both stratigraphic and structural, led to the initial detachment and downslope movement of the Shadow Valley slide sheets. Rock avalanche deposits interspersed with the various slide sheets unequivocally attest to steep topography and rapid, rockfall-type slope collapse. Slide sheet detachment may have been initiated by deformation leading to increased dip of pre-existing surfaces of anisotropy, by increases in topographic relief during crustal extension, and by seismic activity accompanying slip on the Kingston Peak-Halloran Hills detachment fault. It has been shown at numerous historical sites (Radbruch-Hall, 1978; Scott, 1978; Keefer and Wilson, 1989; Keefer, 1993) that earthquakes have induced landsliding at many scales, and it has been postulated by some workers (e.g. Forshee and Yin, 1995) that seismic accelerations of as little as 0.5 g could induce large-scale block sliding.

We consider it improbable that initial detachments of the Shadow Valley slide blocks were triggered by Miocene increases in pore-fluid pressure along favorable horizons. The diverse blocks consist of impermeable Precambrian to Paleozoic crystalline, quartzitic, and carbonate rocks, and the detachment surfaces that underlie them are generally discordant to foliation, bedding, and varied lithologic units.

4.3. Rate of slide sheet movement

The westward emplacement of parts of the Eastern Star allochthon across a saline playa for up to 9 or 10 km is the Shadow Valley slide block phenomenon most difficult to explain by any mechanism. This large "runout" over horizontal playa deposits and bordering, eroded sandflat strata is, perhaps, the most compelling reason to assume that emplacement of this allochthon was rapid, possibly even catastrophic. However, rapid movement, given the great width of the allochthon as measured in the transport direction, would require both very long slopes from an unreasonably high source area and a very steep angle of initial detachment.

Evidence favoring slow rates of movement for the Eastern Star allochthon has already been cited, including the structurally intact nature of its Proterozoic basement and cover units, intra-allochthon tear faults, and the scattered, but widespread deformation of its overridden lacustrine and fluviatile strata; this latter relationship argues for sustained stress transmission between the allochthon and its substrate. Evidence from Shadow Mountain that originally east-vergent Mesozoic thrust faults can be reactivated by west-directed normal faulting or gravity-induced sliding to produce such an allochthon also argues against rapid allochthon displacement.

Perhaps the most compelling evidence that movement of the Eastern Star allochthon occurred at a slow rate–perhaps mm/yr to m/yr–comes from observations at the base of the klippe on the eastern side of hill 988T approximately 3 km ESE of the Eastern Star mine (Davis et al., 1998). Here, crystalline basement rocks rest on gypsiferous lake sediments that contain dilational gypsum veins within 6–8 m of the allochthon base. Planar veins that are subparallel to the base of the slide block (Figs. 10–12) lie within a soft sediment matrix and yet are not disrupted. We inter-pret these veins, which are commonly 2–6 cm wide, as synkinematic features because obliquely inclined gypsum fibers up to 12 cm long within them have trends and shear senses compatible with the direction and sense of southwestward displacement of the overriding allochthon. Gypsum fiber-in-vein growth indicates both fluid presence and high-fluid pressure in the overridden playa sediments—factors facilitating slide block movement across them. The long length of the fibers, indicative of a protracted dilational and fiber growth process, is incompatible with a catastrophic rate of slide block movement.

Studies of modern block landslides are helpful in accepting such slow rates of Shadow Valley allochthon movement as realistic, but may not be relevant to those examples of the Shadow Valley basin that lacked basal stratigraphic or lithologic control. The Portuguese Bend landslide in coastal Los Angeles, for example, has been active since 1956, when block gliding of Miocene marine strata was initiated along a 6° seaward-dipping bentonitic stratum. Total maximum slip during the period 1956–1986 was 215 m (Pipkin and Trent, 1997), for an average of about 7 m/yr. The driving effect of mass at the head of the slide was demonstrated in 1986–1987, when the removal of rock material from the top of the slide block and its transport to the toe reduced the slide's average daily displacement from 17.5 mm/day to 9.0 mm/day.

The Miocene Modelo Fm. hosts landslides in various parts of the Los Angeles basin. Most of these slides occurred above a preconsolidated marine shale detachment, some of which dip as shallowly as 9°. Most slides moved episodically, usually during periods of heavy rain, with varying strain rates. Scott (1978) described a 1969 landslide in Miocene–Pliocene beds of the Modelo Fm.at Mt. Washington, Los Angeles. The slide slipped along a shale horizon dipping 13° at an overall rate of more than 10 m in 60 days. Slip recurrence was extremely regular in both time and magnitude, which Scott attributed to repeated build-ups of shear stress until basal frictional resistance was overcome and sliding occurred. Although in many ways not well constrained, Scott's (1978) data provide rates of episodic sliding from 50 cm/day to less than 1 cm/day. If his observations are applicable to larger gravity-driven block slides, then an average slip rate of 10m/yr is not inconceivable for slide blocks like those in Shadow Valley. Even at half this rate of transport, it would take only 1600 yr for a slide block to move 8 km from its bedrock source.

4.4. Mechanics of slide sheet movement

Slow movement of the Shadow Valley slide sheets over the land surface poses severe mechanical problems of how frictional resistance at the base of the slide sheets was overcome. Localized very high fluid pressures (ca. 1 λ) within gypsiferous playa sediments under the Eastern Star allochthon are indicated by gypsum fibers developed in synkinematic dilational veins. But, in general, geologic relations of the Shadow Valley slide blocks suggest that the role of elevated fluid pressures in facilitating their transport of most of the blocks was probably not substantial. For example, the Francis, 1109, and Valley Wells slide sheets moved across and came to rest on poorly sorted, coarse-grained fan deposits. It is difficult to imagine that high fluid pressures could have been generated in such coarse clastic deposits in the arid or semi-arid environments of block sliding. Furthermore, even if such pressures were generated, the friction carpets at the base of the various blocks should have served as permeable conduits for fluid pressure release. A possible solution to this dilemma exists. Perhaps fluid pressures were built up only in a thin seal of finely comminuted materials beneath the sliding block, not in the underlying coarse clastic sediments or in the overlying carbonate breccias. The possibility of fluid pressure-enhanced sliding of the northern Eastern Star, Valjean, and Carrara-Bonanza King allochthons is even less, given that all overrode eroded land surfaces across tilted volcanic and sedimentary strata. Fluid-pressure buildup in the basal Carrara shales of the latter allochthon is also highly unlikely given their Cambrian age of initial lithification.

Cohesive creep might have been an important slip mechanism for some of the slide sheets (cf. Boyer and Hossack, 1992). Creep occurs in clays under extremely low stresses (Kavazanjian and Mitchell, 1980) that are substantially less than the overburden associated with 100 m of rock. Creep strain rates range from about 2 mm/yr to 0.5 mm/day, within the observed range of modern and ancient slides. The episodic nature of sliding seen in some modern stu-

dies may also be related to episodic creep (Scott, 1978). Given the high deviatoric stresses associated with slide block loading, creep is at least a plausible transport mechanism that has not been thoroughly explored. Experiments in a variety of media more similar to the alluvial fan sediments found below some slide blocks would help to test this mechanism.

The enigma of the mechanics of slide block displacement is most perplexing for the Eastern Star crystalline allochthon with its large runout distance over the Miocene landscape (Fig. 5). We believe that the only explanation for slow displacement of a large allochthon such as the Eastern Star crystalline sheet across a northern eroded landscape and a southern playa is that the sheet had an elevated head that drove its toe across lower slope areas. A possible scenario for the tectonic setting of the Eastern Star allochthon is presented as Fig. 16, illustrating normal fault-generated topography at the eastern edge of a half-graben containing saline playa sediments and intertonguing alluvial fans; the topography of the modern Black Mountain front above the eastern playa deposits of Death Valley is used as as the topographic format for this figure. Tectonic oversteepening of slope in Proterozoic bedrock units by normal faulting could have lead to gravity-induced failure of the slope (cf. Boyer and Hossack, 1992, their Fig. 15), with slump failure controlled by the anisotropy of a pre-existing, basinward-dipping Mesozoic thrust fault (inferred to have been present at Shadow Mountain). The detached landslide block then moved across the bounding normal fault and out onto playa and fan sediments. Subsequent slumpings of the toe of the slide block along intra-allochthon listric normal faults could have extended the allochthon and allowed it to move ever farther away from its source area (Fig. 16). The driving force for movement of the allochthon comes from the gravitational head provided by the original height of the allochthon above the playa surface and by the initial slope of detachment as governed by the dip of the preexisting thrust fault (Davis et al., 1998).

This hypothesis can be considered in light of topographic relief and slope length in modern extensional tectonic settings such as the Black Mountains rising above Death Valley. The average slope angle and approximate relief between the eastern margin of Death Valley and Dantes View in the Black Moun-

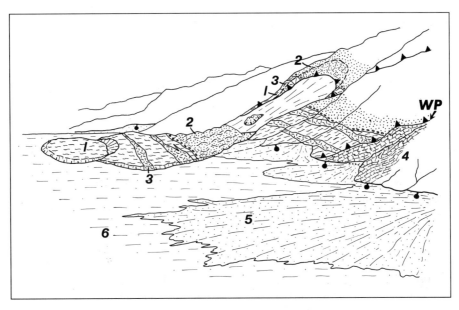

Fig. 16. Possible origin of Eastern Star crystalline slide sheet by southwestward gravitational inversion of the upper plate of a thrust fault in the hanging wall of the Mesozoic Winters Pass thrust (WP). This geologic sketch, based on the pre-Tertiary geology of Shadow Mountain (see text), is superposed on an aerial view of the topography of eastern Death Valley half graben and the adjoining Black Mountains. Units: 1=Proterozoic gneisses; 2=Proterozoic Crystal Spring Fm.; 3=ca 1.1 Ga diabase; 4=Cambrian Carrara Fm. in lower plate of Winters Pass thrust; 5=Miocene alluvial fan deposits; 6=Miocene playa deposits (predominantly siltstones, in part gypsiferous).

tains are respectively 25° and 1.7 km; a single mountain front slump block that would move completely out onto the adjacent valley floor could, thus, not have a runout length greather than the slope length, ca. 4.4 km. Farther south in the Black Mountains, the length of slope between sea level in Death Valley and the 1945 m summit of Funeral Peak is about 10 km, but the average slope is only ca. 12° and perhaps not steep enough to yield to slumping even with preexisting favorably oriented stratigraphic or structural anisotropies.

Although elevated fluid pressures may have been locally generated in water-saturated playa silts and clays of Shadow Valley Member III beneath the Eastern Star sheet, the widespread folding of such overridden sediments to depths up to several tens of meters indicates that the contact between allochthon and its footwall was generally not frictionless. Similar deformation of consolidated footwall fan deposits is even more perplexing, as is the observation that the northern part of the allochthon moved across the lower slopes of an eroded alluvial sandflat environment that probably had topographic asperities (hills) with relief up to several tens of meters. Tear faults in the

allochthon indicating its differential forward transport may have formed between asperity-rich and asperity-poor landscapes.

Parke (1994) has proposed that the friction carpets beneath Shadow Valley slide blocks may facilitate their displacement. This interesting concept is that the thick, intact upper portions of the various slide sheets slip (creep) faster downhill than their basal breccias, which would be shearing and rotating internally, and being slowed by frictional resistance at the base of the block (the analogy of moving large quarried blocks of rock across a substrate of rotating logs or ball-bearings comes to mind). As the intact sheet advances downslope at a rate faster than its basal breccias, upslope breccias are stranded as rubble on the footwall behind the trailing edge of the slide block. This process, which might be considered a very coarse variant of cataclastic flow, diminishes the mass and thickness of the moving allochthon from below. New friction carpet breccias continually form beneath the moving allochthons at the expense of the intact cap. Eventually the allochthon comes to rest, perhaps (1) because it no longer has sufficient head to drive it, (2) the underlying slope is no longer steep enough to generate sufficient shear stresses at the base of the block, (3) the diminished mass (and thickness) of the block due to friction carpet "erosion" lowers shear stresses at its base, or (4) simply because it runs into too large a positive topographic asperity.

5. Conclusions

The Shadow Valley basin slide blocks are diverse and spectacular examples of large-scale gravity tectonics at work in an active supradetachment basin. The extensional and plutonic environment in which the Kingston Range-Halloran Hills detachment fault formed from approximately 13.5 to 7 mybp, spawned steep slopes in diverse structural settings that led to the high energy filling of the basin by debris flows, rocks avalanches, and gravity slide sheets. Although rock avalanche deposits, both in the Miocene basin and in modern settings, form catastrophically, there is considerable geologic evidence that at least some of the large intact slide sheets of the Shadow Mountains area were emplaced slowly, possibly at rates of millimeters to meters per year. High pore-fluid pressures appear not to have played a role in either initial detachments of the slide sheets from their bedrock sites, nor displacements of the slide sheets across both depositional and erosional surfaces within the basin. In some cases, significant footwall deformation below the gravity-driven allochthons indicates that their movement over their substrates was not frictionless. We believe that the strongest explanation for the detachment and slow displacement of the Shadow Valley slide sheets, including the most extensive allochthon—the Eastern Star sheet of Precambrian crystalline rocks—is that they formed as slumps on high, steep bedrock slopes and that their elevated heads drove their toes across lower fan and playa deposits. Initial detachments from bedrock sources were facilitated by pre-existing structural and stratigraphic anisotropies.

Acknowledgements

This paper represents one aspect of a study of the stratigraphy, structure and tectonics of the Shadow

Valley basin by Davis and a team of students at the University of Southern California. Apart from the authors, that team included Kim M. Bishop, T. Kenneth Fowler, Jr., and Mary L. Parke, to whom we extend our appreciation for their efforts and their scientific contributions. The fieldwork by this team was largely supported by the National Science Foundation through its grants EAR-9005588 and EAR-9205711 to Davis. In addition, Friedmann acknowledges the support of NSF grants EAR-9118610 and EAR-9317066 awarded to D.W. Burbank, and the USC Department of Geological Sciences Graduate Research Fund. Brian Darby assisted in preparation of several figures. Helpful manuscript reviews by Darrel Cowan, John Crowell, and Martin Miller are greatly appreciated, as is the invitation and encouragement of James Calzia to participate in this review volume on the geology of the greater Death Valley area, California.

References

Bishop, K.M., 1994. Mesozoic and Cenozoic extensional tectonics in the Halloran and Silurian Hills, eastern San Bernardino County, California. PhD thesis, University of Southern California, Los Angeles.

Bishop, K.M., 1997. Miocene rock-avalanche deposits, Halloran/ Silurian Hills area, southeastern California. Environmental and Engineering Geoscience III, 501–512.

Boyer, S.E., Hossack, J.R., 1992. Structural features and emplacement of surficial gravity-slide sheets, northern Idaho-Wyoming thrust belt. In: Link, P.K., Kuntz, M.A., Platt, L.B. (Eds.), Regional geology of eastern Idaho and western Wyoming, Memoir-Geological Society of America, vol. 179, pp. 197–213.

Braddock, W.A., 1978. Dakota Group rockslides, northern Front Range, Colorado, U.S.A. In: Voight, B. (Ed.), Rockslides and Avalanches: 1. Natural Phenomena, Developments in Geotechnical Engineering, vol. l4A. Elsevier, Amsterdam, pp. 439–480.

Braddock, W.A., Eicher, D.L., 1962. Block-glide landslides in the Dakota Group of the Front Range foothills, Colorado. Geological Society of America Bulletin 73, 317–324.

Burchfiel, B.C., Davis, G.A., 1971. Clark Mountain thrust complex in the Cordillera of southeastern California: geologic summary and field trip guide. University of California Riverside Campus Museum Contributions 1, 1–28.

Burchfiel, B.C., Davis, G.A., 1988. Mesozoic thrust faults and Cenozoic low-angle normal faults, eastern Spring Mountains, Nevada, and Clark Mountains, California. In: Weide, D.C., Faber, M.L. (Eds.), This Extended Land, Geological Journeys in the Southern Basin and Range, Geological Society of America, Cordilleran Section, Field Trip Guidebook, vol. 2. UNLV Dept. of Geoscience Spec. Pub., pp. 87–106.

Calzia, J.D., 1990. Geologic studies in the Kingston Range, southern Death Valley, California. PhD thesis, University of California, Davis, California.

Cladouhos, T.T., 1999. A kinematic model of deformation within brittle shear zones. Journal of Structural Geology 21, 437–448.

Constenius, K.N., 1996. Late Paleogene extensional collapse of the Cordilleran foreland fold and thrust belt. Geological Society of America Bulletin 108, 20–39.

Davis, G.A., Burchfiel, B.C., 1973. The Garlock fault—an intracontinental transform structure. Geological Society of America 84, 1407–1422.

Davis, G.A., Burchfiel, B.C., 1993. Tectonic problems revisited: the eastern terminus of the Miocene Garlock fault and the amount of slip on the southern Death Valley fault zone. Abstracts with Programs-Geological Society of America 25, 28.

Davis, G.A., Parke, M.A., Bishop, K.M., Fowler, T.K., Friedmann, S.J., 1991. Grand scale detachment and emplacement of gravity-driven slide sheets into a Miocene terrestrial basin, eastern Mojave Desert, California. Abstracts with Programs-Geological Society of America 23, A467.

Davis, G.A., Fowler, T.K., Bishop, K.M., Brudos, T.C., Friedmann, S.J., Burbank, D.W., Parke, M.A., Burchfiel, B.C., 1993. Pluton pinning of an active Miocene detachment fault system, eastern Mojave Desert, California. Geology 21, 627–630.

Davis, G.A., Friedmann, S.J., Parke, M.A., 1998. Mode of emplacement of a gravity-driven "crystalline allochthon" in the Late Miocene Shadow Valley basin, eastern Mojave Desert, California. Abstracts with Programs-Geological Society of America 30, 11.

DeWitt, E.H., 1980. Geology and geochronology of the Halloran Hills, southeastern California, and implications concerning Mesozoic tectonics of the southwestern Cordillera. PhD thesis, Pennsylvania State University, University Park, Pennsylvania.

Fleck, R.J, Mattison, J.M., Busby, C.J., Carr, M.D., Davis, G.A., Burchfiel, B.C., 1994. Isotopic complexities and the age of the Delfonte volcanics rocks, eastern Mescal Range, southeastern California: Stratigraphic and tectonic implications. Geological Society of America Bulletin 106, 1242–1253.

Forshee, E.J., Yin, A., 1995. A model for initiation of a mid-Tertiary rock avalanche in the Whipple Mountains area, SE California: Implications for seismicity along low-angle normal faults. Basin Research 7, 347–350.

Fowler Jr., T.K., 1992. Geology of Shadow Mountain and the Shadow Valley basin: Implications for Tertiary tectonics of the eastern Mojave Desert. MS thesis, University of Southern California, Los Angeles.

Fowler, T.K., Friedmann, S.J., Davis, G.A., Bishop, K.M., 1995. Two-phase evolution of the Shadow Valley Basin, south-eastern California: a possible record of footwall uplift during extensional detachment faulting. Basin Research 7, 165–179.

Friedmann, S.J., 1995. The Shadow Valley basin, eastern Mojave Desert, California: the structural and stratigraphic evolution of a supradetachment basin and its implications for extensional tectonics. PhD thesis, University of Southern California, Los Angeles.

Friedmann, S.J., 1997. Rock-avalanche elements of the Shadow Valley basin, eastern Mojave Desert, California: processes and problems. Journal of Sedimentary Research 67, 792–804.

Friedmann, S.J., 1999. Sedimentology and stratigraphy of the Shadow Valley basin, eastern Mojave Desert, California. In: Wright, L.A., Troxel, B.W. (Eds.), Cenozoic Basins of the Death Valley Region, Special Paper-Geological Society of America, vol. 333, pp. 213–243.

Friedmann, S.J., Burbank, D.W., 1995. Rift basins and supradetachment basins: intracontinental extensional endmembers. Basin Research 7, 109–127.

Friedmann, S.J., Davis, G.A., Fowler, T.K., Brudos, T., Burbank, D.W., Burchfiel, B.C., 1994. Stratigraphy and gravity-glide elements of a Miocene supradetachment basin, Shadow Valley, eastern Mojave Desert. In: McGill, S.F., Ross, T.M. (Eds.), Geological Investigations of an Active Margin. Cordilleran Section Guidebook. Geological Society of America, Boulder, CO, pp. 302–320.

Friedmann, S.J., Davis, G.A., Fowler, T.K., 1996. Geometry, paleodrainage, and geologic rates from the Miocene Shadow Valley supradetachment basin, eastern Mojave Desert, California. In: Beratan, K.K. (Ed.), Reconstructing the History of Basin and Range Extension Using Sedimentology and Stratigraphy, vol. 303. Special Paper, Geological Society of America, Boulder, CO, pp. 85–105.

Gucwa, P.R., Kehle, R.O., 1978. Bearpaw Mountain Rockslide, Montana, U.S.A. In: Voight, B. (Ed.), Rockslides and Avalanches: 1. Natural Phenomena, Developments in Geotechnical Engineering, vol. 14A. Elsevier, Amsterdam, pp. 393–421.

Heaman, L.M., Grotzinger, J.P., 1992. 1.08 Ga diabase sills in the Pahrump Group, California: implications for development of the Cordilleran miogeocline. Geology 20, 637–640.

Hewett, D.F., 1955. Structural features of the Mojave Desert region. Special Paper-Geological Society of America 62, 377–390.

Hewett, D.F., 1956. Geology and mineral resources of the Ivanpah quadrangle. U.S. Geological Survey Professional Paper 275.

Hubbert, M.K., Rubey, W.W., 1959. Role of fluid pressure in mechanics of overthrust faulting: I. Mechanics of fluid-filled porous solids and its application to overthrust faulting. Geological Society of America Bulletin 80, 115–166.

Ivins, E.R., Dixon, T.H., Golombek, M.P., 1990. Extensional reactivation of an abandoned thrust: a bound on shallowing in the brittle regime. Journal of Structural Geology 12, 303–314.

Kavazanjian Jr., E., Mitchell, J.K., 1980. Time-dependant deformation behavior of clays. Journal of Geotechnical Engineering, 611–630.

Keefer, D.K., 1993. The susceptibility of rock slopes to earthquake-induced failure. Bulletin of the Association of Engineering Geologists 30.

Keefer, D.K., Wilson, R.C., 1989. Predicting earthquake-induced landslides, with emphasis on arid and semi-arid environments. In: Sadler, P.M., Morton, D.M. (Eds.), Landslides in a Semi-Arid Environment with Emphasis on the Inland Valleys of Southern California, Inland Geological Society, vol. 2, pp. 118–149.

Kehle, R.O., 1970. Analysis of gravity gliding and orogenic translation. Geological Society of America Bulletin 81, 1641–1664.

Lasky, S.G, Webber, B.N., 1949. Manganese resources of the Artillery Mountains region, Mohave County, Arizona. U.S. Geological Survey Bulletin 961, 86.

Miller, M.G., Pavlis, T.L., this volume. The Black Mountains turtle-backs: complicated Rosetta Stones of Death Valley tectonics. Earth-Science Reviews.

Parke, M.L., 1994. Geology and structural evolution of the southern Shadow Mountains, San Bernardino County, California. MS thesis, University of Southern California, Los Angeles.

Pipkin, B.W., Trent, D.D., 1997. Geology and the Environment. West/Wadworth, Belmont, CA.

Radbruch-Hall, D.H., 1978. Gravitational creep of rock-masses on slopes. In: Voight, B. (Ed.), Rockslides and Avalanches: 1. Natural Phenomena, Developments in Geotechnical Engineering, vol. l4A. Elsevier, Amsterdam, pp. 607–657.

Ramsay, J.G., Huber, M.I., 1983. Strain analysis. The Techniques of Modern Structural Geology, vol. 1. Academic Press, London.

Scott, R.F., 1978. Incremental movement of a rockslide. In: Voight, B. (Ed.), Rockslides and Avalanches: 1. Natural Phenomena, Developments in Geotechnical Engineering, vol. l4A. Elsevier, Amsterdam, pp. 659–668.

Topping, D.A., 1993. Paleogeographic reconstruction of the Death Valley extended region: evidence for Miocene large rock-avalanche deposits in the Amargosa Chaos basin, California. Geological Society of America Bulletin 105, 1190–1213.

Wright, L.A., Otton, J.K, Troxel, B.W., 1974. Turtleback surfaces of Death Valley viewed as phenomena of extensional tectonics. Geology 2, 53–54.

Yarnold, J.C., 1993. Rock-avalanche characteristics in dry climates and the effect of flow into lakes: insights from the mid-Tertiary sedimentary breccias near Artillery Peak, Arizona. Geological Society of America Bulletin 105, 345–360.

Yin, A., Dunn, J.F., 1992. Structural and stratigraphic development of the Whipple-Chemehuevi detachment fault system, southeastern California: implications for the geometrical evolution of domal and basinal low-angle normal faults. Geological Society of America Bulletin 104, 659–674.

Available online at www.sciencedirect.com

Earth-Science Reviews 73 (2005) 177–219

www.elsevier.com/locate/earscirev

Late Cenozoic sedimentation and volcanism during transtensional deformation in Wingate Wash and the Owlshead Mountains, Death Valley

Heather Golding Luckow [a,*], Terry L. Pavlis [a], Laura F. Serpa [a], Bernard Guest [a], David L. Wagner [b], Lawrence Snee [c], Tabitha M. Hensley [d], Andrey Korjenkov [e]

[a] *Department of Geology and Geophysics, University of New Orleans, New Orleans, LA 70148, USA*
[b] *California Geological Survey, 801 K St. MS 12-31, Sacramento, CA 95814, USA*
[c] *U. S. Geological Survey, MS 974, Federal Center, Denver, CO 80225-0046, USA*
[d] *Scripps Institution of Oceanography, 9500 Gilman Drive Hall Mail Code 0208, La Jolla, CA 92093, USA*
[e] *Institute of Seismology NAS, Asanbay 52/1, Bishkek 720060, Kyrghyzstan*

Abstract

New 1:24,000 scale mapping, geochemical analyses of volcanic rocks, and Ar/Ar and tephrochronology analyses of the Wingate Wash, northern Owlshead Mountain and Southern Panamint Mountain region document a complex structural history constrained by syntectonic volcanism and sedimentation. In this study, the region is divided into five structural domains with distinct, but related, histories: (1) The southern Panamint domain is a structurally intact, gently south-tilted block dominated by a middle Miocene volcanic center recognized as localized hypabyssal intrusives surrounded by proximal facies pyroclastic rocks. This Miocene volcanic sequence is an unusual alkaline volcanic assemblage ranging from trachybasalt to rhyolite, but dominated by trachyandesite. The volcanic rocks are overlain in the southwestern Panamint Mountains by a younger (Late Miocene?) fanglomerate sequence. (2) An upper Wingate Wash domain is characterized by large areas of Quaternary cover and complex overprinting of older structure by Quaternary deformation. Quaternary structures record ~N–S shortening concurrent with ~E–W extension accommodated by systems of strike-slip and thrust faults. (3) A central Wingate Wash domain contains a complex structural history that is closely tied to the stratigraphic evolution. In this domain, a middle Miocene volcanic package contains two distinct assemblages; a lower sequence dominated by alkaline pyroclastic rocks similar to the southern Panamint sequence and an upper basaltic sequence of alkaline basalt and basanites. This volcanic sequence is in turn overlain by a coarse clastic sedimentary sequence that records the unroofing of adjacent ranges and development of ~N–S trending, west-tilted fault blocks. We refer to this sedimentary sequence as the Lost Lake assemblage. (4) The lower Wingate Wash/northern Owlshead domain is characterized by a gently north-dipping stratigraphic sequence with an irregular unconformity at the base developed on granitic basement. The unconformity is locally overlain by channelized deposits of older Tertiary(?) red conglomerate, some of which predate the onset of extensive volcanism, but in most of the area is overlain by a moderately thick package of Middle Miocene trachybasalt, trachyandesitic, ash flows, lithic tuff, basaltic cinder, basanites, and dacitic pyroclastic, debris, and lahar flows with localized exposures of sedimentary rocks. The upper part of the Miocene stratigraphic sequence in this domain is

---

* Corresponding author.
  E-mail address: lsnee@USGS.GOV (L. Snee).

comprised of coarse grained-clastic sediments that are apparently middle Miocene based on Ar/Ar dating of interbedded volcanic rocks. This sedimentary sequence, however, is lithologically indistinguishable from the structurally adjacent Late Miocene Lost Lake assemblage and a stratigraphically overlying Plio-Pleistocene alluvial fan; a relationship that handicaps tracing structures through this domain. This domain is also structurally complex and deformed by a series of northwest–southeast-striking, east-dipping, high-angle oblique, sinistral, normal faults that are cut by left-lateral strike-slip faults.

The contact between the southern Panamint domain and the adjacent domains is a complex fault system that we interpret as a zone of Late Miocene distributed sinistral slip that is variably overprinted in different portions of the mapped area. The net sinistral slip across the Wingate Wash fault system is estimated at 7–9 km, based on offset of Proterozoic Crystal Springs Formation beneath the middle Miocene unconformity to as much as 15 km based on offset volcanic facies in Middle Miocene rocks. To the south of Wingate Wash, the northern Owlshead Mountains are also cut by a sinistral, northwest-dipping, oblique normal fault, (referred to as the Filtonny Fault) with significant slip that separates the Lower Wingate Wash and central Owlshead domains. The Filtonny Fault may represent a young conjugate fault to the dextral Southern Death Valley fault system and may be the northwest-dipping fault imaged by COCORP studies. Similarly, younger deformation in upper Wingate Wash is probably broadly related to distributed dextral shear along the Panamint Valley fault system. Earlier deformation (Late Miocene?) is more difficult to constrain because of overprinting but appears to be dominated by an E–W extension recognized by a NNW-striking, northeast-dipping, sinistral-oblique normal faults, ~N–S striking normal faults that splay in the northern Owlshead Mountains and include the large west-tilted fault blocks of the northern Owlshead Mountains.

Keywords: Death Valley structure; transtension; Wingate Wash; Owlshead Mountains; argon geochronology; tectonics

1. Introduction

The Death Valley region represents one of the premier sites where strike-slip deformation is occurring contemporaneously with crustal extension. Death Valley itself was first recognized as a pull-apart basin more than 35 yr ago (Burchfiel and Stewart, 1966), and since that time much has been learned about the complex structural interactions that accompany "transtension". In particular, pull-apart basin development is only one of a variety of tectonic processes that accompany transtension (e.g. Tikoff et al., 1994; Dewey, 2001).

Wingate Wash, located in southern Death Valley, is a topographic low between the Panamint and Owlshead Mountains. This area lies in an important transition between the highly extended central Death Valley region and less extended rocks of the northern Mojave Desert (Fig. 1). The extension across the southern Death Valley region was highly oblique, transtensional deformation, and thus, the tectonic evolution involved 3-D interplay of a variety of structures including major detachment faults, folding of faults, and the development of transrotational blocks (e.g. Wernicke, 1988; Wernicke et al., 1988; Wright et al., 1991; Holm et al., 1993; Mancktelow and Pavlis, 1994; Serpa and Pavlis, 1996; Korjenkov et al., 1996; Snow and Wernicke, 2000). Previous attempts to reconstruct this large-scale Neogene transtension have disagreed markedly on the magnitude of extension as well as strike-slip displacement across this region. For example, Snow and Wernicke (2000) restore the Owlshead Mountains and southern Panamint Mountains to a thin (~5 km wide) crustal slice that was dismembered by extension and strike-slip motion to its present form after latest Miocene (8–10 Ma) time. In contrast, Serpa and Pavlis (1996) contended that only minor extension occurred across the Owlshead Mountains and differential motion between the Owlshead block and highly extended terranes to the north implied that a sinistral fault, the Wingate Wash fault, transferred this displacement to the sinistral Garlock fault (Fig. 1) system to the south.

Given the magnitude of the discrepancy between these two tectonic reconstructions and the fact that the Wingate Wash region was known only from reconnaissance mapping (Wagner and Hsu, 1988), we began a detailed mapping program across Wingate Wash between 1996 and 2000. The result of part of that work is reported in this paper with an emphasis on the stratigraphic and structural evolution of the late Cenozoic deposits in Wingate Wash, the northern Owlshead Mountains, and the southern Panamint Mountains. The geologic map (Pavlis et al., in preparation) and a regional tectonic interpretation (Guest et al., submitted for publication) are reported elsewhere.

Fig. 1. LandSat image of southern Death Valley and northern Mojave Desert. Domains labeled as EOWL/LWW: eastern Owlshead Mountains and lower Wingate Wash; NCOWL/CWW: north-central Owlshead Mountains and central Wingate Wash; UWW: upper Wingate Wash; SPMN: southern Panamint Mountains.

We begin with a description of the general geology of the Wingate Wash region and report on new stratigraphic and geochronological data from the syntectonic basin systems developed in the region. These data, together with geologic mapping, reveal a complex history of basin development coincident with multiple stages of faulting that appear to record a complexly evolving transtensional shear system developed across the axis of Wingate Wash. Extensional magnitudes, however, are relatively low and we conclude that any restoration of the southern Panamint and Owlshead Mountains that requires more than ~50% extension after ~12 Ma cannot be reconciled with the field data from the Wingate Wash region. Instead, we propose a tectonic model of sinistral transtension that led to complex fault-block interactions, and transfer of displacements northward into the Panamint Mountains and Death Valley extensional systems.

2. Geologic setting

Wingate Wash is in the southern Basin and Range province, immediately south and west of Death Valley. Prior to Neogene extension, the region was within the Neoproterozoic to Paleozoic passive margin of southwestern North America, and in Mesozoic time, this passive margin was destroyed by contractional structures of the Cordilleran orogen (e.g. Burchfiel et al., 1992). During the Mesozoic, the region was also located at the eastern edge of the Cordilleran magmatic arc, and thus, large parts of the region are underlain by Mesozoic plutonic rocks. All of these rocks were then deeply eroded between latest Mesozoic to middle Cenozoic time to develop a regional unconformity beneath late Cenozoic cover that ranges from a nonconformity to an angular unconformity.

Wingate Wash and the surrounding region were strongly deformed during late Neogene extension and transtension, but the nature of this deformation varies spatially and is debated. To the east, across Death Valley, the Black Mountains have been intensely deformed by late Miocene extension and plutonism (e.g. Wright et al., 1991; Holm et al., 1993; and Miller and Pavlis, 2005—this volume). In the central Black Mountains, low angle detachment fault systems have largely stripped the sedimentary cover and in their footwalls expose a ductilely deformed metamorphic complex as well as Miocene (~11 to 9 Ma) syntectonic plutons that were emplaced at depths of ~10 km or less (Holm et al., 1993; Miller, 1999; Meurer, 1992; Wright et al., 1991). These detachment systems carry brittlely deformed miogeoclinal strata, as well as syntectonic sedimentary and volcanic strata, in their hanging walls; a deformational complex generally referred to as the Amargosa chaos (Noble, 1941; Wright and Troxel, 1973; Wright et al., 1991). Timing relationships of syntectonic cover (e.g. Wright et al., 1991; Topping, 1993) as well as cooling ages from the footwall (e.g. Holm et al., 1993) indicate that extension east of Death Valley was underway by ~11 Ma when the Black Mountains intrusives were emplaced and was largely complete by ~6 Ma when the present pull-apart basin became fully established (e.g. Wright et al., 1991; Topping, 1993; Holm et al., 1993; Holm, 1994; Serpa and Pavlis, 1996). Cooling ages from the Badwater turtleback as well as syntectonic deposits in the hanging-wall suggest that extensional deformation had actu-

ally begun by ~14 Ma (Miller and Prave, 2002). The cover rocks of the central Black Mountains were close to the southern Panamint Range, prior to extension, because basins in the Amargosa chaos contain volcanic rocks that are correlative with rocks in the Wingate Wash area (see below) as well as sedimentary deposits containing clasts derived from the southern Panamint Range (Topping, 1993).

In contrast to the extreme tectonic attenuation of the central Black Mountains, areas immediately to the north, south, and west of the Wingate Wash area are not as complexly deformed by Neogene deformation. To the west of the Owlshead–Panamint block (Fig. 1), the Panamint Valley fault system transfers dextral slip into the Plio-Pleistocene Panamint Valley pull-apart basin; a pull-apart analogous in age and scale to the Death Valley basin, but superimposed on a less deformed pre-Pliocene extensional system (Burchfiel et al., 1992). To the south, the Owlshead Mountains are relatively weakly deformed internally (see below), but are juxtaposed against rocks of the northeast Mojave Desert along a major sinistral fault, the Garlock fault. The Garlock fault has a well-documented sinistral slip of ~64 km (Burchfiel and Davis, 1972), but poses significant displacement complexities at its eastern end where the fault intersects the dextral Death Valley system. Much of this displacement is probably accommodated by transrotation in the northeastern Mojave Desert (e.g. Dokka and Travis, 1990; Pavlis and Serpa, 1997), but transrotation would also have to be transferred into the southern Death Valley region, including the Owlshead Mountains. However, the relative importance of transrotation versus extension in accommodation of the incompatibility is debated (e.g. compare Serpa and Pavlis, 1996; Snow and Wernicke, 2000). Finally, to the north of Wingate Wash the Panamint Mountains block appears relatively intact, but is complexly deformed internally. At the north tip of the Panamint Mountains, two generations of detachment faults-Tucki Wash and Emigrant faults-have displaced what is now the Cottonwood Mountains from an initial position above the northern Panamint Mountains (Wernicke et al., 1988). The connection between these detachment faults and faults to the east, across Death Valley, are controversial. Moreover, the connection between detachment faulting in the northern Panamints and deformation in the southern Panamint Mountains is not well resolved.

In this paper, we present new data to help resolve some of these questions.

3. Methods

The Wingate Wash region has only been examined in reconnaissance fashion by earlier studies (Wagner and Hsu, 1988). In this study, we mapped the region at 1:24,000 using standard U.S.G.S. topographic maps and orthophotos with accompaning 1:24,000 aerial photography for stereographic viewing. Most of our field notes and photographs are captured on handheld computers used the commercial software package Fieldworker. Data were transferred to the public domain software package Fieldlog that operates under the database engine of the commercial software package AutoCAD. Station positions were geo-referenced using Global Positioning System (GPS) units. Most GPS data were captured without differential correction. However, Fieldworker software averages positions for periods of 2–10 min per station. Most of the fieldwork was conducted while selective availability was still in force, and thus, this averaging was necessary to achieve position accuracies to ~10–15 m. Field maps were digitized using the commercial package AutoCAD, and tied to the database using Fieldlog. These data were subsequently exported in arcinfo format and will be released as a digital map through the California Division of Mines and Geology (Pavlis et al., in preparation).

During field mapping, we also measured a series of Tertiary stratigraphic sections where exposure permitted. Most of these sections were measured using standard Jacob staff methods, but two of the sections (upper Wingate Wash, Wingate Wash NW, and Owlsbeak) were measured using sub-meter differential GPS and laser-ranging binoculars. In these sections, contact positions in the measured sections were recorded (or converted) to UTM positions and converted to unit thicknesses using standard trigonometric relationships.

4. General geology of Wingate Wash

The geology of Wingate Wash is dominated by a Middle Miocene volcanic assemblage unconformably overlying Mesozoic granitoids and Neoproterozoic to Paleozoic sedimentary rocks (Figs. 2 and 3). This volcanic assemblage was deposited coeval with syntectonic processes, which is illustrated by early extensional/transtensional structures that grade upward into a complex Late Miocene sedimentary basin system developed within structural basins.

The middle Miocene rocks are primarily an alkaline volcanic complex ranging in composition from basalt to rhyolite, with the majority classified as trachyte, trachybasalt, trachyandesite, and trachy dacite. To the north of Wingate Wash, the volcanic rocks were erupted primarily from a large volcano now represented by an intrusive center located ~5 km north of the axis of Wingate Wash (Wagner and Hsu, 1988). To the south, however, no equivalent volcanic sources have yet been recognized aside from a few localized rhyolitic domes and basaltic cinder cones within the volcanic pile (see below).

The structure of Wingate Wash is complex with cross-cutting relationships with at least three generations of fault systems. The gross structure is a broad syncline with an axis parallel to the topographic axis of the wash, with fault systems parallel to, and oblique to, the axis of the wash. Wingate Wash is not, however, a simple synclinal fold, but rather the result of multiple generations of extensional and strike-slip faulting. In order to decipher this complex structural overprint, we divide the region into structural domains: the Southern Panamint Mountains Domain (SPMN), Upper Wingate Wash Domain (UWW), North-Central Owlshead and Central Wingate Domain (NCOWL/CWW) and the Eastern Owlshead and Lower Wingate Domain (EOWL/LWW) (Fig. 1).

4.1. Southern Panamint Mountains domain (SPMN)

4.1.1. General features

In comparison to adjacent regions in the Death Valley extensional terrane, the geology of the southern Panamint Mountains (labeled on Fig. 1 as SPMN) is remarkably intact, with only minor modification by post-Middle Miocene structures. This domain is essentially a gently south-tilted, Middle Miocene volcanic complex dominated by the eroded remains of a large stratovolcano with an overlying Late Miocene (?) sedimentary basin assemblage in upper Wingate Wash (Fig. 2). This assemblage is cut by only a few small ~N–S-striking faults, and we recognized only two faults with offsets >100 m in this domain. These two faults, however, are minor because neither exposes structural levels deeper than the middle Miocene unconformity beneath the volcanic pile (Fig. 2). Thus, the southern Panamint domain has remained a large, intact crustal slab throughout the Late Miocene and younger extensional history of the Death Valley region.

4.1.2. Middle Miocene volcanic assemblage

The dominant feature of the southern Panamint domain occurs in the eastern half of the southern Panamint Mountains (Fig. 2) where, following the interpretation of Wagner and Hsu (1988) and Wagner (1993, 1994), we infer the presence of a large, exhumed volcanic center. In this area, a 2–3 km wide, crudely circular, region is underlain by shallow-level intrusive rocks ranging from dioritic to granitic in composition. Most of these intrusive rocks have been intensely altered by hydrothermal activity and only the youngest, typically rhyolitic intrusives, retain primary minerals unaffected by hydrothermal alteration (Yawn, 1998). This intrusive center is flanked on all sides by proximal-facies pyroclastic volcanic rocks that dip away from the intrusive center in all directions. Dips are steepest on the south flank of the intrusive center (65–80°), are intermediate on the east and west flank (~30°), and are shallow on the north flank (~10°). Thus, the basic structure of layering around the intrusive center is a cone with a moderately north-plunging axis (Fig. 2). These orientation data suggest an original volcanic cone that has been tilted ~40° south along an ~E–W horizontal axis.

Further evidence that this assemblage represents a volcanic center can be seen in the stratigraphy of the volcanic assemblages. The entire volcanic assemblage within ~2 km of the intrusive center is characterized by very coarse-grained pyroclastic rocks, interlayered with a few flows, and locally interlayed with large masses of rhyodacite (Fig. 3, section G and Table 1). Most of the pyroclastic rocks are diamictites or coarse breccias with boulder size clasts up to tens of meters across. We did not measure a detailed section within this proximal facies assemblage because they are lithologically monotonous, but from map relationships these proximal facies deposits are at least 1500

m thick along the western and southern flanks of the intrusive center.

Farther from the intrusive center, these pyroclastic units become more fine-grained with maximum clast sizes typically in cobble to small boulder range (~10 cm to ~1 m). This characteristic is well displayed in the partial section (Fig. 3, section G and Table 1) that we measured on the west flank of the intrusive center. In this section, the entire assemblage is dominated by pyroclastic units, but the maximum clast size is typically <1 m for the bulk of the pile. Farther west in the southern Panamint Mountains, these pyroclastic units become even finer grained and are interlayered with more diverse lava flows, most of which almost certainly were derived from other volcanic centers or fissure eruptions.

From these observations, we infer that this volcanic assemblage records the initial construction of a large (~15 km in diameter) stratovolcano. Successive generations of shallow-level intrusions into the center of this volcanic edifice led to the production of the intrusive center, and shallow circulation of hydrothermal fluids produced intense alteration within the intrusive center. Late in the construction of this volcanic edifice, large rhyolitic intrusives were emplaced within the core of the stratovolcano, and the magma bodies represented by the intrusives probably also fed rhyodacite dome complexes observed on the flanks of volcanic complex—now represented by large, thick bodies of rhyodacite mapped on the flanks of the intrusive center.

4.1.3. Late Miocene (?) sediments

To the west of the volcanic center, homoclinal, gently south-dipping Middle Miocene volcanics are exposed continuously to at least as far north as Goler Wash (Figs. 1 and 2) where they are cut by fault systems related to the western Panamint range-front faults. Swarms of intermediate to silicic hypabyssal intrusives interrupt the volcanic pile, but otherwise individual flows or pyroclastic units can be traced for several kilometers along strike. In the extreme northwest corner of the mapped area, gently southeast-dipping, Middle Miocene volcanic rocks lie with angular unconformity upon upright, west-dipping, unmetamorphosed Cambrian strata (Zabriski Quartzite to Bonanza King Fm.) just north of Wingate Pass (Fig. 2). Thus, the volcanic pile either thinned

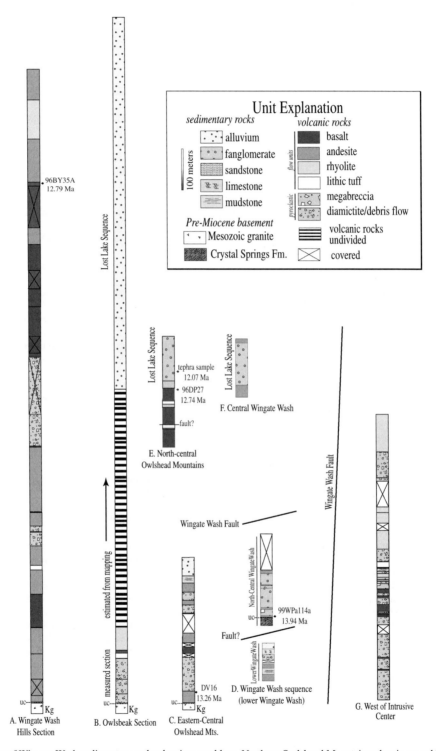

Fig. 3. Stratigraphy of Wingate Wash sedimentary and volcanic assemblage, Northern Owlshead Mountain volcanic assemblage, and intrusive center volcanic and sedimentary sequence. Refer to Table 1 for complete section descriptions and thicknesses.

Table 1
Stratigraphy unit descriptions and thicknesses

Thickness (m)	Unit descriptions
A. Wingate Wash Hills section	
81	Gray weathering, vesicular, basalt–basaltic andesite porphyry. ~3 mm plagioclase phenocrysts in dark gray aphanitic groundmass. Lava flow unit
106.0	Rhyolite. Pink to gray, brown weathering, rhyolite porphyry; 3–4 mm feldspar phenocrysts. Red weathering rhyolitic breccia in upper 2 m probably represents breccia flow top. Cut by 3 m thick dike near center of unit.
116.5	Mostly covered interval. Float dominantly gray, vesicular andesite porphyry with 2–3 mm feldspar phenocrysts. Lower 20–30 m is dark gray vesicular basalt. S1–S5 minus 10 m of exposures
10	Massive dark gray to black basalt. St 5. Sample 96BY35a dated at 12.79 ± 0.09 Ma
112.7	Mostly covered interval. Float of monotonous gray massive andesitic to basaltic andesite, commonly vesicular. Lava flow unit probably contains several flows, S5–S6
45.4	Massive dark gray aphanitic basalt or basaltic andesite. Locally porphyritic with olivine phenocrysts S6–S7
71.0	Vesicular olivine basalt. Package of 2–3 stacked flows S7–S8
47.0	Partially covered interval. Float and scattered outcrop indicates basaltic tephra S8–S9
48.7	Massive dark-gray to black, aphanitic basalt. Flow unit. S9–S10
79.4	Massive dark gray to black basalt. Locally vesicular. Two approximately equal thickness flows within this interval S10–S11
45.5	Partially covered interval. Float and scattered outcrop indicate basaltic tephra. S11–S12
10.0	Massive, partially vesicular basalt. Flow unit
155.0	Covered interval (Quaternary gravels in alluvial valley). No evidence of a fault. Lower 2/3 of covered interval is in slope wash derived from underlying diamictite unit. S12–S13
85.4	Polymict diamictite. Moderately rounded pebble to cobble size clasts of basalt to rhyolite in a muddy brownish gray mud to sand matrix. Clasts are matrix supported. Inferred debris flow deposit. S13–S14
0–0.5	Discontinuous light brown, lithic tuff
0.5–1	Paleosol? (clay rich, multicolored poorly consolidated clay to silt)
175.8	Greenish to greenish gray hydrothermal altered andesitic rock. Less altered near base of the subunit. Represents several altered andesitic flow units. S14–S15
1.5	Polymict diamictite. Debris flow unit
36.8	Greenish, hydrothermally-altered andesitic flow. S15–S16
14.1	Polymict, volcanic diamictite. Gray mud matrix supports moderately rounded pebble to cobble size clasts of basaltic and rhyolitic compositions. Debris flow unit. S16–S17
1.5	Lithic tuff. Brown (top) to white (base) lithic tuff. Abundant pumice fragments near base.
37.7	Polymict, volcanic diamictite. Gray mud matrix supports moderately rounded pebble to cobble size clasts of basaltic and rhyolitic compositions. Debris flow unit. S17–S18
52.6	Greenish gray, hydrothermal altered andesitic rocks. 1 m reddish oxidized horizon at top probably represents a paleosol. Unit consists of altered andesitic flows.
13.5	White lithic tuff. Laminated to massive lithic tuff with >90% white pumice and remainder comprised of polymictic volcanic clasts. Major marker horizon to the east and south in the Owlshead Mountains.
64.9	Gray to greenish-gray aphanitic andesite to basaltic–andesitic flows
91.2	Massive to vesicular and amygdaloidal basalt. Distinctive copper (malachite) alteration filling vugs in lower part of this unit. ~1-m-thick scoria (basaltic breccia) at the top of the sequence. Segment represents a stack of 2–3 flows.
70.9	Altered andesitic and basaltic andesite flows with complex cross-cutting andesitic dikes
68.2	Massive andesitic–dacitic porphyry. Upper 1 m is red, oxidized clay-rich material and may represent a paleosol baked by overlying flows. Copper (malachite) alteration present in veins and vugs. Unit probably represents at least two distinct flows with a breccia horizon approximately half way through the unit at the flow contact.
45.1	Covered interval
19	Brown to greenish brown, altered andesitic to basaltic andesitic rock. Probably represent the beginning of the sequence of flows recognized stratigraphically above the unconformity, across the covered interval
0–1.0	Discontinuous poorly sorted sandstone and conglomerate. Contains clasts of granite. Probable stream channel deposit during initial phases of volcanism.
1–2	Brown, altered andesitic flow. Irregular base suggests deposition on existing topography
1–2	Weathered granite. Paleosol beneath volcanic sequence
	Granite basement
1710	Total thickness of section

Table 1 (*continued*)

Thickness (m)	Unit descriptions
B. Owlsbeak Range section	
?	Lost Lake sequence
575	Upper part of section (thickness estimated from map positions) comprised of andesitic to rhyodacite flows in the lower ~1/3 and basaltic flows in the upper 2/3
65	Top of partial measured section Rhyodacite porphyry with glassy, aphanitic groundmass. Flow unit.
9.5	Vessicular olivine basalt flow unit
0.5	Reddish to light gray sandstone. Fluvial deposit derived from reworking underlying ash bed
17.5	White lithic tuff. Dominantly white pumice with a few lithic rich layers; locally cross-bedded in upper half of the unit. Lithic fragments, pumice fragments, and finer grained ash matrix become mixed toward the top suggesting fluvial reworking. Ash fall deposit that in this section has been partially reworked in the upper 2–3 m.
58	Medium gray andesitic porphyry with 1–3 mm feldspar phenocrysts. Flow unit
40	Polymict diamictite with basaltic to rhyolitic (obsidian) angular to subangular clasts in muddy to silty matrix. Probable debris flow deposit.
32	Matrix supported basaltic breccia and diamictite. Clasts mostly angular to subrounded pebble to large cobble (10 cm) of basaltic rock in an aphanitic gray matrix. Probable debris flow or distal rock avalanche deposit.
–	Nonconformity on megacrystic granite
222.5	Total thickness of measured section
797.5	Total thickness of volcanic section
C. Eastern/central Owlshead Mountain section	
30+	Alluvium: unconsolidated to weakly caliche cemented alluvial gravels and fan gravels. Conglomerate clast composition varies from volcanic to Pahrump Group (lower and middle Crystal Springs), ranging in color from black, green, red, gray to brown. Granite clasts are also present as detritus from the Owlshead Mountains.
20	Mudstone and siltstone
25	Andesite to Trachyandesite; aphanitic, dark gray to blue gray, rich in augite and plagioclase phenocrysts.
20	Volcanogenic debris flow and pyroclastic flows: diamictite with 40–60% muddy to sandy matrix and moderately rounded polymit volcanic clasts, typically ranging in size from pebble to boulder size. Reverse grading common with moderate clast supported rocks near base and matrix supported diamictite near the top of individual units
12.5	Andesite to trachyandesite; aphanitic, dark gray to blue gray, rich in augite and plagioclase phenocrysts.
20	Volcanogenic debris flow and pyroclastic flows: diamictite with 40–60% muddy to sandy matrix and moderately rounded polymit volcanic clasts, typically ranging in size from pebble to boulder size. Reverse grading common with moderate clast supported rocks near base and matrix supported diamictite near the top of individual units
2.5	Andesite to Trachyandesite; aphanitic, dark gray to blue gray, rich in augite and plagioclase phenocrysts.
45	Covered
25	Andesite flow
15	Covered
20	Olivine basalt: black, vesicular olivine basalt.
15	Covered
65–85	Volcanogenic debris flow and pyroclastic flows: diamictite with 40–60% muddy to sandy matrix and moderately rounded polymit volcanic clasts, typically ranging in size from pebble to boulder size. Reverse grading common with moderate clast supported rocks near base and matrix supported diamictite near the top of individual units.
30	Andesite to Trachyandesite; aphanitic, dark gray to blue gray, rich in augite and plagioclase phenocrysts. Sample DV16-00 dated at 13.28 ± 0.39 and 13.26 ± 0.011 Ma
Basement	Alkali-feldspar granite composed of K-feldspar, quartz, orthoclase, and altered biotite.
365	Total thickness
D. Wingate Wash section (lower Wingate Wash)	
10	Alluvium: conglomerate clast composition varies from volcanic to Pahrump Group (lower and middle Crystal Springs), ranging in color from black, green, red, gray to brown.
17	Fossiliferous (organic matter) limestone, local hot spring and/or a saline lake source.
15	Debris flow; dark gray to rusty red color of fresh and weathered surfaces. Contains layered flow andesite.
2	Reworked rhyolitic, lithic tuff.
12	Trachyandesite; dark gray to blue gray, composed of augite and plagioclase phenocrysts.

(continued on next page)

Table 1 (continued)

Thickness (m)	Unit descriptions
15	Siltstone; fine-grained, white to gray, siltstone lacustrine sediments. Units contain sedimentary bedding structures indicating lake bed deposition. Some beds contain reworked ash and black basaltic tephra.
25–30	Trachyandesite; dark gray to blue gray, composed of augite and plagioclase phenocrysts.
110	Alluvial fanglomerate; contains volcaniclastic clast assemblage with minor Pahrump Group clasts.
32	Trachyandesite; dark gray to blue gray, composed of augite and plagioclase phenocrysts.
100	Covered
12	Trachyandesite; dark gray to blue gray, composed of augite and plagioclase phenocrysts.
30	Alluvial fanglomerate; contains volcaniclastic clast assemblage with minor Pahrump Group clasts.
2	Trachyandesite; dark gray to blue gray, composed of augite and plagioclase phenocrysts.
8	Debris flow; dark gray to rusty red color of fresh and weathered surfaces. Contains layered flow andesite.
52	Alluvial fanglomerate; contains volcaniclastic clast assemblage with minor Pahrump Group clasts
5	Trachyandesite; dark gray to blue gray, composed of augite and plagioclase phenocrysts
17	Volcanogenic megabreccia deposit: trachydacite to rhyolitic megabreccia avalanche deposit into the basinal sequence.
5	Trachyandesite; dark gray to blue gray, composed of augite and plagioclase phenocrysts. Sample 99WPa114a dated at 13.94 ± 0.14 Ma
Basement	Crystal Springs and Kingston Peak Fm: composed of diabase, carbonates, and gabbro.
474	Total thickness

E. North-central Owlshead Mountain section

120+	Alluvium: unconsolidated to weakly caliche cemented alluvial gravels and fan gravels. Conglomerate clast composition is volcanic, ranging in color from red, gray and brown. (Lost Lake Sequence). Ash dated at 12.07 Ma
20	Basaltic cinder
37	Olivine basalt: black, vesicular olivine basalt. Sample 96DP27 dated at 12.74 ± 0.09 Ma.
10	Rhyolitic, lithic tuff: white, pumice rich tuff ranging from reworked, water-lain ash deposits, airfall, and tuffaceous mafic debris flows.
12	Andesite to trachyandesite; aphanitic, dark gray to blue gray, rich in augite and plagioclase phenocrysts
45	Olivine basalt: black, vesicular olivine basalt.
10	Rhyolitic, lithic tuff: white, pumice rich tuff ranging from reworked, water-lain ash deposits, airfall, and tuffaceous mafic debris flows.
4	Basaltic cinder
48	Olivine basalt: black, vesicular olivine basalt.
306	Total thickness

F. Central Wingate Wash (Lost Lake sequence)

120+	Alluvium: unconsolidated to weakly caliche cemented alluvial gravels and fan gravels. Conglomerate clast composition is volcanic, ranging in color from red, gray and brown.
20	Basaltic cinder
37	Olivine basalt: black, vesicular olivine basalt.
10	Rhyolitic, lithic tuff: white, pumice rich tuff ranging from reworked, water-lain ash deposits, airfall, and tuffaceous mafic debris flows.
12	Andesite to trachyandesite; aphanitic, dark gray to blue gray, rich in augite and plagioclase phenocrysts
45	Olivine basalt: black, vesicular olivine basalt.
4	Basaltic cinder
150	Total thickness

G. West of intrusive center section

>100? Top not exposed	Massive pink to gray weathering rhyolite porphyry. Flow banded black and red obsidian in lower 10 m. Probable rhyolite dome deposit
53–66	Pink to gray weathering, rhyolitic breccia. Probable flow breccia from overlying unit
9	Matrix-supported, rhyolitic diamictite. Inferred lahar deposit
69	Covered interval; float indicates interval is primarily rhyolitic breccia, but may also include part of the overlying lahar unit
16.1	Massive, gray-weathering rhyodacite porphyry. ~50% phenocrysts of hornblende, biotite, and feldspar. Flow unit

Table 1 (*continued*)

Thickness (m)	Unit descriptions
G. West of intrusive center section	
28.6	Massive, reddish-weather rhyodacite porphyry.<35–40% phenocrysts of hornblende, biotite and feldspar; probably lower part of the same flow represented by overlying unit.
20.5	Partially covered interval. Scattered outcrops and float indicate interval is part of the overlying flow unit and possibly continuous with underlying flow, but two flows are possible
20.5	Massive, reddish weathering rhyodacite porphyry.
20–50	Red, rhyolitic breccia. Probable flow breccia at base of overlying unit. Variable thickness due to uncertainties from minor faulting in the section and depositional variation.
34.4	Brown-weathering, gray rhyolite porphyry. Flow unit
15.8	Matrix-supported andesitic diamictite. Moderately rounded pebble to boulder size andesitic clasts in a brown, muddy matrix. Lahar or debris flow deposit
1.5	Reddish-gray weathering, gray andesitic porphyry. Contains green hornblende and plagioclase phenocrysts ~2 mm in length. Flow unit
0–1	White, lithic tuffaceous rock. Laminated tuffaceous sediment with >50% fragmental pumice
8.2	Polymict pebbly mudstone. Dominantly brown mudstone with ~20% polymict, moderately rounded clasts from 10 mm to 0.5 m in diameter. Contact with underlying unit approximately located (+/ − 1 m) due to small covered interval.
5.9	Clast supported basaltic-breccia with angular clasts of porphyritic basalt (1–4 mm plagioclase and pyroxene phenocrysts) in a sand-gravel size matrix. Probable aa lava flow
9.8	Rhyodacite porphyry. Hornblende and plagioclase phenocrysts ~3–5 mm. Brecciated near top. Probable flow unit
6.0	Polymict, predominantly matrix-supported conglomerate. Moderately rounded andesitic to basaltic cobble to boulder size clasts in a sandy matrix. Locally clast supported. Probable debris flow or alluvial fan deposit
4.3	Porphyritic andesitic-dacitic rock. Large (~1 cm) phenocrysts of plagioclase and hornblende
4.8	Reddish-weathering, matrix supported polymict diamictite. Matrix-supported volcanoclastic sedimentary rock with polymict pebble to boulder size, moderately rounded volcanic clasts in a reddish-brown, muddy matrix. Inferred debris flow deposit.
13.8	Basaltic boulder conglomerate. Moderately rounded basaltic boulders in a basaltic sand-gravel matrix. Probable lahar deposit
11.0	Aphanitic, vesicular basalt. 1–2 m thick breccia layers at top and base suggest aa lava flow
0.7	Matrix-dominant, polymict volcanic diamictite. Mud-dominated (>50%) with moderately rounded, polymict volcanic clasts. May be upper part of underlying unit or a distinct debris flow
6.8	Polymict volcanic diamictite. Moderately rounded cobble to boulder clasts of basaltic to rhyolitic composition locally float in a muddy to sandy matrix, but generally clast supported. Probable debris flow unit.
4.9	Porphyritic gray to dark-gray, non-vesicular olivine basalt. 1–3 mm phenocrysts off olivine. Flow unit.
0–0.5	Aphanitic basalt. Thin, discontinuous flow
18.4	Massive, basaltic breccia. Mostly vesicular basalt clasts 0.1–1 m in a sandy matrix. Probable talus accumulation or rockslide package.
0–1	Vesicular olivine basalt. Thin discontinuous flow. Red, basaltic breccia at base suggest aa flow
18.2	Vesicular, dark-gray to black, porphyritic olivine basalt. Flow unit
3.6	Basaltic diamictite. Moderately rounded pebble to boulder size basaltic clasts floating in a mud to sand size basaltic matrix. Probable lahar deposit
4.9	Porphyritic, dark-gray basalt porphyry. 2–4 mm phenocrysts of plagioclase in a moderately vesicular, aphanitic matrix.
0.5–1	Medium to coarse, buff-colored, volcanoclastic sandstone. Polymict sand volcanic sand grains including a few percent pumice.
3.6	Vesicular olivine basalt. ~1 m thick basaltic breccia at top suggests aa flow deposit
0–1.5	Medium to coarse, gray volcanoclastic sandstone. Moderately to well-rounded, polymict, volcanic sand grains, moderately sorted, locally cross-bedded. Stream deposit on volcanic edifice.
22.3	Polymict volcanic diamictite. Moderately rounded andesitic to basaltic clasts weakly clast supported but partially matrix supported. Matrix is reddish to purplish mud to clay size polymict volcanoclastic material. Inferred debris flow deposit

(continued on next page)

Table 1 (*continued*)

Thickness (m)	Unit descriptions
G. West of intrusive center section	
25.9	Covered interval. Float and small outcrops indicate underlying diamictite, but contact between adjacent units not exposed
62.8	Polymict, matrix-supported volcanic diamictite. Moderately rounded clasts of basaltic to rhyodacitic compositions make up ~50% of rock with remainder comprised of brown to buff colored muddy to sandy matrix. Debris flow deposit
2	Gray andesitic porphyry. Altered hornblende, plagioclase and K-spar phenocrysts ~2 mm make up ~25% of rock, with remainder comprised of a gray, aphanitic groundmass. Flow unit.
7.5	Andesitic diamictite interbedded with gray andesitic porphyry. Composite unit with andesitic diamictite interbedded with 2 discontinuous (0–1 m thick) andesitic flows
5	Clast supported, polymict diamictite. Cobble to large boulder (>1 m) moderately rounded polymict clasts in a mud–sand matrix. Dominantly basaltic clasts, but clasts range from basaltic to rhyolitic
18.1	Polymict, clast-supported volcanic diamictite. Moderately rounded, polymict clasts from cobble to boulder size in a gray to buff, tuffaceous matrix. Debris flow deposit
1.0	Medium to coarse, gray, tuffaceous sandstone. ~30% pumice clasts
0.5	White, lithic tuff. Clay to sand size, dominantly pumice clasts with polymict volcanic clasts from sand to pebble size.
7.0	Polymict, clast-supported volcanic diamictite. Moderately rounded, polymict clasts from cobble to boulder size in a gray to buff, tuffaceous matrix. Debris flow deposit
>73	Stacked sequence of polymict volcanic diamictite. Monotonous package of 2–3-m-thick debris flows continue beyond the area of the measured section. Dominated by rocks with moderately rounded, pebble to cobble size, volcanic clasts (clasts make up 50–70% of rock) supported by a reddish brown muddy to sandy matrix. Upper 30 m contain predominantly basaltic clasts, but rhyolitic clasts become dominant toward base of the observed section.
686+upper unit	Total thickness of partial section

to the west and/or some east tilting occurred near this contact to exhume the angular unconformity.

In the western third of the southern Panamint Mountains, the Miocene volcanic assemblage is unconformably overlain by a gently (4–10°) south-dipping sedimentary assemblage. The unconformity is well exposed just south of Goler Wash for ~5 km along strike and exposes an upward fining sequence of coarse clastic rocks above the unconformity. The lower ~50 m of this sedimentary assemblage are a coarse-grained sedimentary breccia/conglomerate characterized by large (up to 1 m), angular to moderately rounded clasts derived from the underlying volcanic rocks. This lower unit is primarily a clast-supported conglomerate and probably represents proximal facies alluvial fan deposits, possibly including some talus accumulations. These rocks grade upward into a poorly exposed sequence of conglomeratic sediments that are interpreted as fanglomerate deposits. The lower part of this fanglomerate succession is comprised of interlayered clast and matrix supported conglomerates containing moderately rounded pebble to cobble size volcanic clasts. We infer that the matrix supported layers represent debris flow intervals interlayered with fluvial intervals in a typical alluvial fan environment. Finally, this section continues upward into sandstone and conglomeratic sandstone deposits at the top of the exposed section. Due to incomplete exposure, we were unable to measure a complete section of this sedimentary assemblage, but based on cross-section relationships, this sedimentary sequence is 350–500 m thick. It is possible, that the total section is locally thicker beneath Wingate Wash where a significant gravity low is developed along the axis of Wingate Wash (Viana, 2000).

4.1.4. Age data

Topping (1993) reported a biotite $^{40}Ar/^{39}Ar$ date of 13.56 ± 0.85 Ma from a latite flow near the base of the volcanic assemblage near its northern outcrop limit, approximately 10 km north of our study area. To further constrain the age of the volcanic pile, we obtained one new $^{40}Ar/^{39}Ar$ date from the intrusive center. We sampled the youngest large intrusion on the east flank of the intrusive center. The rock is a fresh biotite-bearing rhyolite/microgranite. This sample yielded a simple, clean plateau date on biotite of 13.65 ± 0.14 Ma (sample 99WW12a, Fig. 5, plot A; Tables 2 and 4). Because this is the youngest intrusive

Table 2
Geochronological data collected from CLWW: central and lower Wingate Wash, NWW: north of Wingate Wash, EOWL: eastern Owlshead Mountains, NCOWL: north-central Owlshead Mountains, UWW: upper Wingate Wash, Refer to Table 4 for detailed $^{40}Ar/^{39}Ar$ data

Radiogenic $^{40}Ar/^{39}Ar$ and tephrochronology age data

Sample	99WPa114a	99WW12a	98/99WPa3	DV16-00	96BY35a	96DP27	98WW110
Domain	CLWW	NWW	CLWW	EOWL	EOWL	UWW	NCOWL
UTM(x)	0519557·	0509500	0519224	0519372	0509180	0517061	0515415
UTM(y)	3962029	3971500	3973416	3957978	3961550	3965380	3971292
Method	$^{40}Ar/^{39}Ar$	$^{40}Ar/^{39}Ar$	$^{40}Ar/^{39}Ar$	$^{00}Ar/^{39}Ar$	$^{40}Ar/^{39}Ar$	$^{40}Ar/^{39}Ar$	Tephrochron
Mineral	Biotite	Biotite	Whole rock	Plagioclase	Whole rock	Whole rock	n/a
Material	Rhyolite	Rhyolite	Basalt	Basalt	Basalt	Basalt	Tephra
Age	13.94	13.65	13.61/13.67	13.28/13.26	12.79	12.54/12.74	12.07
2 sigma	0.14	0.14	0.02/0.034	0.39/0.011	0.09	0.011/0.09	n/a

in the area and it is a biotite date from an intrusive rock, we conclude that this date is a reasonable minimum age for the entire stratovolcano complex represented by the intrusive center. The maximum age of the volcanic pile is unclear because our minimum age is concordant, within analytical error, of Topping's (1993) date from the base of the volcanic pile. Thus, until additional data are available, it seems likely that the entire volcanic pile was erupted during a relatively narrow time interval around 13.6 Ma.

The sedimentary section that overlies the volcanic rocks in the southwestern Panamint Mountains has virtually no age control other than relative stratigraphic position above the volcanic pile. To the southeast in lower Wingate Wash, a similar sedimentary section composed of fanglomerates overlie altered volcanic rocks. Some of these sediments are interbedded with 13.5–14 Ma volcanic rocks but others are younger and contain tephra layers estimated at ~12 Ma (see below). However, immediately to the south in the Crystal Hills, similar fanglomerates are locally overlain by 13.5 Ma volcanic rocks (see below), although most of the sedimentary sequence is stratigraphically younger. A major fault separates these two dated sequences from the southern Panamint sedimentary sequence. Thus, although the age of the southern Panamint sedimentary sequence is probably late-Middle or Late Miocene based on tentative correlation, further work is needed to clarify their absolute age.

4.2. Upper Wingate Wash domain (UWW)

4.2.1. General features

Pre-Quaternary rocks in upper Wingate Wash are exposed in four areas (Figs. 1 and 2): (1) Brown

Mountain (BMTN); (2) the Crystal Hills (CSH); (3) the "Radio Tower Range" (RTR)—informally named for the radio transmission tower in the range; and (4) the Wingate Wash Hills (WWHL)—informally named for the isolated low hills within upper Wingate Wash (Fig. 1). Several faults with clear evidence of Late Quaternary slip occur along the axis of upper Wingate Wash. These faults record a combination of Quaternary sinistral-oblique slip and contraction within this segment of the mapped area (Guest, 2000). To the south of this fault system the stratigraphy of the Miocene rocks is distinct from rocks immediately across Wingate Wash to the north. We believe that the stratigraphic discontinuity across upper Wingate Wash reflects a complex history across upper Wingate Wash, that includes both an earlier oblique extension, and Quaternary transpression related to interaction of sinistral-oblique faults along Wingate Wash and dextral faults of the Panamint Valley fault system (e.g. Guest et al., submitted for publication).

4.2.2. Middle Miocene volcanic assemblage

The Middle Miocene volcanic assemblage is exposed at all four sub-areas of the UWW domain, but each locality exposes a distinct sequence. This differentiation is undoubtedly, in part, due to lateral variations in the original volcanic pile, but at least part of this variation is probably due to structural juxtaposition.

4.2.2.1. Wingate Wash Hills section. The most complete section of Middle Miocene volcanic rocks anywhere in the Wingate Wash area is exposed in the Wingate Wash Hills. These low hills expose more than 1700 m of volcanic rocks in a continuous, mod-

erately west-dipping section. In this section (Fig. 3, section A and Table 1), volcanic rocks lie directly on weathered granite with only small, discontinuous layers of volcanogenic conglomerate along the non-conformity. Above the unconformity, the volcanic rocks can be readily divided into three major assemblages: (1) a basal unit (360 m) comprised predominantly of andesitic and basaltic flows with a few interbedded pyroclastic units; (2) a middle, predominantly pyroclastic unit (575 m) with interbedded andesitic flows that begins with a distinctive, thick white lithic tuff unit; and (3) an upper unit (773 m) comprised almost exclusively of basalt flows, and basaltic pyroclastics. The white, lithic tuff unit at the base of the middle unit is a distinctive marker horizon recognizable throughout the northern Owlshead Mountains. However, this tuff unit is absent from the southern Panamint Mountains section implying it was either not deposited on the large volcanic edifice in that region, or was derived from it. In either case, this Wingate Wash Hills section is nearly indistinguishable from sections in the northeastern Owlshead Mountains (see below) implying original paleogeographic affinities between the two areas.

4.2.2.2. Crystal Hills and Radio Tower Range. The Crystal Hills and the Radio Tower Range were examined in detail by Viana (2000) and Guest (2000). In the Crystal Hills, most of the Miocene volcanic rocks have been strongly affected by intense, hydrothermal alteration that largely converted the rocks to multicolored, clay-rich rocks. Indeed, many of the volcanic rocks in the Crystal Hills are sufficiently altered that they could be easily mistaken for fine-grained sediments. This alteration decreases in intensity from east to west within the Crystal Hills, and the limit of intense alteration makes a crude piercing line constraining fault offsets (see below).

The Crystal Hills also contain a distinctive sedimentary assemblage that unconformably overlies the volcanic rocks. These rocks are a package of interbedded diamictites and clast supported conglomerates with clasts derived exclusively from the underlying volcanic units. We interpret these rocks as alluvial fan deposits laid down atop the volcanic units. In the central Crystal Hills, these rocks are locally overlain by volcanic rocks, which were dated by F. Monestero at ~13.5 Ma (Monestero, written communication to

Pavlis, 1999, 2002). Thus, although these fanglomerates overlie the altered volcanics in the Crystal Hills, they are age equivalents of parts of the volcanic assemblage of the Wingate Wash area and presumably record localized basin development within a volcanic terrain.

The Radio Tower Range has been largely stripped of its Miocene cover, and exposes Mesozoic plutonic rocks and small exposures of quartzite and marble that represent highly metamorphosed roof pendants of Paleozoic (?) rocks intruded by Mesozoic granitoids. These Mesozoic crystalline rocks are intruded by swarms of ~N–S striking dikes that range in composition from basaltic to rhyolitic. Many of these dikes were clearly emplaced at very shallow crustal levels as evidenced by vesicular textures and common occurrence of glassy, aphanitic dikes with quenched margins. These dikes almost certainly were feeder dikes to the overlying volcanic pile as they yield Middle Miocene dates indistinguishable from overlying volcanic rocks (see below).

Miocene volcanic rocks are exposed on both the east and west sides of the Radio Tower Range; however, a major fault separates these two exposures (Fig. 4, C–C'). The volcanic rocks generally are not conspicuously layered and are cut by dike swarms, which produce a pseudo-layering that is difficult to distinguish from primary stratification. This problem, combined with faulting, made it impossible to measure a detailed section within this volcanic package. On the southwestern part of the Radio Tower Range, we were able to map the nonconformity beneath the volcanic rocks that dip gently to the west (~12°). Thus, the volcanic rocks along the southwestern part of the range are relatively thin (<200 ms) and have been gently tilted to the west (Fig. 4, E–E'). Along the nonconformity, we also locally observed laminated, fine-grained, siliceous sedimentary rocks that appear to be either tuffaceous rocks or siliceous-cemented siltstones. Bedding in these sedimentary rocks is generally concordant to the nonconformity and they appear to occupy paleotopographic depressions. However, at one locality we observed bedding with much steeper dips than the overlying nonconformity, suggesting either landsliding or some faulting prior to burial by the overlying volcanic rocks.

Lithologically, the volcanic rocks of the Radio Tower Range are dominantly massive andesitic to

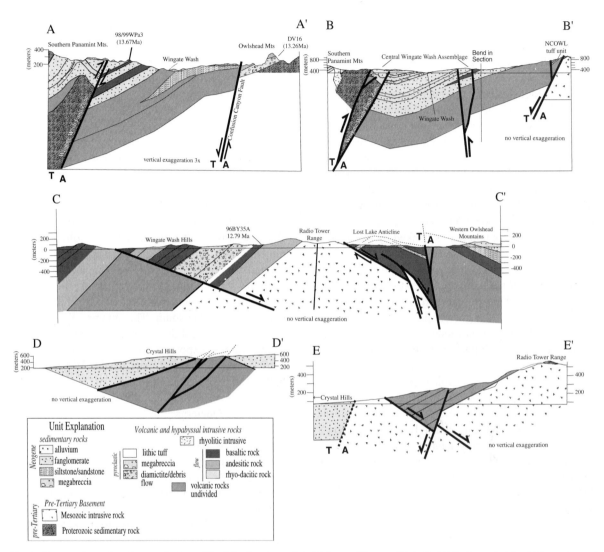

Fig. 4. Cross-sections of study area. Refer to Fig. 2 for locations of A–A′ Lower Wingate Wash, B–B′ Northern Owlshead Mountains and Wingate Wash, C–C′ Central Wingate Wash, D–D′ Crystal Hills, and E–E′ Southern Radio Tower Range transect lines.

dacitic flows with a few interbedded pyroclastic rocks. Pyroclastic rocks are most abundant near the base of the exposed section. As in the Crystal Hills, these rocks are locally and intensely altered, and this alteration decreases in intensity from east to west. Assuming the altered rocks of the Radio Tower Range and the Crystal Hills record the same alteration, the limit of intense alteration is now offset ~3 km sinistrally along the range-front of the Radio Tower Range.

The volcanic rocks in the southwestern part of the Radio Tower Range stratigraphically correlate to the lower unit of the Wingate Wash Hills section, but are lithologically distinct. That is, they appear to be more silica rich and are not as conspicuously layered as the rocks in the Wingate Wash Hills section. Nonetheless, they are probably lateral equivalents with variations produced by a combination of structural juxtaposition and original lateral variation. This tentative correlation is further supported by observations at the southern edge of the mapped area (southwest edge of the Radio Tower Range) where two thick rhyolitic flows appear to stratigraphically overlie the units of the Radio Tower Range and these are in turn overlain by a white lithic tuff that probably correlates with the tuff marker bed in the Wingate Wash Hills section (Fig. 4, C–C′).

Finally, we also conducted a reconnaissance of the northern edge of Brown Mountain at the extreme northwest tip of the Owlshead Mountains. In this area, there are no exposures of the nonconformity at the base of the Miocene section, and thus, the stratigraphic exposure level is uncertain. In the mapped area, this assemblage consists of two distinct assemblages separated by a young, north-directed thrust fault system (Fig. 2). In the hanging wall (south) of this fault system, the rocks are dominated by an elliptical rhyodacite unit that is flanked by pyroclastic units. We infer that these rocks represent a rhyolite dome complex within the volcanic pile, but their association with the rocks to the east is unclear until more work is done on these rocks. To the north, young, north-tilted fan gravels lie unconformably on lacustrine deposits (mudstone and siltstone) with abundant montmorillonitic clay producing a classic popcorn weathering surface. Although these deposits could be highly altered volcanic rocks like those in the Crystal Hills, they are most similar in compositional and physical characteristics to the lacustrine deposits located in lower Wingate Wash Basin (see below). This correlation is highly tentative, but if these are equivalent units, the sinistral displacement of the Wingate Wash fault could be as great as 40 km. Thus, ultimately these deposits need to be examined in more details.

4.2.3. Age data for UWW domain

Age data, at present, are limited from Upper Wingate Wash. We obtained one whole-rock ^{40}Ar/^{39}Ar date from a basalt at the top of the Wingate Wash Hills section (96BY35a) (Fig. 5, plot C; Tables 2 and 4). This sample yielded a relatively clear plateau date of 12.79 ± 0.09 Ma, which is consistent with the date from a basalt in a similar stratigraphic position in central Wingate Wash (sample 96DP27 with isochron age of 12.74 ± 0.09 Ma)(Fig. 5, plot D; Tables 2 and 4). We also obtained two hornblende ^{40}Ar/^{39}Ar dates on dike rocks in the Radio Tower Range (Guest, 2000). Both samples were from dikes that were relatively old in the dike swarm in that they were cut by younger dike sets. Both yielded relatively simple plateau dates of 13.35 ± 0.05 Ma and 13.56 ± 0.07 Ma (Guest, 2000). These dike ages are also indistinguishable from a ^{40}Ar/^{39}Ar date on a trachyandesite flow (F. Monestero, written communication to T. Pavlis, 1999) that was deposited on fanglomerates depositionally overlying the altered volcanic rocks in the Crystal Hills.

4.2.4. Quaternary deformational features

Upper Wingate Wash is deformed with four major groups of structures associated with each of the uplifted areas that expose Miocene and older rocks. The most significant of these structures is an ENE-trending, high-angle fault that we refer to here as the Wingate Wash fault because it can be followed intermittently as a series of fault exposures oriented subparallel to Wingate Wash. The Wingate Wash fault separates the structurally intact rocks of the southern Panamint Mountains from complexly faulted rocks immediately to the south. The Wingate Wash fault is exposed at two localities in upper Wingate Wash: (1) a short bedrock exposure and associated Quaternary scarp in the uppermost part of the Wingate Wash Hills; and (2) a complex zone of faulting in the lower Wingate Wash Hills, which locally also shows evidence of Quaternary activity.

In the upper Wingate Wash Hills, the Wingate Wash fault is poorly exposed in bedrock but juxtaposes rhyolitic breccias on the north against west-dipping basaltic rocks to the south. However, ~1 km along strike from the bedrock exposure, the fault can be traced for ~500 m as a distinct scarp in Quaternary alluvium. The scarp is significantly degraded by erosion but is most easily seen along an ~100-m trace where it is characterized by a 20–30-cm-deep, 1–3-m-wide trough filled with fine-grained mud from ponding water (see photo 99WPa109-3_1324 in Pavlis et al., in preparation). The remainder of the scarp is only recognizable as a low, intermittent S-side down topographic step cutting the Quaternary alluvial fan. No distinct offset Quaternary markers were seen across the fault to clearly constrain its net slip. Nonetheless, the association between the scarp and a high-angle fault in bedrock and the very linear fault trace suggests a high-angle fault that probably represents a strike-slip system.

In the lower exposures of the Wingate Wash Hills, the Wingate Wash fault is well exposed in bedrock on both sides of the present day dry wash. There are

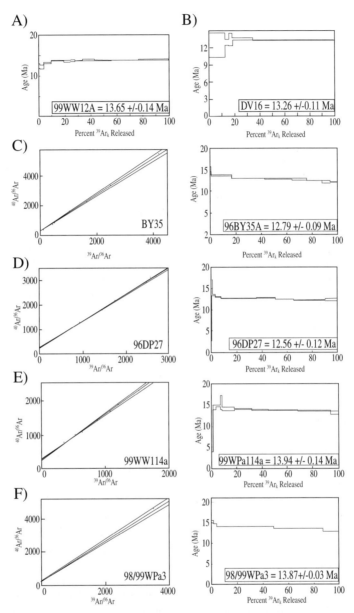

Fig. 5. ^{40}Ar/^{39}Ar geochronological spectra and isochron plots for samples 99WPa114a, 96DP27, 96BY35a, 98/99WPa3, 99WW12a, and DV16-00. Refer to Tables 2 and 4 for geochronological data set.

numerous mappable faults within the fault array, but we recognize one fault within the array (Fig. 2) as the main fault because it juxtaposes distinct stratigraphic assemblages. Specifically, to the south the rocks are variably faulted equivalents of the Wingate Wash Hills section. This truncation includes the tuff marker bed, which is cut by the fault and does not reappear across the fault. To the north, all of the rocks are proximal facies pyroclastic rocks and interbedded flows of the southern Panamint volcanic center.

This main strand of the Wingate Wash fault also appears to have at least some Quaternary slip. Although strongly degraded by erosion, the fault occupies a topographic trough along most of its trace, and drainages are deflected sinistrally at two points along the fault trace. These surface features would be ambiguous in the absence of other evidence for Quaternary slip, but given the clear evidence of Quaternary slip farther west along the fault, we assume these features also record young, presumably sinistral, slip.

To the south of the Wingate Wash fault, there is probably a young, east-dipping normal fault duplicating the upper part of the Wingate Wash Hills section where a playa is now present (Fig. 3, plot A and Fig. 4, C–C'). However, we were unable to recognize a fault scarp in this area and thus, the playa could have been dammed by alluvial fans from the north-side of Wingate Wash.

Quaternary fault scarps are also present along the northern side of Brown Mountain, in the Crystal Hills, and along the northwest side of the Radio Tower Range, indicating recent deformation is not limited to the Wingate Wash fault. At Brown Mountain, a 0 to ~1-m-high north-side-up scarp is developed along the ~E–W striking fault near the base of the north-facing mountain front. Quaternary gravels and underlying units in this area are uplifted and tilted northward immediately to the north of this scarp. Thus, the scarp is actually an uphill-facing scarp. The fault is sub-parallel to the north-dipping bedding of underlying volcanics and overlying gravels, and thus, records localized contraction. We suggest that this geometry probably results from an actively growing synclinal fold above a blind thrust along the front of Brown Mountain, with the fault scarp recording flexural slip above this fold.

The development of a fold at Brown Mountain is consistent with the Quaternary deformation observed immediately to the east in the Crystal Hills. Three sub-parallel, ENE-striking, moderately north-dipping faults represent an en echelon array of oblique sinistral-thrust faults in this area. All three of these faults show evidence of Quaternary slip including scarps in alluvium and damming of cross-cutting drainages. Moreover, bedding dips to the north on the north side of the Crystal Hills and to the south on the south side of the Crystal Hills, indicating the general structure of the hills is a faulted anticline (Fig. 4, D–D').

The Quaternary faults of the Crystal Hills are oblique to both the Wingate Wash fault and a sub-

parallel fault that forms part of the topographic escarpment along the northwest edge of the Radio Tower Range, hereinafter referred to as the Radio Tower Range fault. The Radio Tower Range fault, like the Crystal Hills faults, shows unequivocal evidence of Quaternary slip with a well-developed scarp in alluvium. This scarp shows clear north-side down displacement, but also shows some evidence for a component of sinistral slip. It is important to note, however, that the present Radio Tower Range fault is not coincident with the physiographic front of the entire Radio Tower Range; rather it diverges from the mountain front and projects toward the Crystal Hills (Fig. 2).

Taken together, we suggest that these Quaternary structures record a complex fault interaction that is best considered broadly in the context of a transpressive deformation in upper Wingate Wash. The folding, and associated oblique-slip faults in the Crystal Hills and at Brown Mountain both indicate ~N–S shortening. This shortening is occurring concurrently with both ENE-striking sinistral to sinistral-normal slip faults (Wingate Wash and Radio Tower Range faults) and NW striking dextral faults of the Panamint Valley fault system. We suggest that these strike-slip systems are the primary driver for the localized compression as the sinistral and dextral fault interfere in a general local regime of N–S contraction and ~E–W extension.

4.2.5. Older structure

The Radio Tower Range contains an important cross-cutting relationship indicative of the local structural chronology. Along the base of the western slope of the Radio Tower Range, just east of the Crystal Hills, Mesozoic granitic rocks lie structurally beneath west-dipping Miocene volcanic rocks along a moderately east-dipping contact (Fig. 4, E–E'). Although poorly exposed, this geometry requires that this east-dipping contact is a fault. We suggest this fault is a normal fault that initially was related to west-tilting of the Radio Tower Range and the adjacent Lost Lake Valley to the east. This relationship is important because this fault does not continue to the east, but appears to be truncated by younger west-dipping frontal faults of the Radio Tower Range that have uplifted the range after the earlier normal fault became inactive.

4.3. North-central Owlshead and Central Wingate domain (NCOWL/CWW)

4.3.1. Stratigraphic relationships

The north-central Owlshead Mountains are both stratigraphically and structurally complex, which inhibits reconstruction of the history (Fig. 1). It is possible to construct a composite stratigraphic sequence within this domain from partial sections combined with the large-scale relationship between the Owlsbeak Range and Lost Lake Valley (Fig. 3, section B and Table 1). The Owlsbeak Range and Lost Lake Valley represent a composite west-tilted range block between two east-dipping normal fault systems (RTRF on Fig. 2). Uplift within the range block, together with exhumation along Wingate Wash, has provided a relatively clear view of the composite Tertiary section within this block.

The NCOWL/CWW stratigraphic section is characterized by two distinct stratigraphic assemblages separated by an apparent disconformity: the Middle Miocene volcanic rocks at the base and an Upper Miocene sedimentary assemblage at the top. We informally refer to the upper sedimentary units as the Lost Lake assemblage based on excellent exposures at the northern end of the Lost Lake Valley. In addition, this domain contains a sedimentary sequence that is lithologically similar to the Lost Lake assemblage, but it appears to be older. We refer to this older(?) sequence as the central Wingate Wash assemblage.

4.3.1.1. North-central Owlshead Mountains (NCOWL). The Owlsbeak Range, located in the north-central Owlshead Mountains, exposes a complete section including the Tertiary volcanic pile from the nonconformity to the disconformity at the base of the overlying sedimentary section (Figs. 1 and 2). We were unable to measure this entire section because of logistical constraints, but a generalized section is clear from a partial section and from cross-section relationships (Fig. 3, section E and Table 1). This volcanic section is similar to the upper Wingate Wash section in that it is characterized by a lower section of flows interbedded with pyroclastic units (~130 m thick), a white tuff marker horizon (17.6 m thick), additional flow units (~200 m thick), and an overlying section dominated by basaltic flows (~450 m thick). This entire section is less than half the thickness of the

Wingate Wash Hills section. Farther north into Wingate Wash, the upper basaltic part of the section appears to be much thicker than within the Owlsbeak Range (Fig. 1) and includes three large exposures of buried cinder cones. Thus, the greater thickness in the Wingate Wash Hills section probably represents a primary depositional thickening toward the north. A sample from the basaltic rocks (96DP27) collected directly below the disconformity at the top of the section yielded a whole-rock $^{40}Ar/^{39}Ar$ age of 12.74 Ma (Fig. 5, plot D; Tables 2 and 4), which is only slightly older than a 12.07 Ma tephrochronology age estimate from an ash directly overlying this section (Wagner and Hsu, 1988) (A. Sarna-Wojcicki, written communication to Wagner and Hsu, 1988).

The entire NCOWL volcanic section shows important variations both within local vertical columns and laterally across the mapped area. The basal volcanic units are typically flows interbedded with pyroclastic units that range in composition from trachybasalt to rhyodacite. In this domain, the pyroclastic interbeds are typically relatively distal deposits with maximum clast sizes of a few tens of centimeters, but increase in clast size toward the east in the eastern Owlshead domain (see below). Throughout this domain, two white to light brown, 5–20-m-thick, lithic tuff units lie near the top of this mixed section of pyroclastic rocks interbedded with flows. These ash horizons almost certainly correlate with two tuff horizons recognized in the Wingate Wash Hills section (Fig. 3, section A and Table 1). A few tens of meters above these tuff horizons the rocks become distinctly more mafic, dominated by trachybasalt, basaltic cinder, basaltic andesite, and basanites. These basaltic units locally contain mafic to ultramafic inclusions, suggesting an origin from greater depths or magmatic flow entraining earlier cumulates, a relationship consistent with their geochemistry (see below).

The north-central Owlshead Mountain volcanic section is covered by the Lost Lake assemblage (Fig. 3, section E, Table 1 and Fig. 7a) along an abrupt contact. Throughout this domain, the Lost Lake assemblage consists of a package of 0.5–2-m-thick interbeds of coarse sandstone, sandy clast-supported conglomerate, and muddy to sandy matrix-supported diamictite interpreted as alluvial fan deposits. Clasts within the assemblage vary spatially and vertically, but are dominated by pebble to boulder size volcanic clasts derived from the underlying volcanic complex. The clast composition, together with the abrupt contact, strongly suggests that the base of the sedimentary units is a disconformity.

Time limitations of this project did not allow detailed clast counts, but clast content varies significantly within the Lost Lake assemblage. This relation provides important paleogeographic constraints. For example, at the northern end of the Lost Lake Valley, a large exposure records a distinct unroofing sequence with deposition from both sides of the Lost Lake Valley. Specifically, the lower ~50 m of the section E, Fig. 3 contains clasts derived exclusively from the Middle Miocene volcanic pile, but granitic clasts appear abruptly and become dominant in the upper part of the section in this locality. These granitic clasts increase in abundance toward the west, toward the Radio Tower Range, consistent with derivation from the Radio Tower Range as it was erosionally and tectonically unroofed. Granitic clasts are rare to absent in the easternmost exposures in this locally, which is consistent with derivation of these sediments from the Owlsbeak range to the east.

Farther north, within Wingate Wash, the Lost Lake assemblage is faulted and thus stratigraphic correlation into the eastern Owlshead domain becomes more difficult. Nonetheless, most of the NCOWL area is underlain by a gently north-dipping homocline and throughout most of this homocline, the clasts are derived exclusively from the underlying Middle Miocene volcanic assemblage. The source terrain for this sedimentary assemblage must have contained significant topographic relief because two distinct megabreccia units, within the Lost Lake assemblage, are interbedded within the section (Fig. 2). Both of these megabreccia packages contain large angular rhyolitic fragments indicative of a volcanic source terrain, but it is unclear if the deposits were derived from an active volcano or landsliding from a volcanic edifice.

At the northeast tip of the Radio Tower Range, a volcanic/sedimentary unit occupies the stratigraphic interval between the basaltic upper part of the Middle Miocene volcanic assemblage and the overlying Lost Lake sequence (Fig. 3, section E and Table 1). This interval is ~100 m thick and comprised predominantly of fine-grained claystone and siltstone. These rocks contain high percentages of expanding clays, presum-

ably because of a dominantly volcanic ash source, which gives rise to heaved soils devoid of vegetation. These fine-grained sediments are locally interbedded with thin basalt flows as well as siliceous deposits that occur in lensoidal bodies. We infer that these siliceous deposits are hot spring deposits within a lacustrine system represented by the fine-grained sediments; presumably a lake developed by ponding within a volcanic terrain. Wagner and Hsu (1988) sampled an ash bed from near to top of this local assemblage, which correlated with 12.07 Ma tephras from the region (A. Sarna-Wojcicki, written communication to Wagner and Hsu, 1988). This date is consistent with its stratigraphic position directly above the basalts (12.74 Ma). We dated the lower Lost Lake assemblages as 11.9 Ma and younger (see below).

4.3.1.2. Central Wingate Wash assemblage. The most poorly resolved stratigraphic sequence in the mapped area lies in central and lower Wingate Wash in a triangular shaped area bounded by the Filtonny fault on the east and a complex network of northwest-striking faults that bound the outcrop area of the Lost Lake assemblage to the west, and the Wingate Wash fault to the north (Fig. 2). In this zone, andesitic to dacitic volcanic rocks lie beneath, and are locally interbedded with, gravels that are superficially indistinguishable from the Lost Lake assemblage. However, these gravels may locally be distinguished, from the other sequences, in that they contain relatively fine-grained intervals (sand to siltstone) interbedded with coarse gravels, contain interbedded volcanic units including at least one major volcanic debris flow horizon, and locally contain abundant clasts of gabbro derived from the Proterozoic Crystal Springs Formation. The presence of the Crystal Springs Formation clasts is particularly significant because these clasts were not observed anywhere in the Lost Lake assemblage and the only presently exposed sources for these clasts lie to the north in Burro Wash Highway (Fig. 2) and to the southeast in the northeastern Owlshead Mountains.

Based on stratigraphic position and lithologic similarity to the Lost Lake sequence, we originally assumed, like Wagner and Hsu (1988), that these sedimentary rocks were part of the Lost Lake assemblage. Thus, we attempted to use the interbedded volcanic rocks to date the sequence and assumed these rocks would yield ages younger than basalts below the Lost Lake sequence. Instead, an interbedded rhyolite (99WPa114a) yielded a biotite age of 13.94 Ma and a stratigraphically higher basalt (98/99WPa3) yielded a 13.61 ± 0.02 and 13.67 ± 0.02 Ma whole-rock ages (Fig. 5, plot E and F; Tables 2 and 4), indicating that this section is more than 1 m.y. older than the Lost Lake assemblage. This apparent age is difficult to rationalize from field relationships because these sedimentary units appear to stratigraphically overlie volcanic rocks of the eastern Owlshead domain (Fig. 4, A–A′) which are the same age, or even younger, based on available geochronology (see next section).

We believe that this stratigraphic paradox is unresolvable until further dating is completed on the volcanic rocks in the eastern Owlshead Mountains and lower Wingate Wash. It is possible that the two volcanic units dated within the sedimentary assemblage are actually landslide deposits emplaced into the sedimentary basin, and thus, the dates would only indicate the age of the source. This interpretation is ad hoc because neither sample locality showed evidence of brecciation other than localized fragmentation suggestive of flow-tops or flow-breccias. Alternatively, a mapped fault along the front of the Owlshead Mountains may be more significant than it appears and has juxtaposed two distinct assemblages, but is difficult to trace this structure because it juxtaposes lithologically similar units and is overlapped by the Lost Lake assemblage.

Despite the paradox, we tentatively suggest that the central Wingate Wash assemblage represents an older sedimentary sequence that was being deposited at the same time as the older parts of the Middle Miocene volcanic complex, including the pyroclastic assemblages derived from the volcanic center and associated flows that pre-date the younger basaltic sequence. This tentative conclusion suggests that the faults that bound this assemblage have a significant west-side down component to juxtapose this older sequence against the Lost Lake sequence, that the Lost Lake assemblage has partially overlapped the original structural boundary, or both (Fig. 4, section B–B′). The fault structure in this interpretation is a cryptic system that merges with the larger faults, south of the Wingate Wash axis (Filtonny fault) as well as to the north against the Wingate Wash fault zone.

4.3.2. Age data

The age of the Lost Lake sequence is only partially established and the unconformity may be time transgressive. Just north of the Lost Lake Valley, we collected samples from two ash beds within the assemblage: 98WW110 from just above the unconformity and HG20 from slightly higher in the section. A. Sarna-Wojcicki (written communication to H. Golding Luckow and Pavlis, 2001) tentatively correlated these ash beds to LOVW-18 and LOVW-20, respectively from the Lovell Wash section north of Lake Mead, which suggests ages slightly younger than 11.9 Ma based on a dated unit below these ashes in the Lovell Wash section. These tentative correlations fit both the order of the samples within the Lost Lake sequence as well as their stratigraphic position above the 12.07 Ma tephra collected by Wagner and Hsu (1988). Thus, this correlation appears reasonable and we tentatively conclude that at least the lower part of the Lost Lake assemblage is late Middle to Late Miocene (post 11.9 Ma) in age, but the age of the upper part of the section is unconstrained.

Dates from $^{40}Ar/^{39}Ar$ release spectra for the rhyolite (99WPa114a) were 13.94 ± 0.14 Ma and a basalt (98/99WPa3) were 13.67 ± 0.02 Ma from central Wingate Wash assemblage (Fig. 5, plot E and F; Tables 2 and 4)(described above). Section A–A′ (Fig. 4) shows the approximate structural position of these samples. Sample 96DP27 (Fig. 5, plot D and Tables 2 and 4) is from a basalt near the top of the local volcanic assemblage, and this sample yielded a plateau date of 12.74 Ma.

4.3.3. Structure

The western and central Owlshead Mountains contain the most complex structure in the mapped area. At least two generations of faults are clearly recognizable by cross-cutting relationships, and a young fold system is well developed at the northeast tip of the Radio Tower Range (Fig. 6). Because this deformation varies across the domain, we describe structures from south to north.

The Lost Lake Valley and adjacent ranges are structurally relatively simple in the southern third of the mapped area, but become complex between the north end of the Lost Lake playa and Wingate Wash. These basins and ranges appear to be simple "domino-style" fault blocks with east-dipping faults bounding

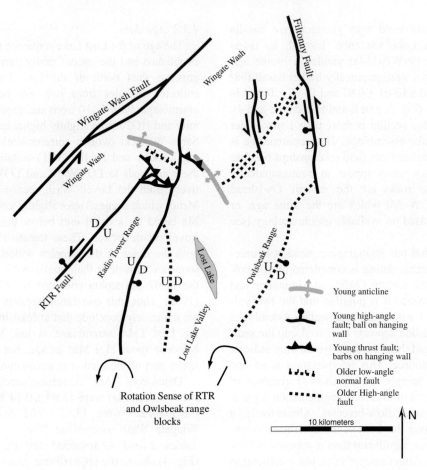

Fig. 6. Simplified geologic map of central Wingate Wash and Northern Owlshead Mountains.

west-tilted layering in the Tertiary rocks (Figs. 2 and 6). Therefore, the RTR, Lost Lake Valley, Owlsbeak Range, and the adjacent unnamed valley to the east are essentially a pair of half-graben with the youngest valley fill represented by the present valley floors and the ranges representing the uplifted blocks. We infer that the Lost Lake sedimentary sequence represents this valley fill, which is supported by the unroofing sequence observed in the sedimentary succession (see above). Thus, because the lowest part of the Lost Lake sequence is Late Miocene, these domino-style faults began to develop by Late Miocene time. However, the duration of this style of faulting has not been determined.

Along the northeast tip of the Radio Tower Range, from the north end of Lost Lake playa to Wingate Wash, we recognize four major groups of structures (Fig. 6): 1) a NW-trending fault-related fold system;

(2) a NNE-striking fault system that connects with the SE range front fault of the Radio Tower Range; (3) a northwest-striking, northeast-dipping normal fault at the NE tip of the Radio Tower Range; and (4) northeast striking, nearly vertical sinistral faults. The first and second group of structures are clearly very young and have been active in the Quaternary whereas the third and fourth groups are clearly older from cross-cutting relationships (Figs. 2 and 6).

The younger structures are probably a linked fault-fold system that developed synchronously on independent structures. The most prominent structure is a northwest-trending, upright, open anticline referred to here as the Lost Lake anticline (Fig. 4, C–C′ and Fig. 6). The Lost Lake anticline exposes Middle Miocene volcanics in its core and warps rocks as young as the upper part of the Lost Lake sequence on its limbs. At the stratigraphic level of Middle Miocene lacustrine sediments, the structure is a doubly plunging anticline. At deeper stratigraphic levels in Middle Miocene basalts, however, the rocks are repeated along an east-dipping thrust fault, indicating the fold is a thrust-related anticline. Dips are lower at higher stratigraphic levels within the Lost Lake sequence indicating a small angular discordance at the unconformity. Dips remain relatively constant, however, through the Lost Lake sequence and sediments form a prominent hill along the axis of the anticline just north of Lost Lake. The hill crest is nearly concordant to bedding. Together these relationships suggest that this anticline developed during two distinct phases of growth: (1) the youngest phase, which is probably continuing today, that produced the prominent hill at the north end of the Lost Lake Valley and the open warp in the Lost Lake sequence; and (2) older warping is indicated by the steeper dips at depth and angular discordance at the base of the Lost Lake sequence, suggesting that anticline was developing in Late Miocene time.

The axis of the Lost Lake anticline is sinistrally offset ~500 m by a NE-striking high-angle fault. This fault cuts the unconformity at the base of the Lost Lake sequence but cannot be unequivocally traced to higher stratigraphic levels. It appears to connect to the NE striking fault that forms the southeastern range front of the Radio Tower Range, but complex mutual cross-cutting relationships along the northeast tip of the range make this correlation tentative. Nonetheless, these structures are presumably part of a younger fault system, and moved at the same time as younger growth of the Lost Lake anticline.

The NE-striking fault system also cross-cuts an older fault system at the northeast tip of the Radio Tower Range (Fig. 6). This older fault is apparently a normal fault because it juxtaposes gently west-dipping basalt and lacustrine sediments along a northeast-dipping contact with Mesozoic granitic rocks. We also recognized a few down-dip slickensided surfaces along the contact, which further supports a conclusion that this structure is a normal fault.

To the east of the Lost Lake anticline we also recognized a significant NE-striking, nearly vertical fault. This fault is marked by a ~1-m-thick, vertical fault gouge with well-developed horizontal striae. This fault cuts at least the lower part of the Lost Lake sequence, but is overlapped by entrenched Quaternary (?) alluvial fans. Moreover, the fault is cut by

~NS-striking, oblique slip faults (Figs. 2 and 6) that are geometrically similar to the Filtonny fault (see below). In addition, similar smaller-scale faults show equivalent cross-cutting relationships within exposures of the Lost Lake assemblage immediately south of Wingate Wash (Fig. 2). Thus, because the Filtonny fault is probably Late Miocene and younger, this NE-striking fault system is probably entirely Late Miocene in age. Based on their geometry and development of horizontal striae locally, we assume these NE-trending faults are sinistral faults.

Exposures of the Lost Lake assemblage along central Wingate Wash appear to be a simple homoclinal sequence dipping 25–30° to the northwest cut be minor faults (Fig. 2). In detail, however, the structure is considerably more complex. Specifically, several distinctive structures are present (Fig. 2): (1) other NE-striking, nearly vertical sinistral faults; (2) an array of NNW-striking normal faults; and (3) near the eastern limit of the Lost Lake sequence, a NE-striking listric normal fault cuts the Lost Lake assemblage displaying a classic roll-over anticline above the curved fault. Finally, the Lost Lake assemblage is truncated to the east by a major structure we refer to here as the Filtonny Fault Zone, which is a sinistral, west-dipping, oblique-slip normal fault. The Filtonny fault is the youngest structure in the NCOWL area because it truncates all other structures, but shows no evidence of Quaternary displacement.

From Wingate Wash, north to the Wingate Wash fault, the structure is difficult to resolve due to a combination of complex faulting, poor stratigraphic control, and partial to complete burial of the Miocene rocks by an unfolded, nearly flat-lying, but now entrenched, alluvial fan. Two major observations can be made:

(1) The Lost Lake sequence is truncated along a complex array of NNW-striking high-angle faults. Although some of the faults within the array show east-side down displacements, the westernmost fault(s) must have a large west-side down component in order to juxtapose the Lost Lake sequence against the central Wingate Wash assemblage.

(2) The Lost Lake sequence clearly terminates to the north against the Wingate Wash fault, even though that fault is not well exposed. This con-

clusion is required because pre-Tertiary basement, consisting of Crystal Springs and Kingston Peak Formations, is exposed in the major wash (Burrow Highway Wash) just north of Wingate Wash (Fig. 2), indicating a large component of north-side-up displacement (Fig. 4, section A–A' and B–B'). NE-trending folds are also prominent beneath the young alluvial fan and represent a fold system with axes subparallel to the Wingate Wash fault contact (Fig. 4, A–A').

4.4. Eastern Owlshead and Lower Wingate domain (EOWL/LWW)

4.4.1. Stratigraphic relationships

The northeastern Owlshead Mountains primarily expose Middle Miocene volcanic rocks unconformably overlying both Crystal Springs Formation and Mesozoic granitoids. In the eastern Owlshead Mountains, older (pre-Middle Miocene?) sedimentary rocks locally lie between this nonconformity/angular unconformity and the overlying volcanic complex. The absolute age of these sediments is unknown and probably varies spatially. Two significant exposures of this assemblage are present: an extensive exposure along Confusion Canyon (Figs. 1 and 2), and a small exposure at the northeast edge of the Owlshead Mountains where the sediments lie on Crystal Springs Formation.

In the Confusion Canyon section, the older sedimentary rocks are comprised of 0 to ~20 m of interbedded red conglomerate and sandstone. In this region, the deposits are spatially limited to lensoidal outcrop bodies. This outcrop pattern strongly suggests the deposits are limited to paleo-channels developed along the erosion surface of the nonconformity. The sediments lie above 5–10 m of red, granitic grus that we interpret as a paleo-weathering zone or paleosol. Lithologically, these sediments are comprised of 0.5–1-m-thick interbeds of cross-bedded arkosic sandstone and clast-supported, pebble to cobble conglomerate. The conglomerates are particularly significant because the clasts are comprised predominantly of well-rounded quartzite and limestone clasts derived from rocks that are not presently exposed in the vicinity of the deposits. Specifically, most of the quartzite clasts appear to be derived from the Stirling Quartzite, Zabriski Quartzite, or both, and the carbonate clasts are derived from a

variety of Paleozoic formations, none of which are exposed within 20 km of the exposure (Fig. 4, section B–B'). The rocks are also a peculiar "broken pebble" conglomerate with highly polished pebbles broken by closely spaced joints near sites of clast–clast contact points. This characteristic is remarkably similar to the well-known broken-pebble conglomerates of the Oligocene Titus Canyon Formation in northern Death Valley. Based on these characteristics, we interpret these deposits as fluvial sandstones and conglomerates deposited in stream channels along the pre-volcanic erosion surface. Based on the exotic clasts and unusual clast types, we tentatively correlate these deposits to the Oligocene Titus Canyon Formation. This implies that these deposits, as well as the weathering zone beneath them, are significantly older than the Miocene volcanic complex.

In contrast to the Confusion Canyon section, the pre-volcanic northeast Owlshead section appears to be a younger deposit laid down during the early phases of volcanism. Section A–A' of Fig. 4 presents a diagrammatic representation of the structural–stratigraphic relationships at this exposure and indicates three key features of these deposits: (1) they are spatially limited to the hanging wall of a small east-side down normal (?) fault that is overlapped by middle Miocene volcanic rocks; (2) the sediments are dominated by coarse sandstones with clasts derived primarily from a volcanic source terrane; and (3) interbedding of a volcanic flow near the top of the sediments. The first observation indicates these deposits are syntectonic to early phases of extension, which contrasts with the channelized deposits of the Confusion Canyon pre-volcanic section. The second and third observations demonstrate a volcanic source terrane and *syn*-depositional volcanism, which is also distinct from the Confusion Canyon section, and these relations further support the conclusion that these deposits are only slightly older than the overlying volcanic pile.

The nonconformity and pre-volcanic sedimentary assemblage are overlain by a series of trachybasalt to trachyandesitic flows followed by pyroclastic flows and surges. We were unable to measure a complete stratigraphic section of the volcanic pile in this area because of complications from faulting, but a generalized section can be constructed from cross-sections and partial sections shown in Fig. 3 (sections C and

D, and Table 1). This section begins with ~300 m of coarse, pyroclastic deposits interpreted as a composite pile of debris and lahar flows interbedded with pyroclastic flows. West of Confusion Canyon, a 10–20-m-thick white to greenish lithic tuff is present near the top of these pyroclastic deposits and is probably correlative with the tuff marker horizon in the central Owlshead Mountains and in the Wingate Wash Hills section. These pyroclastic rocks are capped by a thin (~100 m) assemblage of basaltic flows that are in uncertain contact relationship with volcano-sedimentary deposits exposed in lower Wingate Wash that we refer to as the lower Wingate Wash assemblage (see below). This contact is probably, at least in part, a fault contact, but two local exposures suggest depositional overlap across an irregular topographic surface.

The lower Wingate Wash assemblage begins with a fine-grained diamictite unit that probably represents a major mudflow deposit (Fig. 3, section D and Table 1). This mudflow unit is overlain by interbedded andesitic flows and fine-grained diamictites (mudflows), which are capped by a distinctive terrestrial limestone unit. This limestone contains apparent algael mat structures but is also hydrothermally altered. This suggests the deposits are probably hydrothermal vent deposits within the volcanic pile (Grabyan, 1974). The limestone is overlain by interbedded conglomerate and diamictite that clearly represent fanglomerate deposits. The correlation of these fanglomerates to similar fanglomerates of the NCOWL/CWW domain is uncertain, however, because contact relationships between these two outcrop areas are obscured by an overlying younger alluvial fan.

Finally, the most problematic assemblage in the eastern Owlshead domain is a sequence of horizontal, finely laminated, light to dark gray mudstone interbedded with siltstone. The fine laminations and local occurrence of graded bedding within this unit strongly suggests that these deposits represent lacustrine deposits. Contact relationships between this unit and adjacent units, however, are obscured by Quaternary deposits. This unit is problematic because it crops out on both sides of the projected trace of the Filtonny Fault, yet it appears to be in stratigraphic continuity with the lower Wingate Wash assemblage. Thus, we collected a well-preserved ash bed from near the top of the exposed section. From geochemical characteristics A. Sarna-

Wojcicki (written communication to H. Golding Luckow and T. Pavlis, 2001) correlated this ash with a unit at the top of the Silver Creek volcanic pile that is dated at 6.15 ± 0.15 Ma by Ar/Ar dating. This correlation indicates a latest Miocene age for this lacustrine unit. Based on the age data and the horizontal bedding, we interpret these deposits as lake sediments deposited after much of the deformation in the Wingate Wash area had ended, but prior to much younger alluvial fans that cap ridges north of Wingate Wash.

4.4.2. Age data

$^{40}Ar/^{39}Ar$ analysis of sample DV16-00 from near the base of the eastern central Owlshead Mountain volcanic section (Fig. 3, section C) yielded indistinguishable ages of ~13.28 ± 0.39 and 13.26 ± 0.011 Ma on relatively simple plateau release spectra (Fig. 5, plot B; Tables 2 and 4). These dates indicate that the Middle Miocene volcanic section in the eastern Owlshead Mountains is entirely post 13.28 Ma in age. Paradoxically, these volcanic rocks appear to be stratigraphically overlain by the lower Wingate Wash sequence (Fig. 4, A–A′), which contains interbedded volcanic units that yield $^{40}Ar/^{39}Ar$ ages as old as 13.94 ± 0.14 Ma (sample 99Wpa114a) (Fig. 3, section D). This suggests that (a) there is significantly greater structural complexity in lower Wingate Wash, (b) there are problems in the dates, (c) there are problems in the interpretation of the volcanic interbeds, or (d) some combination of the three. More geochronological work is clearly needed to resolve this paradox.

4.4.3. Structure

Deformation in the eastern Owlshead Mountains is dominated by two N-trending, left-oblique, normal faults. To the west, the Filtonny fault forms the western boundary of the domain and truncates the outcrop area of the Lost Lake assemblage. Farther east the Confusion Canyon fault displaces the nonconformity at the base of the Middle Miocene section, but splays to the south producing both sinistral and dextral shifts of this contact. These Confusion Canyon faults do collectively show a general east-side up displacement. Both of these structures display oblique, south-plunging slickensides that indicate these faults have sinistral-normal oblique-slip displacements.

A third major fault may be present along the northern topographic Owlshead Mountain front where the lower Wingate Wash assemblage is in contact with the volcanic rocks throughout the Owlshead Mountains. However, the authors of this paper are not in agreement on the nature of this contact. Golding et al. (2000) mapped the contact as a buttressing unconformity within the volcanic assemblage whereas Pavlis (unpublished mapping) inferred a fault. If the contact is a fault, however, it was syntectonic with volcanism because it appears to be locally overlapped by volcanic debris flows (Golding et al., 2000). Resolution of this issue ultimately may be important because the lower Wingate Wash assemblage appears to be a significantly older sedimentary sequence, relative to other rocks in the area, and this contact may represent a structure along which this sequence was juxtaposed against other rocks.

5. Geochemistry of Wingate Wash and Owlshead volcanic suites

The primary purpose of performing geochemical analyses on the volcanic sections of Wingate Wash and surrounding areas was to investigate the commonality of the magma source for the volcanic rocks and to better understand the volcanic and sedimentary stratigraphic unit continuity. The evolution of the Wingate Wash and Owlshead Mountain volcanic province has

been a manifestation of at least three volcanic episodes and/or separate bimodal magmatic systems, which are also spatially distinct as three separate volcanic provinces. Samples from the three volcanic regions were chemically analyzed (Table 3) and selected samples were dated: the eastern Owlshead, central and lower Wingate Wash, and north-central Owlshead volcanic provinces are delineated in Fig. 7a. Assuming all the geochronological data are accurate representations of the absolute ages, the $^{40}Ar/^{39}Ar$ ages show the oldest region to be the southern Panamints (pre-13.65 Ma) and lower Wingate Wash at 13.94–13.56 Ma, followed by the Eastern Owlshead at 13.4–13.26 Ma, and ending with the basalts in the upper part of the volcanic section in the north-central Owlshead and upper Wingate Wash region at 12.79–12.56 Ma.

Geochemically, the southern Panamint, Wingate Wash and Eastern Owlshead Volcanics are chemically indistinguishable, whereas the north-central Owlshead region shows a different major element trend. Fig. 7a depicts total alkali vs. silica for all samples with rock names superimposed. The north-central Owlshead samples range from basalt to trachybasalts and basaltic trachyandesites, and overall are distinctively more mafic than Wingate Wash or Eastern Owlshead rocks. In contrast, these volcanic rocks in Wingate Wash and the Eastern Owlshead Mountains have elevated silica and total alkalis and are mostly rhyolites to trachyte and

Table 3

Geochemical data collected from EOWL: Eastern Owlshead Mountains, WW: central and lower Wingate Wash, NCOWL: north-central Owlshead Mountains

Sample	G35	G58	G108	G116	G53b	G79	G135b	G135a	G146	G148
Domain	EOWL	EOWL	EOWL	EOWL	WW	WW	WW	WW	WW	WW
SiO_2	54.63	63.66	56.52	62.35	54	58.49	65.85	55.44	53.32	73.28
TiO_2	1.34	0.8	0.84	0.8	1.34	1.19	0.62	1.02	1.2	0.47
Al_2O_3	15.22	14.31	15.47	15.47	15.68	17.35	14.79	15.09	16.76	14.4
FeO	7.61	4.16	6.37	4.46	6.48	6.24	3.7	5.77	5.95	2.99
MnO	0.066	0.042	0.098	0.058	0.13	0.14	0.076	0.074	0.094	0.068
MgO	2.08	1.22	4.19	2.65	3.44	0.95	2.13	4.32	2.14	0.37
CaO	5	3.96	5.51	5.2	5.84	4.55	3.86	6.73	2.46	2.36
Na_2O	8.03	7.55	8.19	4.99	8.62	3.45	4.81	7.86	7.36	2.25
K_2O	3.99	3.4	2.26	2.96	3.31	6.22	4.18	2.4	9.26	4.38
Total Alk	12.02	10.95	10.45	7.95	11.93	9.67	8.99	10.26	16.62	6.63
P_2O_5	0.96	0.34	0.29	0.49	0.68	1.03	0.48	0.6	0.69	0.23
SO_3	0.41	0.18	0.026	0.056	0.026	0.026	0.14	0.056	0.16	0.081	0.051
Cl	0.023	0.023	0.029	0.026	0.023	0.026	0.027	0.031	0.027	0.018
Mg #	35.9	37.5	57.4	54.9	52.1	23.8	54.1	60.5	42.4	20.2
Total	99.359	99.645	99.793	99.51	99.569	99.776	99.495	100.579	99.342	100.867

All chemicals are in wt.%. Spectro energy dispersive XFR analyses completed at Tulane University's CIF Inorganic Laboratory.

trachyandesite, with one foidite. Fig. 7b isolates K_2O and further emphasizes the highly alkaline nature of the Wingate Wash and Eastern Owlshead provinces, with 2 samples, G146 and G79, having K_2O greater than 6 wt.%. Presumably the samples from these regions represent a source highly contaminated with crustal melts, which then underwent crystal fractionation in a shallow-depth magma chamber. Fig. 7d displays a fairly typical, linear fractionation trend of all three assemblages with the K_2O vs. MgO plot. Wingate Wash and Eastern Owlshead samples group with low MgO and intermediate to high K_2O in a linear trend. In contrast, more fractionated samples from north-central Owlshead Mountain sections exhibit the low MgO and higher K_2O values expected in evolved volcanic rocks.

The younger north-central Owlshead volcanic province is separated from the Eastern Owlshead province by the Filtonny Fault. Samples from the north-central Owlshead volcanic provinces are considerably more mafic in composition and all groups have high MgO, low K_2O (Fig. 7c) and elevated TiO_2 (Fig. 7d). Petrographically, the north-central Owlshead samples contain mafic and ultramafic inclusions, which push some of the MgO contents to high levels in samples P62, G152, and P47d. A plot of the Mg number vs. TiO_2 wt.% (Fig. 7e) shows the north-central Owlshead samples to have higher Mg number and elevated TiO_2 contents (>1%) relative to the other two assemblages. The Wingate Wash and Eastern Owlshead volcanic province have a larger range of Mg numbers and exhibit variable enrichments in TiO_2, further suggesting internal fractionation and a crustally contaminated source. Fig. 7d shows the clearest example of a fractionation trend of lower TiO_2 with increasing SiO_2 content. North-central Owlshead samples represent a more primary mantle melt depleted in easily melted components and enriched in incompatible elements like Ti.

These patterns, in addition to the presence of ultramafic inclusions in north-central Owlshead samples and younger eruption times, suggest the magma source for the north-central Owlsheads was deeper and hotter relative to the source of the southern Panamint Mountains, Wingate Wash and the Eastern Owlshead Mountains volcanic suites. In addition, basaltic cinder cones, suggestive of a deeper and more mafic source, are distributed solely within the north-central Owlshead volcanic sequence.

6. Discussion and conclusions

6.1. Paleogeographic evolution of the Wingate Wash Basin derived from Tertiary stratigraphy of the Wingate Wash area

The lower Wingate Wash basin is a composite feature generated by two distinct processes: an early history of volcanic and volcanogenic sediment deposi-

G170	G163	P87	G165	G152	G158	G158b	G173	P47a	P47d	P62	P71
WW	WW	WW	WW	NCOWL	NCOWL	NCOWL	NCOWL	NCOWL	NCOWL	NCOWL	NCOWL
62.16	73.79	69.8	54.93	50.64	51.16	50.77	48.26	48.06	49.97	48.58	48.97
0.72	0.33	0.58	1.68	1.14	1.11	1.11	1.3	1.28	1.13	1.6	1.57
14.97	11.65	13.92	15.74	13.98	16.14	15.78	13.97	16.92	16.09	14.4	16.71
4.06	2.42	2.95	7.24	8.25	7.55	7.52	8.13	8.11	7.8	8.46	8.78
0.035	0.076	0.055	0.11	0.15	0.13	0.13	0.13	0.13	0.13	0.014	0.14
1.2	0.73	1.43	3.32	10.08	7.38	7.3	7.82	6.92	9.2	10.66	5.71
3.81	2.99	3.49	6.89	12.33	10.11	10.27	9.63	10.49	9.79	10.17	10.75
7.78	3.49	3.87	5.69	2.5	4.5	5.16	8.41	6.53	3.53	2.8	5.57
3.67	3.46	4.26	3.49	0.48	1.16	1.15	1.4	0.74	0.88	1.74	0.83
11.45	6.95	8.13	9.18	2.98	5.66	6.31	9.81	7.27	4.41	4.54	6.4
0.48	0.4	0.29	1.13	0.84	0.45	0.48	0.56	0.51	0.64	0.97	0.48
0.35	0.26	0.039	0.56	0.021	0.017	0.021	0.031	0.01	0.029	0.02	0.3
0.022	0.032	0.031	0.023	0.061	0.021	0.019	0.018	0.015	0.018	0.017	0.018
37.7	38.2	49.8	48.4	71.5	66.7	66.6	66.3	63.6	70.7	72.0	57.1
99.257	99.628	100.715	100.803	100.472	99.7279	99.71	99.659	99.715	99.207	99.431	99.828

tion and a later history of fanglomerate deposition in a closed basin developed by *syn*-sedimentary faulting. Based on our geochronology data we can further divide the volcanic history into two basic periods: an early interval (~14–13.5 Ma) dominated by pyroclastic rocks and alkaline volcanic flows with a large compositional range, and a later phase (~13–12.5 Ma) of basaltic eruptions localized in what is now the central portion of the mapped area. Contradictions between absolute age estimates and stratigraphic position suggest that all of the early volcanic history occurred in a sufficiently short time interval (<0.5 m.y.) and that a few scattered dates cannot resolve the details of this history. Thus, until a larger geochronological data base is available, the best record of the volcanism is provided by the basic stratigraphic sequences of the volcanic rocks and the spatial relationships among these sections.

Pre-mid-Middle Miocene (pre-13.94 Ma) sedimentary deposition began on an irregular erosional surface presently exposed extensively in the Owlshead Mountains, and locally in the southern Panamint Mountains. This erosional surface was marked by deep weathering recorded now as a thick paleo-weathering zone developed on the granite basement. The oldest deposits above the unconformity are only locally exposed, but the presence of conglomerates with well rounded clasts derived from sources now well removed from the area suggest fluvial systems were present along the erosion surface prior to deposition of the Middle Miocene volcanic rocks above the unconformity. These conglomerates also contain volcanic clasts, but it is unknown if these clasts are derived from the Middle Miocene volcanic complex; i.e. these conglomerates may be much older, possibly as old as the Oligocene Titus Canyon Formation.

The early phase of middle Miocene volcanism began after deposition of these conglomerates, although the character of this volcanism varies throughout the mapped area. In the southern Panamint domain, north of what is now the Wingate Wash fault, a large volcanic edifice was constructed at the site of the present intrusive center and deposited thick accumulations of lava flows interbedded with debris flows, flow breccias, pyroclastic surges and lahars. Similar, but more distal facies of the pyroclastic deposits and lava flows accumulated in the northern Owlshead Mountains, and Wingate Wash Hills. In the northeast-

ern Owlshead Mountains, more proximal-facies pyroclastic rocks were deposited during this stratigraphic interval suggesting that this region of the Owlshead Mountains was closer to the volcanic source. Nonetheless, this region is now separated from the most likely source in the southern Panamint Range by a basinal section that appears to be the same age as the early volcanic series, a relationship implying either geochronological problems (see above) or significant younger structural complications.

Early in the volcanic history, sedimentation began, in what is now lower Wingate Wash, with the accumulation of fine-grained lacustrine, silty sands as well as episodes of coarser sedimentation. Flows interbedded with the sedimentary packages indicate that the sedimentation was interrupted was by periods of volcanism. Near the end of this cycle, non-marine limestones formed on saline playa deposits, hot spring deposits or both. These carbonates were localized within a larger internally drained basin system. During volcanism, uplift on either the north or south side of the basin exhumed the Crystal Springs from beneath the volcanic pile and exposed it to erosion so that clasts from these rocks were deposited in fanglomerates now exposed along the northern edge of the basin. The closest likely source for these clasts lies nearby, immediately to the north, across the Wingate Wash fault. This structure clearly has significant post-12.5 Ma displacement, and thus the paleogeography at the time of fanglomerate sedimentation (~13.6 Ma) is unclear. The spatial association between this basinal, fanglomerate assemblage and the Wingate Wash fault implies that it may have been active at the time of deposition of these sediments.

Late in deposition of the volcanic and sedimentary system (~13–12.5 Ma), the thick mafic volcanic pile of basanites and basalts in the central Owlshead domain were deposited. These deposits may have, in fact, been partially responsible for the development of localized closed basins recorded by the lacustrine sediments at the northern end of the Lost Lake Valley.

Volcanism had largely ceased by approximately 12 Ma, and was followed by a period of erosion, nondeposition, or both that occurred throughout the Wingate Wash region. The length of this interval of nondeposition cannot be determined until more geochronological data become available, but if the tephra correlations reported here are correct, this interval was <1 m.y.

Following this interval, the Lost Lake sequence was deposited through much of the present area of lower Wingate Wash. These fanglomerates are derived from predominantly local sources, but exposures are spatially limited within Wingate Wash. That is, they occur only to the west of the north–south Filtonny fault and to the east of the normal fault at the northeast tip of the Radio Tower Range. In the Owlsbeak Range, the Lost Lake sequence is deposited disconformably on Middle Miocene basalts. The Lost Lake sequence can be traced continuously southwestward where it merges with correlated sediments of the Lost Lake Valley. Indeed, a complete cross-section of the Lost Lake Sequence sedimentary basin is exposed just north of Lost Lake where headward erosion from Wingate Wash and uplift along the Lost Lake anticline has exhumed the section. This observation strongly suggests that the Lost Lake sequence records the position of an ancestral Lost Lake Valley that predates exhumation by deformation and erosion throughout the Wingate Wash region. This basin is clearly a half graben to the south in Lost Lake Valley proper, but was probably a graben in the north, given the truncation of the Lost Lake fanglomerates to the east against a fault.

During deposition of the fanglomerates, some displacement apparently continued(s) on the cryptic ENE-trending fault, just north of Wingate Wash. This conclusion is based on the presence of folded fanglomerates just to the east and south of the exposures of Pahrump Group in the Burro Highway Wash. Deformation apparently ceased prior to deposition because of young fan gravels that buried this fault. This cryptic fault also connects in some way to the Wingate Wash fault system imaged in COCORP data (Serpa et al., 1988), but how these faults connect in the subsurface is unknown.

The lack of evidence for Quaternary motion along the Wingate Wash fault in lower Wingate Wash is surprising given the clear evidence of Quaternary slip along strike to the southwest in upper Wingate Wash. This suggests that present motion of the Wingate Wash fault represents a local reactivation of an older fault that may record some type of displacement transfer between the Owlshead and southern Panamint blocks.

Table 4
$^{40}Ar/^{39}Ar$ data from Wingate Wash and the Owlshead Mountains samples[a]

Temp (°C)	$^{40}Ar_R$[b]	$^{39}Ar_K$	$^{38}Ar_{Cl}$	$^{37}Ar_{Ca}$	$^{36}Ar_{At}$	% Rad	% $^{39}Ar_K$	Age+Error (Ma)[c]
99WW114a, Biotite, Total-gas date: 13.87 ± 0.26 Ma; Plateau date: 13.94 ± 0.14 Ma; J = 0.006071, 0.12%; wt. 29.9 mg								
600	0.0197	0.028	0.00036	0.00447	0.00198	3.3	1.7	8 ± 4
700	0.03872	0.0291	0.00020	0.00242	0.00016	45.1	1.8	14.5 ± 0.6
800	0.08124	0.05949	0.00083	0.00177	0.00011	70.5	3.6	14.9 ± 0.2
850	0.0200	0.0139	0.00025	0.00035	0.00004	62.9	0.8	16 ± 2
950p	0.21052	0.16324	0.00351	0.00314	0.00018	79.3	9.9	14.1 ± 0.4
1050p	0.28437	0.22024	0.00497	0.00532	0.00018	83.6	13.4	14.09 ± 0.11
1100p	0.27499	0.21575	0.00485	0.00571	0.00025	78.7	13.1	13.91 ± 0.06
1150p	0.48813	0.38345	0.00843	0.00898	0.00035	81.9	23.4	13.89 ± 0.03
1200p	0.55115	0.43304	0.01032	0.02005	0.00028	86.4	26.4	13.89 ± 0.11
1300	0.11727	0.09553	0.00392	0.02366	0.00015	72.2	5.8	13.4 ± 0.4
99WW12a, Biotite, Total-gas date: 13.56 ± 0.14 Ma; Plateau date: 13.65 ± 0.14 Ma; J = 0.006813, 0.12%; wt 44.1 mg								
650	0.0277	0.02733	0.00028	0.00142	0.00079	10.6	0.9	12.4 ± 0.6
750	0.0924	0.0939	0.00128	0.00585	.00081	27.8	2.9	12.0 ± 0.5
850	0.1932	0.18057	0.00360	0.00318	.00086	43.1	5.6	13.1 ± 0.2
900p	0.35363	0.31622	0.00702	0.00385	.00052	69.1	9.9	13.69 ± 0.06
950p	0.17782	0.16149	0.00361	0.00251	.00024	71.3	5.0	13.5 ± 0.3
1000p	0.12105	0.10905	0.00255	0.00206	.00019	67.6	3.4	13.6 ± 0.2
1050p	0.20591	0.18547	0.00441	0.00375	.00034	67.0	5.8	13.59 ± 0.09
1100p	0.33920	0.30286	0.00716	0.00588	.00046	70.9	9.4	13.7 ± 0.2
1150p	0.46031	0.41475	0.00969	0.00495	.00039	79.6	12.9	13.59 ± 0.14
1200p	0.75089	0.67356	0.01541	0.01452	.00040	85.9	21.0	13.65 ± 0.05
1350p	0.83012	0.74263	0.01726	0.00777	.00048	84.7	23.2	13.69 ± 0.12

(continued on next page)

Table 4 (continued)

Temp (°C)	$^{40}Ar_R$[b]	$^{39}Ar_K$	$^{38}Ar_{Cl}$	$^{37}Ar_{Ca}$	$^{36}Ar_{At}$	% Rad	% $^{39}Ar_K$	Age+Error (Ma)[c]
98/99WPa3, Whole rock, Total-gas date: 13.86±0.04 Ma; No plateau; Isochron date: 13.61±0.02 Ma (all steps);								
J = 0.006289, 0.12%; wt 248.8 mg								
600	0.09630	0.0615	0.00059	0.03298	0.00493	6.2	0.7	17.7±0.2
700	0.57656	0.43323	0.00054	0.15927	0.00408	32.3	5.2	15.04±0.04
800	2.6545	2.1074	0.00000	0.63461	0.00059	93.2	25.5	14.23±0.02
900	3.0253	2.4623	0.00022	0.69454	0.00036	95.9	29.8	13.89±0.02
1000	1.5634	1.2760	0.00224	0.38106	0.00036	92.9	15.5	13.85±0.03
1100	1.2258	1.0484	0.00371	0.46911	0.00052	88.2	12.7	13.22±0.05
1200	0.80457	0.71680	0.00354	0.84870	0.00091	74.5	8.7	12.69±0.02
1400	0.18367	0.15187	0.00091	0.29054	0.00028	68.5	1.8	13.67±0.4
98/99WPa3, Whole rock, Total-gas date: 13.81±0.03 Ma; No plateau; Isochron date: 13.67±0.34 Ma (all steps);								
J = 0.006490, 0.12%; wt 246 mg								
700	0.20310	0.1547	0.00097	0.06551	0.00769	8.2	1.9	15.3±0.3
800	0.26102	0.20749	0.00019	0.07590	0.00079	52.6	2.6	14.67±0.04
900	4.2168	3.4938	0.00019	1.02333	0.00017	92.9	43.9	14.08±0.02
1050	3.6013	3.0685	0.00473	0.94607	0.00015	92.9	38.5	13.69±0.02
1300	1.1510	1.20405	0.00434	1.06738	0.00016	80.3	13.1	12.90±0.02
DV16-00, Plagioclase, Total-gas date: 13.28±0.39 Ma; Plateau date: 13.40±0.42 Ma; J = 0.005040, 0.12%; wt 267.5 mg								
700	0.14617	0.1062	0.00254	0.28372	0.00955	4.9	12.4	12±2
800	0.03634	0.02531	0.00009	0.08134	0.00028	30.8	2.9	13.0±0.5
900p	0.03916	0.02631	0.00008	0.10436	0.00006	68.9	3.1	13.5±1.0
1000p	0.20316	0.13566	0.00003	0.58353	0.00006	91.3	15.8	13.6±0.1
1450p	0.83463	0.56594	0.00234	2.50357	0.00118	70.3	65.9	13.36±0.06
DV16-00, Plagioclase, Total-gas date: 12.06±0.46 Ma; Plateau date: 13.26±0.11 Ma; J = 0.005025, 0.12%; wt 252.1 mg								
700	0.02618	0.0864	0.00119	0.26306	0.00852	1.0	9.6	2.8±0.6
900	0.04615	0.04132	0.00004	0.14477	0.00033	32.2	4.6	10±1.3
1000	0.13414	0.09391	0.00000	0.39822	0.00006	88.3	10.5	12.9±0.4
1150p	0.48687	0.33270	0.00053	1.47502	0.00013	92.2	37.1	13.22±0.12
1250p	0.34577	0.23569	0.00040	1.02289	0.00049	70.4	26.3	13.25±0.11
1350p	0.15781	0.10638	0.00049	0.48729	0.00023	69.4	11.9	13.40±0.11
96BY35a, Whole rock, Total-gas date: 12.81±0.08 Ma; Plateau date: 12.79±0.09 Ma; Isochron date: 12.67±0.12 Ma (all steps);								
J = 0.005765, 0.12%; wt 248.5 mg								
600	0.04636	0.0318	0.00155	0.03278	0.00127	11.0	1.1	15.1±0.5
700	0.63682	0.48113	0.00127	0.29446	0.00047	81.5	15.9	13.71±0.08
800p	0.96936	0.78094	0.00028	0.37251	0.00019	94.0	25.7	12.86±0.04
900p	0.80360	0.65294	0.00044	0.47798	0.00026	90.5	21.5	12.75±0.04
1000p	0.43528	0.35583	0.00067	0.46403	0.00022	86.4	11.7	12.68±0.19
1100	0.44355	0.37086	0.00108	0.57983	0.00044	77.1	12.2	12.40±0.02
1200	0.22575	0.19500	0.00113	0.99070	0.00047	61.4	6.4	12.00±0.3
1400	0.18900	0.16496	0.00152	2.99565	0.00035	64.6	5.4	11.88±0.02
96DP27, Whole rock, Total-gas date: 12.62±0.11 Ma; No plateau; Isochron date: 12.54±0.11 Ma (all steps);								
J = 0.006685, 0.12%; wt 249.2 mg								
600	0.00892	0.0109	0.00106	0.02259	0.00040	7.0	0.4	10±7
700	0.07133	0.06367	0.00315	0.09922	0.00034	41.7	2.6	13.46±0.11
800	0.11395	0.10491	0.00209	0.16704	0.00011	76.6	4.3	13.05±0.09
900	0.71873	0.67802	0.00112	0.94276	0.00039	85.7	28.0	12.74±0.03
1000	0.39176	0.36746	0.00017	0.58791	0.00014	89.7	15.2	12.81±0.06
1100	0.36477	0.34910	0.00037	0.49836	0.00014	89.1	14.4	12.56±0.04
1200	0.62635	0.61075	0.00183	2.80995	0.00023	89.7	25.2	12.33±0.05
1400	0.24758	0.23787	0.00185	3.65128	0.00016	83.0	9.8	12.51±0.3

Tables 4 (*continued*)

Temp (°C)	$^{40}Ar_R$[b]	$^{39}Ar_K$	$^{38}Ar_{Cl}$	$^{37}Ar_{Ca}$	$^{36}Ar_{At}$	% Rad	% $^{39}Ar_K$	Age + Error (Ma)[c]
96DP27, Whole rock, Total-gas date: 12.67 ± 0.07 Ma; Plateau date: 12.74 ± 0.09 Ma; Isochron date: 12.66 ± 0.15 Ma (all steps); J = 0.006847, 0.12%; wt 250.5 mg								
800p	0.79930	0.77095	0.00625	1.03890	0.00084	75.8	33.2	12.76 ± 0.09
900p	0.32950	0.31710	0.00049	0.44910	0.00011	90.5	13.7	12.79 ± 0.15
1100p	0.51478	0.50021	0.00068	0.83997	0.00073	70.1	21.6	12.67 ± 0.03
1350	0.74357	0.73108	0.00419	5.82374	0.00030	88.5	31.5	12.52 ± 0.04

[p] Fraction included in plateau date. Plateaus determined according to the method of Fleck et al. (1977).

[a] Ninety-nine percent pure biotite or plagioclase was separated from rock samples by standard mineral separation procedures; phenocrysts of pyroxene and olivine were removed from whole-rock concentrates. Grain sizes were between 250 and 125 μm. Whole-rock samples were treated with dilute HCl to remove carbonate. All samples were cleaned with reagent-grade acetone, alcohol, and deionized water and air-dried. All samples were wrapped in aluminum foil packages. All samples were encapsulated in silica glass vials with fluence standards adjacent to each unknown. The standard for this experiment is Fish Canyon Tuff (FCT) sanidine with age of 28.03 Ma calibrated internally against MMhb1 hornblende standard (percent K = 1.555, $^{40}Ar_R = 1.624 \times 10^{-9}$ mole/gm, Samson and Alexander, 1987, and K–Ar age = 523.1 ± 2.6 Ma; age reevaluated by Renne et al., 1998). For irradiation, an aluminum canister was loaded with the silica vials. Samples were irradiated in two different irradiation packages and three samples were analyzed twice. Each irradiation was for 30 h in the TRIGA reactor at the U.S. Geological Survey in Denver, Colorado. The samples and standards were analyzed in the Denver Argon Geochronology Laboratory of the U.S. Geological Survey using a Mass Analyser Products 215 rare-gas mass spectrometer on a Faraday-cup collector. All samples were heated in a double-vacuum low-blank resistance furnace for 20-min heating intervals at each temperature, in a series of 4 to 12 steps, to a maximum of 1400 °C, and analyzed using the standard stepwise heating technique described by Snee (2002). Each standard was degassed to release argon in a single step at 1350 °C. For every argon measurement, five isotopes of argon (^{40}Ar, ^{39}Ar, ^{38}Ar, ^{37}Ar, and ^{36}Ar) are measured. Detection limit at the time of this experiment was 2×10^{-17} moles of argon.

[b] Abundance of all isotopes including $^{40}Ar_R$ (Radiogenic ^{40}Ar), $^{39}Ar_K$ (irradiation-produced K-derived ^{39}Ar), $^{38}Ar_{Cl}$ (irradiation-produced Cl-derived ^{38}Ar), $^{37}Ar_{Ca}$ (irradiation-produced Ca-derived ^{37}Ar), and $^{36}Ar_{At}$ (atmospheric ^{36}Ar) is measured in volts and calculated to five decimal places. Voltage may be converted to moles using 1.160×10^{-12} moles argon per volt signal. The "*F*" value ($^{40}Ar_R/^{39}Ar_K$) may be directly calculated from $^{40}Ar_R$ and $^{39}Ar_K$ in the table. All isotopic abundances have been corrected for mass discrimination. Mass discrimination was determined by calculating the $^{40}Ar/^{36}Ar$ ratio of aliquots of atmospheric argon pipetted from a fixed pipette on the extraction line; the ratio during this experiment was 298.9, which was corrected to 295.5 to account for mass discrimination. Final isotopic abundances were corrected for all interfering isotopes of argon including atmospheric argon. ^{37}Ar and ^{39}Ar, which are produced during irradiation, are radioactive and their abundances were corrected for radioactive decay. Abundances of interfering isotopes from K and Ca were calculated from reactor production ratios determined by irradiating and analyzing pure CaF_2 and K_2SO_4; the K_2SO_4 was degassed in a vacuum furnace prior to irradiation to release extraneous argon. Corrections for Cl-derived ^{36}Ar were determined using the method of Roddick (1983). Production ratios for this experiment are available in Snee (2002).

[c] Apparent ages and associated errors were calculated from analytical data then rounded using associated analytical errors. Apparent ages of each fraction include the error in *J* value (0.12%), which was calculated from the reproducibility of splits of the argon from several standards. Apparent ages were calculated using decay constants of Steiger and Jäger (1977). All apparent age errors are cited at 1 sigma. Uncertainties in the calculations for apparent age of individual fractions were calculated using equations of Dalrymple et al. (1981).

6.2. Tectonic synthesis of Wingate Wash and northern Owlshead Mountains

6.2.1. Lower Wingate Wash

Approximately north-south-striking, sinistral-normal oblique-slip faults are the youngest and most prominent structures throughout the northeastern Owlshead Mountains. These faults may accommodate some of the left-lateral displacement for the region as suggested by Serpa and Pavlis (1996), although their orientation is distinctly different than the ENE trend of the more cryptic Wingate Wash fault. The sense of displacement along this sinistral-normal fault system is indicated by: (1) slickenside measurements on the faults; (2) the sinistral shift of the two major north-dipping unconformities in the northern Owlshead Mountains; (3) the observation that the younger (12.07–12.74 Ma) sedimentary basin assemblage is limited to the area west of the major N–S fault of this system; and (4) the sinistral offset of a NNE-striking, ESE-ipping normal fault in the southern part of the mapped area—a shift that requires a significant component of left-lateral motion, given the sinistral offset of other, north-dipping contacts.

The geometry of these oblique-normal faults is important from a regional perspective, because their strike is ~50–60° from the southern Death Valley fault system. That orientation is significant because it is approximately the predicted orientation for a conjugate left-lateral fault array to the dextral Southern Death Valley fault system, but is also close to predicted orientations for normal faults in a dextral transtensional array that might develop along Southern

Death Valley fault. We interpret these faults as a composite of these two effects: a secondary conjugate system to the main dextral fault system of southern Death Valley producing the sinistral component, and the normal component accommodating the extension in a system that is fundamentally transtensional.

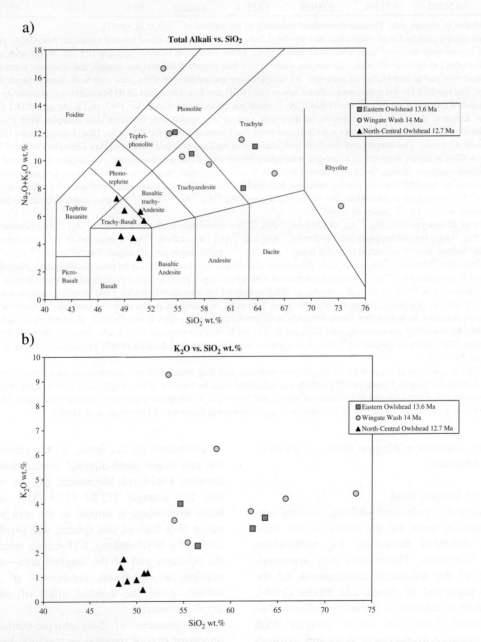

Fig. 7. (a) Geochemical plot of total alkali vs. SiO_2. (b) Geochemical plot of K_2O vs. SiO_2 wt.%. (c) Geochemical plot of K_2O vs. MgO wt.%. (d) Geochemical plot of TiO_2 vs. SiO_2 wt.%. (e) Geochemical plot of Mg # vs. TiO_2 wt.%.

c)

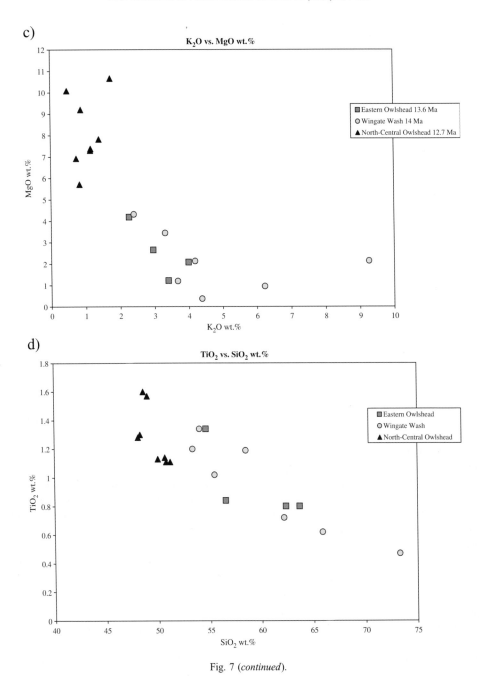

Fig. 7 (*continued*).

It is noteworthy that this fault system may provide a partial explanation for the normal-slip fault system that Serpa et al. (1988) recognized in COCORP data. Specifically, the strike of the fault recognized by Serpa et al. (1988) was not well constrained and could easily have a significantly more northerly strike than the NE strike inferred. Thus, if a fault similar to the sinistral oblique faults recognized in lower Wingate Wash were present in the subsurface beneath southern Death Valley, the fault could produce a feature similar to that recognized by Serpa et al. (1988).

e)

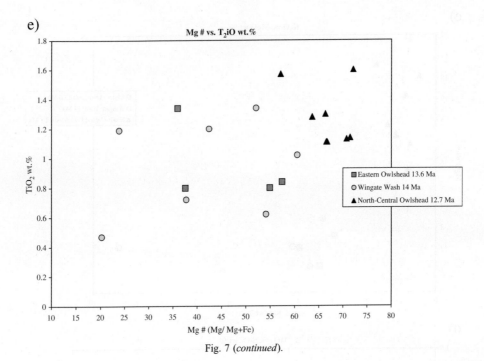

Fig. 7 (*continued*).

In line with the observations of the conjugate geometry of the N–S-striking oblique-slip faults above, a series of NNW-striking normal faults in central Wingate Wash and ENE-trending folds can be placed in a broader context of distributed left-lateral transtension. That is, projecting the ENE trend of the Wingate Wash fault system through the region places the NNW-striking normal faults in the position predicted for en echelon normal fault arrays along a transtensional sinistral system. In addition, this interpretation also provides an explanation for the unusual ENE-trending folds because in a similar manner, these folds lie approximately in the predicted shortening field for sinistral shear system sub-parallel to Wingate Wash (Fig. 8a and b).

6.2.2. Younger structures of Upper Wingate Wash and the Radio Tower Range

Unlike lower Wingate Wash, where evidence of Quaternary deformation is lacking, upper Wingate Wash continues to be actively deformed as evidenced by numerous fault scarps and fold-related hills. This Quaternary deformation appears to be occurring on three major groups of structures: (1) NE-striking, sinistral-normal (?) oblique faults including the Win-

gate Wash fault, the northwestern range front fault of the Radio Tower Range, and possibly the SE boundary fault of the Radio Tower Range; (2) N–S contractional structures in the Crystal Hills and Brown Mountain represented by oblique thrust faults and fault-related folds; and (3) NE-directed contraction at the northeast tip of the Radio Tower Range represented by the Lost Lake Anticline and related thrust faults.

Aside from the Lost Lake Anticline, the Quaternary structures in upper Wingate Wash can be broadly interpreted in the context of a zone of distributed dextral shear to weakly transpressional dextral shear. The active fault motions produce north–south contraction through interaction of dextral and sinistral faults with E–W-trending thrust faults and fault-related folds, and ~E–W extension through simultaneous motion on the conjugate strike-slip faults. Quaternary motions also appear to record predominantly vertical extensional strain. Specifically, the actively growing folds (Crystal Hills and Brown Mountain) record localized vertical thickening, which represents a larger-scale lengthening of a vertical crustal column or a bulk vertical extensional strain. In addition, the basic geology of upper Wingate Wash also argues for ver-

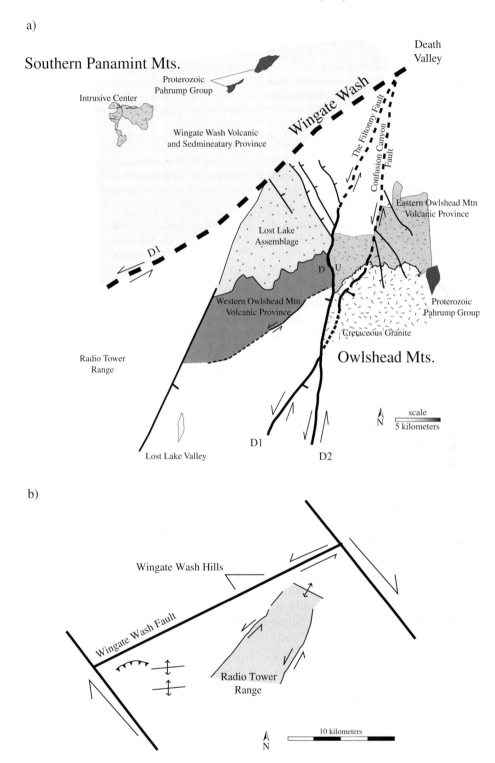

Fig. 8. (a) Tectonic reconstruction of southern Panamint Mountains, central and lower Wingate Wash, and the northern Owlshead Mountains. (b) Tectonic reconstruction of upper Wingate Wash and Radio Tower Range.

tical thickening (extension) in the broadly synclinal structure along the axis of the wash, paralleled by a linear topographic depression immediately north of contractional structures along Brown Mountain and Crystal Hills, a geometry suggestive of foreland basin type depression at the front of an active thrust. This apparent Quaternary strain pattern suggests active 3D strain with two elongation axes (vertical and E–W-horizontal) and a N–S, subhorizontal contraction axis, which represents a bulk flattening strain. In transcurrent systems, such as the Wingate Wash area, flattening strains require a transpressional system (e.g. Teyssier et al., 1995), which is surprising in an area most known for extensional tectonics. We explain this relationship, however, as a local complication of fault interactions between regional dextral shear combined with localized sinistral displacement transfer across the Garlock and Wingate Wash fault systems. This young deformation contrasts with the older structure, which appears to accommodate both transcurrent motion and simultaneous extension (i.e. transtension), suggesting motion has changed with time.

The origin of the Lost Lake anticline is more problematic. This structure shows evidence of both Miocene and Quaternary periods of growth, and these two episodes probably occurred under very different circumstances. The fold probably originated as an extensional roll-over (see below), but in the Quaternary the structure appears to record NE contraction. Following Guest (2000), we tentatively suggest that the younger growth of the fold records contraction related to localized transrotation of the Radio Tower Range. That is, sinistral shear along the Wingate Wash system may be forcing the Radio Tower Range to rotate counterclockwise about a vertical axis, driving shortening at the northern tip of the range. Alternatively, we note here that the transrotation could be driven by the same distributed dextral shear inferred for the regional deformation of upper Wingate Wash. In this case, the Radio Tower Range would represent a clockwise rotating block, bounded by sinistral faults, in a zone of distributed dextral shear. This hypothesis is attractive in that it could account for the lack of significant Quaternary deformation to the east of the Lost Lake Anticline, and is broadly consistent with other young structures. Indeed, the present NE strike of the range bounding faults may represent clockwise rotation from faults that were originally ~N–S, parallel

to the oblique slip faults of the northeastern Owlshead Mountains. Testing of either hypothesis will require paleomagnetic data, further dating of the rock sequences, or both.

6.3. Older (pre-Quaternary) structure of the Wingate Wash area

The most prominent older structure in the mapped area is the Wingate Wash fault. The concept of a fault running through central and lower Wingate Wash was suggested by COCORP data, which indicated a normal fault dipping to the north at 25–30° with strike sub-parallel to Wingate Wash (Serpa et al., 1988). Later, Serpa and Pavlis (1996) inferred a sinistral-oblique fault system along Wingate Wash. This study confirms the existence of a Wingate Wash fault, although in a different form than either of these previous studies suggested.

The Wingate Wash fault is an active structure in upper Wingate Wash, but is overlapped by older, entrenched alluvial fans in lower Wingate Wash. Thus, the motion of the fault in upper Wingate Wash is almost certainly a reactivation of an older structure.

The existence of an older Wingate Wash fault is evidenced by several observations. First, there is a dramatic change in structural style at the projected position of the fault from the intact, gently south-tilted volcanic rocks of the southern Panamint Domain into the complexly faulted rocks along the axis of Wingate Wash. Second, there is dramatic structural relief along the fault in lower Wingate Wash with Crystal Springs and Kingston Peak formation exhumed on the north and juxtaposed against the youngest part of the Lost Lake sequence on the south (Fig. 4, A–A' and B–B')—a stratigraphic throw of nearly 3 km. Third, the emplacement of large megabreccia sheets of rhyolite into the Lost Lake sequence occurred long after the volcanic center became inactive (post ~12 Ma) implying local large-scale structural relief-the most obvious source is to the north, across the nearly 3 km of structural relief associated with the Wingate Wash fault. Fourth, NW-striking Pahrump Group strata occur directly below the Tertiary unconformity in both the northeastern Owlshead Mountains and southern Panamints, including the newly recognized exposure in Burro Wash highway immediately north of the buried part of the Wingate Wash fault. Piercing lines are difficult to define because of differences in strike between the two exposures, but currently exposures are sinistrally separated by 7–9 km. Finally, the most definitive evidence for sinistral offset is juxtaposition of different volcanic and sedimentary facies across the fault. In central Wingate Wash, proximal facies volcaniclastic rocks on the north are juxtaposed against a stratigraphic section (Fig. 3, section D and Table 1) that contains a thinner, more distal facies pyroclastic assemblage. Rocks to the south of the fault are indistinguishable from the section in the Owlshead Mountains with a thin distal facies pyroclastic assemblage and distinctive white tuff marker bed near the top of the pyroclastic assemblage and an overlying basaltic assemblage. We assume that the lower pyroclastic assemblage south of the fault was derived from the same volcanic center as the pyroclastic rocks to the north, but the truncation of the tuff marker bed at the fault, and difference in facies of the pyroclastic rocks implies the two assemblages have been transported laterally from their original depositional positions. Similarly, in upper and lower Wingate Wash, the younger upper Miocene sediments (Lost Lake sequence and upper sedimentary units of upper Wingate Wash) are truncated at the fault. In addition, older (Middle Miocene) fanglomerates with interbedded volcanic rocks appear to be truncated at the fault.

Together these relationships clearly establish the existence of a Wingate Wash fault, but slip on the structure remains elusive. The structure is almost certainly a strike-slip fault because: (1) it is high-angle in all areas we have seen subsidiary faults related to the structure; (2) subsidiary faults along the fault show a predominance of strike-slip slickensides; (3) vertical throw varies dramatically along the strike of the fault-i.e. north-side up in lower Wingate Wash, south-side up in central and upper Wingate Wash. Offset of the two Pahrump Group exposures across the fault probably provides the best minimum slip estimate of 7–9 km of sinistral slip. However, other markers are present. Golding et al. (2000) noted that the volcanic facies boundary within the northern Owlshead Mountains between proximal facies rocks to the east and distal facies to the west may represent an offset piercing line. Correlation of this boundary to its approximate equivalent north of the fault (western side of the volcanic center) indicates 8–15 km of sinistral slip across the fault. Finally, the most

extreme amount of lateral offset (~19 km) would be inferred from correlation of the outcrop belt of the Lost Lake sequence with the upper Miocene sedimentary deposits of upper Wingate Wash. We consider this last case less likely, however, because of uncertainties in the absolute age of the southern Panamint sediments.

The Wingate Wash fault is not the only older structure in the mapped area. The NW-striking normal fault at the northeast tip of the Radio Tower Range is clearly relatively old because it is crosscut by other faults, and reactivated(?) by younger thrust systems. In addition, the east-dipping normal faults that have produced the west-tilted ranges of the Radio Tower and Owlsbeak ranges both originated at ~12 Ma with deposition of the Lost Lake sequence. These east-dipping normal faults and the Wingate Wash fault were probably active at the same time because they both affect the same rocks and bound the depositional area of the Lost Lake sequence. Thus, the normal faults and the Wingate Wash fault system are presumably linked.

This correlation is significant because aside from the northeast tip of the Radio Tower Range, these normal faults are NNE-striking and oblique to the NE-trending Wingate Wash system; a pattern consistent with en echelon normal fault arrays developed along a sinistral fault. Even the fault at the NE tip of the Radio Tower Range may have originated with this same orientation if the northern tip of the range has experienced significant vertical axis rotation. In any case, we suggest these faults are part of a general Wingate Wash system characterized by sinistral transtension across what is now Wingate Wash. This deformation ceased, however, by the time of deposition of the overlapping and undeformed Quaternary alluvial fans in lower Wingate Wash.

Acknowledgements

We would like to thank Christopher Menges and Emily K. Rose for comments that greatly improved the paper. This research was funded by the National Science Foundation Grant EAR-9706233 and United States Geologic Survey EDMAP grant #99HQAG0028. We thank Andre Sarna for providing the tephrochronology services. We thank Frank Monastero, Al Katzenstein and other members of the geothermal program staff at China Lake who helped us extensively with logistics to gain access to the field area through the China Lake Naval Weapons Center. We thank Bennie Troxel and Lauren Wright for logistical assistance on the project, and for discussions on the local geology.

References

Burchfiel, B.C., Davis, G.A., 1972. Garlock fault of southern California; an intracontinental transform structure. Geological Society of America Abstract with Program 4, 461–462.

Burchfiel, B.C., Stewart, J.H., 1966. "Pull-apart" origin of the central segment of Death Valley, California. Geological Society of America Bulletin 77, 142–439.

Burchfiel, B.C., Cowan, D.S., Davis, G.A., 1992. Tectonic overview of the Cordilleran Orogen in the Western United States. In: Burchfiel, B.C., Lipman, P.W., Zoback, M.L. (Eds.), The Cordilleran Orogen; Conterminous U.S. Decade of North American Geology. Geologic Society of America, pp. 407–479.

Dalrymple, G.B., Alexander Jr., E.C., Lanphere, M.A., Kraker, G.P., 1981. Irradiation of Samples for ^{40}Ar/^{39}Ar Dating using the Geological Survey TRIGA Reactor. U.S. Geological Survey Professional Paper, vol. 1176. 55 pp.

Dokka, R.K., Travis, C.J., 1990. Late Cenozoic strike-slip faulting in the Mojave Desert, California. Tectonics 9, 311–340.

Dewey, J.F., 2001. Transtension in arcs and orogens. The Bulletin of the Houston Geological Society 44, 15.

Fleck, R.J., Sutter, J.F., Elliott, D.H., 1977. Interpretation of discordant ^{40}Ar/^{39}Ar age spectra of Mesozoic tholeiites from Antarctica. Geochimica et Cosmochimica Acta 41, 15–32.

Golding, H.R., Pavlis, T., Serpa, L., 2000. Syntectonic volcanism and sedimentation in a transtensional environment, Wingate Wash Death Valley. Geological Society of America Abstract with Program 32, 7.

Grabyan, R.J., 1974. Investigations of the geology and mineralization of the Wingate Wash Mine area, Death Valley, California. M.S. Thesis, Los Angeles, University of Southern California.

Guest, B., 2000. The geology of the northwestern Owlshead Mountains, Death Valley, CA. M.S. Thesis, University of New Orleans, New Orleans, LA.

Guest, B., Pavlis, T.L., Serpa, L., Luckow, H.G., submitted for publication. Tectonic response of the Owlshead Mountain, Garlock Fault and southern Death Valley Fault Zone to rotation of the N.E. Mojave. Geology.

Holm, D.K., 1994. Relation of Miocene deformation and multiple intrusion in the Black Mountains crystalline core, Death Valley, California. Eos Transactions, American Geophysical Union 74, 609.

Holm, D.K., Geissman, J.W., Wernicke, B., 1993. Tilt and rotation of the footwall of a major normal fault system; paleomagnetism of the Black Mountains, Death Valley extended

terrane, California. Geological Society of America Bulletin 105, 1373–1387.

Korjenkov, A.M., Pavlis, T., Serpa, L., Viana, R., 1996. Distributed left-lateral motion along an intercontinental transform: Quaternary left-lateral slip along Wingate Wash, Death Valley Area, California, linked to the Garlock Fault. The 30th IGC, Beijing, China, p. 206.

Mancktelow, N.S., Pavlis, T.L., 1994. Fold-fault relationships in low-angle detachment systems. Tectonics 13, 668–685.

Meurer, K.J., 1992. Tectonic evolution of Smith Mountain–Gold Valley region, Death Valley, California. M.S. Thesis, University of New Orleans, New Orleans, LA.

Miller, M.G., 1999. Implications of ductile strain on the Badwater Turtleback for pre-14 Ma extension in the Death Valley region, California. In: Wright, L.A., Troxel, B.W. (Eds.), Cenozoic Basins of the Death Valley Region. Geological Society of America Special Paper, 333, 115–126.

Miller, M.G., Prave, A.R., 2002. Rolling hinge or fixed basin? A test of continental extensional models in Death Valley, California, United States. Geology 30, 847–850.

Miller, M.B., Pavlis, T.L., 2005. The Black Mountain turtlebacks: Rosetta stones of Death Valley tectonics. Earth-Science Reviews 73, 115–138. doi:10.1016/j.earscirev.2005.04.007.

Noble, L.F., 1941. Structural features of the Virgin Spring area, Death Valley, California. Geological Society of America Bulletin 52, 941–999.

Pavlis, T.L., Serpa, L., 1997. Structural consequences of intersecting dextral–sinistral strike slip fault systems; the eastern termination of the Garlock fault zone. Geological Society of America Abstract with Program 29, 234.

Pavlis, T.L., Wagner, D.L., Serpa, L.F., Luckow, H.G., Guest, B., in preparation. Geologic Map of Wingate Wash and Owlshead Mountains region, Death Valley, California Geological Survey.

Renne, P.R., Swisher, C.C., Deino, A.L., Karner, D.B., Owens, T.L., DePaolo, D.J., 1998. Intercalibration of standards, absolute ages and uncertainties in ^{40}Ar/^{39}Ar dating. Chemical Geology 145, 117–152.

Roddick, J.C., 1983. High precision intercalibration of 40Ar–39Ar standards. Geochimica et Cosmochimica Acta 47, 887–898.

Samson, S.D., Alexander Jr., E.C., 1987. Calibration of the interlaboratory 40Ar–39Ar dating standard, MMhb-1. Chemical Geology 66, 27–34.

Serpa, L., Pavlis, T.L., 1996. Three-dimensional model of the late Cenozoic history of the Death Valley region, southeastern California. Tectonics 15, 1113–1128.

Serpa, L., de Voogd, B., Wright, L., Willemin, J., Oliver, J., Hauser, E., Troxel, B., 1988. Structure of the central Death Valley pull-apart basin and vicinity form COCORP profiles in the southern Great Basin. Geologic Society of America Bulletin 100, 1437–1450.

Snee, L.W., 2002. Argon thermochronology of mineral deposits—A review of analytical methods, formulations, and selected applications. U.S. Geological Survey Bulletin 2194, 39 pp.

Snow, J.K., Wernicke, B., 2000. Cenozoic tectonism in the central Basin and Range; magnitude, rate, and distribution of upper crustal strain. American Journal of Science 300, 659–719.

Steiger, R.H., Jäger, E., 1977. Subcommission on geochronology: convention on the use of decay constants in geo- and cosmochronology. Earth and Planetary Science Letters 36, 359–362.

Teyssier, C., Tikoff, B., Markley, M., 1995. Oblique plate motions and continental tectonics. Geology 23, 447–450.

Tikoff, B., Teyssier, C., Markley, M., Mckinnon, S., 1994. The role of plate motion on strike-slip partitioning at zones of oblique convergence/divergence. Geological Society of America Abstract with Program 26, 110.

Topping, D.J., 1993. Paleogeographic reconstruction of the Death Valley extended region; evidence from Miocene large rock-avalanche deposits in the Amargosa Chaos Basin, California. Geological Society of America Bulletin 105, 1190–1213.

Viana, R.F., 2000. Late Cenozoic deformational history of Wingate Wash, Death Valley Region, California. M.S. Thesis, University of New Orleans, New Orleans, LA.

Wagner, D.L., 1993. Extension and Syntectonic Volcanism in Wingate Wash, Southwestern Death Valley, California. Geological Society of America, Cordilleran Section, vol. 26, p. 101.

Wagner, D.L., 1994. Evidence for Late Cenozoic extension across Wingate Wash, Death Valley Region, Southeastern California. Geological Society of America Abstract with Program 20, 240.

Wagner, D.L., Hsu, E.Y., 1988. Reconnaissance geologic map of parts of the Wingate Wash, Quail mountains and Manly Peak quadrangles, Inyo and San Bernardino Counties, Southeastern California. California Department of Conservation, Division of Mines and Geology Open-File Report.

Wernicke, B.P., 1988. Magnitude and timing of extreme continental extension, the central Death Valley region, California. Abstracts with Programs, Geological Society of America 29, 381.

Wernicke, B.P., Douglas, J., Kip, V., 1988. Detachment surfaces in the southern Great Basin; field guide to the northern part of the Tucki Mountain fault system, Death Valley region, California. In: Weide, D.L., Faber, M.L. (Eds.), This Extended Land Geological Journeys in the Southern Basin and Range, Field Trip Guidebook, Geological Society of America, Cordilleran Section, and University of Las Vegas, Nevada, pp. 58–63.

Wright, L.A., Troxel, B.W., 1973. Shallow-Fault Interpretation of Basin and Range Structure, Southwestern Great Basin, Gravity and Tectonics. John Wiley & Sons, New York, pp. 397–407.

Wright, L.A., Thompson, R.A., Troxel, B.W., Pavlis, T.L., DeWitt, E., Otton, J.K., Ellis, M.A., Miller, M.G., Serpa, L.F., 1991. Cenozoic Magmatism and Tectonic Evolution of the East-Central Death Valley Region, California, Geological Excursions in Southern California and Mexico, pp. 93–127.

Yawn, B., 1998. Petrogenesis of Wingate Wash volcanic rocks, Death Valley. M.S. Thesis, University of New Orleans, New Orleans, LA.

Available online at www.sciencedirect.com

Earth-Science Reviews 73 (2005) 221–243

www.elsevier.com/locate/earscirev

Miocene rapakivi granites in the southern Death Valley region, California, USA

James P. Calzia [a,*], O. Tapani Rämö [b]

[a] U.S. Geological Survey, Menlo Park, CA, USA
[b] Department of Geology, P.O. Box 64, FI-00014 University of Helsinki, Finland

Abstract

Rapakivi granites in the southern Death Valley region, California, include the 12.4-Ma granite of Kingston Peak, the ca. 10.6-Ma Little Chief stock, and the 9.8-Ma Shoshone pluton. All of these granitic rocks are texturally zoned from a porphyritic rim facies, characterized by rapakivi textures and miarolitic cavities, to an equigranular aplite core. These granites crystallized from anhydrous and peraluminous to metaluminous magmas that were more oxidized and less alkalic than type rapakivi granites from southern Finland. Chemical and isotope (Nd–Sr–Pb) data suggest that rapakivi granites of the southern Death Valley region were derived by partial melting of lower crustal rocks (possibly including Mesozoic plutonic component) with some mantle input as well; they were emplaced at shallow crustal levels (4 km) in an actively extending orogen.
Published by Elsevier B.V.

Keywords: rapakivi granite; Death Valley; California; petrology; geochemistry; isotope geology

1. Introduction

In 1891, J.J. Sederholm introduced rapakivi granites to the international geological community in his pioneering paper on Proterozoic granites of southern Finland; since then, southern Finland has been considered the type locality of rapakivi granites. These "type" rapakivi granites are characterized by textures consisting of relatively large crystals of potassium feldspar (K-feldspar) rimmed by plagioclase in a finer grained matrix of quartz and mafic minerals (Fig. 1) and are typically found as discordant plutons intruded into a metamorphic crust; that crust was differentiated from the mantle a few hundred million years earlier (Haapala and Rämö, 1990). Geochemically, the rapakivi granites, crystallized from alkalic, anhydrous, and slightly metaluminous to peraluminous magmas, have high contents of the light rare earth elements (LREE) except Eu, show high Zr, Rb, Ba, Y, F, Fe/Mg, and Ga/Al, low Ca, Mg, P, and Sr, and vary from reduced to oxidized in character (e.g., Anderson and Bender, 1989; Rämö, 1991; Rämö and Haapala, 1995; Frost and Frost, 1997). Rapakivi granites are often associated with bimodal magmatic suites

* Corresponding author. Tel.: +1 650 329 5538; fax: +1 650 329 5130.
E-mail address: jpcalzia@usgs.gov (J.P. Calzia).

0012-8252/$ - see front matter. Published by Elsevier B.V.
doi:10.1016/j.earscirev.2005.07.006

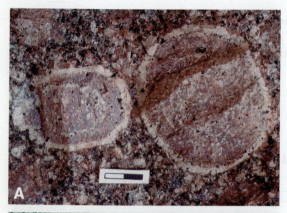

Fig. 1. Photographs showing rapakivi texture in (A) wiborgite (hornblende granite with alkali feldspar megacrysts mantled by sodic plagioclase) from the classic 1.65- to 1.62-Ga Wiborg rapakivi granite batholith of southeastern Finland, and (B) rapakivi porphyry from the 9.76-Ma Shoshone pluton, southern Death Valley region, California. Length of scale bar is 2 cm in (A) and ~3 cm in (B). Photos: O.T. Rämö.

such as anorthosite–mangerite–charnockite–granite complexes as well as spatially and temporally associated mafic dike swarms; this magmatic association has been described in detail from southern Wyoming (Frost et al., 2001), Labrador (Emslie, 1991), Ukraine (Amelin et al., 1994), and Fennoscandia (Rämö, 1991; Neymark et al., 1994; Amelin et al., 1997), and has led many researchers to advocate magmatic underplating as the probable mechanism for generating rapakivi granites (Emslie, 1978; Anderson, 1983; Haapala, 1985; Rämö, 1991). Magmatic underplating may result in partial melting of the lower crust and mixing of lower crust and more mafic melts (Eklund et al., 1994; Salonsaari, 1995).

In the 20th century, rapakivi granites are described from around the world, especially Ukraine, South

Greenland, eastern Canada, Brazil, and the United States (see reviews in Rämö and Haapala, 1995; Haapala and Rämö, 1999). In the United States, rapakivi granites are well exposed in areas of pronounced crustal extension and syntectonic magmatism, such as the Colorado River extensional corridor in southernmost Nevada and adjacent Arizona (Volborth, 1973; Falkner et al., 1995; Haapala et al., 2005) as well as the southern Death Valley region in southeastern California (Calzia, 1990). In this paper, we describe three late Cenozoic rapakivi granites from the southern Death Valley region; we compare and contrast their geology, petrography, and geochemistry with Proterozoic rapakivi granites and note the remarkable similarity of these fascinating rocks.

2. Geologic setting

The southern Death Valley region is bounded on the west by the Panamint Range and the south by the Providence and the New York Mountains (Fig. 2). The pre-Cenozoic stratigraphy in this region consists of Early Proterozoic cratonic rocks and Middle Proterozoic to Triassic sedimentary rocks. The cratonic rocks include Early Proterozoic paragneiss, schist, and quartzite intruded by ca. 1700-Ma orthogneiss and 1400-Ma anorogenic granites (Lanphere et al., 1964; Wooden and Miller, 1990; Rämö and Calzia, 1998). The Middle Proterozoic Pahrump Group unconformably overlies the cratonic rocks and consists of approximately 2100 m of conglomerate, sandstone, shale, and carbonate rocks divided into the Crystal Spring Formation, Beck Spring Dolomite, and Kingston Peak Formation (Hewett, 1940); the Crystal Spring Formation is intruded by 1068- and 1087-Ma (Heaman and Grotzinger, 1992) diabase sills. The Pahrump Group is overlain by more than 7500 m of Late Proterozoic to Triassic rocks. The Late Proterozoic and Paleozoic rocks include a basal dolomite overlain by widespread units of quartzite, conglomeratic quartzite, siltstone, shaly siltstone, limestone, and dolomite; the Triassic rocks include metasedimentary and metavolcanic rocks west and south of the Owlshead Mountains.

Most of the miogeoclinal rocks are intruded by Mesozoic and Cenozoic plutons. The Mesozoic plutons vary in composition from granite to gabbro

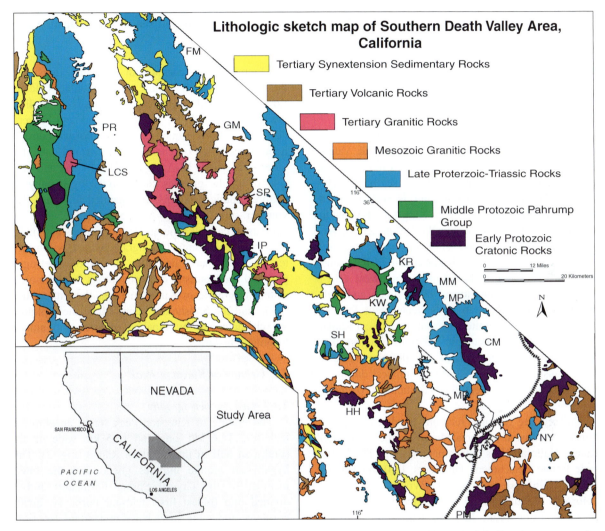

Fig. 2. Index and lithologic sketch map of the southern Death Valley region, California. CM=Clark Mountains, GM=Greenwater Mountains, IP=Ibex Pass, FM=Funeral Mountains, HH=Halloran Hills, KR=Kingston Range, KW=Kingston Wash, LCS=Little Chief stock, MM=Mesquite Mountains, MP=Mesquite Pass, MR=Mescal Range, NY=New York Mountains, OM=Owlshead Mountains, PM=Providence Mountains, PR=Panamint Range, SH=Silurian Hills, SP=Shoshone pluton. Modified from Calzia and Rämö (2000).

(Rämö et al., 2002); Cenozoic plutons vary from granite to quartz monzonite (Calzia and Rämö, 2000). All of these rocks are unconformably overlain by younger Tertiary sedimentary and volcanic rocks and Quaternary alluvial deposits.

The oldest Cenozoic extensional faults in the southern Death Valley region include the Harrisburg and Kingston Range–Halloran Hills fault systems. The Harrisburg fault system along the crest and eastern side of the Panamint Range consists of a basal detachment fault (the Harrisburg fault) and high-angle

transfer faults; the Harrisburg fault cuts the Cretaceous Skidoo monzogranite pluton and is cut by late Miocene normal faults (Hodges et al., 1990). Wernicke et al. (1988) reported that the Harrisburg fault is synchronous with fanglomerate and monolithic breccia along the north and northeast side of the range; they tentatively correlated these deposits with the middle Miocene Bat Mountain Formation at the southern end of the Funeral Mountains (Fig. 2). Cemen et al. (1999) redefined the Bat Mountain Formation and concluded that the fanglomerate deposits at the north end of the Panamint Mountains may be the same age as the basal conglomerate member of their Amargosa Valley Formation; tuff near the base of the conglomerate member yields a K–Ar age of 24.7 ± 0.3 Ma. If this correlation is correct, these data suggest that Cenozoic extension began during the late Oligocene in the Panamint Range.

The Kingston Range–Halloran Hills fault system defines the eastern boundary of the Death Valley extended terrain (Burchfiel et al., 1983) and is divided into northern and southern segments. The Kingston Range detachment fault is well exposed in the northern and eastern Kingston Range to about the latitude of Kingston Wash (Fig. 2). Reconstruction of fault and depositional contacts suggests that maximum horizontal displacement of the hanging wall is about 6 km to the southwest (B.C. Burchfiel, written commun., 1989; Fowler and Calzia, 1999). The Halloran Hills detachment fault is best exposed in Mesquite Pass and the Mescal Range (Fig. 2). Regional geologic relations suggest that rocks in the hanging wall of this detachment fault were transported 5–9 km to the southwest during at least two phases of west-directed sliding (Fowler et al., 1995).

Crosscutting geologic relations bracket the age of crustal extension associated with the Kingston Range–Halloran Hills detachment system. The Kingston Range detachment fault cuts 16.0-Ma (Friedmann, 1996) ash and is cut (Wright, 1968) and deformed (Calzia et al., 1986) by the 12.4-Ma granite of Kingston Peak; 12.5-Ma syntectonic andesite flows are present in the upper plate of the Kingston Range detachment fault (Calzia, 1990). The Halloran Hills detachment fault cuts a 13.4-Ma hypabyssal sill; 13.1-Ma volcanic breccia was deposited on the subsiding hanging wall of this detachment fault (Friedmann et al., 1996). Subhorizontal basalt flows and the undeformed Tecopa lake beds unconformably overlie east-tilted strata related to crustal extension south and west, respectively, of the Kingston Range. The basalt flows yield K–Ar ages of 4.48 Ma (Dohrenwend et al., 1984) and 5.12 Ma (Turrin et al., 1985); the lake beds may be as old as a 7-Ma tuff (Louie et al., 1992). These data indicate that crustal extension in and around the Kingston Range began between 13.4 and 13.1 Ma and stopped by 5 Ma.

3. Rapakivi granites of the southern Death Valley region

Magmatic rocks coeval with Cenozoic extension in the southern Death Valley region include volcanic flows, tuffs, and volcaniclastic deposits as well as granitic plutons, dikes, and sills. Calzia and Finnerty (1984) divided the granitic plutons into two petrogenetic suites based on chemical and textural data. The older suite yields K–Ar ages of 12–14 Ma and consists of alkalic granites characterized by hypabyssal textures, mafic xenoliths, and mantled feldspars. The younger suite yields K–Ar ages of 10–6.5 Ma and consists of calc-alkaline quartz monzonites with medium- to coarse-grained equigranular and porphyritic textures. This paper describes the geology, petrography, and geochemistry of three plutons from the older suite: the granite of Kingston Peak, the Little Chief stock, and the Shoshone pluton. All of these plutons are characterized by rapakivi textures and similar magmatic histories.

3.1. Granite of Kingston Peak

3.1.1. Granite petrography

The granite of Kingston Peak forms an elliptical batholith, 14.6 km long and 10.5 km wide, in the southwest half of the Kingston Range (Fig. 2). This hypabyssal granite intrudes gneiss and the lower member of the Crystal Spring Formation and is divided into (oldest first) feldspar porphyry, quartz porphyry, and aplite facies based on textural variations and intrusive relations. Aplite dikes and quartz veins are common in all three facies; rhyolite porphyry dikes and mafic xenoliths are common only in the feldspar porphyry and quartz porphyry facies. The modes of these rocks are listed in Table 1 and shown in Fig. 3.

The feldspar porphyry facies consists of gray fine- to medium-grained biotite hornblende granite porphyry characterized by seriate feldspar phenocrysts up to 15 mm long, rapakivi textures, and miarolitic cavities. Subhedral to anhedral K-feldspar phenocrysts, 2–4 mm long, are turbid, perthitic, and locally altered to sericite; quartz, plagioclase, and zircon inclusions are common. K-feldspar in the core of rapakivi textures is partially rounded, locally embayed, and bounded by a discontinuous rim of

Table 1
Modal analyses (in volume percent) of Miocene rapakivi granites, southern Death Valley region, CA

Percent phenocrysts and groundmass divided into different minerals and normalized to 100%

Spl no.	Phenocrysts (<2.0 mm)			Groundmass >(>2.0 mm)							
	Percent phenocrysts	Percent total feldspar		Percent groundmass	Quartz	Plagioclase	K-feldspar	Mafics[a]	Opaque	Vugs	Accessory
		Plagioclase	K-feldspar								
Granite of Kingston Peak											
Feldspar porphyry facies											
4A	35.3	50.4	49.6	64.7	24.6	4.5	68.9	2.0	2.0–3.0		
6A	32.1	68.8	31.2	67.1	37.7	6.7	50.3	5.3	1.0–2.0	0.8	
190.0	19.2	69.8	30.2	80.8	35.4	10.8	49.6	4.2	1.0	0.1	
Quartz porphyry facies											
36.0	6.5	24.6	75.4	93.2	32.8	8.2	51.2	7.8	<1	0.3	
42.0	13.9	60.4	39.6	85.7	21.7	20.1	50.2	8.1	1.0–2.0	0.4	
107.0	39.4	46.3	53.7	60.6	26.8	5.0	61.8	6.5	2.0–3.0		
Aplite facies											
H3	3.2		100.0	96.6	41.6	0.2	56.5	1.7	<1	0.2	
34.0	1.4	57.1	42.9	98.6	34.0	9.0	54.9	2.1	<1	0.3	
Little Chief stock											
Southern facies											
DV74	24.0	50.0	50.0	76.0	31.6	9.9	48.7	7.2	1.3		1.3
Topping A	35.0	14.3	85.7	65.0	46.2	23.1	23.1	3.1*	4.6		
Northern facies											
DV56	43.5	77.0	23.0	56.5	34.5		51.3	9.7	2.7	0.9	
DV128	33.5	9.0	91.0	66.5	37.6	1.5	52.6	6.0	1.5		
DV151	38.0	15.8	84.2	62.0	46.0	1.6	42.7	5.6	1.6	0.8	1.6
Shoshone pluton											
Rim facies											
D2	30.3	76.2	23.8	69.3	22.8	10.7	60.8	3.0*	2.0		
A13	9.9	75.8	24.2	90.1	24.0	10.0	62.5	2.0*	0.6		
Topping D	20.0	65.0	35.0	80.0	28.8	18.8	45.0	6.3*	1.3		
Core facies											
G2	16.3	85.3	14.7	83.8	27.9	7.5	61.0	2.6*	0.5		
E3	25.4	69.3	30.7	74.6	23.1	13.9	57.5	4.0*	0.9		
Aplite facies											
Apl				100.0	12.0	18.0	66.0	3.0*	0.5		

[a] Includes biotite and amphibole.

* biotite only.

quartz mantled by oligoclase; a few mafic minerals occur within the plagioclase mantles. The abundance of mafic minerals in the plagioclase mantle and their absence in the K-feldspar core indicate that these rapakivi textures developed synchronous with or slightly later than crystallization of the mafic minerals. Subhedral to anhedral oligoclase phenocrysts are embayed and fractured; the fractures are filled with opaque oxides.

The fine-grained aplitic groundmass of the feldspar porphyry facies consists of subhedral quartz, K-feldspar, and plagioclase as well as mafic and accessory minerals; graphic granite and granophyric intergrowths are common. Groundmass plagioclase is albite and is more varied in composition than oligoclase phenocrysts. Mafic minerals include biotite and hornblende in nearly equal proportions. Subhedral and anhedral biotite crystals, generally 0.3–0.5 mm and rarely 1 mm across, are pleochroic and poikilitic with apatite, opaque oxides, and zircon inclusions; molar Fe/(Fe+Mg) ratios vary from 0.17 to 0.25. Subhedral to euhedral hornblende crystals are 0.5–0.8 mm long,

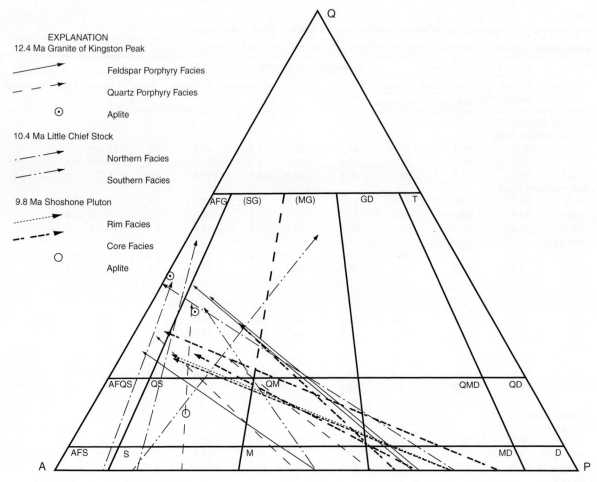

Fig. 3. Modal mineral classification diagram of Miocene rapakivi granites from the southern Death Valley region, California. Arrows represent approximate feldspar tie lines from phenocrysts to groundmass compositions. Q=quartz, A=alkali feldspar, P=plagioclase; Q+A+P=100%. Field boundaries and nomenclature from Streckeisen (1973).

usually altered to chlorite and opaque oxides, and often covered by a reddish brown opaque dust; molar $Mg/(Mg+Fe)$ ratios from hornblende vary from 0.71 to 0.73. Clots of biotite, titanite, and opaque oxides form pseudomorphs after amphibole; these clots probably represent hornblende that has reacted with the melt during crystallization. Accessory minerals include anhedral zircon inclusions in K-feldspar, plagioclase, and biotite, subhedral to euhedral apatite that forms blunt equant crystals 0.2–0.3 (locally up to 0.5) mm long, and subhedral titanite with opaque oxides, apatite, and zircon inclusions. Euhedral and anhedral opaque oxides are 0.2–0.8 mm across and contain apatite inclusions.

The quartz porphyry facies consists of pale gray to white fine- to medium-grained granite porphyry characterized by anhedral (subrounded) quartz phenocrysts, up to 2 or 3 mm across, and seriate feldspar phenocrysts up to 12 mm long; perthite, graphic granite, granophyric intergrowths, and rapakivi textures are common. The quartz phenocrysts locally contain subhedral K-feldspar inclusions. Subhedral to anhedral K-feldspar phenocrysts are generally 1–2 mm long (maximum 5–8 mm) and contain many inclusions of quartz and plagioclase. K-feldspar in the core of rapakivi textures is rounded and bounded by a discontinuous quartz rim mantled by plagioclase; anti-rapakivi textures are rare but are more common in this facies than in the feldspar porphyry facies. Anhedral to subhedral oligoclase phenocrysts are generally 1–4 mm long and rarely up to 12 mm long. These phenocrysts are poikilitic with abundant anhedral quartz, K-feldspar, and mafic inclusions.

The groundmass of the quartz porphyry facies is aplitic and consists of anhedral (rarely subhedral) very fine-grained quartz, perthitic K-feldspar, and plagioclase; miarolitic cavities are common and are often filled by secondary quartz. Biotite occurs as individual anhedral and subhedral crystals, 0.5–1.0 mm across, or in clots, up to 3 mm long, with opaque oxides. Biotite crystals are pleochroic and poikilitic with euhedral apatite and zircon inclusions; $Fe/(Fe+Mg)$ varies from 0.26 to 0.3. Hornblende is absent although most of the clots of biotite and opaque oxides (locally with zircon and feldspar) form pseudomorphs after amphibole. Accessory minerals include large (0.4–0.6 mm) euhedral opaque oxides, subhedral to euhedral apatite crystals up to 0.2 mm long, and anhedral zircon; titanite is rare or absent. The quartz porphyry facies contains about as much opaque oxides as does the feldspar porphyry facies; the abundance of opaque oxides decreases, however, toward contacts with the older rocks. Smaller (=0.1 mm) opaque oxide crystals and zircon occur as inclusions in biotite.

The aplite facies is generally fine-grained and equigranular but does contain a few ($\leq 3\%$) K-feldspar phenocrysts; miarolitic cavities, graphic granite, and granophyric intergrowths are common. Anhedral to subhedral quartz crystals are generally 0.2–0.3 mm across; K-feldspar phenocrysts, up to 2 mm long, are subhedral and perthitic. Plagioclase occurs as euhedral crystals of albite and oligoclase in the groundmass and as anhedral inclusions of calcic-rich oligoclase in the K-feldspar phenocrysts. Subhedral to euhedral biotite crystals are poikilitic with apatite inclusions; amphibole is absent. Accessory minerals include subhedral apatite, characterized by rounded terminations, and rare zircon. Titanite is nearly absent; a single crystal was observed in one aplite plug in the feldspar porphyry facies. The aplite facies contains less than one percent subhedral to euhedral opaque oxides uniformly distributed throughout the rock.

3.1.2. Mafic xenoliths

Mafic xenoliths are found as rounded to irregular-shaped clots (up to 10 mm across) or teardrop-shaped inclusions (up to several centimeters long) in the feldspar porphyry and quartz porphyry facies. The xenoliths are widely scattered throughout the batholith but are most common near the contact between these two facies. Dark, teardrop-shaped vesicular mafic xenoliths form sharp contacts with the feldspar porphyry facies; thin discontinuous gray zones around the margin of a few xenoliths are caused by small apophyses of the granite in the xenolith.

The mafic xenoliths consist mainly of fine-grained subhedral andesine, hypersthene mostly altered to chlorite, and abundant opaque oxides; green hornblende, up to 6 mm long, is common. Spherulitic biotite forms a reaction rim around an unidentified mafic mineral that is completely altered to chlorite or actinolite; the shape of the biotite reaction rim suggests that the original mineral was amphibole. A few embayed and skeletal crystal clots have clinopyroxene habit but are completely altered to chlorite, biotite, and opaque oxides; their original composition could not be identified.

The feldspar porphyry facies adjacent to the mafic xenoliths is generally finer grained and contains less plagioclase that is more sodic than normal for this rock; green hornblende, locally with blue pleochroism, is common. The change in pleochroism indicates that the hornblende is more sodic than other hornblendes in this facies. Rare orthopyroxene crystals in the granite are completely surrounded by opaque oxides and are altered to talc or actinolite. The abundance and texture of the orthopyroxene crystals suggest that they are xenocrysts not in equilibrium with the adjacent melt during crystallization.

3.1.3. Dikes and veins

Rhyolite porphyry and aplite dikes as well as quartz veins are common in and adjacent to the granite of Kingston Peak. The rhyolite porphyry dikes cut all but the aplite facies of the granite of Kingston Peak and are cut by aplite dikes. Quartz veins fill joints and faults that crosscut the dikes and are the youngest rocks coeval with the granite of Kingston Peak.

Gray, white, and purple biotite rhyolite porphyry dikes consist of quartz and K-feldspar phenocrysts in an aphanitic groundmass; euhedral to subhedral oligoclase and albite phenocrysts are rare. The groundmass consists of very fine-grained subhedral to anhedral quartz and turbid K-feldspar. Subhedral hornblende crystals with tan to green pleochroism are rare and are restricted to the coarser-grained cores of the larger dikes. Most hornblende crystals are replaced by clots of subhedral brown biotite altered to chlorite or actinolite, opaque oxides, and red brown opaque stain. Accessory minerals include fine-grained subhedral to euhedral opaque oxides, very fine-grained anhedral zircon, and equant or elongate apatite crystals up to 0.2 mm long.

Fine-grained equigranular aplite dikes consist of subhedral to anhedral quartz, subhedral perthitic K-feldspar, albite, and less than one percent opaque oxides. Subhedral biotite crystals with apatite inclusions are generally less abundant than opaque oxides but locally constitute 1% to 3% of the rock; amphibole(?) completely altered to chlorite is rare.

White quartz veins are most common in the eastern half of the batholith. Although the veins are generally massive, color bands and vesicles are common; the vesicles are locally filled with amethyst. Locally, the vesicles are so common that the veins develop a "honeycomb" or crude stockwork texture on weathered surfaces; these stockwork veins contain anomalous concentrations of lead and barium but no sulfide or oxide minerals (Calzia et al., 1986). The "honeycomb" texture is caused by erosion and leaching of soluble inclusions (probably carbonate material).

3.1.4. Age

The age of the granite of Kingston Peak was determined by conventional K–Ar and ^{40}Ar/^{39}Ar methods (Calzia, 1990). Biotite and hornblende from the feldspar porphyry facies yield concordant K–Ar ages of 12.1 ± 0.9 and 12.4 ± 0.3 Ma, respectively. Incremental heating of hornblende from the same sample yields a plateau age of 12.42 ± 0.04 Ma for more than 95% of the ^{39}Ar released during the experiment, and a recombined total fusion age of 12.36 ± 0.18 Ma. The concordant K–Ar and ^{40}Ar/^{39}Ar ages indicate that the granite of Kingston Peak is middle Miocene.

3.1.5. Granite of Ibex Pass

The fine- to medium-grained granite of Ibex Pass in the Sperry Hills (Fig. 2) is characterized by rapakivi textures and is modally, chemically, and isotopically equivalent to the 12.4-Ma granite of Kingston Peak (Calzia et al., 1991); biotite from the granite of Ibex Pass yields a K–Ar age of 12.6 ± 0.3 Ma (J.P. Calzia, unpublished data). Although Burchfiel et al. (1985) suggested that granitic rocks in the Sperry Hills once rested on the granite of Kingston Peak, L.A. Wright (written commun., 1993) and McMackin (1997) proposed that the granite of Ibex Pass is autochthonous bedrock. The granite of Ibex Pass has no exposed intrusive contacts; it tectonically overlies miogeoclinal rocks south of Tecopa Peak, and is tectonically interleaved with late Miocene China Ranch beds in the Sperry Hills. Isostatic gravity data indicate that this granite forms a tabular body approximately one km thick and is rootless (Calzia et al., 1991). These geologic and geophysical data, combined with chemical, isotopic, and geochronological observations, support the Burchfiel et al. hypothesis, as refined by Topping (1993) and Holm et al. (1994), and suggest that the granite of Ibex Pass is a large landslide deposit derived from the granite of Kingston Peak.

Disaggregated hypersthene occurs in vesicular mafic xenoliths in the granite of Ibex Pass. Hypersthene is surrounded by, and isolated from, the xenolith host by an assemblage of quartz, feldspar, biotite, titanite, and zircon characterized by a medium-grained granitic texture. Modal and textural data suggest that these granitic assemblages represent blebs of granitic melt that invaded the xenolith host; the abundance and shape of these blebs, as well as their close spatial association with disaggregated hypersthene, suggest that the granitic melt filled vesicles and reacted with hypersthene in the mafic xenolith. These observations, combined with the abundance of orthopyroxene xenocrysts in the granite of Kingston Peak, indicate that some type of magmatic and chemical interaction occurred between Cenozoic granitic and mafic melts in the southern Death Valley region.

3.2. Little Chief stock

3.2.1. Granite petrography

The Little Chief stock is a dome-shaped hypabyssal stock along the crest of the Panamint Range south of Telescope Peak (Fig. 2). This stock, 28 km^2 in outcrop, intrudes the Pahrump Group as well as the oldest miogeoclinal rocks, with sharp discordant contacts; a chill zone approximately 0.3 m wide is common between the stock and the country rocks (Albee et al., 1981). Regional cross sections suggest that the Little Chief stock has a flat floor at the contact between gneiss and carbonate rocks in the Pahrump Group (McDowell, 1967).

The Little Chief stock is divided into southern and northern facies (McDowell, 1967). The contact between facies is sharp; apophyses within the chill zone indicate the southern facies is intruded by the northern facies. Inclusions of an earlier magmatic phase are common in both facies.

The southern facies consists of medium gray hornblende–biotite granite with plagioclase and sanidine phenocrysts, 2–5 mm across, in a very fine-grained (0.2–0.5 mm) groundmass of K-feldspar, quartz, and plagioclase; megacrysts of K-feldspar and plagioclase, up to 15 mm across, are present but rare. Plagioclase phenocrysts are characterized by embayed calcic cores with sodic plagioclase rims; compositions vary from andesine to oligoclase from core to rim, respectively. The rims are characterized by one to three compositional reversals and locally yield very sodic compositions. Subhedral sanidine phenocrysts are usually embayed and bounded by rims of cloudy or mottled oligoclase; quartz blebs along the core/rim boundary are common. The groundmass consists of quartz, sanidine, plagioclase, and less than seven percent mafic minerals. Groundmass sanidine forms equant crystals, locally with zoned plagioclase cores. Plagioclase in the groundmass occurs as glomeroporphyritic aggregates, up to 10 mm across, associated with hornblende and biotite.

The northern facies is divided into exterior and interior phases; contact relations indicate that the exterior phase was intruded by the interior phase (McDowell, 1967). The exterior phase consists of pinkish gray biotite–hornblende granite with plagioclase and sanidine phenocrysts up to 7 mm across; the interior phase is similar and consists of white to very light gray hornblende–biotite granite with phenocrysts up to 10 mm across. Euhedral to subhedral plagioclase phenocrysts consist of embayed labradorite cores with oligoclase rims. Albite twins are common in the cores of these phenocrysts; the rims are zoned but not twinned. Euhedral to subhedral sanidine phenocrysts have partially embayed cores and oligoclase rims 0.5–1 mm wide; inclusions of euhedral plagioclase with albite twins are common in the core of sanidine phenocrysts. The inclusions occur as laths aligned

with the long dimension of the sanidine host, as disoriented inclusions near the outer edge of sanidine phenocrysts, or as patch perthite; graphic granite is common. The groundmass of the exterior and interior phases consists of quartz, sanidine, plagioclase, and <10% mafic minerals 1–6 mm long; megacrysts of plagioclase aggregates and sanidine up to 20 mm across are common. Vugs up to 5 mm in diameter as well as pegmatite and aplite dikes are common in the northeast part of the northern facies; the vugs constitute less than one percent of the rock.

A contact facies surrounds the northern facies as well as the east side of the southern facies and consists of anhedral sanidine and plagioclase phenocrysts in a fine grained groundmass; vermicular to graphic granite textures are common in the groundmass of the contact facies. Sanidine phenocrysts are characterized by lamellar perthite; plagioclase phenocrysts are complexly zoned with andesine cores and oligoclase rims; the outer edge of the rims includes a thin compositional reversal to andesine.

Mafic and accessory minerals, including hornblende, biotite, titanite, magnetite, and apatite, are common in all facies of the Little Chief stock. Hornblende generally forms euhedral prisms, 0.2–5 mm long; Fe/ (Fe+Mg+Mn) is approximately 0.5. The core of hornblende crystals is often skeletal and replaced by biotite, quartz, and K-feldspar. Calcium- and magnesium-rich augite is present but rare in the core of several hornblende crystals in the southern facies; reaction rims of biotite and diopside around hornblende are common in the contact facies. Euhedral to anhedral biotite occurs as scattered plates up to 1 mm across or in clots of hornblende, magnetite, and plagioclase in the groundmass; the biotite plates yield lower Mg/(Mg+Fe) than groundmass biotite. Biotite pseudomorphs after pyroxene are common throughout the stock. Titanite crystals, 0.1–0.6 mm long, are euhedral but skeletal. Magnetite is anhedral and less than 1 mm across. Apatite occurs as a replacement product after Na-plagioclase and K-feldspar.

3.2.2. Magmatic inclusions

Round inclusions of an earlier magmatic phase are concentrated in an east–west zone near the contact between the southern and northern facies. The inclusions consist of green very fine-grained igneous rock with sanidine phenocrysts in a groundmass of quartz and plagioclase. The sanidine phenocrysts are rimmed by cloudy plagioclase with quartz blebs; plagioclase laths in the groundmass are 0.2–1 mm long and rimmed by K-feldspar. The igneous inclusions consist of 20–25% mafic and accessory minerals including hornblende, biotite, magnetite, titanite, apatite, and secondary chlorite; hornblende is locally rimmed by diopside. The inclusions include rapakivi and anti-rapakivi textures and are not flow banded. Some inclusions include euhedral plagioclase phenocrysts with numerous quartz blebs and albite twinning.

3.2.3. Dikes

Felsic dikes and sills that are pre- and synmagmatic with the Little Chief stock are divided into two dike swarms (McDowell, 1967). North-trending porphyritic rhyolite dikes that extend from the northeast corner of the stock are truncated by the northern facies and are older than the Little Chief stock. Many of these dikes trend into breccia that consists of angular fragments of country rocks in a matrix of dike rock. A southern dike swarm consists of a vertical 100-m-thick dike that divides into numerous subparallel dikes. This dike swarm follows pre-existing faults and is 100–300 m wide approximately 0.5 km from the stock.

A dike sheath injected between the Little Chief stock and the country rocks consists of very fine-grained aplite characterized by spherulites and gray laminations; inclusions of dolomite, approximately 30 m long, are stoped into the dike sheath. The dike sheath is older than the southern facies, younger than the northern dike swarm, and is synmagmatic with the northern facies of the Little Chief stock (McDowell, 1967).

3.2.4. Age

K-feldspar from a monzonite boulder at the mouth of Hanaupah Canyon yields a conventional K–Ar age of 12.5 ± 1.3 Ma (Stern et al., 1966). Hodges et al. (1990) reported a three-point Rb–Sr isochron of 10.6 ± 0.2 Ma using plagioclase, biotite, and hornblende. Topping (1993) reported a zircon fission-track age of 11.2 ± 0.5 Ma for the southern facies and 9.8 ± 0.5 Ma for the northern facies. Although discordant, the isotopic data suggest that the Little Chief stock is middle Miocene and may be as old as 12.5 Ma.

3.3. Shoshone pluton

3.3.1. Granite petrography

The Shoshone pluton is the youngest intrusive rock in the southern Greenwater Mountains (Fig. 2). This elliptical pluton, 3 km wide and 4 km long, is tilted to the east and has no exposed intrusive contacts; it is unconformably overlain by Miocene volcanic rocks.

The Shoshone pluton consists of gray porphyritic quartz monzonite with phenocrysts of plagioclase and K-feldspar in pink groundmass. The pluton is texturally zoned into rim, core, and aplite facies. The rim facies is intruded by the core facies; both are characterized by miarolitic cavities and rapakivi textures.

The rim and core facies consist of phenocrysts of plagioclase, microcline, quartz, and biotite as well as rare pseudomorphs of brown chlorite, hematite, and calcite after amphibole. Euhedral to subhedral oligoclase phenocrysts are 0.2–1.0 cm across and are zoned and sericitized. Microcline occurs as ovoid phenocrysts or as micrographic intergrowths with quartz; the ovoid phenocrysts are slightly perthitic and rimmed by oligoclase (Fig. 1B). The interface between microcline and oligoclase is uneven to serrate; the oligoclase rim is slightly more Ab-rich than coexisting plagioclase phenocrysts. Subhedral quartz phenocrysts are strongly embayed. Euhedral biotite phenocrysts are rimmed by aggregates of magnetite, zircon, and ilmenite needles; $FeO^*/(FeO^*+MgO)$ of biotite varies from 0.48 to 0.5. Euhedral to anhedral magnetite occurs as reaction rims around dark reddish-brown biotite phenocrysts; the magnetite is oxidized to hematite (locally limonite), Mn-rich ilmenite, and pseudobrookite.

The rim and core facies are differentiated by aphanitic and granophyric groundmass, respectively (Almashoor, 1980). The aphanitic groundmass consists of an equigranular aggregate of microcline, quartz, and plagioclase; the granophyric groundmass is characterized by resorbed quartz microphenocrysts, 1.5 mm in diameter, that include cuneiform, vermicular, and radiating textures. The groundmass of each facies contains less than five percent biotite and magnetite; apatite inclusions and red brown rutile needles are common in groundmass biotite.

The white fine-grained aplite facies is equigranular and consists of anhedral plagioclase and granophyric intergrowths of quartz and K-feldspar. Although the contact with the core facies is poorly exposed, Almashoor (1980) concluded that the aplite facies is a late differentiate of the core facies.

3.3.2. Xenoliths, inclusions, and dikes

The Shoshone pluton includes numerous green xenoliths as well as green and gray inclusions and is cut by several generations of dikes. The green xenoliths are igneous in origin, felsic to intermediate in composition, and are rounded, up to 15 m in diameter, or elongated north–south. Dark green quartz diorite inclusions consist of fine- to coarse-grained hornblende in a groundmass of white feldspar; the randomly oriented hornblende crystals are 1–1.5 cm long and are characterized by acicular to lath-shaped textures. Gray felsic inclusions consist of 20–25% plagioclase, rare K-feldspar, one to two percent hornblende, and are characterized by granophyric textures and mafic inclusions.

Latite, andesite, and lamprophyre dikes cut the Shoshone pluton. Pink fine-grained quartz latite dikes consist of plagioclase and K-feldspar phenocrysts, up to 5 mm long, in a microcrystalline groundmass. These dikes are 1.5–5 m wide and locally emplaced along faults that are synmagmatic with the Shoshone pluton. Green andesite dikes and plugs consist of plagioclase, augite, hypersthene, and magnetite; plagioclase crystals are characterized by concentric compositional zoning. One andesite dike cuts a latite dike and co-mingles with the quartz monzonite. Andesite plugs are elliptical and as much as 25 m long. Dark gray to greenish gray porphyritic lamprophyric dikes consist of dark green hornblende and rare biotite phenocrysts in a holocrystalline groundmass of hornblende, andesine, and minor biotite.

3.3.3. Age

The Shoshone pluton is disconformably overlain by 7.5- to 8.5-Ma (Wright et al., 1991) dacite and rhyolitic flows and tuffs of the Shoshone Volcanics; biotite from the rim facies yields a $^{40}Ar/^{39}Ar$ total fusion age of 9.76 ± 0.25 Ma (Holm et al., 1992). These data suggest that the Shoshone pluton is late Miocene.

4. Geochemistry

The chemical compositions of rapakivi granite samples from the southern Death Valley region are listed in Table 2 and shown on Figs. 4 and 5. Samples from these granites span a relatively large range in SiO_2 (69.8 to 77.6 wt.%), straddle the peraluminous–metaluminous boundary (Fig. 4A), are moderately enriched in Fe relative to Mg ($FeO^*/[FeO^*+MgO]$ varies from 0.75 to 0.95; Fig. 4B), have high alkali metal and intermediate high field strength element (HFSE) contents (Fig. 4C), and mostly plot in the within-plate field on Nb vs. Y tectonomagmatic discrimination diagrams (Fig. 4D). The Shoshone pluton is less enriched in Fe relative to Mg (Fig. 4B) and has lower contents of HSFE than the granite of Kingston Peak, granite of Ibex Pass, and Little Chief stock. Thus the Shoshone pluton is less like typical A-type (Fig. 4C) and within-plate (Fig. 4D) granites than other rapakivi granites in the southern Death Valley region.

Chemical analyses of the ca. 1.6 Ga rapakivi granites of Finland and 1.88 Ga anorogenic granites from the Amazon craton are also shown on Fig. 4. Rapakivi granites in the southern Death Valley region are similar to the Finnish rapakivi granites in terms of aluminum saturation and within-plate geochemical character (Fig. 4A and D), but are, in general, not as strongly enriched in Fe relative to Mg and HFSE (Fig. 4B and C). These data indicate that rapakivi granites in the southern Death Valley region are more oxidized than the "type" rapakivi granites (cf. Frost and Frost, 1997); they more nearly resemble the more oxidized anorogenic granites of the Amazon craton (Fig. 4B; Dall'Agnol et al., 1999).

The rare earth element (REE) content of rapakivi granites in the southern Death Valley region is shown in chondrite-normalized diagrams in Fig. 5. The overall patterns of these three plutons are similar and show a slightly less pronounced enrichment in heavy REE relative to rapakivi granites from Finland (Fig. 5A–C). Relative enrichment in REE contents may be caused by a lower degree of partial melting of the source rocks of the Death Valley granites.

The initial Nd, Sr, and Pb isotope compositions of the Miocene rapakivi-type granites in the southern Death Valley region are listed in Table 3 and shown in Figs. 6 and 7. Fig. 6 shows the initial Nd and Sr isotope compositions of these rapakivi granites compared to those of Mesozoic granitoids from the Sierra Nevada and Peninsular Ranges batholiths, Cretaceous granitoids of the Mojave Desert, and Cretaceous batholiths of the southern Death Valley region. The rapa-

Table 2
Geochemistry of Miocene rapakivi granites, southern Death Valley region, California

Column groupings:

- **Kingston Range**
  - *Granite of Kingston Peak*
    - Feldspar porphyry facies: 4A, 6A
    - Quartz porphyry facies: 190, 42, 36, 107
    - Aplite facies: H3, 34
  - *Mafic xenolith*: 170
  - *Dikes*: Mafic (164), Rhy. porp. (43), Aplite (166)
- **Sperry Hills**
  - *Syenite Peak 4611*: 47
  - *Granite of Ibex Pass*: 52, 276
- **Panamint Mtns** — *Little Chief stock, Southern phase*: LCS, 3B, Topping A
- **Greenwater Mtns** — *Shoshone pluton*: 10, 126, Topping D

Sample no.	4A	6A	190	42	36	107	H3	34	170	164	43	166	47	52	276	LCS	3B	Topping A	10	126	Topping D
Major element geochemistry (wt.%)																					
SiO_2	72.90	71.10	71.80	69.70	71.90	71.40	77.60	74.40	54.30	60.00	70.70	69.00	68.70	72.60	72.20	69.80	67.98	74.64	70.50	70.80	71.16
TiO_2	0.28	0.35	0.33	0.40	0.33	0.36	0.11	0.14	1.15	0.73	0.42	0.46	0.38	0.34	0.40	0.42	0.43	0.21	0.34	0.36	0.34
Al_2O_3	13.60	14.10	14.20	15.00	13.90	14.40	12.20	12.40	16.50	17.80	15.50	4.17	15.50	13.50	13.90	14.50	15.90	12.94	14.40	13.80	14.33
Fe_2O_3	0.77	1.37	1.50	1.71	1.52	1.55	0.63	0.42	3.08	3.00	2.02	0.24	1.84	1.26	1.31	1.72	1.69	1.33*	1.96	2.79*	2.2*
FeO	0.30	0.44	0.37	0.38	0.25	0.27	0.07	2.26	2.65	2.07	0.10	1.09	0.13	0.50	0.49	0.74	0.56		0.31		
MnO	0.02	0.04	0.03	0.08	0.06	0.02	0.02	0.03	0.57	0.03	0.09	0.03	0.05	0.06	0.08	0.10	0.06	0.12	0.05	0.12	0.05
MgO	0.21	0.49	0.40	0.58	0.35	0.36	0.10	0.16	5.19	1.73	0.12	7.59	0.18	0.40	0.45	0.68	1.88	0.14	0.60	0.73	0.62
CaO	0.82	1.08	1.03	0.95	0.61	0.78	0.33	0.38	4.81	3.82	0.70	10.50	0.54	1.11	0.82	1.41	0.96	0.25	1.90	1.56	1.44
Na_2O	3.78	4.01	4.16	4.43	3.88	4.16	3.99	3.57	4.78	5.23	8.81	1.32	3.99	3.82	4.00	4.65	5.25	4.46	3.64	3.95	4.02
K_2O	5.45	5.09	4.98	5.01	5.61	4.95	4.56	4.89	3.36	3.89	0.07	1.09	7.47	5.15	5.42	4.70	4.65	4.98	4.70	4.63	5.15
H_2O^+	0.32	0.35	0.33	0.80	0.40	0.44	0.16	0.40	1.42	0.49	0.34	0.95	0.32	0.24	0.19	0.36	0.52	0.38*	0.37	0.75*	0.81*
H_2O^-	0.33	0.38	0.10	0.56	0.24	0.09	0.06	0.15	0.37	0.12	0.35	0.57	0.18	0.23	0.10	0.26	0.37		0.24		
P_2O_5	0.08	0.10	0.10	0.19	0.08	0.13	0.05	0.05	0.74	0.52	0.08	0.11	0.07	0.10	0.20	0.17	0.58	0.08	0.09	0.09	0.15
CO_2	0.10	0.10	0.09	0.20	0.26	0.10	0.24	0.28	0.31	0.13	0.40	2.97	0.13	0.26	0.06	0.30	0.10		0.35		
Cl	0.12	0.06	0.13	0.04	0.03	0.03	0.02	0.03	0.02	0.05	0.01	0.02	0.01	0.06	0.02	0.04	0.02		0.01		
F	0.04	0.07	0.07	0.08	0.06	0.06	0.01	0.03	0.64	0.07	0.02	0.04	0.04	0.06	0.06	0.08	0.06		0.04		
Total	99.12	99.13	99.62	100.11	99.48	99.10	100.15	99.59	99.89	99.68	99.73	100.15	99.53	99.65	99.70	99.93	101.01	97.82	99.50	96.04	97.26
CIPW normative minerals																					
Q	28.2	25.6	26.0	22.4	26.3	26.4	36.3	32.4	0.0	5.7	17.8	33.8	15.7	29.2	26.9	22.6	16.8	29.4	28.1	26.4	24.7
C	0.2	0.3	0.3	1.0	0.5	1.1	0.2	0.7	0.0	0.0	0.0	0.0	0.0	0.6	0.7	0.5	2.2	0.0	1.0	0.0	0.0
OR	32.2	30.1	29.4	29.6	33.1	29.2	26.9	28.9	0.0	0.0	0.4	6.4	44.1	30.7	32.2	30.0	27.5	29.7	28.1	27.7	30.6
AB	32.0	33.9	35.2	37.5	32.8	35.2	33.8	30.2	40.4	44.3	74.5	11.2	33.8	32.4	33.9	39.3	44.2	38.1	31.1	33.8	3.0
AN	3.5	4.7	4.5	3.5	2.5	3.0	1.3	1.6	13.6	13.6	2.5	2.2	2.2	2.9	2.1	3.5	0.3	0.6	6.5	6.3	34.2
DI	0.0	0.0	0.0	0.0	0.0	0.0	0.0	0.0	4.2	6.1	0.0	38.4	0.0	0.0	0.0	0.0	0.0	0.0	0.0	0.1	0.0
HY (EN+FS)	0.5	1.2	1.0	1.4	0.9	0.9	0.0	0.0	6.1	3.8	0.3	2.0	0.4	1.0	1.1	1.7	4.7	0.4	1.5	1.8	1.6
MT	0.2	0.5	0.3	0.3	0.5	0.3	0.2	0.0	4.4	4.4	0.3	0.3	0.4	0.8	0.7	1.5	0.8	0.4	0.2	0.0	0.0
HM	0.6	1.0	1.3	1.5	1.5	1.6	0.6	0.6	0.0	0.0	1.8	0.0	1.8	0.7	0.8	0.7	1.2	1.3	1.9	2.8	2.2
IL	0.5	0.7	0.6	0.8	0.6	0.6	0.3	0.3	1.4	1.4	0.4	0.9	0.4	0.7	0.8	0.8	0.8	0.3	0.7	0.3	0.1
TN	0.0	0.0	0.0	0.0	0.0	0.0	0.0	0.0	0.0	2.2	0.0	0.0	0.0	0.0	0.0	0.0	0.0	0.1	0.0	0.0	0.0
RU	0.0	0.0	0.0	0.0	0.0	0.0	0.0	0.0	0.0	0.0	0.1	0.0	0.2	0.0	0.0	0.0	0.0	0.0	0.6	0.6	0.2
AP	0.2	0.2	0.2	0.4	0.2	0.3	0.1	0.1	1.2	1.2	0.2	0.3	0.2	0.2	0.5	0.4	1.4	0.2	0.2	0.2	0.4

Trace element geochemistry (ppm)

Element																					
Nb	48	30	30	28	30	30	60	32	14	26	60	10	60	28	38	34	31	114	16	14	29
Rb	186	150	168	195	168	166	237	202	247	95	10	24	314	170	170	150	139	17	120	120	
Sr	206	237	219	312	155	210	8	56	867	967	28	46	88	160	166	290	231	322	290	229	240
Zr	260	250	280	310	260	300	150	116	230	380	510	150	520	270	275	330	301	32	200	188	163
Y	30	30	32	32	34	38	16	14	30	28	34	20	40	34	46	32	29	32	20	19	18
Ba	630	750	710	960	690	710	40	200	1250	1850	640	320	480	500	620	780	700	86	1050	809	1150
Cu	20	20	20	20	20	20	20	36	20	20	20	20	20	20	20	20		15	20		
Ni	20	20	20	20	20	20	20	26	20	20	20	20	20	20	20	20		21	20		
Zn	26	60	40	80	48	50	20	32	98	45	60	30	54	64	46	90		78	56		
U																				4	12
Th																				17	32

Rare earth element geochemistry (ppm)

Element																					
La	37.60	44.00	60.50	71.80	58.50	65.10	39.10	36.00	82.30	81.40	140.00	6.00	124.00	65.20	70.00	71.00	91.00	56.38	46.00	44.10	41.49
Ce	87.90	91.90	124.00	134.00	122.00	119.00	59.10	46.00	151.00	156.00	270.00	18.00	240.00	132.00	130.00	130.00	160.00	102.56	76.00	69.60	69.53
Pr	11.10	10.20	13.30	15.60	13.50	14.10	4.40	4.90	18.30	17.30	26.00	2.80	25.70	15.00	13.00	14.00	17.00	29.19	7.40	8.40	21.41
Nd	39.00	36.70	44.90	53.80	45.60	46.80	10.10	16.00	70.10	60.40	89.00	12.00	82.10	50.00	46.00	49.00	62.00		26.00	24.60	
Sm	6.70	6.40	7.50	9.00	8.10	7.80	1.20	2.30	11.60	8.80	13.00	2.70	12.30	8.06	7.00	6.90	8.80	5.66	3.70	4.10	3.96
Eu	0.91	0.93	1.03	1.33	1.00	1.16	0.08	0.35	2.89	2.19	1.60	0.44	1.55	0.97	1.10	0.91	1.00	0.43	0.71	0.81	0.68
Gd	5.70	5.30	6.00	7.40	6.20	5.80	1.20	1.70	8.60	6.10	7.30	2.40	8.30	6.07	4.40	4.70	7.00	4.70	2.90	3.30	3.16
Tb	1.00	1.00	1.00	1.00	1.00	1.00	0.30	0.30	1.00	1.00	0.97	0.34	1.00	1.00	0.67	0.83	1.10		0.47	0.50	
Dy	4.90	4.60	5.00	5.80	4.40	4.40	0.50	1.60	5.30	3.90	5.80	2.20	6.40	5.50	4.30	4.60	5.90	4.29	2.40	2.90	2.69
Ho	0.90	0.90	0.95	1.08	0.99	0.80	0.19	0.36	0.97	0.66	1.20	0.42	1.21	1.03	0.90	0.87	1.10	0.82	0.52	0.59	0.66
Er	2.40	2.40	2.70	2.90	2.60	2.10	0.70	1.20	2.30	1.50	3.30	1.40	3.20	2.60	2.20	2.40	3.20		1.60	2.00	
Tm	0.40	0.40	0.42	0.47	0.43	0.30	0.14	0.18	0.36	0.24	0.49	0.20	0.53	0.40	0.35	0.35	0.52		0.25	0.40	
Yb	3.00	2.30	2.70	2.60	2.50	1.90	1.00	1.60	2.00	1.30	3.40	1.80	3.50	2.50	3.20	2.70	3.30	3.17	1.40	2.10	1.72
Lu	0.41	0.32	0.37	0.37	0.36	0.27	0.18	0.00	0.00	0.00	0.00	0.00	0.51	0.35				0.49	0.28	0.28	0.27

Note: Analytical methods described in Calzia (1990). Sample 126 was analyzed at the XRAL laboratories (Toronto) using methods described in Rämö et al. (2002).

* Total iron or total water.

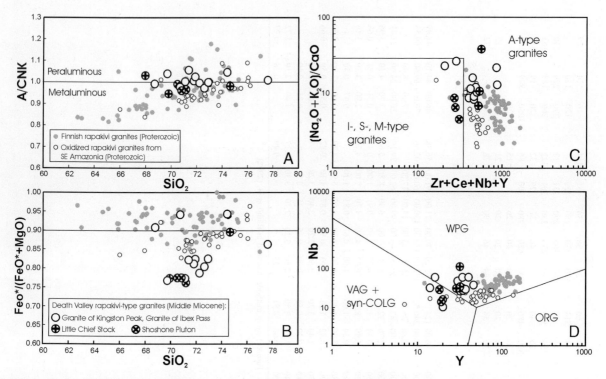

Fig. 4. Chemical composition of Miocene rapakivi granites from the southern Death Valley region compared with "type" 1.6-Ga Finnish rapakivi granites (Rämö and Haapala, 1995), and oxidized 1.88-Ga anorogenic granites from the southeastern flank of the Amazonian craton, Brazil (Dall'Agnol et al., 1999). (A) A/CNK vs SiO_2; (B) $FeO^*/(FeO^*+MgO)$ vs SiO_2; (C) $(Na_2O+K_2O)/CaO$ vs $Zr+Ce+Nb+Y$; and (D) Nb vs Y diagrams. A/CNK=molecular $Al_2O_3/(CaO+Na_2O+K_2O)$, FeO^*=total Fe expressed as FeO. Fields in (C) and (D) from Whalen et al. (1987) and Pearce et al. (1984), respectively: WPG—within plate granites, ORG—ocean ridge granites, VAG—volcanic arc granites, syn-COLG—syn-collision granites.

kivi granites are less influenced by upper crustal contamination (higher ε_{Nd}, lower Sr_i) than the Cretaceous plutons from Death Valley (Fig. 6; Rämö et al., 2002), but were probably derived from more evolved source rocks (lower ε_{Nd}, higher Sr_i) than most of the Peninsular Ranges and Sierra Nevada batholiths. These data indicate that the rapakivi granites in the southern Death Valley region include a juvenile or mantle component in their source rocks. In terms of Nd isotopes, the three plutons in the southern Death Valley region are similar, with initial ε_{Nd} values of about -6 to -7; this variation barely exceeds the experimental error of ε_{Nd}. In contrast, the initial Sr isotope composition shows more scatter (Sr_i=0.7059–0.7126) and could reflect contamination of the granitic magmas from source or country rocks with varying Sr isotope composition.

The Pb isotope composition of the Miocene rapakivi-type granites in the southern Death Valley region is shown in Fig. 7. The samples show relatively little variation and, in both diagrams, fall close to the boundary between the Precambrian Mojave and Arizona crustal provinces (MCP and ACP, respectively) of Wooden and Miller (1990). Pb isotopes of the rapakivi granites are also more radiogenic than those of Cretaceous batholiths of the southern Death Valley region. This is compatible with the hypothesis of a marked post-Cretaceous mantle component in the rapakivi granites of the southern Death Valley region.

5. Emplacement and crystallization history

Stratigraphic reconstruction of Proterozoic, Paleozoic, and Miocene country rocks in the Kingston Range as well as regional tectonic relations suggest that the granite of Kingston Peak was emplaced at shallow (≤ 4 km) crustal levels in an actively extend-

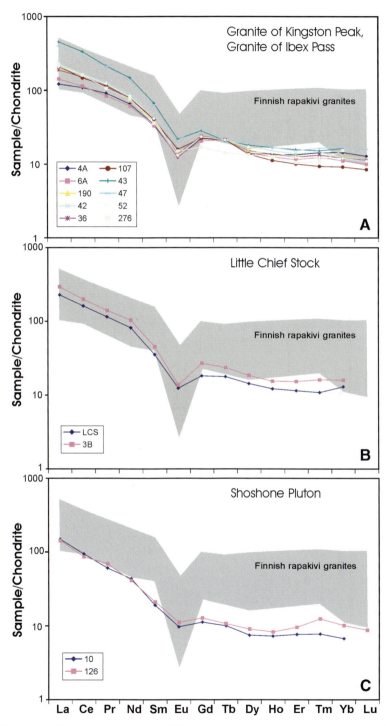

Fig. 5. Chondrite-normalized rare earth element patterns of rapakivi-type granite plutons from the southern Death Valley region: (A) Granite of Kingston Peak, Granite of Ibex Pass; (B) Little Chief stock; (C) Shoshone pluton. Overall field for the Finnish rapakivi granites (Haapala et al., 2005) is also shown.

Table 3
Isotopic data of Miocene rapakivi granites, southern Death Valley, California

Area	Kingston Range									Sperry Hills		Panamint Mtns	Greenwater Mtns	
Name	Granite of Kingston Peak (12.4 Ma)							Mafic xenolith	Syenite, Peak 4611	Granite of Ibex Pass		Little Chief stock (10.6 Ma)	Shoshone pluton (9.8 Ma)	
Description	Feldspar porphyry facies			Quartz porphyry facies			Aplite facies					Southern phase		
Sample no.	4A	6A	190	42	36	107	H3	173	47	52	7A	LCS	10	126
Lead isotopic data														
$^{206}Pb/^{204}Pb$	19.236	19.183	19.209	19.177	19.194	19.162	19.103	18.903	18.982	19.211	19.210	19.101	18.782	18.821
$^{207}Pb/^{204}Pb$	15.695	15.680	15.694	15.675	15.678	15.688	15.671	15.646	15.672	15.699	15.683	15.662	15.648	15.661
$^{208}Pb/^{204}Pb$	39.674	39.541	39.631	39.503	39.559	39.563	39.337	38.786	39.143	39.675	39.655	39.405	39.425	39.354
Strontium isotopic data														
Rb (ppm)	155	150	168	159	168	166	237	247	274	179	176	155	126	128
Sr (ppm)	182	237	219	274	155	210	8	867	77	156	143	288	291	238
Rb/Sr	0.852	0.633	0.767	0.580	1.080	0.790	29.620	0.285	3.580	1.150	1.230	0.538	0.433	0.538
$^{87}Rb/^{86}Sr$	2.470	1.830	2.220	1.670	3.140	2.290	85.760	0.825	10.370	3.320	3.560	1.560	1.250	1.550
$^{87}Sr/^{86}Sr$	0.7130	0.7088	0.7091	0.7090	0.7099	0.7091	0.7211	0.7075	0.7123	0.7102	0.7108	0.7081	0.7085	0.7082
Sr_i	0.7126	0.7085	0.7087	0.7087	0.7094	0.7087	0.7059	0.7074	0.7105	0.7096	0.7102	0.7079	0.7083	0.7080
Neodymium isotopic data														
Sm (ppm)	6.469	7.646		7.808		10.490	1.345	9.501	11.460			7.867		3.658
Nd (ppm)	37.40	46.26		45.14		63.55	10.84	62.72	78.45			48.93		23.05
$^{147}Sm/^{144}Nd$	0.10360	0.09994		0.10460		0.09977	0.07502	0.09157	0.08832			0.09720		0.09593
$^{143}Nd/^{144}Nd$	0.512236	0.512271		0.512270		0.512269	0.512340	0.512312	0.512345			0.512308		0.512308
$\varepsilon_{Nd\,i}$	−7.7	−7.0		−7.0		−7.0	−5.6	−6.2	−5.5			−6.3		−6.3
T_{DM} (Ma)	1118	1037		1081		1038	781	918	854			966		956

Note: for analytical methods used, see Wooden et al. (1988) for Pb and Sr isotopes, and Rämö et al. (2002) for Nd isotopes as well as Pb and Sr isotopes for sample 126. Sr and Nd initial compositions corrected for age (Miocene) of crystallization.

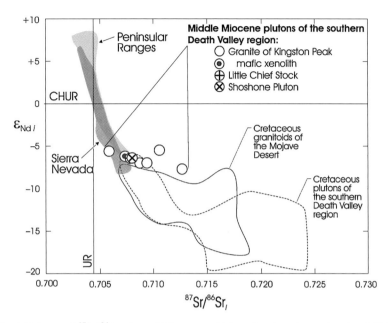

Fig. 6. Diagram comparing initial ε_{Nd} and $^{87}Sr/^{86}Sr$ values of Miocene rapakivi granites from the southern Death Valley region (this study) to Cretaceous granitoids in the southern Death Valley region, western and eastern Mojave Desert, and the Sierra Nevada and Peninsular Ranges batholiths (data from Kistler and Peterman, 1973, 1978; DePaolo, 1981; Kistler et al., 1986; Miller and Wooden, 1994; Allen et al., 1995; Barth et al., 1995, Gerber et al., 1995; Mahood et al., 1996; Rämö et al., 2002). CHUR is Chondritic Uniform Reservoir; UR is Sr uniform mantle reservoir (DePaolo and Wasserburg, 1976).

ing orogen (Calzia, 1990). The abundance of randomly oriented oligoclase inclusions near the cores of K-feldspar phenocrysts and K-feldspar inclusions near the cores of plagioclase phenocrysts indicate that plagioclase and K-feldspar began to crystallize early and at about the same time; the relative size and abundance of the phenocrysts suggest, however, that plagioclase was the liquidus phase. Quartz also occurs as randomly oriented inclusions or as blebs mantled by K-feldspar near the cores of the feldspar phenocrysts. The coexistence of quartz, K-feldspar, and(or) plagioclase inclusions suggests that the granitic melt quickly reached eutectic conditions early in its crystallization history. The abundance of biotite and hornblende in the groundmasses of the feldspar and quartz porphyry facies and as inclusions near the margins of the feldspar phenocrysts suggests that the mafic minerals began to crystallize after plagioclase, K-feldspar, and quartz.

Comparison of the interpreted crystallization sequence with experimental studies of granitic systems, combined with the abundances of titanite, magnetite, and quartz, suggest that the feldspar porphyry

facies crystallized at 675 °C and at high oxygen fugacity (f_{O_2}>NNO but <HM buffer curves). Late crystallization of hydrous minerals indicates that the volatile content (primarily H_2O) varied from less than two weight percent near the liquidus to saturation near the solidus; the abundance of miarolitic cavities in the groundmass of all three facies indicates that a volatile phase was exsolved during late-stage crystallization of the granitic melt (Calzia, 1990).

Granitic magma that crystallized to form the Little Chief stock was probably injected into a narrow, vertical east-trending feeder dike, then expanded laterally and upward to form a laccolith-like plug along the contact between gneiss and the Pahrump Group (McDowell, 1967, 1974). Assuming that the magmatic inclusions were incorporated at this time, the feeder dike may have been located along the zone of inclusions near the contact between the southern and northern facies; this zone of inclusions coincides with a zone of faulting, breccia, and small plugs that are synmagmatic with the stock. Initial doming and extension of the country rocks caused formation of west-dipping normal faults north of the Little Chief stock

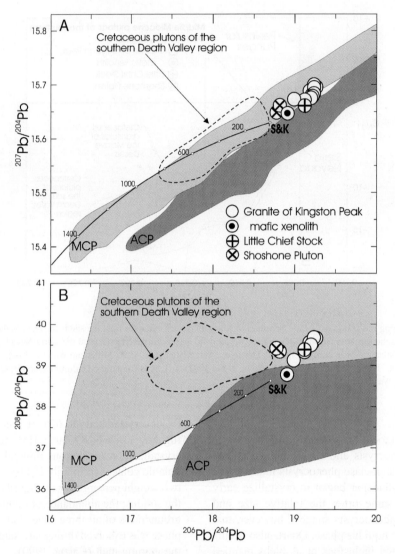

Fig. 7. Pb isotope composition of Miocene rapakivi granites from the southern Death Valley region. Fields for Cretaceous granitoids of the southern Death Valley region from Rämö et al. (2002); MCP and ACP represent Precambrian Mojave and Arizona crustal provinces, respectively (Wooden and Miller, 1990). S and K denotes the growth curves of average crustal Pb (Stacey and Kramers, 1975); numbers on the growth curves are model ages in Ma.

followed by intrusion of the northern dike swarm. Continued injection of magma resulted in formation of a trapdoor structure consisting of a nearly rectangular block of sedimentary country rocks bounded by vertical faults. As magma was injected, the western half of the trapdoor structure was tilted east; differential uplift of the trapdoor resulted in offset of the northern dike swarm along the trapdoor faults. McDowell concluded that the southern facies was at its present position during formation of the trapdoor

structure; intrusion of the northern facies resulted in truncation and deformation of the northern dike swarm and produced sharply discordant contacts with the country rocks.

Petrographic data suggest that plagioclase, augite, and magnetite were the early liquidus phases of the Little Chief stock (McDowell, 1967, 1978). Plagioclase phenocrysts, however, are zoned from labradorite to andesine and oligoclase; many of these phenocrysts are partially resorbed and reversely zoned. With continued crystallization, sanidine phenocrysts began to crystallize with oligoclase; augite became unstable and hornblende began to crystallize with biotite and magnetite. McDowell reported that these changes in phenocryst composition and stability were caused by progressive changes in water content and P_{H_2O} as the magma crystallized. With crystallization of feldspar and augite, water content and P_{H_2O} increased. Eventually P_{H_2O} exceeded P_{load}, and the magma moved up in the crust to a level where $P_{H_2O} = P_{load}$ and crystallization continued. During ascent, plagioclase phenocrysts were partially resorbed and zoned with more Na-rich compositions; with continued crystallization, a more Ca-rich plagioclase became the stable phase, resulting in reversals of zoned-plagioclase compositions. The process was repeated until both plagioclase and sanidine were stable. McDowell (1967) concluded that sanidine began to crystallize at 710–680 °C and 1.0–2.5 kbar; assuming a crustal density of 2.85 gm/cm^3, two feldspar crystallization probably began at a depth of 6–6.5 km.

Intrusion of the magma into the Pahrump Group resulted in assimilation of dolomitic or calcareous country rocks relatively rich in H_2O and CO_2 (McDowell, 1967, 1974). With a sudden increase in P_{fluid} (where $P_{fluid} = P_{H_2O} + P_{CO_2}$) or decrease in P_{load}, the feldspars become unstable. Sanidine phenocrysts reacted with the melt and were partially resorbed and rimmed by oligoclase, resulting in the formation of rapakivi textures; oligoclase rims may also have formed on plagioclase phenocrysts at this time. Assimilation of dolomitic country rocks also resulted in formation of diopside rims around hornblende; cumulophyric aggregates of plagioclase, biotite, and hornblende probably resulted from assimilation of calcareous greywacke from the upper Pahrump Group.

Episodic increase of water content followed by relatively rapid loss of P_{fluid} probably resulted in biotite pseudomorphs after hornblende as well as simultaneous crystallization of K-feldspar and quartz in the groundmass. Continued crystallization of the groundmass resulted in high volatile concentrations and formation of a vapor phase as indicated by the abundance of vugs, pegmatite dikes, and graphic-granite textures in the northeast part of the northern facies. With the bulk of the groundmass crystallized, biotite

reacted with a vapor phase to form magnetite or hematite; magnetite reacted with the volatile-rich phase to form hematite. The appearance of these minerals indicates high oxygen fugacity during final crystallization. Based on this crystallization sequence, McDowell (1967) concluded that the Little Chief stock finally crystallized at 670–620 °C and 0.4–0.9 kbar.

The emplacement history of the Shoshone pluton is enigmatic. This pluton has no exposed intrusive contacts and is disconformably overlain by the volcanics of Deadman Pass and the Miocene Shoshone Volcanics. Fragments of the quartz monzonite are found in tuff of the Shoshone Volcanics at the contact; these relations show that the Shoshone pluton formed the erosional surface beneath the Miocene tuff.

The rim and core facies have similar whole-rock and feldspar compositions but different groundmass textures; the groundmass textures were controlled by water contents (Almashoor, 1980). The abundance of miarolitic cavities suggests that the rim facies contained more water than the core facies and crystallized more slowly, resulting in the coarser-grained granophyric groundmass. The abundance of biotite relative to hornblende suggests, however, that the rim facies initially contained less than three percent H_2O. With crystallization, the magma became saturated in water. Eventually $P_{H_2O} > P_{total}$, the wall rock was fractured, and magma intruded to higher crustal levels. This process was repeated until the magma was emplaced in a subvolcanic environment and crystallized rapidly, resulting in the aphanitic groundmass of the core facies. Almashoor estimated that the magma finally crystallized at a depth of 2.5–2.0 km (P is approximately 0.8–0.5 kbar). Assuming $P \leq 1.0$ kbar and using two-feldspar geothermometry, he concluded that the Shoshone pluton crystallized at 750–850 °C.

6. Concluding remarks

The Miocene rapakivi granites of the southern Death Valley region include the 12.4-Ma granite of Kingston Peak, the ca. 10.6-Ma Little Chief stock, and the 9.8-Ma Shoshone pluton; the granite of Ibex Pass is a landslide breccia of the granite of Kingston Peak. Textural variations and intrusive relations show that each of these plutons is zoned inward from a porphyritic rim facies to an equigranular aplite core; all porphyritic facies are characterized by rapakivi texture and miarolitic cavities. Mafic xenoliths and inclusions are common; they probably represent earlier magmatic phases that were partially resorbed during crystallization. Regional stratigraphic and tectonic data indicate that all three plutons were emplaced at shallow crustal levels in an actively extending orogen.

Petrographic data suggest that each rapakivi granite in the southern Death Valley region evolved through a long and complex crystallization history of anhydrous magma. Crystallization of anhydrous minerals, primarily feldspars and pyroxenes, increased water content and P_{H_2O}. Eventually P_{H_2O} will exceed P_{load}, the country rock fractured, and the magma moved to a higher crustal level. Minor amounts of normative anorthite (cf. Table 3) are indicators of early crystallization of plagioclase and K-feldspar in equilibrium with melt along the cotectic between plagioclase and K-feldspar fields in the Q–Or–Ab–H_2O system (Tuttle and Bowen, 1958). A sudden loss of volatiles caused an abrupt decrease in pressure and shifted the cotectic toward the Or apex. The melt compensated for this shift by crystallizing plagioclase; K-feldspar was resorbed and did not crystallize until the composition of the melt reached the new cotectic. Plagioclase nucleated and crystallized either as new crystals or as mantles on existing feldspars. The overgrowth of plagioclase on partially resorbed K-feldspar crystals resulted in rapakivi texture.

Geochemical data show that the Miocene rapakivi granites in the southern Death Valley region straddle the peraluminous–metaluminous boundary and are more oxidized and less clearly A-type than rapakivi granites from Finland. Overall, the Pb and Sr isotope data are compatible with the idea that the rapakivi granites from Death Valley were derived by partial melting of Mesozoic plutonic rocks; Nd isotope data, however, suggest that all these three Miocene plutons include a juvenile (mantle) component. These data, combined with the volume of coeval volcanic rocks in the region (Fig. 2), suggest that heat required for melting may be transported into the crust by magmatic underplating. If so, then the Miocene rapakivi granites in the southern Death Valley region were generated by similar magmatic processes as Proterozoic rapakivi granites of southern Finland and the Amazonian craton. Although previous Sr isotope data suggest that the granite of

Kingston Peak could have been derived by partial melting of Mesozoic plutonic rocks (Calzia, 1990), our new Sr, Nd, and Pb isotope data do not support this hypothesis. The Nd isotope composition suggests that all three plutons included a juvenile (mantle) component in their source area; Pb isotope data imply that the Pb in these rocks may have been derived from metamorphosed Early Proterozoic supracrustal rocks and granitoids of the Mojave crustal province. Seismic modeling implies that the lower crust beneath Death Valley consists of relatively low-velocity ductily deformed crustal material interlayered with and underplated by numerous relatively high-velocity subhorizontal mafic sills probably derived from the mantle (McCarthy and Thompson, 1988; Serpa et al., 1988). Combined, our isotope data and the seismic models suggest that rapakivi granites in the southern Death Valley region may be derived by partial melting of lower crustal rocks; the heat required for melting may be transported into the crust by magmatic underplating. Thus the Miocene rapakivi granites in the southern Death Valley region could have been generated by similar magmatic processes as Proterozoic rapakivi granites of southern Finland and the Amazonian craton.

Acknowledgments

The authors thank B.W. Troxel for providing field accommodations in Shoshone, CA, J.L. Wooden, USGS, for Pb and Sr analyses of selected samples, and the staff of Isotope Geology, Geological Survey of Finland, for help while performing the isotopic analyses. We also thank R.L. Christiansen and E.H. McKee, Jr., both from the USGS, for critical and thoughtful reviews of the manuscript. Rämö's research in the southern Death Valley region was funded by the Academy of Finland (Project SA 36002).

References

Albee, A.L., Labotks, T.C., Lanphere, M.A., McDowell, S.D., 1981. Geologic map of the Telescope Peak Quadrangle, California. U.S. Geological Survey Geologic Quadrangle Map 1532.

Almashoor, S.S., 1980. The petrology of a quartz monzonite pluton in the southern Greenwater Range, Inyo County, California. Ph.D. Thesis, Pennsylvania State University.

Allen, C.M., Wooden, J.L., Howard, K.A, Foster, D.A., Tosdal, R.M., 1995. Sources of the Early Cretaceous plutons in the Turtle and West Riverside Mountains, California: anomalous Cordilleran interior intrusions. Journal of Petrology 36, 1675–1700.

Amelin, Yu.V., Heaman, L.M., Verchogliad, V.M., Skobelev, V.M., 1994. Geochronological constraints on the emplacement history of an anorthosite–rapakivi granite suite: U–Pb zircon and baddeleyite study of the Korosten complex, Ukraine. Contributions to Mineralogy and Petrology 116, 411–419.

Amelin, Yu.V., Larin, A.M., Tucker, R.D., 1997. Chronology of multiphase emplacement of the Salmi rapakivi granite–anorthosite complex, Baltic Shield: implications for magmatic evolution. Contributions to Mineralogy and Petrology 127, 353–368.

Anderson, J.L., 1983. Proterozoic anorogenic granite plutonism of North America. In: Medaris Jr., L.G., Byers, C.W., Mickelson, D.M., Shanks, W.C. (Eds.), Proterozoic Geology: Selected Papers from an International Proterozoic Symposium, Geological Society of America, Memoir 161, 133–154.

Anderson, J.L., Bender, E.E., 1989. Nature and origin of Proterozoic A-type granitic magmatism in the southwestern United States of America. Lithos 23, 19–52.

Barth, A., Wooden, J.L., Tosdal, R.M., Morrison, J., 1995. Crustal contamination in the petrogenesis of a calc-alkalic rock series: Josephine Mountain intrusion, California. Geological Society of America Bulletin 107, 201–212.

Burchfiel, B.C., Walker, J.D., Davis, G.A., 1983. Kingston Range and related detachment faults — a major "breakaway" zone in the southern Great Basin. Abstracts with Programs—Geological Society of America 15, 536.

Burchfiel, B.C., et al., 1985. The Kingston Range detachment system: structures at the eastern edge of the Death Valley extensional zone, southeastern California. Abstracts with Programs—Geological Society of America 17, 345.

Calzia, J.P., 1990. Geologic studies in the Kingston Range, southern Death Valley region, California. Ph.D. Thesis, University of California, Davis.

Calzia, J.P., Finnerty, A.A., 1984. Geologic and geochemical reconnaissance of Late Tertiary granitic plutons, Death Valley, California. Abstracts with Program—Geological Society of America 16, 461.

Calzia, J.P., Rämö, O.T., 2000. Late Cenozoic crustal extension and magmatism. Southern Death Valley region, California. In: Lagerson, D.R., Peters, S.G., Lahren, M.M. (Eds.), Great Basin and Sierra Nevada, Field Guide—Geological Society of America 2, 135–164.

Calzia, J.P., Verosub, K.L., Jachens, R.C., 1986. Tilting and faulting of the late Miocene Kingston Peak granite, southern Death Valley, California. EOS 67, 1189.

Calzia, J.P., Blakely, R.J., Jachens, R.C., 1991. Miocene magmatism and extension in Ibex Pass, southern Death Valley, California. EOS 72, 469.

Cemen, I., Wright, L.A., Prave, A.R., 1999. Stratigraphy and tectonic implications of the latest Oligocene and early Miocene sedimentary succession, southwesternmost Funeral Mountains, Death Valley region, California. In: Wright, L.A., Troxel, B.W. (Eds.), Cenozoic Basins of the Death Valley Region, Special Paper—Geological Society of America 333, 65–86.

Dall'Agnol, R., Rämö, O.T., de Magalhães, M.S., Macambira, M.J.B., 1999. Petrology of the anorogenic, oxidized Jamon and Musa granites, Amazonian Craton: implications for the genesis of Proterozoic A-type granites. Lithos 46, 431–462.

DePaolo, D.J., 1981. A neodymium and strontium isotopic study of the Mesozoic calc-alkaline granitic batholiths of the Sierra Nevada and Peninsular Ranges, California. Journal of Geophysical Research 86, 10470–10488.

DePaolo, D.J., Wasserburg, G.J., 1976. Nd isotopic variations and petrogenetic models. Geophysical Research Letters 3, 249–252.

Dohrenwend, J.C., McFadden, L.D., Turrin, B.D., Wells, S.G., 1984. K–Ar dating of the Cima volcanic field, eastern Mojave Desert, California: Late Cenozoic volcanic history and landscape evolution. Geology 12, 163–167.

Eklund, O., Fröjdö, S., Lindberg, B., 1994. Magma mixing, the petrogenetic link between anorthositic suites and rapakivi granites, Åland, SW Finland. Contributions to Mineralogy and Petrology 50, 3–19.

Emslie, R.F., 1978. Anorthosite massifs, rapakivi granites, and late Proterozoic rifting of North America. Precambrian Research 7, 61–98.

Emslie, R.F., 1991. Granitoids of rapakivi granite — anorthosite and related associations. In: Haapala, I., Condie, K.C. (Eds.), Precambrian Granitoids — Petrogenesis, Geochemistry and Metallogeny, Precambrian Research 51, 173–192.

Falkner, C.M., Miller, C.F., Wooden, J.L., Heizler, M.T., 1995. Petrogenesis and tectonic significance of the calc-alkaline, bimodal Aztec Wash pluton, Eldorado Mountains, Colorado River extensional corridor. Journal of Geophysical Research 100, 10453–10476.

Fowler, T.K., Calzia, J.P., 1999. Kingston range detachment fault, southeastern Death Valley region, California: relation to Tertiary deposits and reconstruction of initial dip. In: Wright, L.A., Troxel, B.W. (Eds.), Cenozoic Basins of the Death Valley Region, Special Paper—Geological Society of America 333, 245–257.

Fowler, T.K., Friedmann, S.J., Davis, G.A., Bishop, K.M., 1995. Two-phase evolution of the Shadow Valley basin, southeastern California: a possible record of footwall uplift during extensional detachment faulting. Basin Research 7, 165–179.

Friedmann, S.J., 1996. Miocene strata below the Shadow Valley basin fill, eastern Mojave Desert, California. In: Reynolds, R.E., Reynolds, J. (Eds.), Punctuated Chaos in the Northeastern Mojave Desert, San Bernardino County Museum Association Quarterly 43, 123–126.

Friedmann, S.J., Davis, G.A., Fowler, T.K., 1996. Geometry, paleodrainage, and geologic rates from the Miocene Shadow Valley supradetachment basin, eastern Mojave Desert, California. In: Beratan, K.K. (Ed.), Reconstructing the History of Basin and Range Extension Using Sedimentology and Stratigraphy, Special Paper—Geological Society of America 303, 85–105.

Frost, C.D., Frost, B.R., 1997. Reduced rapakivi-type granites: the tholeiitic connection. Geology 25, 647–650.

Frost, C.D., Bell, J.M., Frost, B.R., Chamberlain, K.R., 2001. Crustal growth by magmatic underplating: isotopic evidence from the northern Sherman batholith. Geology 29, 515–518.

Gerber, M.E., Miller, C.F., Wooden, J.L., 1995. Plutonism at the interior margin of the Jurassic magmatic arc, Mojave Desert, California. In: Miller, D.M., Busby, C. (Eds.), Jurassic Magmatism and Tectonics of the North American Cordillera, Special Paper—Geological Society of America 299, 351–374.

Haapala, I., 1985. Metallogeny of the Proterozoic rapakivi granites of Finland. In: Taylor, R., Strong, D.F. (Eds.), Granite-Related Mineral Deposits. Canadian Institute of Mining and Metallurgy, pp. 123–131.

Haapala, I., Rämö, O.T., 1990. Petrogenesis of the rapakivi granites of Finland. In: Stein, H.J., Hannah, J.L. (Eds.), Ore-Bearing Granite Systems: Petrogenesis and Mineralizing Processes, Special Paper—Geological Society of America 246, 275–286.

Haapala, I., Rämö, O.T., 1999. Rapakivi granites and related rocks: an introduction. Precambrian Research 95, 1–7.

Haapala, I., Rämö, O.T., Frindt, S., 2005. Comparison of Proterozoic and Phanerozoic rift-related basaltic–granitic magmatism. In: Rämö, O.T. (Ed.), Granitic Systems—Ilmari Haapala Special Issue, Lithos 80, 1–32.

Heaman, L.M., Grotzinger, J., 1992. 1.08 Ga diabase sills in the Pahrump group, California: implications for the development of the Cordilleran miogeocline. Geology 20, 637–640.

Hewett, D.F., 1940. New formation names to be used in the Kingston Range, Ivanpah Quadrangle, California. Journal of the Washington Academy of Sciences 30, 239–240.

Hodges, K., McKenna, L.W., Harding, M.B., 1990. Structural unroofing of the central Panamint Mountains, Death Valley region, southeastern California. In: Wernicke, B.P. (Ed.), Basin and Range Extensional Tectonics at the Latitude of Las Vegas, Nevada, Special Paper—Geological Society of America 176, 377–390.

Holm, D.K., Snow, J.K., Lux, D.R., 1992. Thermal and barometric constraints on the intrusive and unroofing history of the Black Mountains: implications for timing, initial dip, and kinematics of detachment faulting in the Death Valley region, California. Tectonics 11, 507–522.

Holm, D.K., Pavlis, T.L., Topping, D.J., 1994. Black Mountains crustal section, Death Valley region, California. In: McGill, S.F., Ross, T.M. (Eds.), Geological Investigations of an Active Margin. Cordilleran Section Guidebook—Geological Society of America, pp. 31–54.

Kistler, R.W., Peterman, Z.E., 1973. Variations in Sr, Rb, K, Na, and initial Sr^{87}/Sr^{86} in Mesozoic granitic rocks and intruded wall rocks in Central California. Geological Society of America Bulletin 84, 3489–3512.

Kistler, R.W., Peterman, Z.E., 1978. Reconstruction of crustal blocks of California on the basis of initial strontium isotopic compositions of Mesozoic granitic rocks. U.S. Geological Survey Professional Paper 1071.

Kistler, R.W., Chappell, B.W., Peck, D.L., Bateman, P.C., 1986. Isotopic variation in the Tuolumne Intrusive Suite, central Sierra

Nevada, California. Contributions to Mineralogy and Petrology 94, 205–220.

Lanphere, M.A., Wasserburg, G.J.F., Albee, A.L., Tilton, G.R., 1964. Redistribution of strontium and rubidium isotopes during metamorphism, World Beater Complex, Panamint Range, California. In: Craig, H., Miller, S.L., Wasserburg, G.J.F. (Eds.), Isotopic and Cosmic Chemistry. North Holland Publishing Company, Amsterdam, pp. 269–320.

Louie, J., et al., 1992. 7 Ma age of Tecopa Lake and tectonic quiescence from seismic stratigraphy east of Death Valley, California. EOS 73, 548.

Mahood, G.A., Nibler, G.E., Halliday, A.N., 1996. Zoning patterns and petrologic processes in peraluminous magma chambers: Hall Canyon pluton, Panamint Mountains, California. Geological Society of America Bulletin 108, 437–453.

McCarthy, J., Thompson, G.A., 1988. Seismic imaging of extended crust with emphasis on the western United States. Geological Society of America Bulletin 100, 1361–1374.

McDowell, S.D., 1967. The intrusive history of the Little Chief granite porphyry stock, Panamint Range, California. Ph.D. Thesis, California Institute of Technology.

McDowell, S.D., 1974. Emplacement of the Little Chief stock, Panamint Range, California. Geological Society of America Bulletin 85, 1535–1546.

McDowell, S.D., 1978. Little Chief granite porphyry: feldspar crystallization history. Geological Society of America Bulletin 89, 33–49.

McMackin, M.R., 1997. Late Tertiary evolution of the southern Death Valley fault system: the origin of the Tecopa hump, a tectonic dam on the Amargosa River. In: Reynolds, R.E., Reynolds, J. (Eds.), Death Valley: The Amargosa Route, San Bernardino County Museum Association Quarterly 44, 37–42.

Miller, C.F., Wooden, J.L., 1994. Anatexis, hybridization, and the modification of ancient crust: Mesozoic plutonism in the Old Woman Mountains area, California. Lithos 32, 111–133.

Neymark, L.A., Amelin, Yu., Larin, A.M., 1994. Pb–Nd–Sr isotopic constraints on the origin of the 1.54–1.56 Ga Salmi rapakivi granite–anorthosite batholith (Karelia, Russia). In: Haapala, I., Rämö, O.T. (Eds.), IGCP Project 315 Publication 12, pp. 173–194.

Pearce, J.A., Harris, N.B.W., Tindle, A.G., 1984. Trace element discrimination diagrams for the tectonic interpretation of granitic rocks. Journal of Petrology 25, 956–983.

Rämö, O.T., 1991. Petrogenesis of the Proterozoic rapakivi granites and related basic rocks of southeastern Fennoscandia: Nd and Pb isotopic and general geochemical constraints. Geological Survey of Finland, Bulletin 355.

Rämö, O.T., Calzia, J.P., 1998. Nd isotopic composition of cratonic rocks in the southern Death Valley region: evidence for a substantial Archean source component in Mojavia. Geology 26, 891–894.

Rämö, O.T., Haapala, I., 1995. One hundred years of rapakivi granite. Mineralogy and Petrology 52, 129–185.

Rämö, O.T., Calzia, J.P., Kosunen, J., 2002. Geochemistry of Mesozoic plutons, southern Death Valley region, California: insights into the origin of Cordilleran interior magmatism. Contributions to Mineralogy and Petrology 143, 416–437.

Salonsaari, P.T., 1995. Hybridization in the bimodal Jaala–Iitti complex and its petrogenetic relationship to rapakivi granites and associated mafic rocks of southeastern Finland. Geological Society of Finland, Bulletin 67.

Sederholm, J.J. 1891. Ueber die finnländischen Rapakiwigesteine: Tshermak's Mineralogische und Petrographische Mitteilungen 12, 1–31.

Serpa, L., deVoogd, B., Wright, L., Willemin, J., Oliver, J., Hauser, E., Troxel, B., 1988. Structure of the central Death Valley pull-apart basin and vicinity from COCORP profiles in the southern Great Basin. Geological Society of America Bulletin 100, 1437–1450.

Stacey, J.S., Kramers, J.D., 1975. Approximation of terrestrial lead isotope evolution by a two-stage model. Earth and Planetary Science Letters 26, 207–221.

Stern, T.W., Newell, M.F., Hunt, C.B., 1966. Uranium–lead and potassium–argon ages of parts of the Amargosa thrust complex, Death Valley, California. U.S. Geological Survey Professional Paper 550-B, 142–147.

Streckeisen, Albert, chairman, 1973. Plutonic rocks: classification and nomenclature recommended by the IUGS subcommission on the systematics of igneous rocks. Geotimes 18, 26–30.

Topping, D.J., 1993. Paleographic reconstruction of the Death Valley extended region: evidence from Miocene large rock–avalanche deposits in the Amargosa chaos basin, California. Geological Society of America Bulletin 105, 1190–1213.

Turrin, B.D., Dohrenwend, J.C., Drake, R.E., Curtis, G.H., 1985. K–Ar ages from the Cima volcanic field, eastern Mojave Desert, California. Isochron–West 44, 9–16.

Tuttle, O.F., Bowen, N.L., 1958. Origin of granite in the light of experimental studies in the system $NaAlSi_3O_8$–$KalSi_3O_8$–SiO_2–H_2O. Geological Society of America, Memoir 74 (153 pp.).

Volborth, A., 1973. Geology of the granite complex of the El dorado, Newberry, and northern Dead Mountains, Clark County, Nevada. Nevada Bureau of Mines and Geology, Bulletin 80, 40.

Whalen, J.B., Currie, K.L., Chappell, B.W., 1987. A-type granites: geochemical characteristics, discrimination, and petrogenesis. Contributions to Mineralogy and Petrology 95, 407–419.

Wernicke, B., Wlaker, J.D., Hodges, K., 1988. Field guide to the northern part of the Tucki Mountain fault system, Death Valley region, California. In: Weide, D.L., Faber, M.L. (Eds.), This Extended Land. Cordilleran Section Field Trip Guidebook—Geological Society of America, pp. 58–63.

Wooden, J.L., Miller, D.M., 1990. Chronologic and isotopic framework for Early Proterozoic crustal evolution in the eastern Mojave Desert region, SE California. Journal of Geophysical Research 95, 20133–20146.

Wooden, J.L., Stacey, J.S., Howard, K.A., 1988. Pb isotopic evidence for the formation of Proterozoic crust in the southwestern United States. In: Ernst, W.G. (Ed.), Metamorphism and crustal evolution, western conterminous United States, Prentice-Hall Ruby, vol. 7, pp. 68–86.

Wright, L.A., 1968. Talc deposits of the southern Death Valley–Kingston Range region, California. Special Report—California Division of Mines and Geology 95, 79.

Wright, L.A., Thompson, R.A., Troxel, B.W., Pavlis, T.L., DeWitt, E.H., Otton, J.K., Ellis, M.A., Miller, M.G., Serpa, L.F., 1991. Cenozoic magmatic and tectonic evolution of the east–central Death Valley region, California. In: Walawender, M.J., Hanan, B.B. (Eds.), Geological Excursions in Southern California and Mexico. Cordilleran Section Field Trip Guidebook—Geological Society of America, pp. 93–127.

Available online at www.sciencedirect.com

Earth-Science Reviews 73 (2005) 245–270

www.elsevier.com/locate/earscirev

Upper Neogene stratigraphy and tectonics of Death Valley — a review

J.R. Knott [a,*], A.M. Sarna-Wojcicki [b], M.N. Machette [c], R.E. Klinger [d]

[a]*Department of Geological Sciences, California State University Fullerton, Fullerton, CA 92834, United States*
[b]*U. S. Geological Survey, MS 975, 345 Middlefield Road, Menlo Park, CA 94025, United States*
[c]*U. S. Geological Survey, MS 966, Box 25046, Denver, CO 80225-0046, United States*
[d]*Technical Service Center, U. S. Bureau of Reclamation, P. O. Box 25007, D-8530, Denver, CO 80225-0007, United States*

Abstract

New tephrochronologic, soil-stratigraphic and radiometric-dating studies over the last 10 years have generated a robust numerical stratigraphy for Upper Neogene sedimentary deposits throughout Death Valley. Critical to this improved stratigraphy are correlated or radiometrically-dated tephra beds and tuffs that range in age from >3.58 Ma to <1.1 ka. These tephra beds and tuffs establish relations among the Upper Pliocene to Middle Pleistocene sedimentary deposits at Furnace Creek basin, Nova basin, Ubehebe–Lake Rogers basin, Copper Canyon, Artists Drive, Kit Fox Hills, and Confidence Hills. New geologic formations have been described in the Confidence Hills and at Mormon Point. This new geochronology also establishes maximum and minimum ages for Quaternary alluvial fans and Lake Manly deposits. Facies associated with the tephra beds show that ~3.3 Ma the Furnace Creek basin was a northwest–southeast-trending lake flanked by alluvial fans. This paleolake extended from the Furnace Creek to Ubehebe. Based on the new stratigraphy, the Death Valley fault system can be divided into four main fault zones: the dextral, Quaternary-age Northern Death Valley fault zone; the dextral, pre-Quaternary Furnace Creek fault zone; the oblique–normal Black Mountains fault zone; and the dextral Southern Death Valley fault zone. Post −3.3 Ma geometric, structural, and kinematic changes in the Black Mountains and Towne Pass fault zones led to the break up of Furnace Creek basin and uplift of the Copper Canyon and Nova basins. Internal kinematics of northern Death Valley are interpreted as either rotation of blocks or normal slip along the northeast–southwest-trending Towne Pass and Tin Mountain fault zones within the Eastern California shear zone.
© 2005 Elsevier B.V. All rights reserved.

Keywords: Neogene; stratigraphy; tectonics; tephrochronology

1. Introduction

For most of the 20th century, the Upper Neogene (Pliocene through Quaternary) stratigraphy of Death Valley was relatively straightforward and unrefined. The first geologists working in Death Valley estab-

* Corresponding author. Tel.: +1 714 278 5547; fax: +1 714 278 7266.
    E-mail addresses: jknott@fullerton.edu (J.R. Knott), asarna@usgs.gov (A.M. Sarna-Wojcicki), machette@usgs.gov (M.N. Machette), rklinger@do.usbr.gov (R.E. Klinger).

lished a very useful relative-age stratigraphy for the Upper Neogene that consisted of seven main geologic units (Fig. 1): (1) Furnace Creek Formation; (2) Nova Formation; (3) Funeral Formation (or QTg1); (4) Quaternary gravel 2 (Qg2); (5) Qg3; (6) Qg4; and (7) Lake Manly deposits (Noble, 1934; Noble and Wright, 1954; Drewes, 1963; Hunt and Mabey, 1966; Denny, 1967; Hooke, 1972).

Recently, several independent, but coordinated, studies have reinforced and expanded this stratigraphic framework utilizing a variety of geochronologic techniques (Holm et al., 1994; Snow and Lux, 1999; Knott et al., 1999b; Klinger, 2001b; Klinger and Sarna-Wojcicki, 2001; Machette et al., 2001c; Sarna-Wojcicki et al., 2001). This has resulted in defining new geologic units and helps elucidate the complex tectonic history of the Death Valley pull-apart basin, which has long been a fundamental proving ground for extensional tectonics (Burchfiel and Stewart, 1966; Wright and Troxel, 1967; Wright et al., 1974; Wright, 1976; Wernicke, 1992).

The purpose of this paper is to review the recent refinements to the Upper Neogene stratigraphy of Death Valley. Although an abbreviated review of the stratigraphy is presented in Knott (1999), new research (Knott et al., 1999b; Klinger, 2001b; Klinger and Sarna-Wojcicki, 2001; Machette et al., 2001c; Sarna-Wojcicki et al., 2001) necessitates a more extensive update. In addition, in the past 10 yr, theories regarding the Late Cenozoic tectonic framework of Death Valley have evolved as the region is considered within the broader tectonic framework of the Eastern California shear zone (Knott et al., 1999b; Klinger and Sarna-Wojcicki, 2001; Lee et al., 2001).

2. Upper Neogene stratigraphy

Our understanding of the Upper Neogene stratigraphy of Death Valley has evolved greatly in the past decade (Fig. 1). Geologic units such as the Furnace Creek, Funeral and Nova Formations—all initially

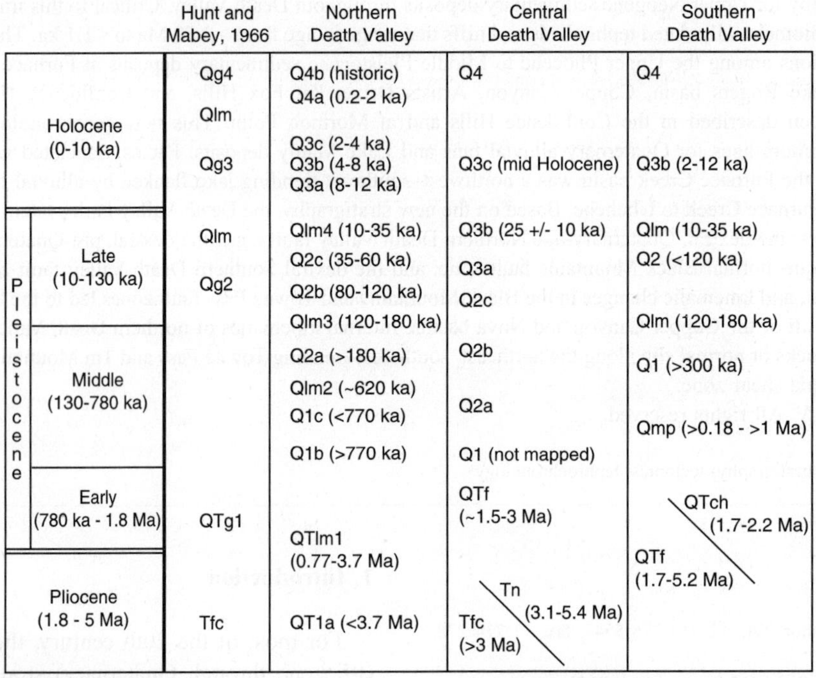

	Hunt and Mabey, 1966	Northern Death Valley	Central Death Valley	Southern Death Valley
Holocene (0-10 ka)	Qg4	Q4b (historic) Q4a (0.2-2 ka)	Q4	Q4
	Qlm			
	Qg3	Q3c (2-4 ka) Q3b (4-8 ka) Q3a (8-12 ka)	Q3c (mid Holocene)	Q3b (2-12 ka)
Pleistocene — Late (10-130 ka)	Qlm	Qlm4 (10-35 ka) Q2c (35-60 ka) Q2b (80-120 ka)	Q3b (25 +/- 10 ka) Q3a Q2c	Qlm (10-35 ka) Q2 (<120 ka)
	Qg2	Qlm3 (120-180 ka)		Qlm (120-180 ka)
Middle (130-780 ka)		Q2a (>180 ka) Qlm2 (~620 ka) Q1c (<770 ka)	Q2b Q2a	Q1 (>300 ka) Qmp (>0.18 - >1 Ma)
		Q1b (>770 ka)	Q1 (not mapped)	
Early (780 ka - 1.8 Ma)	QTg1	QTlm1 (0.77-3.7 Ma)	QTf (~1.5-3 Ma)	QTch (1.7-2.2 Ma) QTf (1.7-5.2 Ma)
Pliocene (1.8 - 5 Ma)	Tfc	QT1a (<3.7 Ma)	Tn (3.1-5.4 Ma) Tfc (>3 Ma)	

Fig. 1. Chart comparing Late Neogene geologic formations and map units of Hunt and Mabey (1966) for central Death Valley to later studies of northern (after Klinger, 2001b), central (after Machette et al., 2001c), and southern Death Valley (after Knott et al., 1999a,b; Beratan et al., 1999; Klinger and Piety, 2001; Sarna-Wojcicki et al., 2001). Abbreviations for formations and map units are: Quaternary gravels (Qg2, Qg3 and Qg4 of Hunt and Mabey; Q1, Q2, Q3 and Q4 for other studies); Lake Manly (Qlm and QTlm); Mormon Point Formation (Qmp); Funeral Formation (QTf); Confidence Hills Formation (QTch); Furnace Creek Formation (Tfc); Nova Formation (Tn).

described at least 50 years ago (see Hunt and Mabey, 1966)—now have radiometric- or correlated-age control. Conversely, the Confidence Hills (Beratan et al., 1999; Beratan and Murray, 1992) and Mormon Point Formations (Knott et al., 1999b) are newly described and the Ubehebe–Lake Rogers and Furnace Creek depocenters (Fig. 2) are the subject of several new studies (Klinger and Sarna-Wojcicki, 2001; Liddicoat, 2001; Machette et al., 2001c). The ages of the classic Quaternary alluvial-fan deposits, whose ages have long been problematic (McFadden et al., 1991), now have an incipient numeric-age framework (e.g., Nishiizumi et al., 1993) and are the subject of ongoing cosmogenic isotope investigations (Machette, pers. commun., 2003). In the following sections, a brief summary of each geologic formation or unit is provided followed by a description of the new age control and correlation to other deposits in Death Valley. The age control is mainly the result of the correlation of several key tuff and tephra marker beds summarized in Fig. 3.

2.1. Furnace Creek Formation

The Furnace Creek Formation, as defined by Noble (1934) and mapped in detail by McAllister (1973), consists of interbedded siltstones, sandstones, conglomerates and basalts in the Furnace Creek basin (Figs. 2 and 4). The Furnace Creek Formation and the overlying Funeral Formation mark the final depositional phases that began during the Late Miocene. The Furnace Creek basin is between the Black and Funeral Mountains. The Furnace Creek basin is bounded by the pre-Quaternary Furnace Creek and Grandview fault zones on the northwest and southwest, respectively (Wright et al., 1999). Hunt and Mabey (1966) and Wright and Troxel (1993) extended the Furnace Creek Formation northwest to include fine-grained deposits near the Kit Fox Hills (Fig. 2); however, the correlation between the Furnace Creek and Kit Fox Hills areas is lithostratigraphic and thus tentative.

Early studies of diatom and plant-fossil assemblages indicated a Pliocene age for the Furnace Creek Formation (Hunt and Mabey, 1966). The first radiometric age control for the Furnace Creek Formation was a 4.0 Ma K/Ar (whole-rock) age from a basalt flow in the overlying Funeral Formation

(McAllister, 1973). These data supported a Pliocene age for the underlying upper part of the Furnace Creek Formation.

More recently, Machette et al. (2001c) found the 3.1–3.35 Ma Mesquite Spring tuffs in the upper Furnace Creek Formation near the southern margin of the Furnace Creek basin (Fig. 3). The Mesquite Spring tuffs are important marker beds in Late Neogene sediments of Death Valley. Originally thought to be a single tuff (Snow and White, 1990), Knott et al. (1999b) showed that there are at least two biotite phenocryst tuffs with similar glass shard composition. The stratigraphic, paleomagnetic and geochronologic data indicate that the lower and upper Mesquite Spring tuffs have ages of 3.1 and 3.35 Ma, respectively (Snow and White, 1990; Holm et al., 1994; Knott et al., 1999b; Knott and Sarna-Wojcicki, 2001a).

About 7.5 km to the southeast, Liddicoat (2001) interpreted paleomagnetic reversals within the upper part of the Furnace Creek Formation to be correlative with either between 3.04 and 3.33 Ma or 2.67 and 2.81 Ma. Either of these paleomagnetically-determined age ranges are consistent with the tephrostratigraphy of Machette et al. (2001c). All of these data indicate that the age of the upper part of the Furnace Creek Formation is <3.5 Ma.

Machette et al. (2001c) noted that the <3.5 Ma age for the upper part of the Furnace Creek Formation conflicts with the 4.0 Ma K/Ar age in the overlying Funeral Formation reported by McAllister (1973). One possible explanation for the age discrepancy is that 4.0 Ma age is flawed. Alternatively, if the K/Ar age is correct, then a number of hypotheses must be entertained to explain the various geologic relations. These include, but are not limited to, (1) post-Pliocene, northwest-down, slip on northeast–southwest-trending faults across the Furnace Creek basin (Wright et al., 1999) or (2) that the Funeral Formation containing the basalt is a proximal, coarse-grained facies of the Furnace Creek Formation to the northwest.

As a result, Machette et al. (2001c) placed the contact between the Furnace Creek Formation and overlying Funeral Formation at the facies transition where mudstones grade upward into conglomerates. These conglomerates contain Paleozoic clasts that Wright et al. (1999) interpret to record the progradation of alluvial fans due to uplift and denudation of the Funeral Mountains on the opposite side of the basin.

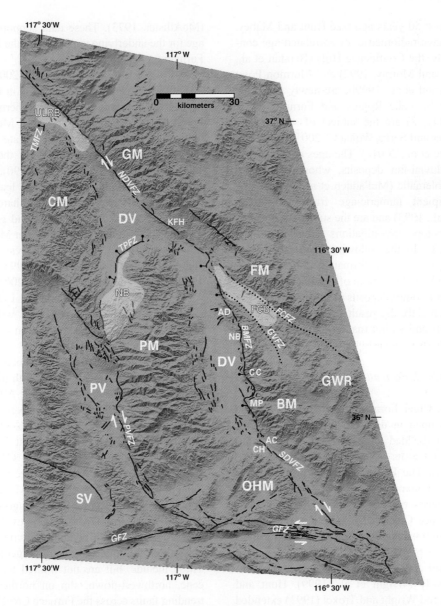

Fig. 2. Map showing major geographic features and Quaternary faults (black lines) of Death Valley. Arrows indicate motion on strike-slip faults; bar and ball indicates downthrown side of normal faults. Major geographic features are: Black Mountains (BM), Cottonwood Mountains (CM), Death Valley (DV), Funeral Mountains (FM), Grapevine Mountains (GM), Greenwater Range (GWR), Owlshead Mountain (OHM), Panamint Mountains (PM), Panamint Valley (PV), and Searles Valley (SV). Major fault zones are: the Black Mountains (BMFZ), Furnace Creek (FCFZ), Garlock (GFZ), Grandview (GVFZ), Northern Death Valley (NDVFZ), Panamint Valley (PVFZ), Southern Death Valley (SDVFZ), Tin Mountain (TMFZ), and Towne Pass (TPFZ). Locations of sedimentary deposits discussed in text are: Artists Drive (AD), Ashford Canyon (AC), Confidence Hills (CH), Copper Canyon (CC), Furnace Creek Basin (FCB), Kit Fox Hills (KFH), Mormon Point (MP), Natural Bridge (NB), Nova Basin (NB), and Ubehebe–Lake Rogers Basin (ULRB). Shaded relief base map derived from digital elevation model.

This interpretation places the Furnace Creek–Funeral contact slightly higher in the section, but still generally concordant with the interpretation of McAllister (1973). Machette et al. (2001c) have argued that this higher facies transition is a more appropriate upper contact for the Furnace Creek because of the associa-

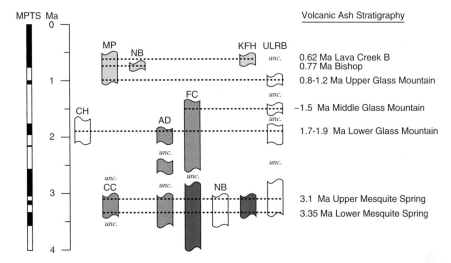

Fig. 3. Diagram showing correlation of Late Neogene deposits of Death Valley using selected volcanic ash and tuff marker beds. Locations shown on diagram are: Artists Drive (AD), Confidence Hills (CH), Copper Canyon (CC), Furnace Creek (FC), Kit Fox Hills (KFH), Mesquite Flat (MF), Mormon Point (MP), Natural Bridge (NB), Nova basin (NB), and Ubehebe–Lake Rogers (ULRB). Mormon Point Formation is light shade; Funeral Formation is intermediate shade; Furnace Creek Formation is dark shade. Unconformities (unc.) between stratigraphic sequences are shown as well. Discussion of other stratigraphic units is found in text. Magnetic Polarity Time Scale (MPTS) is shown at left. Black segments of MPTS are normal polarity and white segments are reversed polarity.

tion with the tectonic uplift of the Funeral Range. In contrast, the lower facies transition preferred by McAllister is indistinguishable from other facies changes in the section. Blair and Raynolds (1999) described similar facies in the upper Furnace Creek Formation adjacent to the northern margin of the Furnace Creek basin and Furnace Creek fault zone. These likely record uplift and progradational events; however, the lack of age control makes correlation to studies on the southern margin by Machette et al. (2001c) difficult.

Identification of a Mesquite Spring tuff in the Furnace Creek Formation allows correlation of the upper Furnace Creek with deposits at Copper Canyon, Artists Drive, and the Nova basin (Figs. 2 and 3).

2.2. Ubehebe–Lake Rogers deposits

The Ubehebe–Lake Rogers deposits are found in the Cottonwood Mountains and on the eastern Cottonwood Mountains piedmont southeast of the Tin Mountain fault zone (Fig. 2). This Miocene–Quaternary sedimentary sequence has only recently been described and the interpretation is rapidly evolving (Snow and White, 1990; Snow and Lux, 1999; Klin-

ger and Sarna-Wojcicki, 2001). As a result, we believe that while the data collected to date are substantive, the formation designations for stratigraphic sequences remain tentative. Thus, we prefer Ubehebe–Lake Rogers deposits rather than describe these data under formal formational designations.

Snow and White (1990) defined the Ubehebe basin deposits as a sequence of interbedded conglomerates, tuffs, mudstones and basalts ranging in age from 3.17 ± 0.09 to 23.87 ± 0.23 Ma. Snow and Lux (1999) interpreted the gently dipping (<20°) conglomerate, marl and sandstone beds overlying the 3.7 ± 0.2 Ma Basalt of Ubehebe Hills as Nova Formation. This correlation to the Nova Formation is based largely on the presence of the 3.1–3.35 Ma Mesquite Spring tuff in both the Nova Formation and the sediments above the 3.7 Ma Basalt of Ubehebe Hills.

Snow and Lux (1999) noted that the Mesquite Spring tuff grades to the northeast into a tuffaceous (altered) marl as the surrounding facies grade from conglomerate to mudstone. Snow and Lux (1999) interpreted these relations to show that a playa lake, bound on the southwest by alluvial fans existed along Death Valley Wash ~3.2 Ma. To the east, between Death Valley Wash and the Northern Death Valley

Fig. 4. Oblique aerial view looking southeast at the north-dipping Furnace Creek Formation in the Furnace Creek basin. The light colored beds in the right center are lacustrine mudstones with breccias and conglomerates comprising the ridge in the foreground. In the lower right is the Furnace Creek alluvial fan with a dashed line along the trend of the Black Mountains fault zone. The arrow indicates the location of the 3.1–3.35 Ma Mesquite Springs tuff in the upper part of the Furnace Creek Formation.

fault zone, ~3.2 Ma breccia/conglomerates are interpreted as proximal alluvial-fan deposits (Klinger and Sarna-Wojcicki, 2001). Thus, the ~3.2 Ma playa lake inferred by Snow and Lux (1999) would have been narrowly constrained to the area of the present Death Valley Wash with flanking alluvial-fan deposits to the east and west.

One of the best exposures of the upper Ubehebe basin beds is in Death Valley Wash (Fig. 5) where Klinger (2001a,b) dated and correlated several tephra beds. Here, the 3.7 Ma Basalt of Ubehebe Hills is

overlain by a Mesquite Spring tuff. This tuff is interbedded with mudstones that grade upward to conglomerates (Klinger and Sarna-Wojcicki, 2001). Intercalated with the conglomerates and separated by unconformities are tuffs from the lower (1.7–2.0 Ma), middle (~1.5) and upper (0.76–1.2 Ma) Glass Mountain tephra groups, all suggesting that the basin persisted well into the Early Quaternary (Klinger and Sarna-Wojcicki, 2001).

The 1.7–1.9 Ma lower Glass Mountain tuffs are interbedded with fine-grained sediments, which Klinger and Sarna-Wojcicki (2001) designated as Qlm1 (Fig. 1) or an early phase of Lake Manly. The Qlm1 deposits of Klinger and Sarna-Wojcicki (2001) are overlain in angular unconformity by breccia/conglomerates that contain 0.8–1.2 Ma upper Glass Mountain ash beds.

Overlying the gently dipping upper Ubehebe basin sediments are the flat-lying mudstones and evaporites of Lake Rogers. Klinger and Sarna-Wojcicki (2001) defined the Lake Rogers basin as a Quaternary-age structural depression located northwest of the Tin Mountain fault zone and west of the Northern Death Valley fault zone. These flat-lying sediments overlie the Ubehebe basin deposits in angular unconformity. Clements (1952) found mammoth fossils in the mudstones and sparse evaporite beds and interpreted the age as Quaternary.

Based on tephrochronology, the upper Ubehebe basin deposits are correlative with those from the upper part of the Nova Formation; the upper part of the Furnace Creek Formation; the upper part of the Funeral Formation at Copper Canyon; and the lower part of the Funeral Formation at Artists Drive. Identification of the lower Glass Mountain tephra beds indicates that the upper Ubehebe basin deposits are temporally correlative with the Funeral Formation in the Furnace Creek basin, the upper Funeral Formation at Artists Drive and the Confidence Hills Formation. The presence of upper Glass Mountain tephra beds indicates a correlation with the Mormon Point Formation.

Based on the presence of the Mesquite Springs tuff, Snow and Lux (1999) correlated the upper Ubehebe basin deposits with the Nova Formation. However, unlike the Nova Formation, deposition in the Ubehebe basin persisted into the Quaternary. It is obvious that the upper part of the Ubehebe basin

Fig. 5. View to the north of the Ubehebe–Lake Rogers basin deposits. Dark bed in the canyon wall at right is the 3.7 Ma Basalt of Ubehebe Hills, which marks the base of the upper part of the Ubehebe basin deposits. In the center are the Quaternary-age, flat-lying Lake Rogers deposits.

deposits are significantly younger from the Nova Formation. Also, the fine-grained facies of the upper part of the Ubehebe basin deposits are more similar to the upper part of the Furnace Creek Formation. We believe that these discrepancies make the correlation to the Nova Formation tentative. The presence of unconformities within the upper part of the Ubehebe basin deposits (Klinger and Sarna-Wojcicki, 2001) suggests that the depositional and tectonic history of this area is not sufficiently resolved and may warrant a new formation.

2.3. Funeral Formation

Since being described in the Furnace Creek area by Thayer in 1897, the Funeral Formation has been mapped throughout the Black Mountains of central and southern Death Valley (see Hunt and Mabey, 1966). Funeral Formation outcrops range from broad expanses to small isolated outcrops. In general, the Funeral Formation consists of tilted and uplifted alluvial-fan conglomerates containing sparse intercalated basalts. The basalts are most prominently exposed in the upper reaches of the present Furnace Creek drainage basin (Fig. 2). The most extensive

deposits of the Funeral Formation are found in the Furnace Creek basin (Fig. 2); Artists Drive and Copper Canyon; (Drewes, 1963; Hunt and Mabey, 1966; McAllister, 1970; McAllister, 1973; Wright and Troxel, 1984); and Kit Fox Hills (Wright and Troxel, 1993).

Both Noble (1934) and Hunt and Mabey (1966) hypothesized that the discontinuous outcrops prohibited correlation of the Funeral Formation from place to place. In addition, they both noted that there was no evidence that the Funeral Formation deposits were contemporaneous.

McAllister (1970, 1973) published the most comprehensive map and descriptions of the Funeral Formation. He also provided the initial numeric age control using a K/Ar date of 4.0 Ma on a basalt flow near the southeastern margin of the Furnace Creek basin. However, as described above, the basalt's age is not supported by ~3 Ma tephrochronologic and paleomagnetic ages in the underlying Furnace Creek Formation (Liddicoat, 2001; Machette et al., 2001c; Sarna-Wojcicki et al., 2001). The correlation of lower Glass Mountain (1.7–1.9 Ma) and middle Glass Mountain (~1.5 Ma) tephra beds within the Funeral Formation several kilometers northeast of the

Furnace Creek type locality supports the younger age for the Funeral Formation (Sarna-Wojcicki et al., 2001).

Knott et al. (1999a,b) mapped the Funeral Formation in the Artists Drive block, which is located west of the Black Mountains across the Black Mountains fault zone. Near the basal unconformity, Knott et al. (1999a) found two tuffs above and below the 3.28 Ma Nomlaki Tuff that have identical glass shard compositions. They named these tuffs the lower and upper Mesquite Spring tuffs because of their similarity to the Mesquite Spring tuff of Snow and White (1990). Based on the exposures and data from the Artists Drive locality, the ages of the lower and upper Mesquite Springs tuffs are interpreted to be 3.1 and 3.35 Ma (Knott and Sarna-Wojcicki, 2001a,b).

Below the 3.35 Ma lower Mesquite Spring tuff at Artists Drive is the lower Nomlaki Tuff. The lower Nomlaki tuff has a shard composition general similar to the Nomlaki Tuff, but outcrop data clearly shows this is not the Nomlaki Tuff (Knott et al., 1999b). Knott and Sarna-Wojcicki (2001a,b) used paleomagnetic and stratigraphic data to infer the age of the lower Nomlaki tuff to be >3.58 Ma (Fig. 6). Also below the lower Mesquite Spring tuff at Artists Drive is the tuff of Curry Canyon (Knott and Sarna-Wojcicki, 2001b). The age of this tuff is estimated to be between 3.35 and 3.58 Ma. The tuff of Curry Canyon is also found in the upper part of the Furnace Creek Formation (Machette et al., 2001c).

The upper part of the Funeral Formation at Artists Drive also contains a lower Glass Mountain ash bed. The lower Glass Mountain family of ash beds are important stratigraphic markers in Death Valley. This family of ash beds have an age range of 1.7–1.9 Ma (Sarna-Wojcicki et al., 2001). The reversed paleomagnetic polarity of the lower Glass Mountain ash bed at Artists Drive narrows the age of that ash bed to 1.8–1.9 Ma (Knott, 1998).

The tephrochronology shows that the lower part of the Funeral Formation at Artists Drive is time equivalent to the upper part of the Furnace Creek Formation. The upper part of the Funeral Formation at Artists Drive, however, is time equivalent to the Furnace Creek Formation in the Furnace Creek basin.

A Mesquite Spring tuff is also found in the upper part of the Funeral Formation at Copper Canyon (Knott et al., 1999b). This Mesquite Spring tuff has

Fig. 6. Funeral Formation breccias and conglomerates interbedded with the 3.35 Ma lower Mesquite Spring tuff (LMS), 3.28 Ma Nomlaki Tuff (N), and ~3.1 Ma upper Mesquite Spring tuff (UMS) in Hunt Canyon at Artists Drive. LMS is approximately 1 m thick.

a biotite ^{40}Ar/^{39}Ar age of 3.1 ± 0.2 Ma (Holm et al., 1994), and thus could be either the upper or lower Mesquite Spring tuff.

Topping (1993) obtained a zircon fission-track age of 5.2 Ma for a tuff within the Funeral Formation of the southern Black Mountains. This age makes these breccias and conglomerates the oldest deposits mapped as Funeral Formation yet dated. The 5.2 Ma age makes this southern outcrop of Funeral Formation time equivalent to the middle part of the Furnace Creek Formation at its type locality and the lower Nova Formation.

Wright and Troxel (1993) mapped poorly sorted conglomerates in the Kit Fox Hills northeast of Furnace Creek as belonging to the Funeral Formation. Underlying the Funeral Formation just 2 km northwest, Knott and Sarna-Wojcicki (2001a,b) found a Mesquite Spring tuff interbedded with mudstones. If the interpretation of Machette et al. (2001c) is extended to the Kit Fox Hills, then the mudstones underlying the Funeral Formation would be temporally and lithologically equivalent to the upper Furnace Creek Formation.

Wright and Troxel (1984) also mapped the Funeral Formation near Ashford Canyon in the southern Black Mountains northeast of the Confidence Hills (Fig. 2). The Funeral Formation here is comprised of uplifted and tilted alluvial-fan conglomerates. Wright and Troxel (1984) inferred that a basalt flow exposed at the base of the section is the same as the 1.7 Ma basalt (K/Ar) that forms Shoreline Butte in the northern Confidence Hills (Fig. 2). The 1.7 Ma age is consistent with the correlation of 1.7–1.9 Ma lower Glass Mountain tephra beds in the overlying conglomerates (Knott, 1998). This section of Funeral Formation is contemporaneous with the Funeral Formation in the Furnace Creek basin (Sarna-Wojcicki et al., 2001); the upper part of the Funeral Formation at Artists Drive (Knott et al., 1999b); the Confidence Hills Formation (Sarna-Wojcicki et al., 2001); and the upper part of the Ubehebe–Lake Rogers deposits (Klinger, 2001a). Given its location only a few kilometers from the Confidence Hills, this deposit is probably more likely and appropriately part of the Confidence Hills Formation; however, this is based on limited data and should be confirmed by additional studies.

The more recent tephrochronologic studies have enabled correlation of the Funeral Formation from place to place throughout the Black Mountains. In addition, the tephrochronology allows correlation of the Funeral Formation to other deposits in Death Valley. The tephrochronology also has shown that the age of the Funeral Formation is time transgressive (Fig. 3).

2.4. Nova Formation

The Nova Formation was originally mapped northwest of the Panamint Mountains (Hunt and Mabey, 1966). The Nova Formation is composed of conglomerates, breccias and intercalated basalt flows. According to Hunt and Mabey (1966), the base of the Nova Formation (Fig. 1) in Death Valley is not exposed, but is in fault contact with the low-angle Emigrant Canyon fault. To the north and west, the Nova Formation is uplifted by the high-angle Towne Pass fault (Fig. 1). The age of the Nova Formation ranges from 5.4 ± 0.4 Ma (whole rock K/Ar Hodges et al., 1989) to 3.35 ± 0.13 Ma (Snow and Lux, 1999).

The 3.35 Ma age is on a Mesquite Spring tuff found in the upper part of the Nova Formation (Fig. 7). The Mesquite Spring tuffs are found in the Ubehebe basin and in isolated outcrops in the Cottonwood Mountains (Snow and White, 1990; Snow and Lux, 1999). Based on the presence of the 3.1–3.35 Ma Mesquite Spring tuffs, Snow and Lux (1999) extended the Nova Formation to include the Cottonwood Mountain sediments. In the Cottonwood Mountains, the base of the Nova Formation is delineated by the 3.7 ± 0.2 Ma Basalt of Ubehebe Hills (Snow and Lux, 1999). As mentioned above, new research by Klinger and Sarna-Wojcicki (2001) indicate the Ubehebe basin deposits are significantly younger and lithologically different. For these reasons, we believe that the extension of the Nova Formation to the upper Ubehebe basin deposits is tentative at this time and requires additional research (see Section 2.2).

The Nomlaki Tuff is also found in the Nova Formation northwest of the Panamint Mountains (J. Tinsley, pers. commun., 1994). In addition, the Nomlaki Tuff is found in alluvial-fan deposits along the eastern Cottonwood Mountain piedmont, south of Ubehebe (Knott, 1998).

Based on the tephrochronology (Fig. 3), the upper part of the Nova Formation is temporally equivalent to the upper part of the Furnace Creek Formation of Machette et al. (2001c); the lower part of the Funeral Formation at Artists Drive; and the upper part of the Copper Canyon (Knott et al., 1999b). These correlations are different from Hunt and Mabey (1966) who equated the Nova Formation with the Funeral Formation based on lithology.

2.5. Confidence Hills Formation

In southern Death Valley, evaporite, mudstone and conglomerate beds are uplifted and exposed in the Confidence Hills adjacent the Southern Death Valley fault zone (Wright and Troxel, 1984). Troxel et al. (1986) identified three ash beds within these sediments, including the Huckleberry Ridge ash bed, which was erupted from the Yellowstone Caldera at

Fig. 7. Typical exposure of Nova Formation breccias and conglomerates interbedded with the 3.35 Ma lower Mesquite Spring tuff. Total thickness of tuff is about 6 m. The upper 5 m is tuffaceous debris flows.

about 2.07 Ma. Paleomagnetic data show that this sedimentary section ranges in age from <1.79 to >2.15 Ma (Pluhar et al., 1992). Based on these unique lithologies and temporal qualities, Beratan et al. (1999) named these playa and alluvial-fan deposits the Confidence Hills Formation.

Subsequent tephrochronologic studies of the Confidence Hills Formation have identified the Blind Springs Valley tuff (formerly the tuff of Taylor Canyon; 2.22 Ma); the tuff of Confidence Hills (1.95–2.09 Ma); the tuffs of Emigrant Pass (1.95–2.09 Ma); the lower Glass Mountain tuffs (1.79–1.95 Ma); and the Huckleberry Ridge ash bed (Sarna-Wojcicki et al., 2001).

The lower Glass Mountain tuffs are a key stratigraphic marker for correlation of the Confidence Hills Formation to other deposits in Death Valley (Fig. 3). The Confidence Hills Formation is temporally equivalent to the upper Funeral Formation at Artists Drive, the lower Funeral Formation at Furnace Creek and the upper Ubehebe–Lake Rogers deposits (Fig. 3). Based on the presence of a lower Glass Mountain tuff and proximity to the type locality, we have tentatively extended the Confidence Hills Formation to include tilted basalt flows and conglomerates across Death Valley at Ashford Canyon (Fig. 2). These deposits

were previously mapped as Funeral Formation (see Section 2.3).

2.6. Mormon Point Formation

The Mormon Point Formation (formally described herein) consists of interbedded mudstones, conglomerates and tephra beds, which are well exposed in their type locality at Mormon Point (Knott et al., 1999b). Based on lithostratigraphy, Noble and Wright (1954) assigned a Quaternary age to these sediments. Subsequent studies show that the Mormon Point Formation contains the ~0.5 Ma Dibekulewe, 0.62 Ma Lava Creek B, 0.77 Ma Bishop and 0.8–1.2 Ma Upper Glass Mountain ash beds (Knott et al., 1999a,b; Hayman et al., 2003). The base of the Mormon Point Formation is in fault contact with Precambrian rock (Fig. 8). The fault is one of the low-angle normal faults often referred to as turtlebacks (Wright et al., 1974). The top is defined by an angular unconformity, on which 120–180 ka Lake Manly gravels are deposited.

The Mormon Point Formation is also found at Natural Bridge, just north of Badwater (Fig. 2). At Natural Bridge, coarse-grained breccias, interpreted to be alluvial-fan deposits, are interbedded with the 0.77

Fig. 8. Oblique aerial photograph of Mormon Point in the southern Black Mountains. The solid line is along the trace of the Mormon Point low-angle or turtleback fault. This fault separates Mormon Point Formation (Qmp) and metamorphosed Precambrian rocks (PC). The dashed line at the base of the Black Mountains shows the trace of the Black Mountains fault zone. Relatively upthrown (U) and downthrown (D) blocks across faults are also indicated.

Ma Bishop ash bed (Hayman et al., 2003). Like Mormon Point, the base of the Mormon Point Formation at Natural Bridge is in fault contact with a low-angle normal fault and overlain by the 120–180 ka deposits of Lake Manly (Knott et al., 1999a,b; Hayman et al., 2003).

The Mormon Point Formation is time equivalent to the upper part of the Ubehebe basin deposits and mudstones in the southern Kit Fox Hills (Figs. 2 and 3). In the Kit Fox Hills, Klinger (2001b) identified the Lava Creek B ash bed interbedded with fine-grained mudstones (See Section 2.8).

2.7. Lake Manly deposits

Lake Manly is the name used for the lake that intermittently occupied Death Valley during pluvial periods. In 1924, Levi Noble, W. M. Davis and H. E. Gregory described the first evidence of a lake in Death Valley (Noble, 1926). Means (1932) was the first to apply the name Lake Manly, honoring William Manly who lead a group of pioneers through Death Valley in 1849. Blackwelder (1933, 1954) and Clements and Clements (1953) described and speculated on the age of seven known outcrops known at that time. They all

assumed that these outcrops were all related to a single lake stand of Lake Manly. The compilation of Machette et al. (2001a) nearly 50 years later identified 30 localities where evidence of Lake Manly (e.g., bars, spits or shoreline features) is found (Fig. 9); and this listing is probably incomplete.

Given the size of Death Valley, the geomorphic and stratigraphic evidence of various high stands of Lake Manly is relatively sparse and discontinuous, yet convincing where preserved (Figs. 9 and 10). These deposits are mappable (Hunt and Mabey, 1966), but have never been assigned formal formation status. This is probably due to the fact that the ages of discrete lake stands are poorly defined (Machette et al., 2001a) and confusion arising from distinguishing between shorelines and fault scarps (Klinger, 2001b; Machette et al., 2001c; Knott et al., 2002).

Drilled cores recovered from the valley floor have found evidence of two pluvial lakes. Hooke (1972) reported radiocarbon ages between 11 and 26 ka for lake-bottom sediments retrieved from shallow cores near Badwater. Lowenstein et al. (1999) found evidence of lakes at 10–35 ka and 120–186 ka in their 126-m-deep core near Badwater. Anderson and Wells (2003) found evidence of several small lakes that

Fig. 9. South-dipping foreset beds in a spit of Lake Manly at Desolation Canyon (Loc. 17 on Fig. 10). Beds are composed of subrounded to rounded platey gravel and sand. Such outcrops are convincing evidence of Lake Manly.

occupied only the lowest parts of Death Valley between 10 and 35 ka. The 10–35 ka and 120–186 ka time frames are periods traditionally associated with large pluvial lakes in the western North America (i.e., Bonneville, Lahonton and Tecopa) and correlative with marine oxygen isotope stages 2 and 6, respectively (Smith, 1991).

Age dating of outcrops has yielded definitive evidence of only the 120–186 ka lake. Hooke and Lively (1979), cited in Hooke and Dorn (1992), used uranium-series disequilibrium to determine a preferred age range of 60 to 225 ka for tufa interbedded with near-shore facies gravels ~90 m above mean sea level (amsl) along the western Black Mountains piedmont. Ku et al. (1998) used uranium-series disequilibrium to date tufa deposits associated from this same ~90 amsl shoreline and found an age range between 128 and 216 ka. Ku et al. (1998) interpreted the clustering of shoreline tufa ages between 120 and 180 ka to correspond with the oxygen isotope stage 6 lake sediments found in the core.

Mormon Point Formation at Mormon Point and equivalent age sediments at the Kit Fox Hills and Ubehebe basin indicate at least two Pleistocene lake phases older than those found in the drilled cores (Knott, 1997). At Mormon Point, green, massive to laminated mudstones with reversed paleomagnetic polarity (>0.78 Ma) and containing 0.8–1.2 Ma Upper Glass Mountain ash beds are interpreted as lake deposits (Knott et al., 1999b). Lake facies are also found between the 0.77 Ma Bishop and the 0.62 Ma Lava Creek B ash beds at Mormon Point. In the Kit Fox Hills, Klinger (2001b) found the Lava Creek B ash bed interbedded with lacustrine facies. Fine-grained facies with an age of <0.77 Ma are also found in the Ubehebe–Lake Rogers basin. These lacustrine facies of Early to Middle Pleistocene age are referred to as Lake Manly phase 2 (Qlm2 on Fig. 1) by Klinger (2001b) and are equivalent to marine oxygen isotope stages 16 and 18.

Klinger (2001b) also extended usage of Lake Manly to include Pliocene (<3.7 and >0.77 Ma) lacustrine facies in the Ubehebe–Lake Rogers basin as well. Klinger designated these Pliocene to Pleistocene age deposits Lake Manly phase 1 (Qlm1 on Fig. 1). Blair and Raynolds (1999) referred to lacustrine facies in the northeast Furnace Creek basin as paleolake Lake Zabriske sediments. McAllister (1970) mapped these same sediments as the Furnace Creek Formation; however, the age of these sediments is poorly defined.

The two names (Lake Manly phase 1 and Lake Zabriske) for the Pliocene (?) to Pleistocene (?) paleolake could be referring to the same lake phase and illustrates the problem of attempting to name lake phases. Given the lack of exposure and limited age control for the Pliocene lake sediments in Death Valley, it may be prudent to resist naming each new sequence of fine-grained deposits until better age control and mapping is completed. Similarly, given the rapid uplift of the Sierra Nevada Mountains in Pliocene time and incumbent climatic changes (Smith, 1991) and the post-Pliocene deformation throughout Death Valley, we now believe it is inappropriate to apply the term "Lake Manly" to deposits older than Pleistocene.

2.8. Alluvial-fan deposits

Numeric age control on the spectacular alluvial-fan deposits of Death Valley has remained sparse because of a lack of datable material (i.e., ^{14}C) in the fan deposits themselves and in the older formations (e.g., Funeral Formation). Early studies consistently inferred Quaternary ages for the alluvial fans based on limited fossil data and their relative lack of deforma-

tion (Noble and Wright, 1954; Hunt and Mabey, 1966; Denny, 1967; Hooke, 1972).

Rock varnish methods were used to estimate ages of Middle Pleistocene to Holocene for alluvial fans of the eastern Panamint Mountains piedmont (Dorn, 1988; Dorn et al., 1990; Hooke and Dorn, 1992). Rock varnish methods, however, have been challenged and their validity remains in question (e.g., Bierman and Gillespie, 1994; Beck et al., 1998; Watchman, 2000). As a result many questions regarding the age of Death Valley alluvial-fan deposits remained unanswered. These include: Are morphologically similar alluvial-fan units the same age?; What are the stratigraphic relations between the alluvial fans and Lake Manly?; What roles do climate and tectonics play in development of these alluvial fans?

The four-fold (Q1 [oldest], Q2, Q3 and Q4 [youngest]) relative stratigraphic framework that Hunt and

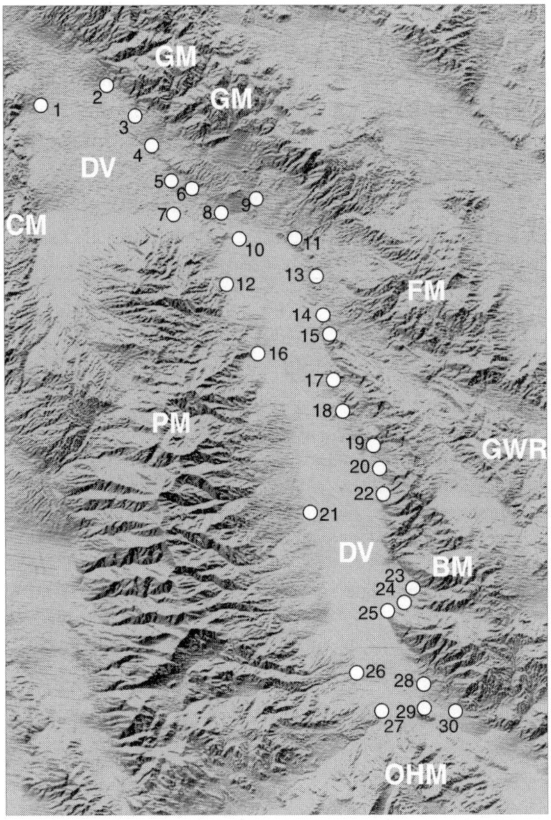

Fig. 10. Map showing selected locations where Lake Manly deposits or landscape feature have been identified. See Fig. 2 for abbreviations to geographic elements. After Machette et al. (2001a,b,c).

1. Niter beds
2. Titus Canyon
3. E. Mesquite Flat
4. Triangle Spring
5. Mud Canyon
6. NPS Rte 5 @ Hwy 190
7. Stovepipe Wells
8. Salt Creek anticline
9. Beatty Junction
10. Salt Creek
11. Three Bare Hills
12. North of Salt Spring
13. Park Village Ridge
14. Road to NPS landfill
15. Tea House
16. Unnamed ridge
17. Desolation Canyon
18. Manly terraces
19. Natural bridge
20. Nose Canyon
21. Tule Springs–Hanaupah Cyn
22. Badwater
23. Sheep Canyon
24. Willow Wash
25. Mormon Point
26. Warm Springs Canyon
27. Wingate Delta
28. East of Cinder Hill
29. Shoreline Butte
30. East of Ashford Mill

Mabey (1966) developed using empirical geomorphic observations has remained relatively unchallenged and viable. However, several subsequent workers (e.g., Hooke, 1972; Dorn, 1988; Klinger, 2001b; Menges et al., 2001) have subdivided the four original units of Hunt and Mabey (1966) (Fig. 1).

Studies of soil development and tephrochronology (Klinger, 2001b), fan morphology (Klinger, 2001b; Menges et al., 2001) and cosmogenic radionuclide surface exposure dating (Nishiizumi et al., 1993) are providing relative and numeric ages for some alluvial-fan deposits (Table 1). In addition, the stratigraphic relations between Lake Manly (Ku et al., 1998) and alluvial-fan deposits have improved age control for the alluvial fans as well.

In order to facilitate a review of the stratigraphic framework, the four main units of Hunt and Mabey are described below with subdivisions noted as identified where appropriate.

2.8.1. Q1

In the four-fold nomenclature of Hunt and Mabey (1966), the oldest alluvial-fan unit is the QTg1, or the Funeral Formation. However, the Funeral Formation lacks alluvial-fan morphology and the map symbol is unconventional (QTg1 instead of QTf). As a result,

we recommend that the Funeral Formation should no longer be regarded as an alluvial-fan gravel, but should remain as a formation rank geologic unit. Later studies have described an alluvial-fan unit on the eastern Panamint piedmont that has fan morphology and is older than Q2 (Hooke, 1972; Dorn, 1988; Jayko and Menges, 2001; Klinger, 2001b). We recommend that this alluvial-fan unit along with morphologically similar deposits (described below) be identified as Q1.

Our recommended Q1 is found near the mouth of Hanaupah Canyon along the eastern Panamint piedmont (Hooke, 1972). Here, Hooke (1972) described an alluvial-fan unit (older surface facies) that is older than Q2 and, unlike the Funeral Formation, has alluvial-fan morphology. Dorn (1988) labeled this unit Q1. Jayko and Menges (2001) used remote sensing and surface morphology to map this unit (their QTa) and its equivalents along the eastern piedmont of the Panamint Mountains, thereby showing the regional extent of the unit. As a result, we recommend that this upper alluvial-fan deposit along the eastern Panamint piedmont be Q1.

Menges et al. (2001) described the surface morphology of Q1 as highly degraded, having none to weak or no desert pavement development and weakly

Table 1
Compilation of age control and geochronologic methods used on alluvial-fan deposits of Death Valley

Alluvial-fan unit/location	Age	Method	Reference
Q1			
Hanaupah Fan	≥ 0.3 Ma	Cosmogenic radionuclides	Nishiizumi et al. (1993)
Kit Fox Hills	≤ 0.62 Ma	Tephrochronology	Klinger (2001b)
Six Springs Canyon	≤ 3.1–3.35 Ma	Tephrochronology	Knott et al. (2000)
Q2			
Northern Death Valley	30–180 ka	Soil development	Klinger (2001b)
Hanaupah Fan	260–318 ka	Cosmogenic radionuclides	Nishiizumi et al. (1993)
Mormon Point	<120–180 ka	Stratigraphic relations	Knott et al. (1999a,b)
Q3a			
Hanaupah Fan	117 ka	Cosmogenic radionuclides	Nishiizumi et al. (1993)
Central Death Valley	<125 ka	Archeology	Hunt and Mabey (1966)
Northern Death Valley	<12 ka	Soil development	Klinger (2001a)
Q4a			
Northern Death Valley	<1.2 ka	Tephrochronology	Klinger (2001a)
Q4b			
Northern Death Valley	0.14–0.30 ka	[14]C	Klinger (2001a)

preserved varnish with chips of pedogenic carbonate commonly found on the surface. Soils developed on the Q1 have Stage III–V petrocalcic horizons (See Machette, 1985 for nomenclature) and are commonly dissected (Menges et al., 2001).

Numeric age control on the Q1 unit is sparse (Table 1). Nishiizumi et al. (1993) determined a minimum age surface exposure age of ~0.3 Ma for clasts on the Q1c surface using cosmogenic ^{10}Be and ^{26}Al. In the Kit Fox Hills, Klinger (2001b) mapped Q1 deposits unconformably overlying the 0.62 Ma Lava Creek B ash bed (Fig. 11). This relation shows that the maximum age for Q1 in the Kit Fox Hills is 0.62 Ma or younger than the Mormon Point Formation.

At Six Springs Canyon, Knott et al. (2000) found a Mesquite Springs tuff (3.1–3.35 Ma) in a terrace deposits at Six Springs Canyon along the eastern Panamint Mountains piedmont. The upper surfaces of the terraces, which have Stage IV–V petrocalcic horizons, are the upstream extension of the Q1 surface on the piedmont. Based on the soil development and tephrochronology, Knott et al. (2000) inferred a Plio-cene age for the Q1 deposits at Six Spring Canyon and along the eastern Panamint piedmont. Thus, the Q1 at Six Springs Canyon is equivalent to the Furnace Creek Formation at Furnace Creek, the upper part of the Nova Formation and the lower part of the Funeral Formation at Artists Drive.

The broad range in ages (>0.3 to <3.35 Ma) for Q1 (Table 1) and sparse data illustrates the challenge that obtaining reliable numeric ages has been so far. The age range may also show the time-transgressive char-acter of this morphological unit. These ages and the correlation of Q1 to other formations and units in Death Valley should be considered tentative and clearly shows that additional research is warranted.

2.8.2. Q2

The Q2 alluvial fans are some of the most distinc-tive geomorphic landforms in Death Valley. This extensive fan unit has a well developed desert pave-ment (Fig. 12) and is comprised of darkly varnished (2.5YR4/8), tightly packed clasts (Hunt and Mabey, 1966). Soils on Q2c deposits (youngest of three sub-units) in northern Death Valley show Avkz/Btkz/2Bkz horizon profiles. The carbonate dominated profiles demonstrates significant pedogenesis mainly from the addition and redistribution of airborne materials such as silt, calcium carbonate and salt (Klinger, 2001b).

Based on soil development and relative strati-graphic position, Klinger (2001b) estimated that Q2 has an age range between 35–180 ka (Table 1). The maximum age is assumed because the Q2 deposits appear to be overlain by the Lake Manly deposits that are dated at 120–180 ka by Ku et al. (1998). Along the

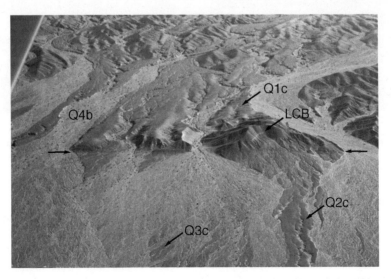

Fig. 11. Oblique aerial photograph of southern Kit Fox Hills showing relations between 0.62 Ma Lava Creek B ash bed (LCB) and younger alluvial fan units. See text for unit definitions and ages. Horizontal arrows show location of Northern Death Valley fault zone (after Klinger, 2001b).

Fig. 12. Typical alluvial fan morphology for units Q2 (A) and Q3 (B). The Q2 surface has a characteristic level, well-developed desert pavement. Traces of the oblique normal Black Mountains fault zone offset the surface. Q3 shows the characteristic remnant bar-and-swale morphology.

eastern Panamint piedmont, Nishiizumi et al. (1993) measured minimum surface exposure ages of 260 ± 9 and 318 ± 12 ka (^{10}Be and ^{26}Al) for clasts on Q2 surfaces; however, the clasts on which these exposure ages were analyzed may have had a complex prior exposure history (i.e., inherited cosmogenic radionuclides), potentially leading to a greater exposure age for the clast than the age of the surface.

2.8.3. Q3

Alluvial fan unit Q3 is also found throughout Death Valley, albeit less extensively than Q2 (Hunt and Mabey, 1966). Q3 typically displays various forms of bar and swale morphology. Surface clasts are only partially coated with a light colored (2.5YR4/8 to 5YR6/6) desert varnish (Fig. 12). Soils have Avk/Bkz/2C horizons with a profile ranging from 20 cm (Q3c) to 72 cm (Q3a) thick (Klinger, 2001b).

Based on archeological evidence from the western Black Mountains piedmont, Hunt and Mabey (1966) estimated that the age of unit Q3 is latest Pleistocene to Holocene. Based on soil development and morphology, Klinger (2001b) suggested an age of 2–12 ka for Q3. Nishiizumi et al. (1993) determined a minimum exposure age of 117 ± 4 ka for clasts on the Q3a surface along the eastern Panamint piedmont. The surface exposure age, however, may be too old due to inherited cosmogenic radionuclides (Nishiizumi et al., 1993). This explanation seems plausible because unpublished mapping by Machette and Janet Slate (USGS) suggest that the sampled surface is underlain by Q2.

2.8.4. Q4

Q4 deposits are found in both the active stream-channels and flood plain. The Q4 deposits have surfaces that have prominent bar-and-swale topography. Thin (4 cm), poorly developed soils with Av/2C horizons are developed on these deposits (Klinger, 2001b). Based on the soil development, Klinger (2001b) suggested that Q4 was several hundred years old or less.

Age control on Q4 is sparse as well. In northern Death Valley, Klinger (2001a) correlated an ash bed within the Q4a (older) unit with a <1200-yr-old Mono Craters ash. In the overlying Q4b alluvium, Klinger (2001a) obtained a ^{14}C age of 140–300 cal. yr. B.P. from a fragment of charcoal. The charcoal fragment underlies the Ubehebe Craters tephra bed, which also provides a maximum age for the most recent eruption of Ubehebe Crater.

3. Tectonic implications

The Late Neogene tectonics of Death Valley is a topic of great interest, particularly since 1966. In that year, Burchfiel and Stewart (1966) proposed the term "pull-apart" basin for Death Valley. They envisioned the Furnace Creek, Northern Death Valley, Black Mountains, and Southern Death Valley fault zones as the main structural components (see Machette et al., 2001b for discussion). In addition, Hill and Troxel (1966) suggested that the regional extensional stress

field had a NW–SE orientation. In this same year, Hunt and Mabey (1966) published their seminal geophysical and geologic study, which provided a more detailed map of the Late Neogene faults and described Quaternary fault scarps as well.

Interest in the Late Neogene tectonics of Death Valley was reinvigorated by the Eastern California shear zone hypothesis (Dokka and Travis, 1990). The Eastern California shear zone appears to transfer ~25% of the San Andreas right-lateral plate boundary motion to strike slip fault systems in the southwestern Basin and Range. The Death Valley fault system is thought to be the easternmost fault system of the Eastern California shear zone.

The tephrostratigraphy and alluvial-fan stratigraphy provide age control for the deformation and translocation of basinal deposits uplifted throughout Death Valley. They also provide age control for Pleistocene and younger faulting events. In the following sections, we describe some of the more recent hypotheses and interpretations of Late Neogene tectonics that have taken advantage of the numeric stratigraphy.

3.1. Death Valley fault system

Machette et al. (2001b) propose a revision to the nomenclature for the Death Valley fault system based on Quaternary structural/stratigraphic studies during the last decade. They propose that the Fish Lake Valley, Northern Death Valley, Black Mountains and Southern Death Valley fault zones be collectively referred to as the Death Valley fault system (Fig. 13). This also suggests limiting the use of the term Furnace Creek fault zone to the largely pre-Quaternary fault southeast of Furnace Creek. Further, they recommend returning to the original name, Black Mountains fault zone. Black Mountains fault zone was originally proposed by Noble and Wright (1954) for the oblique–normal fault along the western piedmont of the Black Mountains.

The north–south to northeast–southwest trending, dextral slip fault zone that bounds the northeast margin of northern Death Valley has had several names including Northern Death Valley fault zone (Wesnousky, 1986); Furnace Creek fault zone (Hill and Troxel, 1966; Hunt and Mabey, 1966; Wright and Troxel, 1993); and Death Valley–Furnace Creek fault zone (Burchfiel and Stewart, 1966; Wright and Troxel,

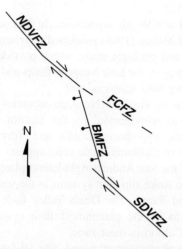

Fig. 13. Schematic diagram showing proposed nomenclature for the Death Valley fault system. Northern Death Valley fault zone (NDVFZ); Furnace Creek fault zone (FCFZ); Black Mountains fault zone (BMFZ); and Southern Death Valley fault zone (SDVFZ); after Machette et al. (2001b).

1967). The different names for the same fault zone have been confusing. In an attempt to clarify this issue, Machette et al. (2001b) divided this fault zone into pre-Quaternary and Quaternary sections. The dividing line is roughly at Furnace Creek where the Black Mountains fault zone abuts this strike-slip fault from the south (Machette et al., 2001b).

Machette et al. (2001b) advocate the name Furnace Creek fault zone for the major fault zone southeast of Furnace Creek Ranch. The Furnace Creek fault zone was an important structural element in the development of Death Valley and the Furnace Creek basin during the Late Miocene and Pliocene. The Furnace Creek fault zone, however, has been largely inactive during the Quaternary (Burchfiel and Stewart, 1966; Hamilton, 1988; Klinger and Piety, 1996).

Northwest of Furnace Creek, Machette et al. (2001b) proposed the name Northern Death Valley fault zone for the dextral oblique fault zone that exhibits Late Pleistocene–Holocene slip. The Northern Death Valley fault zone merges with the Fish Lake Valley fault zone, which is essentially a northward extension of the Death Valley fault system into Nevada.

Machette et al. (2001b) recommended using Furnace Creek–Northern Death Valley fault zone when discussing pre-Pliocene tectonics of the region. These two fault zones appear to have formed the tectoni-

cally-active northeast margin of Death Valley during that time.

Machette et al. (2001b) also propose that the oblique–normal fault zone found at the western foot of the Black Mountains be called the Black Mountains fault zone. This returns to the name originally proposed on the map of Noble and Wright (1954). Likewise, Machette et al. (2001b) proposed that the names Death Valley fault zone (now system) and Central Death Valley fault zone (now Black Mountains) be discontinued with respect to this fault.

Although this change is nomenclature seems cumbersome and unnecessary, we think that this change will provide clarity and consistency for future studies. The recommended fault zone definitions are especially important in terms of defining seismic hazards associated with these fault zones and the overall fault system. In the case of the Northern Death Valley and Furnace Creek fault zones, the nomenclature proposed by Machette et al. (2001b) provides a useful discrimination between the pre-Pliocene and Quaternary faults, potential fault hazards and regional tectonics.

3.2. Breakup of the Furnace Creek basin

The Miocene–Pleistocene Furnace Creek basin had a northwest–southeast trend and was located between the dextral–oblique Grandview and Furnace Creek fault zones (Wright et al., 1999). Clast provenance and paleocurrent data from the lower part of the Furnace Creek Formation indicate that a northwest-to southeast-flowing fluvial system occupied the Furnace Creek basin during the Late Miocene (Hunt and Mabey, 1966; Prave and Wright, 1996; Wright et al., 1999). In contrast, provenance and sedimentary structures in the upper part of the Furnace Creek Formation record southerly and northerly progradation of alluvial fans from the Funeral and Black Mountains, respectively, into a perennial lake/playa (Hunt and Mabey, 1966; Blair and Raynolds, 1999).

The exact timing of the end of the Furnace Creek basin as well as the extent of the basin has been problematic because of the lack of age control on the Furnace Creek Formation, Funeral Formation and deposits of the Kit Fox Hills and Ubehebe–Lake Rogers areas to the northwest. Correlation of the Mesquite Spring tuffs in each of these locations has helped to resolve this problem. The Mesquite Spring tuffs are interbedded with perennial lake/playa deposits at Furnace Creek, in the Kit Fox Hills, and in the Ubehebe/Lake Rogers basin. Identification of the Mesquite Spring tuff and Nomlaki-like tuffs interbedded with alluvial fan deposits in the Nova Formation of the northern Panamint Mountains, Funeral Formation at Artists Drive and along the eastern Cottonwood Mountains piedmont provides broad limits on the playa dimensions. This is interpreted as evidence of either a continuous basin or multiple, contemporaneous basins along a northwest–southeast trend separated by intervening alluvial fans, much like the present Death Valley playa (Fig. 14). The basin axis appears to roughly parallel the trend of the Northern Death Valley–Furnace Creek fault zone.

Correlation of the Mesquite Spring tuff and the underlying lower Nomlaki tuff at Artists Drive (Fig. 2) shows that alluvial-fan deposition began there >3.58 Ma (Knott et al., 1999b; Knott and Sarna-Wojcicki, 2001a). Knott et al. (1999a,b) interpreted

the onset of alluvial-fan deposition at Artists Drive coincided with uplift of the Black Mountains and downdropping of the Artists Drive block. They hypothesized that this uplift was related to along-strike growth of the Black Mountains fault zone from south to north. Knott (1998) also inferred that this along-strike growth of the Black Mountains fault zone uplifting the Furnace Creek basin and the northernmost part of the Black Mountains. This may have lead to the eventual deactivation of the Furnace Creek fault zone as well.

Northward growth of the Black Mountains fault zone generated a zone of compression as it impinged on the Northern Death Valley–Furnace Creek fault zone (Machette et al., 2001b). This is expressed as the NNW-trending Texas Spring syncline and Echo Canyon thrust fault on the north side of Furnace Creek and a series of en-echelon faults that transfer slip from the Furnace Creek fault zone to the Black Mountains fault zone (Klinger and Piety, 2001; Machette et al., 2001c).

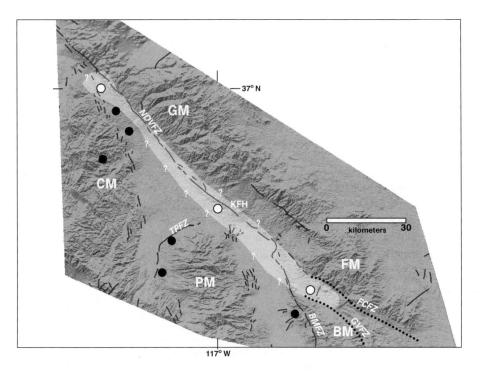

Fig. 14. Map of northern Death Valley showing the distribution of the Furnace Creek lake about 3.2 Ma along with Quaternary fault traces (black lines). White circles indicate locations where Late Pliocene tuffs are interbedded with lake sediments. Black circles indicate where Late Pliocene tuffs are interbedded with alluvial-fan deposits. The shaded area is the aerial extent of the postulated lake. See Fig. 2 for key to abbreviations to geographic elements and fault zones. Shaded relief base map derived from digital elevation model.

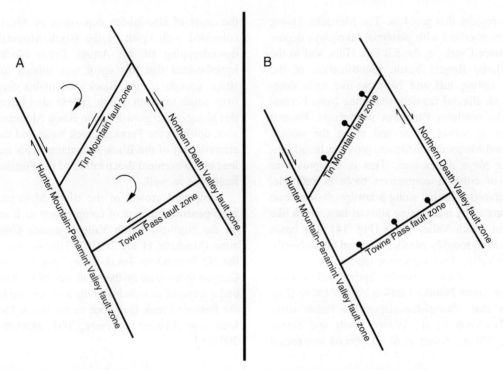

Fig. 15. Schematic diagram showing two dextral shear models proposed for the Late Neogene tectonic evolution of northern Death Valley. A. Following McKenzie and Jackson (1983, 1986), Klinger and Sarna-Wojcicki (2001) propose oblique left-lateral slip on the Towne Pass and Tin Mountain fault zones resulting in clockwise rotation of discrete blocks. (B) After Oldow et al. (1994), Lee et al. (2001) propose pure normal slip on the Towne Pass and Tin Mountain fault zones.

3.3. Tectonic development of Northern Death Valley

Klinger and Sarna-Wojcicki (2001) proposed that regional dextral shear in Northern Death Valley between the Northern Death Valley and Hunter Mountain–Panamint Valley fault zones is accommodated by a set of discrete rotating blocks bounded by the Towne Pass and Tin Mountain fault zones (Fig. 15). Klinger and Sarna-Wojcicki's model is similar to models proposed by McKenzie and Jackson (1983, 1986) where blocks are rotating between two strike-slip faults and

are either rotating about a fixed or shifting vertical axis. The Tin Mountain and Towne Pass faults would be left-lateral accommodating faults and both show evidence of Quaternary movement.

According to Klinger and Sarna-Wojcicki (2001), the northeast corners of the rotating blocks generate compressive structures along the Northern Death Valley fault zone where the rotating block impinge on the bounding Funeral Mountains blocks. In contrast, in the southeast corners (or regions) are extensional, generating deep structural depressions, such as the

Fig. 16. Schematic map showing Death Valley during the Late Neogene. (A) ~3.3 Ma or the time of eruption of the Mesquite Spring tuffs. A lake (lined region) occupied the Furnace Creek basin from Furnace Creek (FC) to the Ubehebe–Lake Rogers (ULRB) area. Alluvial fan deposits of the Nova basin (NB), Artists Drive (AD) and Copper Canyon (CC), flank the lake. Solid thick black lines show active faults at the time. Dotted lines show the future locations of faults that are inactive 3.3 Ma. (B) ~1.8 Ma or the time of eruption of the lower Glass Mountain ash beds. Black Mountain (BMFZ) fault zone has extended along strike to the north cutting through the Furnace Creek basin and making the Grandview (GVFZ) and Furnace Creek (FCFZ) fault zones inactive. stepped basinward uplifting the Copper Canyon basin. The Towne Pass (TPFZ) fault zone has cut and is uplifting the Nova basin. Alluvial fans are depositing in both the Ubehebe and Furnace Creek basin. The playa and alluvial fan deposits of the Confidence Hills Formation (CH) are being deposited along the Southern Death Valley fault zone (SDVFZ). (C) Present. The SDVFZ has stepped to the northeast uplifting the Confidence Hills Formation. The Northern Death Valley fault zone (NDVFZ) is now active causing uplift of the Kit Fox Hills. Compression between the NDVFZ and the BMFZ creates the Salt Creek anticline and Texas Spring syncline. Playas are formed in both Northern and central Death Valley.

Mesquite Flat basin. This interpretation is supported by gravity data from Blakely et al. (1999) who estimates that the Mesquite Flat basin is ≥5 km deep.

In contrast, Lee et al. (2001) observed that movement on the accommodating faults further north in this region, such as the Deep Springs further north and Towne Pass fault zones, is predominantly dip slip with no oblique component. Based on this observation, Lee et al. (2001) proposed a model modified from Oldow et al. (1994), in which displacement on the accommodating faults is normal or dip slip with little rotation.

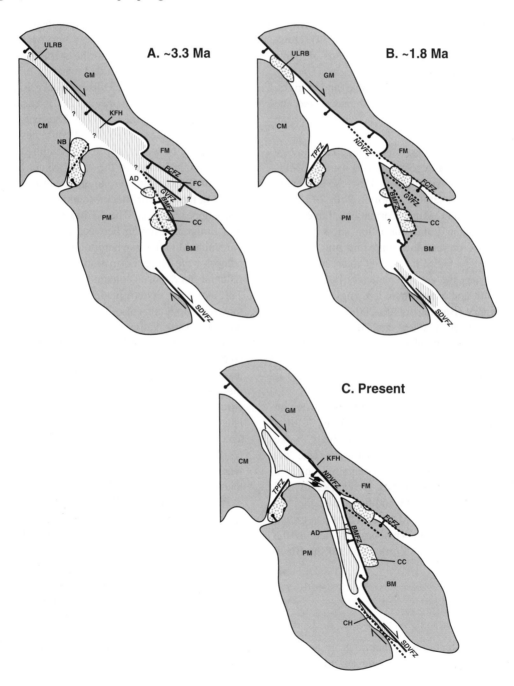

3.4. Spatial and temporal development of the Black Mountains fault zone

The Black Mountains faults zone (BMFZ) is the oblique–normal fault zone found at the base of the Black Mountains. Slip on the BMFZ has created a created a mountain-front morphology of triangular facets, wineglass canyons and fault scarps indicative of active faulting (Bull and McFadden, 1977). However, the fault zone is a complex mixture of faults dipping both at high-angles and low-angles (Drewes, 1963; Hunt and Mabey, 1966; Noble and Wright, 1954). In addition, a paucity of earthquakes and age control left the level of Quaternary activity in question (Knott et al., 1999b).

Knott et al. (1999a,b) established a geochronology for the Late Neogene deposits faulted by the BMFZ that resulted in two main findings. First, the low-angle normal or turtleback fault at Mormon Point offset the Quaternary Mormon Point Formation. Lake beds within the Mormon Point Formation show that the fault was at its present low-angle dip and had not tilted. High-angle faults that offset the 120–180 ka Lake Manly deposits are kinematically linked to the low-angle fault and demonstrate low-angle slip in the Late Quaternary (Hayman et al., 2003). These same temporal and fault relations are found at Natural Bridge above the Badwater turtleback fault as well (Hayman et al., 2003).

Another main finding of Knott et al. (1999a,b) is that the BMFZ is a dynamic structure. After 3.35 Ma, the BMFZ stepped basinward at Copper Canyon and grew northerly along strike northerly Natural Bridge. The along strike growth effectively broke up the Furnace Creek basin. After ~1.8 Ma, the BMFZ changed from a normal mountain front to a graben-bounded front at Artists Drive (Knott and Sarna-Wojcicki, 2001a,b). In addition, over the last ~1 Ma, the BMFZ west of Mormon Point has grown along strike to the north. This has resulted in the uplift of the Mormon Point Formation into the Black Mountains footwall.

4. Summary

Correlations of tephra beds that range in age from Late Pliocene to Holocene provide an unparalleled opportunity to reconstruct the tectonics and paleogeography of the Death Valley pull-apart basin. In particular, the 3.1–3.35 Ma Mesquite Springs group tuffs and 1.7–1.9 Ma lower Glass Mountain ash beds are keys to understanding the breakup of the Furnace Creek basin and development of the Death Valley fault system.

The 3.1–3.35 Ma time line provided by the Mesquite Springs tuffs shows that a northwest–southeast trending lake occupied northeastern Death Valley (Fig. 16). This area included what is presently the Furnace Creek, Kit Fox Hills and Lake Rogers areas (Klinger and Sarna-Wojcicki, 2001; Knott et al., 1999b; Machette et al., 2001c). The margins of this lake are broadly constrained and could have extended to the south. However, any lake extending to the south was probably limited to the valley center because the Mesquite Springs tuffs are found in alluvial fan deposits along the western Black Mountains at Artists Drive and Copper Canyon. The ~3 Ma lake did not extend into Emigrant Canyon between the Cottonwood and Panamint Mountains where alluvial-fans, which now comprise the Nova Formation were being deposited. Alluvial-fan deposits containing the 3.28 Ma Nomlaki Tuff along the eastern piedmont of the northern Cottonwood Mountains also limit the lake to the center of northern Death Valley.

The presence of a lake in Death Valley ~3 Ma is consistent with data from Searles Valley. A ~3.1 Ma tephra layer is found in the Searles Lake core during the longest lake phase identified in the 3.5 Ma long core record (Smith, 1991). Knott et al. (1997) correlated this tephra layer with the Mesquite Springs tuffs. The location of the lake along the Furnace Creek–Northern Death Valley fault zone suggests that there was a vertical slip component to slip on this fault zone that created this elongate lake.

The time line provided by the 1.7–1.9 Ma lower Glass Mountain tephra beds shows the dynamic climatic and tectonic history of Death Valley. In the Furnace Creek basin, the lake that existed ~3 Ma has given way to alluvial fans, which are prograding from both the Black and Funeral Mountains (Fig. 16). This progradation and uplift is interpreted to be coincident with a decrease in activity on the Furnace Creek fault zone and along-strike growth of the Black Mountains fault zone (Machette et al., 2001b). Alluvial fans are also being deposited in the Kit Fox

Hills and to the north in the Ubehebe–Lake Rogers areas (Klinger and Sarna-Wojcicki, 2001; Knott and Sarna-Wojcicki, 2001b). To the south in the Confidence Hills, alluvial-fan and playa sediments are being deposited (Beratan et al., 1999). The alluvial fans are interpreted to record a relatively dry climatic period, which is consistent with a dry climate at Searles Valley (Smith, 1991).

Data on younger deposits remains sparse, but continues to improve. Uplift and tilting of the Confidence Hills has occurred entirely since the end of the Pliocene. During this same time period, the Black Mountains fault zone has stepped basinward at Mormon Point and Natural Bridge and developed a graben at Artists Drive (Knott et al., 1999a,b). At the transpressive zone between the Black Mountains and Northern Death Valley fault zone, uplift of and development of the Texas Spring syncline has continued through the Quaternary (Machette et al., 2001c). In northern Death Valley, the data are consistent with either oblique left lateral slip or normal slip on cross faults within a right lateral shear zone (Klinger and Sarna-Wojcicki, 2001; Lee et al., 2001).

Acknowledgements

The authors have benefited from entertaining and fruitful discussions with many individuals through the years including Lauren Wright and Bennie Troxel, Janet Slate, Angela Jayko, Chris Menges, John Tinsley, Steve Wells and a host of others. The participants of the 2001 Friends of the Pleistocene field trip stimulated and challenged many of our hypotheses as well. Support for work by Knott has been from NSF (EAR 94-06029) and the U.S. Geological Survey. We thank Matthew Kirby and Lewis Owen for constructive reviews of the manuscript.

References

Anderson, D.E., Wells, S.G., 2003. Latest Pleistocene lake highstands in Death Valley, California. In: Enzel, Y., Wells, S.G., Lancaster, N. (Eds.), Paleoenvironments and Paleohydrology of the Mojave and Southern Great Basin Deserts. Geological Society of America, Boulder, Colorado.

Beck, W., Donahue, D.J., Jull, A.J.T., Burr, G., Broecker, W.S., Bonani, G., Hajdas, J., Malotki, E., 1998. Ambiguities in direct dating of rock surfaces using radiocarbon measurements. Science 280, 2132–2139.

Beratan, K.K., Murray, B., 1992. Stratigraphy and depositional environments, southern Confidence Hills, Death Valley, California. San Bernardino County Museum Association Quarterly 39 (2), 7–11.

Beratan, K.K., Hsieh, J., Murray, B., 1999. Pliocene–Pleistocene stratigraphy and depositional environments, southern Confidence Hills, Death Valley, California. In: Wright, L.A., Troxel, B. (Eds.), Cenozoic Basins of the Death Valley Region. Geological Society of America, Boulder, CO, pp. 289–300.

Bierman, P.R., Gillespie, A.R., 1994. Evidence suggesting that methods of rock-varnish cation-ratio dating are neither comparable nor consistently reliable. Quaternary Research 41, 82–90.

Blackwelder, E., 1933. Lake Manly, an extinct lake of Death Valley. Geographical Review 23, 464–471.

Blackwelder, E., 1954. Pleistocene lakes and drainage in the Mojave region, southern California. In: Jahns, R.H. (Ed.), Geology of Southern California. California Division of Mines and Geology, Sacramento, CA, pp. 35–40.

Blair, T.C., Raynolds, R.G., 1999. Sedimentology and tectonic implications of the Neogene synrift hole in the wall and wall front members, Furnace Creek Basin, Death Valley, California. In: Wright, L.A., Troxel, B. (Eds.), Cenozoic Basins of the Death Valley Region. Geological Society of America, Boulder, CO, pp. 127–168.

Blakely, R.J., Jachens, R.C., Calzia, J.P., Langenheim, V.E., 1999. Cenozoic basins of the Death Valley extended terrane as reflected in regional-scale gravity anomalies. In: Wright, L.A., Troxel, B. (Eds.), Cenozoic Basins of the Death Valley Region. Geological Society of America, Boulder, CO, pp. 1–16.

Bull, W.B., McFadden, L.D., 1977. Tectonic geomorphology north and south of the Garlock fault, California. In: Doehring, D.O. (Ed.), Geomorphology in Arid Regions; Proceeding of the 8th Annual Geomorphology Symposium. State University of New York, Binghamton, Binghamton, NY, pp. 115–138.

Burchfiel, B.C., Stewart, J.H., 1966. "Pull-apart" origin of the central segment of Death Valley, California. Geological Society of America Bulletin 77, 439–442.

Clements, T., 1952. Lake Rogers, a Pleistocene lake in the north end of Death Valley, California. Geological Society of America Bulletin 63, 1324.

Clements, T., Clements, L., 1953. Evidence of Pleistocene man in Death Valley, California. Geological Society of America Bulletin 64 (1), 1189–1204.

Denny, C.S., 1967. Fans and pediments. American Journal of Science 265, 81–105.

Dokka, R.K., Travis, C.J., 1990. Role of the eastern California shear zone in accommodating Pacific-North American plate motion. Journal of Geophysical Research 17 (9), 1323–1326.

Dorn, R.I., 1988. A rock varnish interpretation of alluvial-fan development in Death Valley, California. National Geographic Research 4 (1), 56–73.

Dorn, R.I., Jull, A.J.T., Donahue, D.J., Linick, T.W., Toolin, L.J., 1990. Latest Pleistocene lake shorelines and glacial chronology in the western Basin and Range Province, U.S.A.: insights from AMS radiocarbon dating of rock varnish and paleoclimatic implications. Palaeogeography, Palaeoclimatology, Palaeoecology 78, 315–331.

Drewes, H., 1963. Geology of the Funeral Peak Quadrangle, California, on the east flank of Death Valley. United States Geological Survey Professional Paper 413.

Hamilton, W.B., 1988. Detachment faulting in the Death Valley Region, California and Nevada. U S G S Bulletin 1790, 51–85.

Hayman, N.W., Knott, J.R., Cowan, D.S., Nemser, E., Sarna-Wojcicki, A.M., 2003. Quaternary low-angle slip on detachment faults in Death Valley, California. Geology 31 (4), 343–346.

Hill, M.L., Troxel, B.W., 1966. Tectonics of Death Valley region, California. Geological Society of America Bulletin 77, 435–438.

Hodges, K.V., McKenna, L.W., Stock, J., Knapp, J., Page, L., Sternlof, K., Silverberg, D., Wust, G., Walker, J.D., 1989. Evolution of extensional basins and basin and range topography west of Death Valley, California. Tectonics 8 (3), 453–467.

Holm, D.K., Fleck, R.J., Lux, D.R., 1994. The Death Valley turtlebacks reinterpreted as Miocene–Pliocene folds of a major detachment surface. Journal of Geology 102, 718–727.

Hooke, R.L., 1972. Geomorphic evidence for Late-Wisconsin and Holocene tectonic deformation, Death Valley, California. Geological Society of America Bulletin 83, 2073–2098.

Hooke, R.L., Dorn, R.I., 1992. Segmentation of alluvial fans in Death Valley, California: new insights from surface exposure dating and laboratory modelling. Earth Surface Processes and Landforms 17, 557–574.

Hooke, R.B., Lively, R.S., 1979. Dating of Late Quaternary Deposits and Associated Tectonic Events by U/Th Methods, Death Valley, California. National Science Foundation. EAR-7919999.

Hunt, C.B., Mabey, D.R., 1966. Stratigraphy and structure Death Valley, California. United States Geological Survey Professional Paper, 494-A.

Jayko, A.S., Menges, C.M., 2001. A short note on developing digital methods for regional mapping of surficial basin deposits in arid regions using remote sensing and DEM data — example from central Death Valley, California. In: Machette, M.N., Johnson, M.L., Slate, J.L. (Eds.), Quaternary and Late Pliocene Geology of the Death Valley Region: Recent Observations on Tectonics, Stratigraphy, and Lake Cycles (Guidebook for the 2001 Pacific Cell — Friends of the Pleistocene Fieldtrip). U.S. Geological Survey, Denver, CO, pp. 167–172.

Klinger, R.E., 2001a. Late Quaternary volcanism of Ubehebe crater. In: Machette, M.N., Johnson, M.L., Slate, J.L. (Eds.), Quaternary and Late Pliocene Geology of the Death Valley Region: Recent Observations on Tectonics, Stratigraphy, and Lake Cycles (Guidebook for the 2001 Pacific Cell — Friends of the Pleistocene Fieldtrip). U.S. Geological Survey, Denver, CO, pp. 21–24.

Klinger, R.E., 2001b. Road log for day A, northern Death Valley. In: Machette, M.N., Johnson, M.L., Slate, J.L. (Eds.), Quaternary and Late Pliocene Geology of the Death Valley Region: Recent Observations on Tectonics, Stratigraphy, and Lake Cycles (Guidebook for the 2001 Pacific Cell — Friends of the Pleistocene Fieldtrip). U.S. Geological Survey, Denver, CO, pp. 6–20.

Klinger, R.E., Piety, L.A., 1996. Late Quaternary activity on the Furnace Creek fault, northern Death Valley, California.

Geological Society of America Abstracts with Programs 28 (7), A-193.

Klinger, R.E., Piety, L.A., 2001. Late Quaternary flexural-slip folding and faulting in the Texas Spring syncline, Death Valley. In: Machette, M.N., Johnson, M.L., Slate, J.L. (Eds.), Quaternary and Late Pliocene Geology of the Death Valley Region: Recent Observations on Tectonics, Stratigraphy, and Lake Cycles (Guidebook for the 2001 Pacific Cell — Friends of the Pleistocene Fieldtrip). U.S. Geological Survey, Denver, CO, pp. 185–192.

Klinger, R.E., Sarna-Wojcicki, A.M., 2001. Active tectonics and deposition in the Lake Rogers basin. In: Machette, M.N., Johnson, M.L., Slate, J.L. (Eds.), Quaternary and Late Pliocene Geology of the Death Valley Region: Recent Observations on Tectonics, Stratigraphy, and Lake Cycles (Guidebook for the 2001 Pacific Cell — Friends of the Pleistocene Fieldtrip). U.S. Geological Survey, Denver, CO, pp. 25–31.

Knott, J.R., 1997. An Early to Middle Pleistocene pluvial Death Valley lake, Mormon Point, central Death Valley, California. San Bernardino County Museum Association Quarterly 44 (2), 85–88.

Knott, J.R., 1998. Late Cenozoic tephrochronology, stratigraphy, geomorphology, and neotectonics of the western Black Mountains piedmont, Death Valley, California: implications for the spatial and temporal evolution of the Death Valley fault zone. Ph.D. Thesis, University of California, Riverside, Riverside, CA, 407 pp.

Knott, J.R., 1999. Quaternary stratigraphy and geomorphology of Death Valley. In: Slate, J.L. (Ed.), Proceeding of Conference on Status of Geologic Research and Mapping in Death Valley National Park, Las Vegas, Nevada, April 9–11, 1999. U.S. Geological Survey, Denver, CO, pp. 90–96.

Knott, J.R., Sarna-Wojcicki, A.M., 2001a. Late Pliocene tephrostratigraphy and geomorphic development of the Artists Drive structural block. In: Machette, M.N., Johnson, M.L., Slate, J.L. (Eds.), Quaternary and Late Pliocene Geology of the Death Valley Region: Recent Observations on Tectonics, Stratigraphy, and Lake Cycles (Guidebook for the 2001 Pacific Cell — Friends of the Pleistocene Fieldtrip). U.S. Geological Survey, Denver, CO, pp. 80–87.

Knott, J.R., Sarna-Wojcicki, A.M., 2001b. Late Pliocene to Early Pleistocene paleoclimate and tectonic reconstructions of Death Valley. AAPG Bulletin 85 (6), 1128.

Knott, J.R., Sarna-Wojcicki, A.M., Montanez, I.P., Geissman, J.W., 1997. Differentiating the Upper Pliocene Mesquite spring tuffs from the Middle Pleistocene Bishop ash bed, Death Valley, California: implications for reliable correlation of the bishop ash bed. EOS, Transactions of the American Geophysical Union 78 (46), F760.

Knott, J.R., Sarna-Wojcicki, A.M., Klinger, R.E., Tinsley, J.C., Troxel, B.W., 1999a. Late Cenozoic tephrochronology of Death Valley, California — new insights into stratigraphy, paleogeography, and tectonics. In: Slate, J.L. (Ed.), Proceeding of Conference on Status of Geologic Research and Mapping in Death Valley National Park, Las Vegas, Nevada, April 9–11, 1999. U.S. Geological Survey, Denver, Colorado, pp. 115–116.

Knott, J.R., Sarna-Wojcicki, A.M., Meyer, C.E., Tinsley, J.C.I., Wells, S.G., Wan, E., 1999b. Late Cenozoic stratigraphy and tephrochronology of the western Black Mountains piedmont, Death Valley, California: implications for the tectonic development of Death Valley. In: Wright, L.A., Troxel, B.W. (Eds.), Cenozoic Basins of the Death Valley Region. Geological Society of America, Boulder, Colorado, pp. 345–366.

Knott, J.R., Jayko, A.S., Sarna-Wojcicki, A.M., 2000. Late Pliocene alluvial fan deposits, eastern Panamint Mountains piedmont, Death Valley, California. Geological Society of America Abstracts with Programs 32 (7), 183.

Knott, J.R., Tinsley, J.C., Wells, S.G., 2002. Are the benches at Mormon Point, Death Valley, California, USA, scarps or shorelines? Quaternary Research 58, 352–360.

Ku, T.L., Luo, S., Lowenstein, T.K., Li, J., Spencer, R.J., 1998. U-series chronology of lacustrine deposits in Death Valley, California. Quaternary Research 50, 261–275.

Lee, J., Rubin, C.M., Calvert, A., 2001. Quaternary faulting history along the Deep Springs fault, California. Geological Society of America Bulletin 113 (7), 855–869.

Liddicoat, J.C., 2001. Paleomagnetism of the upper part of the Furnace Creek Formation, Death Valley, California. In: Machette, M.N., Johnson, M.L., Slate, J.L. (Eds.), Quaternary and Late Pliocene Geology of the Death Valley Region: Recent Observations on Tectonics, Stratigraphy, and Lake Cycles (Guidebook for the 2001 Pacific Cell — Friends of the Pleistocene Fieldtrip). U.S. Geological Survey, Denver, CO, pp. 137–141.

Lowenstein, T.K., Li, J., Brown, C., Roberts, S.M., Ku, T., Luo, S., Yang, W., 1999. 200 k. y. paleoclimate record from Death Valley salt core. Geology 27 (1), 3–6.

Machette, M.N., 1985. Calcic Soils in the Southwestern United States. Geological Society of America, Boulder, Colorado, pp. 1–21.

Machette, M.N., Klinger, R.E., Knott, J.R., 2001a. Questions about Lake Manly's age, extent, and source. In: Machette, M.N., Johnson, M.L., Slate, J.L. (Eds.), Quaternary and Late Pliocene Geology of the Death Valley Region: Recent Observations on Tectonics, Stratigraphy, and Lake Cycles (Guidebook for the 2001 Pacific Cell — Friends of the Pleistocene Fieldtrip). U.S. Geological Survey, Denver, CO, pp. 143–149.

Machette, M.N., Klinger, R.E., Knott, J.R., Wills, C.J., Bryant, W.A., Reheis, M.C., 2001b. A proposed nomenclature for the Death Valley fault system. In: Machette, M.N., Johnson, M.L., Slate, J.L. (Eds.), Quaternary and Late Pliocene Geology of the Death Valley Region: Recent Observations on Tectonics, Stratigraphy, and Lake Cycles (Guidebook for the 2001 Pacific Cell — Friends of the Pleistocene Fieldtrip). U.S. Geological Survey, Denver, CO, pp. 173–183.

Machette, M.N., Menges, C.M., Slate, J.L., Crone, A.J., Klinger, R.E., Piety, L.A., Sarna-Wojcicki, A.M., Thompson, R.A., 2001c. Field trip guide for day B, Furnace Creek area. In: Machette, M.N., Johnson, M.L., Slate, J.L. (Eds.), Quaternary and Late Pliocene Geology of the Death Valley Region: Recent Observations on Tectonics, Stratigraphy, and Lake Cycles (Guidebook for the 2001 Pacific Cell — Friends of the Pleistocene Fieldtrip). U.S. Geological Survey, Denver, CO, pp. 51–87.

McAllister, J.F., 1970. Geology of the Furnace Creek Borate Area, Death Valley, Inyo County, California. California Division of Mines and Geology.

McAllister, J.F., 1973. Geologic map and sections of the Amargosa Valley borate area, southeast continuation of the Furnace Creek area, Inyo County, CA. U.S. Geological Survey Miscellaneous Map I-782.

McFadden, L.D., Bull, W.B., Wells, S.G., 1991. Stratigraphy and geomorphology of Quaternary piedmont deposits. In: Morrison, R.B. (Ed.), Quaternary Nonglacial Geology: Conterminous U.S. Geology of North America. Geological Society of America, Boulder, CO, pp. 327–331.

McKenzie, D., Jackson, J., 1983. The relationship between strain rates, crustal thickening, paleomagnetism, finite strain and fault movements with a deforming zone. Earth and Planetary Science Letters 65, 182–202.

McKenzie, D., Jackson, J., 1986. A block model of distributed deformation by faulting. Journal of the Geological Society of London 143, 349–353.

Means, T.H., 1932. Death Valley. Sierra Club Bulletin 17, 67–76.

Menges, C.M., Taylor, E.M., Workman, J.B., Jayko, A.S., 2001. Regional surficial-deposit mapping in the Death Valley area of California and Nevada in support of ground-water modeling. In: Machette, M.N., Johnson, M.L., Slate, J.L. (Eds.), Quaternary and Late Pliocene Geology of the Death Valley Region: Recent Observations on Tectonics, Stratigraphy, and Lake Cycles (Guidebook for the 2001 Pacific Cell — Friends of the Pleistocene Fieldtrip). U.S. Geological Survey, Denver, CO, pp. 151–166.

Nishiizumi, K., Kohl, C.P., Arnold, J.R., Dorn, R., Klein, J., Fink, D., Middleton, R., Lal, D., 1993. Role of in situ cosmogenic nuclides ^{10}Be and ^{26}Al in the study of diverse geomorphic processes. Earth Surface Processes and Landforms 18, 407–425.

Noble, L.F., 1926. Note on a colemanite deposit near Shoshone, Calif., with a sketch of the geology of a part of Amargosa Valley. U.S. Geological Survey Bulletin 785, 63–73.

Noble, L.F., 1934. Rock formations of Death Valley, California. Science 80 (2069), 173–178.

Noble, L.F., Wright, L.A., 1954. Geology of the central and southern Death Valley region, California. In: Jahns, R.H. (Ed.), Geology of Southern California. California Division of Mines and Geology, Sacramento, CA, pp. 143–160.

Oldow, J.S., Kohler, G., Donelick, R.A., 1994. Late Cenozoic extensional transfer in the Walker Lane strike-slip belt, Nevada. Geology 22, 637–640.

Pluhar, C.J., Holt, J.W., Kirschvink, J.L., 1992. Magnetostratigraphy of Plio–Pleistocene lake sediments in the Confidence Hills of southern Death Valley, California. San Bernardino County Museum Association Quarterly 39 (2), 12–19.

Prave, A.R., Wright, L.A., 1996. Pebbles and cross-beds: data to constrain the Miocene tectonic evolution of the Furnace Creek basin, Death Valley, California. Geological Society of America Abstracts with Programs 28 (7), 309.

Sarna-Wojcicki, A.M., Machette, M.N., Knott, J.R., Klinger, R.E., Fleck, R.J., Tinsley, J.C., Troxel, B., Budahn, J.R., Walker, J.P., 2001. Weaving a temporal and spatial framework for the Late Neogene of Death Valley — correlation and dating of Pliocene and Quaternary units using tephrochronology, ^{40}Ar/^{39}Ar dating, and other dating methods. In: Machette, M.N., Johnson, M.L., Slate, J.L. (Eds.), Quaternary and Late Pliocene Geology of the Death Valley Region: Recent Observations on Tectonics, Stratigraphy, and Lake Cycles (Guidebook for the 2001 Pacific Cell — Friends of the Pleistocene Fieldtrip). U.S. Geological Survey, Denver, CO, pp. 121–135.

Smith, G.I., 1991. Stratigraphy and chronology of Quaternary-age lacustrine deposits. In: Morrison, R.B. (Ed.), Quaternary Nonglacial Geology: Conterminous U.S. Geological Society of America, Boulder, Colorado, pp. 339–352.

Snow, J.K., Lux, D.R., 1999. Tectono-sequence stratigraphy of Tertiary rocks in the Cottonwood Mountains and northern Death Valley area, California and Nevada. In: Wright, L.A., Troxel, B. (Eds.), Cenozoic Basins of the Death Valley Region. Geological Society of America, Boulder, CO, pp. 17–64.

Snow, J.K., White, C., 1990. Listric normal faulting and synorogenic sedimentation, northern Cottonwood Mountains, Death Valley region, California. In: Wernicke, B.P. (Ed.), Basin and Range Extensional Tectonics near the Latitude of Las Vegas, Nevada. Geological Society of America, Boulder, Colorado, pp. 413–445.

Topping, D.J., 1993. Paleogeographic reconstruction of the Death Valley extended region: evidence from Miocene large rock-avalanche deposits in the Amargosa chaos basin, California. Geological Society of America Bulletin 105, 1190–1213.

Troxel, B.W., Sarna-Wojcicki, A.M., Meyer, C.E., 1986. Ages, correlations, and sources of three ash beds in deformed Pleistocene beds, Confidence Hills, Death Valley, California. In: Troxel, B.W. (Ed.), Quaternary Tectonics of Southern Death Valley. B W Troxel, Shoshone, CA, pp. 29–30.

Watchman, A., 2000. A review of the history of dating rock varnishes. Earth-Science Reviews 49, 261–277.

Wernicke, B., 1992. Cenozoic extensional tectonics of the U. S. Cordillera. In: Burchfiel, B.C., Lipman, P.W., Zoback, M.L. (Eds.), The Cordilleran Orogen: Conterminous U. S. The Geology of North America. Geological Society of America, Boulder, Colorado, pp. 553–581.

Wesnousky, S.G., 1986. Earthquakes, Quaternary faults, and seismic hazard in California. Journal of Geophysical Research 91 (B12), 12587–12631.

Wright, L., 1976. Late Cenozoic fault patterns and stress fields in the great basin and westward displacement of the Sierra Nevada block. Geology 4, 489–494.

Wright, L.A., Troxel, B.W., 1967. Limitations on right-lateral, strike-slip displacement, Death Valley and Furnace Creek fault zones, California. Geological Society of America Bulletin 78, 933–950.

Wright, L.A., Troxel, B.W., 1984. Geology of the North 1/2 Confidence Hills 15' quadrangle, Inyo County, California. California Division of Mines and Geology, Sacramento, CA.

Wright, L.A., Troxel, B.W., 1993. Geologic map of the central and northern Funeral Mountains and adjacent areas, Death Valley region, southern California. U.S. Geological Survey, Washington, D.C.

Wright, L.A., Otton, J.K., Troxel, B.W., 1974. Turtleback surfaces of Death Valley viewed as phenomena of extensional tectonics. Geology 2, 53–54.

Wright, L.A., Greene, R.C., Cemen, I., Johnson, F.C., Prave, A.R., 1999. Tectonostratigraphic development of the Miocene–Pliocene Furnace Creek Basin and related features, Death Valley region, California. In: Wright, L.A., Troxel, B. (Eds.), Cenozoic Basins of the Death Valley Region. Geological Society of America, Boulder, CO, pp. 87–114.

Available online at www.sciencedirect.com

Earth-Science Reviews 73 (2005) 271–289

www.elsevier.com/locate/earscirev

Late Quaternary denudation, Death and Panamint Valleys, eastern California

A.S. Jayko *

U.S. Geological Survey, White Mountain Research Station, 3000 East Line St., Bishop, CA 93514, USA

Accepted 1 April 2005

Abstract

Late Quaternary denudation rates are constrained from alluvial fans and tributary watersheds in central Death and Panamint Valleys. Preliminary results suggest that the denudation rate is in part a function of the mean watershed elevation. Rainfall increases semi-logarithmically with higher elevation to about 2500 m where it becomes limited by the regional average maximum moisture content of the air mass. The fan volumes show a power-law relation to the watershed areas. The fan volumes ranged from about 250,000 to 4000 km^3 and the watershed areas range from about 60,000 to 2000 km^2. The upper limit of the denudation rates estimated from small Death Valley fans restricted to the east side of the basin along the Black Mountain frontal scarp range between about 0.03 to 0.18 mm/yr. The maximum is made by assuming most of the clastic accumulation in these fans followed the last highstand of Lake Manly around 24,000 yr which is the least conservative condition. The upper limit of the denudation rates from the Panamint fans range from 0.04 to 0.20 mm/yr assuming the accumulation mainly postdates OIS-4 ~60,000 yr or OIS-2 ~20,000 yr based on the presence or absence of inset shorelines from the last glacial–pluvial maximum. The greater denudation rate associated with the higher mean watershed elevations can mainly be attributed to the greater rainfall at higher elevation. Denudation rates are about a third or less of the Neogene dip-slip rates reported from nearby active faults consistent with relief increasing during dryer periods.
© 2005 Elsevier B.V. All rights reserved.

Keywords: Death Valley; alluvial fans; Quaternary; denudation rates; arid processes; Panamint Valley

1. Introduction

Some of the most extreme relief and arid land in the co-terminus United States lies in the southwestern part of the Great Basin, which includes Panamint and Death Valleys. Here, elevations range from 4342 m at Mt Whitney in the Sierra Nevada to −85 m at Badwater in Death Valley. Relief from ridge to valley floor in the fan watersheds ranges from 1980 m near Badwater to 3100 m near Ballarat in Panamint Valley (Fig. 1). Denudation rate most strongly correlates with relief in most regions (Schumm, 1963; Ahnert, 1970). Likewise, the higher relief regions are also

* Tel.: +1 760 873 7040; fax: +1 760 873 7830.
 E-mail address: ajayko@usgs.gov.

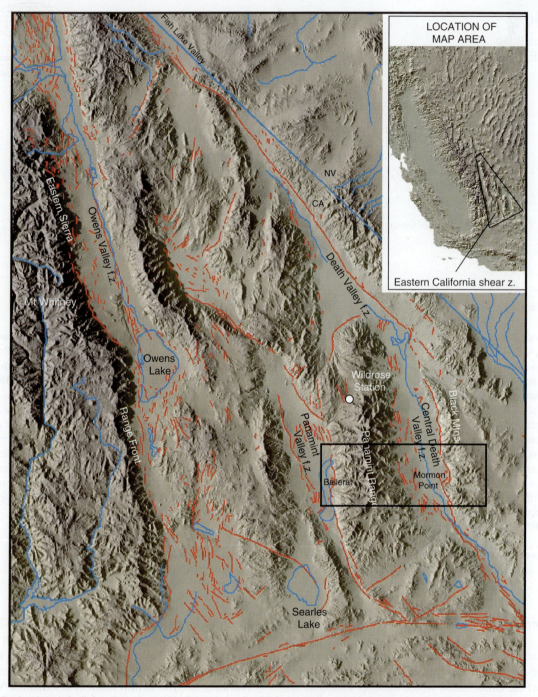

Fig. 1. Regional shaded relief map showing location of study area, major faults and topographic features described in the text. Quaternary faults from California Division of Mines and Geology, Quaternary fault Map of California, 1 : 750,000 shown in red, major drainages in blue.

generally higher mean elevation (Summerfield and Hulton, 1994). Denudation rates can be estimated for an area by measuring the amount of eroded debris transported from the watershed, which is commonly determined from the sediment load measured at gaging stations in perennial and intermittent streams

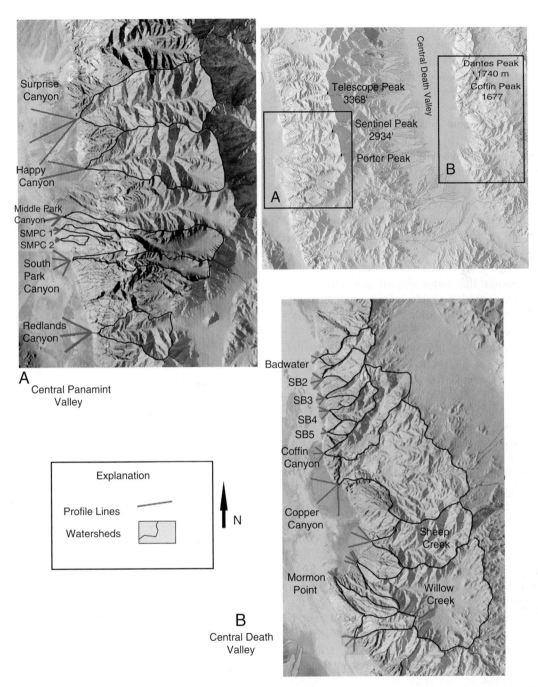

Fig. 2. Map showing location of fan profiles and watershed areas used to calculate denudation rates from Death and Panamint Valleys. The profiles adjacent to the labeled canyons were used.

(Ahnert, 1970). Such gaging stations are not typically maintained in arid regions where stream flow can be so intermittent that it occurs on the order of once every several years or decades rather than annually. Therefore, the volume of material in alluvial fans exiting from adjoining watersheds can be used to approximate the material removed from the watershed area (for example, Granger and others, 1996). The geomorphic parameters measured in this study occur along active fault-bounded range front escarpments (Fig. 2).

The Death Valley region is noted for its extreme aridity. Mean annual precipitation on the valley floor ranges from 0 to 4 cm/yr and at higher elevations ranges from 9.0 to 20 cm/yr (Hunt and Mabey, 1966; Smith et al., 1983). Data from gaging stations in the Amargosa, northern Mojave and Death Valley areas show elevation dependent variation of mean annual precipitation for the region over the last 30 years (Fig. 3). Studies of hillslope processes associated with extreme rainfall events in arid regions suggest that much of the transport of hillslope colluvium that forms alluvial fans occurs during the infrequent but highly intense rainfall events that frequently induce flash floods (Lustig, 1965; Hooke, 1968, 1972).

Mean annual precipitation (M.A.P.) from the Wildrose Ranger Station in the Death Valley area at 1250 m elevation is 19 cm/yr, similar to adjacent arid regions of equivalent elevation including the Amargosa and Mojave basins (Fig. 3). Rainfall in the Death Valley region tends to occur during infrequent, irregularly distributed cloudbursts, usually from 2 to 6 storm events per year. The rainfall events tend to be very energetic and frequently catastrophic with enough power to modify the landscape by erosion, incision, and hillslope failure thereby transporting significant amounts of sediment (Hooke, 1972). Single, large magnitude storms are capable of releasing one or two times the mean annual precipitation. For example, during July, 1984, a weather station at Wildrose recorded 30.5 cm of rain within a 24-h period.

Death and Panamint Valleys lie within the eastern California shear zone, a tectonically and seismically active region (Reheis and Sawyer, 1997). The basin-range blocks are fault bounded and undergo relative uplift with rates on the order of 0.5 to 1.5 mm/yr (Table 1) (Klinger and Piety, 1996; Reheis and Sawyer, 1997; Klinger, 1999). Dip-slip rates along part of the central Death Valley fault zone have been estimated to be as great as 3–5 mm/yr near Mormon Point (Klinger and Piety, 1996; Klinger, 1999) although a more conservative slip rate of 1.6 mm/yr has been

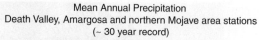

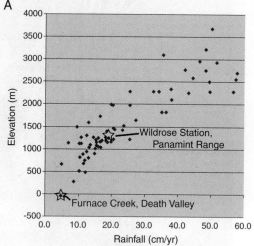

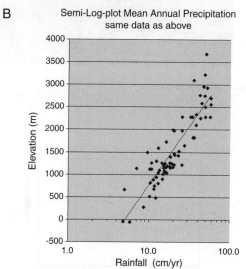

Fig. 3. Mean annual precipitation versus elevation from stations located in the Death Valley, Amargosa and northern Mojave area. Length of record about 30 yr. (A) Linear plot; (B) semi-log plot of data.

suggested (Knott et al., 2001). Dip-slip displacement on normal faults bounding the eastern Sierran range front, where the relief is comparable to the west side of the Panamint Range, is on the order of 0.3 to 0.8 mm/yr (Martel et al., 1987; Lubetkin and Clark, 1988; Beanland and Clark, 1994; Zehfuss et al., 2001). Individual seismic events can cause considerable uplift, for example, the Owens Valley fault zone

Table 1
Displacement rates tabulated from published reports

Fault or location	Reference	Dip-slip rate (mm/yr)	Average (mm/yr)	Duration or age of contraining horizon	Name of contraining horizon
Dip-slip rates, greater Death Valley area					
Deep Springs Fault	Reheis and Sawyer, 1997	0.2–0.5	0.37		
Emigrant Peak FZ, Fish Lake V.	Reheis and Sawyer, 1997	0.2–0.5	0.37		
Dyer Section, Fish Lake V.	Reheis and Sawyer, 1997	0.1–0.3	0.2		
Chiatovich Creek, Fish Lake V.	Reheis and Sawyer, 1997	0.2–0.4	0.3		
Ash Hill Fault, Panamint Valley	Densmore and Anderson, 1997	0.1	0.1	Pliocene basalt	
Panamint Fault, Panamint Valley	Vogel and others, 2002	0.34	0.34	Pleistocene alluvium	
		Mean	0.28		
Dip-slip rates, Death Valley area					
DV FZ, Willow Creek, central Death Valley	Klinger, 1996	3.0–5.0	4	Holocene	
Dip-slip rates, eastern Sierra					
Owens Valley Fault zone	Beanland and Clark, 1994	0.1–0.8	0.45	300 ka	
Fish Springs Fault	Martel and others, 1987	0.1–0.3	0.2	314 ka	
Fish Springs Fault	Zehfuss and others, 2001	0.24	0.24	300 ka	basalitc cinder cone
Lone Pine area	Martel and others, 1987	0.4	0.4		bedrock offset
Big Pine area	Martel and others, 1987	0.3	0.3		bedrock offset
Sierra escarpment, Mono Quad'	Kistler, 1966	0.41	0.4	0.84 ka	Sherwin till?
		Mean	0.33		

Uplift rates eastern Sierra

Fault or location	Reference	Uplift rate (mm/yr)	Average (mm/yr)	Age (Ma)	Horizon
Owens Gorge	Curry, 1971	0.3	0.3	3.2	Owens Gorge basalt
San Joaquin Crest	Curry, 1971	0.25	0.25	2.7	Two Teats quartz latite
San Joaquin Crest	Curry, 1971	0.34	0.34	0.76	Bishop Tuff
San Joaquin Crest	Curry, 1971	0.37	0.37	3.1	Andesite of Deadman Pass
Mono Basin	Gilbert et al., 1968	0.3	0.3	3	Mono Lake Rocks
Sierra escarpment, Mono Quad'	Kistler, 1966 1966–1971	0.41	0.41 0.33	0.94	Sherwin till?
East of Big Pine	Bachman, 1974, 1978	1 (0.5?)	1	2.3	Waucobi Lake Beds
Bloody Cnyn, Mono Basin	Phillips et al., 1990; Burke and Birkeland, 1979	0.7–1.0	0.85		Mono Basin moraines
Long Valley	Clark and Gillespie, 1981	1.1	1.1		Hilton Creek fault
White Mtns	Depolo, 1989 1978–1990	0.9	0.9 Mean 0.56		Marble Creek

locally developed a 4.4-m vertical scarp from the 1872 earthquake, inferred to be at least a 7.5 to 7.7 moment magnitude with a recurrence interval of about 3000 yr (Beanland and Clark, 1994). Similar estimates of vertical displacement and rupture frequency have been made along the central Death Valley fault zone at the base of the Black Mountain escarpment (Brogan et al., 1991; Klinger and Piety, 1996; Klinger, 1999; Frankel et al., 2001).

Extensive studies of the sediment record from cores drilled in Searles and Owens Lakes provides a framework for interpreting the paleoclimate and physiographic history of the greater Owens River watershed (Smith, 1968, 1997; Smith et al., 1983; Bischoff et al., 1997) which drained into Death Valley during extreme pluvial maximums. Cores from Panamint Valley (Smith and Pratt, 1957; Jannick and others, 1991) and Death Valley (Hooke, 1972; Lowenstein et al., 1999) provide additional essential constraints on the paleoclimate and related sedimentation record. Sedimentation rates from Death Valley, Pana-

mint, Searles and Owens basin core data obtained from the basin depocenters (Table 2) generally range between 0.2 and 0.6 mm/yr between about 600,000

Table 2
Sedimentation rates tabulated from published reports. Sedimentation rates determined from cores, Death Valley, Panamint Valley, Searles Valley and Owens Valley

Location	Age* (1000)	Rate (mm/yr)	Age range	Reference
Searles	3500	0.22	3.5 Ma	Smith et al., 1983
Owens V.	780	0.41	780 ka	Smith and Bischoff, 1997
Searles	660	0.32	651–683 ka	Liddicoat et al., 1980; Smith et al., 1983
Searles	645	0.21	643–651 ka	Liddicoat et al., 1980; Smith et al., 1983
Searles	630	0.3	616–643 ka	Liddicoat et al., 1980; Smith et al., 1983
Searles	570	0.21	522–616 ka	Liddicoat et al., 1980; Smith et al., 1983
Searles	485	0.1	378–398 ka	Liddicoat et al., 1980; Smith et al., 1983
Searles	480	0.2	456–522 ka	Liddicoat et al., 1980; Smith et al., 1983
Searles	450	0.43	447–456 ka	Liddicoat et al., 1980; Smith et al., 1983
Searles	430	0.28	424–447 ka	Liddicoat et al., 1980; Smith et al., 1983
Searles	415	0.55	408–424 ka	Liddicoat et al., 1980; Smith et al., 1983
Searles	400	0.07	398–408 ka	Liddicoat et al., 1980; Smith et al., 1983
Panamint	392	0.4	392 ka	Jannik et al., 1991
Searles	330	0.22	221–378 ka	Liddicoat et al., 1980; Smith et al., 1983
Searles	215	0.23	204–221 ka	Liddicoat et al., 1980; Smith et al., 1983
Searles	200	0.25	130–310 ka	Smith et al., 1983
Searles	195	0.12	185–204 ka	Liddicoat et al., 1980; Smith et al., 1983
Death V.	192	0.95		Ku et al., 1998
Death V.	186	0.87		Ku et al., 1998
Panamint	175	0.39	175 ka	Jannik et al., 1991
Death V.	166	0.92		Ku et al., 1998
Death V.	146	0.95		Ku et al., 1998
Death V.	128	0.99		Ku et al., 1998
Death V.	120	0.91		Ku et al., 1998
Death V.	98.6	0.78		Ku et al., 1998
Death V.	97.1	0.89		Ku et al., 1998
Owens V.	80	0.68	80 ka	Bischoff and others, 1997
Searles	75	0.32	32–130 ka	Smith et al., 1983
Death V.	59.3	0.95		Ku et al., 1998
Death V.	57.5	1.04		Ku et al., 1998
Owens V.	55	0.65	55 ka	Bischoff and others, 1997
Death V.	30.1	0.54		Ku et al., 1998
Owens V.	30	1.01	30 ka	Bischoff and others, 1997
Searles	26	1.6	24–32 ka	Smith et al., 1983
Searles	24	0.36	10–24 ka	Smith et al., 1983
Death V.	21	1.11	12,980–21,500	Hooke, 1972
Death V.	21	1.39	12,980–21,500	Hooke, 1972
Death V	21	0.62	12,980–21,500	Hooke, 1972
Death V.	21	0.77	12,980–21,500	Hooke, 1972
Death V.	9.6	0.8		Ku et al., 1998
Searles	6	0.97	6 ka	Smith et al., 1983
Searles	6	3.5	6–10 ka	Smith et al., 1983

31,130–30,670 ka 1.02 mm/yr.
55,100–80,000 ka 0.69 mm/yr.

 * Used to calculate rate, source cited or mean age if range provided.

and 30,000 yr. Rates tend to be higher during the last 30,000 yr ranging between 0.6 and 1.6 mm/yr (Smith, 1968; Smith et al., 1983; Bischoff et al., 1998).

2. Methods

Geomorphic parameters were quantified on fifteen conical alluvial fans from the east sides of Panamint and Death Valleys (Fig. 2) where they accumulated adjacent to active escarpments as opposed to the surfaces of rotated structural mountain blocks which are exposed on the west side of the valleys. The denudation rate for the watersheds was determined from the fan volume, watershed area and estimated fan age constrained by the lacustrine history of the basins. The fans overlie and interfinger with playa and lacustrine deposits of late Pleistocene age (Smith and Pratt, 1957; Hooke, 1972; Fitzpatrick et al., 1993). They emanate from westward-draining watersheds lying within the east side range fronts which are escarpments bounding the basins. Discrete, generally small conical fans with radii from 4 to 18 km were selected for the volume estimates. Small non-coalescing or very minimally coalescing fans simplify the volume estimate and help constrain the potential age variability. The conical volume estimate is considered a minimum value, as the fans were deposited concurrently with range front faulting, which may be associated with tilting and (or) partial tectonic burial of the fans. In addition, some material removed from the watershed chemically dissolves and is deposited beyond the fan area, and some very fine grained material reaches the playa and interfingers with the playa deposits.

The fans selected for analysis were identified from topographic maps, SPOT images and 30-m digital elevation models (DEMs) based on their simple and discrete geometric forms, as well as minimal overlap between adjacent fans. Parameters of the tributary watersheds were determined from the DEMs. The watershed areas extend from the head of a fan to the first drainage divide. The fan surface extends from 0° to 1° slope transition on the playa surface to the pour point (exit from the confined watershed) of the bedrock-confined canyon at the fan head along the range front. Most of the analyzed fans developed concentrically away from the range front. For simplicity, the base of the fan was estimated to be the elevation of the valley floor. West of the range front in the central playa area and along the toes of the west-side fans 20,000–25,000 yr sediment lies about 10–15 m below the playa surface (Hooke, 1972; Ku et al., 1998). Conical geometry was used to determine the fan volume (Figs. 2 and 4).

Discrete radial fan symmetry is notable on topographic maps and can be delineated geomorphically and from the distribution of fine grained sediment (Fig. 2). The conical fan volume is calculated using the percent of the circumference in the horizontal and assuming a vertical abutment against the range front (Fig. 5). This assumption results in a small overestimate because the range front escarpment dips 50° to 70° east. The volume estimate was also simplified by assuming that the fans are flat at the base (0° slope), although late Pleistocene and Holocene dip-slip displacement and tilting has occurred (Maxson, 1950; Hunt and Mabey, 1966; Hooke, 1972). This assumption results in a small underestimate of the fan volume that tends to be larger than the abutment overestimate. The volume calculation is dominated by the radius of the fan, which is generally an order of magnitude or more greater than the fan height. Increasing the fan height by about 10 m to allow for late Pleistocene and Holocene buried sediment increases the denudation rate by about 2–10% depending on the size of the fan. Thus, the simplifying assumptions regarding geometry of the basal and range front boundaries of the fan tend to cancel each other out, and if anything underestimate the rate giving conservative values (Fig. 5).

The greatest weakness in the determination is assuming the entire conical fan above the playa surface lies within the best estimated age range. In this study the fans selected in Death Valley are assumed to be younger than the last pluvial maximum and the mixed-age fans selected in Panamint Valley are assumed to be either less than OIS-4 (~60,000 yr) or the last glacial/pluvial maximum, OIS-2 (~24,000 yr). The fans are most likely younger than OIS-6 (~160,000 yr). Several relations suggest the approach may give useful constraints in spite of age uncertainties. If the fans are of composite age, which is reasonable, then these denudation rates are maximum rates. However, as the fan volume and watershed dimensions are provided herein, the rates can be refined or adjusted as better age information becomes available in the future.

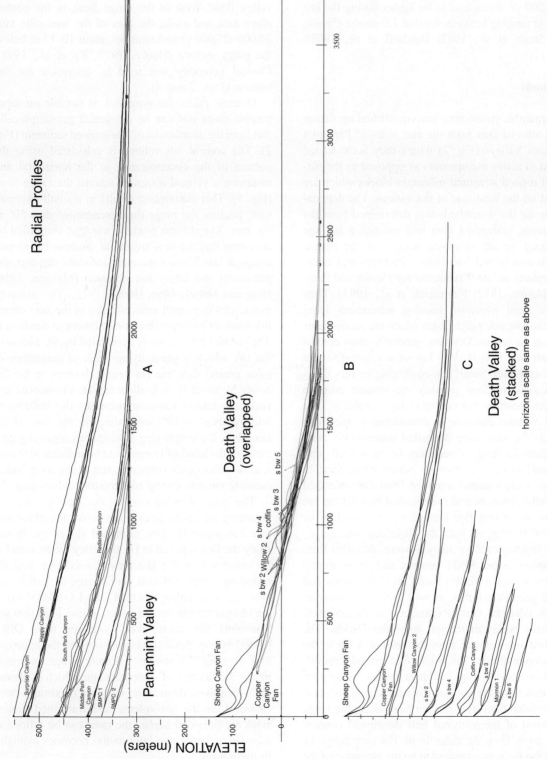

Fig. 4. Radial topographic profiles drawn from fan head to toe in Panamint and Death Valleys. Horizontal and vertical scale constant in 4 A, B, and C.

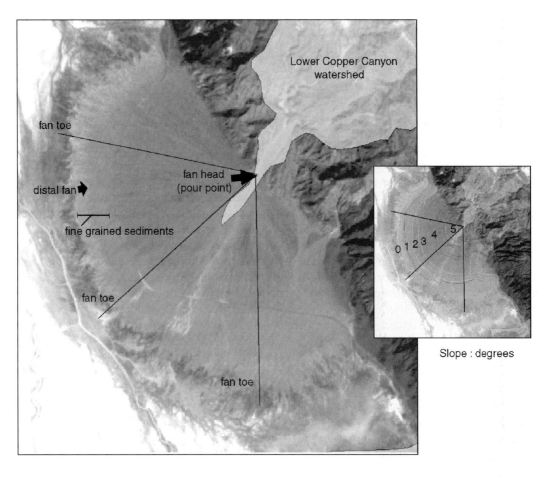

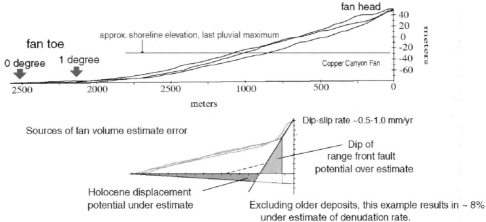

Fig. 5. Enlarged view of Copper Canyon fan showing major features including fan head, pour point from the canyon mouth, fan toe, distribution of fine grained sediments characterized by lighter spectral characteristics at the distal fan fringes and 0–1° slope area defining the distal edge of the fan.

3. Age constraints

3.1. Death Valley

The conical fans along the Black Mountain Range front in Death Valley are inferred to be at least younger than the penultimate highstand of Lake Manly ~186,000–128,000 yr (Hooke, 1972; Lowenstein et al., 1999) and may be as young as the last highstand ~30,000–15,000 yr (Hooke, 1972; Lowenstein et al., 1999) based on the absence of evidence for paleoshorelines on the fan surfaces, absence of buried shorelines along the range front, and extremely youthful fan surface morphology (Brogan and others, 1991; Klinger and Piety, 1996). The presence of a late Wisconsin lake around 26,000 to 11,900 yr was established by Hooke (1972) from ^{14}C ages on black mudstone from shallow cores. Hooke (1972) noted that the absence of oxidized sediments indicated a persistent shallow lacustrine environment between 21,500 and 12,980 yr. Uranium-series dating of sediment from a 183 m deep core drilled in the central part of the playa west of Badwater also records late Pleistocene lacustrine sediments ranging in age from about 10,000 to 30,000 yr at depths of about 8 to 16 m (Ku et al., 1998). The drill hole data combined with tufa uranium-series dates of 24,700 and 18,000 yr from − 22 to − 30 m elevation near Badwater indicate a shallow lake about 64–78 m deep (Ku et al., 1998). The presence of a late Wisconsin lake in Death Valley is consistent with lake occurrences in adjacent basins. For example Searles Lake (Stuiver, 1964; Smith, 1968; Smith et al., 1983) persisted during this time and also overflowed at about 15,000 and 22,000–24,000 yr (Smith, 1968; G.I. Smith, personal communication 2001). A large late Wisconsin lake is also documented to the south at the termination of the Mojave River in Lake Mojave, which overflowed prior to about 14,500, and developed shoreline features between 13,750 and 12,000 (Ore and Warren, 1971). The outlet for Lake Mojave trends northward and joins the southern Amargosa drainage system into southern Death Valley (Ore and Warren, 1971).

Uranium-series dating of tufa from the 90 m highstand yields 160,000–185,000 yr ages (Hooke et al., 1978; Lowenstein et al., 1999). Likewise, uranium-series dating of saline units overlying and underlying black mud from a deep core indicates a perennial lake was present between 186,000 and 128,000 yr (Hooke et al., 1978; Lowenstein et al., 1999, 1994). These tufa altitudes combined with core data suggest water depths of about 160 m for the older Blackwelder lake stand, although the estimate is complicated by the active faulting along the Black Mountain front.

3.2. Panamint Valley

The Surprise and Happy Canyon fans in Panamint Valley are twice the size of the Copper Canyon Fan studied in Death Valley and apparently older. Both fans have inset strandlines and (or) overlying cobble beach deposits around 420 m that are inferred to be recessional from the last pluvial highstand based on preliminary ^{14}C ages on tufa from shoreline and beach deposits (Jayko et al., 2001). The fan heads at 550 and 475 m lie at lower elevation than the abrasion platform and strandline, Gale High Stand (Smith, 1976), inferred to be OIS-6, although poorly constrained U-series isochrons ranging between 54 and 91,000 yr on shoreline tufa are also consistent with a OIS-4 age (Fitzpatrick and Bischoff, 1993). Individual U-series tufa analysis gives results that mainly lie between 75,000 and 175,000 yr, with some ~275,000+ yr ages (Fitzpatrick and Bischoff, 1993). The elevation of the Gale strandline is limited by the spillway elevation via Wingate Wash (603 m) to Death Valley.

The Middle Park Canyon fan and 3 to the south may be age equivalent to the east-central Death Valley fans in the study area. The fan heads at ~400 m lie below the Late Pleistocene to Holocene strandline of the last pluvial lake and prominent shorelines recognized at ~384 m (Smith, 1976). Denudation rate for these fans are calculated with respect to the last pluvial maximum, whereas the Surprise and Happy Canyon fans are with respect to approximately OIS-4.

Major overflows from Searles Lake into Panamint Valley occurred between 120,000 and 150,000 yr (Jannik et al., 1991) and 52,000 to 67,000 yr (Smith, 1968). The most prominent, but not the highest, shoreline in Panamint Valley occurs at about the elevation of the Wingate Pass spillway into Death Valley (Gale, 1914; Smith, 1976). ^{14}C results from shells and tufa from that shoreline exceed the ^{14}C range (Davis, 1970; Smith, 1976, 1978). Lower elevation strandline tufa associated with beach and wetland deposits suggest only a 130–200 m deep lake during the last pluvial maximum (Smith, 1976; Jayko et al., 2001). The prominent and topographically irregular shoreline controlled by the Wingate Pass spillway, the Gale strandline, has tentatively been correlated with the Tahoe glaciation and the highstand (Blackwelder stand) in Death Valley (Blackwelder, 1933; Smith, 1976).

4. Results

Nine small late Pleistocene and (or) Holocene fans and adjoining watersheds were measured along the Black Mountain range front between Badwater and Mormon Point on the east side of Death Valley. The crest of the watersheds ranges from 1600 to 1900 m with mean watershed elevations ranging from 400 to 1200 m. The range front is an escarpment bound by the active central Death Valley fault zone. In the Black Mountains area, this fault zone has at least 3 scarps of middle and late Holocene age that cut the fan deposits (Brogan et al., 1991; Klinger and Piety, 1996; Frankel et al., 2001). The fans lie on the Panamint block which extends from the Panamint Valley fault on the west side of the Panamint Mountains to the Black Mountain escarpment and is inferred to have a component of eastward tilt associated with displacement on the central Death Valley fault (Hooke, 1972).

In central Death Valley, the playa surface underlying the study area lies at about − 84 m below sea level near Badwater at the north end of the study area (Fig. 3) and rises slightly to about − 78 m near Mormon Point. The fan heads range from about − 30 to +50 m elevation and have 50 to 130 m relief from head to toe. The elevation of the 6 northern Death Valley fan heads (Badwater to Coffin Canyon) at the canyon mouth pour points ranges from − 15 to 0 m. The Copper Canyon fan head reaches about +50 m and the others to the south are +80 m elevation and lower. The shoreline elevation inferred for the last pluvial maximum of Lake Manly (~24,000 to 15,000 yr) lies at about − 22 to − 30 m as suggested by U-series dating of tufa deposited on the bedrock escarpment at Badwater (Ku et al., 1998).

The 6 small fans situated at low elevation abut bedrock and have relatively simple smooth slope profiles with very low curvature or slightly concave

upward (Fig. 3). The construction of these fans is inferred to postdate the last pluvial maximum based on the age of surficial deposits, fan morphology, and lack of shoreline features on the fans (Brogan et al., 1991; Klinger, 1999). The surface of the fans is presently active.

The Copper Canyon and Sheep Canyon fans are considerably larger and have higher relief than the other fans to the north (Fig. 3). The distal parts of these fan profiles are characterized by similar morphology to the northern fans with constant low-angle, smooth slopes that have low surface relief (0–5 m). The fans are cut by Quaternary faults with active scarps and small-scale horst and graben features that contribute to the upper surface irregularity and concavity. The graben surface on the Sheep Canyon fan is apparently rotated clockwise relative to the range front forming oversteepened slopes, whereas horsts on the upper Copper Canyon fans are flattened by counter clockwise rotation.

The distal toes of all the fans show very subdued steps and/or slope breaks (Fig. 3) in response to erosional activity associated with intermittent lakes on the valley floor or fluvial activity along the meandering axial wash which flows near the escarpment. Erosion at the fan toe undercuts the slope causing slightly steeper up-slope surfaces that gradually merge with the depositionally controlled surface which is generally 2–3° more gently inclined than the erosional surfaces. The middle and upper parts of the 6 small northern fans have highly constant surface slopes. The slopes near the tops or heads of the fans tend to flatten near the range front. This smaller slope angle is inferred to be controlled by erosional excavation along the faulted contacts with bedrock because the wash channels emerging from the canyon mouths tend to be captured by the fault zones near the fan heads. Thus, the resulting fan surfaces have an overall slightly concave upward form due to erosional processes at the top and base of the fans. However, these erosional and tectonically controlled variations in fan slope morphology are more chronologically limited that the segmentation of many older, larger fan complexes that occur in a variety of geomorphic regions (for example, Bull, 1964; Hooke, 1972; Hooke and Dorn, 1992) Seven of the fans (Badwater to Copper Canyon, Fig. 2) occur along a continuous northwest trending section of the central

Death Valley fault, one, Sheep Canyon, occurs at the bend and two, Willow Creek and Mormon Point, along a short right step in the fault. Topographic profiles indicate that four fans (SBW-2, -3, -4, and -5 on Fig. 4) along the main trace of the escarpment (Fig. 5) are not strongly truncated. These three fans yield denudation rates of between 0.07 and 0.15 mm/yr (calculated with respect to the last pluvial period ~24,000 yr; Table 3). The Coffin Canyon drainage has recently captured a large, about 6 km², watershed area from the Copper Canyon drainage, increasing its area by about 80%. The Coffin Canyon fan also yields a low rate of about 0.06 mm/yr with its present expanded watershed indicating the fan has not yet equilibrated to this capture event. The denudation rate calculated from the former watershed area with the present fan size yields about 0.57mm/yr. The captured area is only about 25% of the Copper Canyon watershed; the relatively high average denudation rate of 0.18 mm/yr combined with the youthful geomorphic expression and low incision at the capture gap suggests that the capture occurred recently. The Sheep Canyon fan occurs at the bend in the fault step-over. In spite of the structurally complicated location, the fan yields 0.17 mm/yr rate.

Willow Creek and Mormon Point fans occur along a east step-over of the fault and yield low rates of 0.03 and 0.06 mm/yr. These low rates suggest that the fans could be tectonically denuded and undersized relative to their adjacent watersheds. The tectonic denudation would be caused by intermittent fault-controlled down-dropping of the fan surface below the lake or playa level. This area has the highest vertical slip-rate estimates for the Death Valley fault (Klinger, 1996). Projecting the fan area from the up-thrown side of the fault and keeping the present fan radius constant (minimal estimate) yields a rate of 0.24 mm/yr more similar to the northern fans.

In central Panamint Valley the playa surface lies at about 317 m, the crest of the higher watersheds lies around 2500 to 2900 m, and the mean elevation of the watersheds lie between 690 and 1950 m. The older fans are generally 2 to 3 times as large as the moderate sized fans in the Death Valley part of the study area with relief between 140 and 210 m. Likewise the watersheds are 2 to 10 times as large as the medium and small fans in Death Valley.

Fans from Panamint Valley including the Surprise, Happy and South Park Canyon fans were also examined. These fans have slightly concave upward sur-

Table 3

Denudation rates derived for central death and Panamint Valleys. Denudation rates, Death Valley area

Fan	Fan	Basin	Fan volume (m³)	Wshd area (m²)	Rate* (mm/yr)	Mean watershed elevation (m)	Approximate age (ka)
2	S Badwater 2	D V	18,477,439	5,225,226	0.15	1041	1
3	S Badwater 3	D V	6,447,842	3,872,495	0.07	1109	1
4	S Badwater 4	D V	22,303,292	6,118,007	0.15	917	1
5	S Badwater 5	D V	4,052,533	1,934,704	0.09	728	1
6	Coffin Canyon	D V	16,382,163	11,074,864	0.06	409	1
7	Copper Canyon	D V	247,548,847	58,858,996	0.18	1221	1
8	Sheep Canyon	D V	48,260,643	58,900,847	0.03	1234	1
9	Willow Creek	D V	111,048,665	27,376,080	0.17	1215	1
10	Mormon Pt	D V	8,112,058	5,411,591	0.06	747	1
11	Surprise Canyon	P V	656,599,320	47,942,100	0.23	1953	2
12	Happy Canyon	P V	396,049,975	47,652,300	0.14	1912	2
13	Middle Park Canyon	P V	24,000,067	12,077,100	0.08	1366	1
14	S of MPC 1	P V	6,504,797	6,084,900	0.04	1272	1
15	S of MPC 2	P V	2,526,710	2,942,100	0.04	691	1
16	South Park	P V	100,476,054	8,321,400	0.20	1555	2
17	Redlands Canyon	P V	101,392,259	16,148,386	0.10	1492	2

Approximate age: 1 calculated with respect to 24,000; 2 calculated with respect to 60,000.

DV, Death Valley, mean 0.11 mm/yr.

PV, Panamint Valley, mean 0.12 mm/yr.

* Calculated with respect to approximate age.

faces that are truncated in midsections by Holocene fault scarps and at the toes by Holocene strandlines. Like the fans situated along the Black Mountain range front, these fans sit on a downthrown block, whereas the watersheds lie on the hanging-wall block. The Surprise and Happy Canyon fans are characterized by well developed pavements and heavy varnish suggesting pre-Holocene aged surfaces (Brogan et al., 1991; Klinger and Piety, 1996). In addition, the fans are more deeply dissected, have larger variation in horizontal curvature, and are less concentric; these morphologies suggest that the fans are more modified and likely older than the studies

of fans in Death Valley. These two fans yield denudation rates of 0.1–0.23 mm/yr with respect to the last 60,000 yr, however Happy Canyon fan lies in a similar step-over position as the Willow Creek fan in Death Valley its lower rate may also be caused by tectonic denudation of the fan complex. The three smaller fans south of Ballarat, the Middle Park Canyon fan and two unnamed fans to the south, are more geomorphically similar to the fans along the Black Mountain escarpment. They are inferred to be younger that the Surprise and Happy Canyon fans, thus their rates have been calculated with respect to the last pluvial maximum yielding ~0.04 to 0.08

Alluvial Fan Geomorphic Parameters
Death and Panamint Valleys

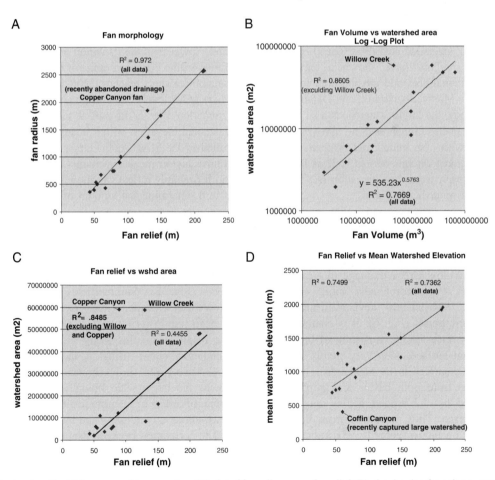

Fig. 6. Plots showing alluvial fan geomorphic parameters: (A) plot of fan radius versus fan relief; (B) plot showing fan volume versus watershed area; (C) plot showing fan relief versus watershed area; (D) plot showing fan relief versus mean watershed elevation.

mm/yr, although they also are locally modified by tectonic and erosional (mainly Holocene lacustrine shoreline) features. The mean elevation of watersheds for these drainages is 600–1200 m higher than those of the Black Mountain escarpment (Table 3) with greater total relief.

The fans show consistent geomorphic trends among fan dimensions and adjacent watershed parameters (Fig. 6 A,B,C, and D) even though, in both basins the heads of the fans are modified by Quaternary faults with Holocene scarps. Many of the fan surfaces adjacent to the range front are modified by active faults. There is a very strong correlation ($R^2=0.972$) between fan relief and radius in both basins suggesting that modification of the fans by range front faults is acting similarly to all the fans and primarily modifying a small area near the fan head, or not significant for this duration. There is a power-law relation and moderate correlation between fan volume and watershed area ($R^2=0.860$). Jannson et al. (1993) similarly have shown a correlation between depositional area and erosion area for landscape within the Mojave, Amargosa and Death Valley area. In this study, the larger watersheds also tend to have higher mean elevations, resulting in a systematic correlation between fan relief and watershed area ($R^2=0.845$ for most catchments) as well as between fan relief and mean watershed elevation ($R^2=0.736$) in nondisturbed watersheds (Fig. 6).

The correlation between fan relief and mean watershed elevation is good and significantly improves by excluding the watersheds that show geomorphic evidence for recent capture and abandonment. The Willow Creek drainage in Death Valley, which has a geomorphically more complex catchment area, consistently deviates from most the morphometric trends (Fig. 6). It is characterized by a bimodal surface terrain consisting of the range front escarpment surface, plus very gently sloping paleotopographic surface in its highlands. The other catchment areas from both basins are morphologically dominated by the eroded active range front escarpment surface. Thus, deviation from the morphologic trends on Fig. 6 seems largely indicative of disturbances in the discharge history of the watershed, for example, stream capture and abandonment, or fundamentally complex morphologic character.

5. Discussion

The denudation rates are only first order estimates, as the age of the fans studied are only loosely constrained by stratigraphic and geomorphic associations with the last pluvial maximum in Death Valley and the last two pluvial events in Panamint Valley. Analysis of conical fans and associated watersheds in central Death Valley suggest denudation rates of about 0.06–0.18 mm/yr are not exceeded and, likewise about 0.04 to 0.23 mm/yr are maximums for central Panamint Valley.

Elevation is commonly used as a surrogate to estimate mean annual precipitation (MAP) where rainfall has not been directly measured. In this region, metered rainfall increases semi-logarithmically to elevations of around 2500 m where it levels off to about 50–60 cm/yr (Fig. 3). The mean watershed elevation of the Black Mountain fans around 950 m, is significantly lower than the Panamint watersheds at 1463 m. However, the elevation range of the watersheds in the Panamint Range reaches 3300 m well in excess of where MAP levels off. Preliminary results suggest the denudation rate is in part a function of the mean watershed elevation which is apparently mainly controlled by the MAP. As the larger watersheds also tend to have higher mean elevations, the trends in Fig. 7A and B probably reflect variation in precipitation which in turn determine the amount of power available for erosion and transport.

A relation between denudation rate and mean relief of watersheds has been noted for the large drainage basins associated with the major continental river systems (Ahnert, 1970; Summerfield and Hulton, 1994) (Fig. 8). Summerfield and Hulton (1994) show mean elevation (m.e.) is roughly proportional to mean relief (m.r.) by a factor of about 0.5 to 0.67. The results from this study best fit Ahnert (1970) denudation curve using a factor of 0.667 (m.r. ~0.67*m.e.) and show similar relief to denudation rate as the larger-scale fluvially dominated systems inspite of orders of magnitude difference in watershed areas, climatic conditions and methods.

The correlation between denudation rate and mean watershed elevation holds well for the fans in Death Valley, except for the Willow Creek fan. The watersheds of the Death Valley fans except Willow Creek occupy the steep, west-facing escarpment of the range

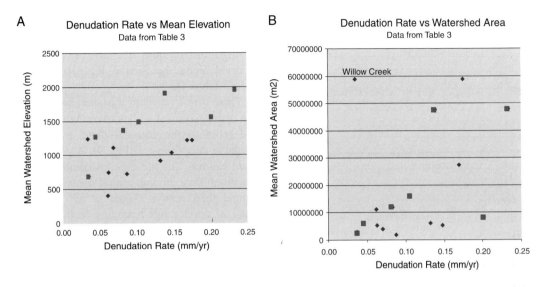

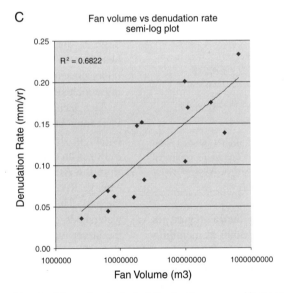

Fig. 7. Plot showing the denudation rate determined for each watershed relative to other geomorphic parameters. Plot A shows mean watershed elevation versus denudation rate; Plot B shows mean watershed area versus denudation rate; Plot C shows fan volume versus denudation rate.

and have fairly uniform slope distributions with mean watershed slopes ranging between 20° and 32° (average 27°). The Willow Creek watershed includes the escarpment surface as well as a broad, gently sloping upland basin at the top of the range which makes up almost two-thirds of the watershed and has an average watershed elevation of 17°. The Willow Creek

watershed has the lowest mean slope of all the catchment areas studied, and also has the highest mean elevation for the Death Valley catchments. Thus, the low denudation rate for this watershed is inferred to be due to detainment of colluvium on the relatively flat highland surface that is not being eroded as rapidly as from the steeper escarpment-surface dominated basins.

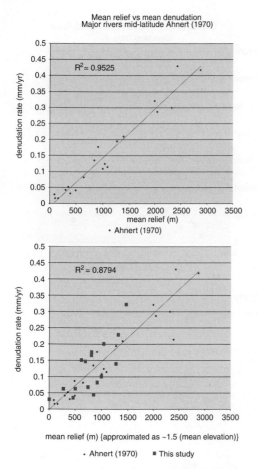

Fig. 8. Denudation rate relative to estimated mean watershed relief for the study area (~1.5 × Mean Watershed Elevation) compared with denudation rate relative to mean watershed relief of major continental river systems from Ahnert (1970).

Relief in the greater eastern Sierra region is commonly on the order of 2500 to 3000 m including along much of the Sierran Range Crest, White, Inyo and Panamint Mountains. Relief on the Black Mountain Range, in the Death Valley part of the study area is on the order of 1900 m with elevations ranging between − 50 to 1800 m along the ridgecrest between Dantes Peak, 1740 m and Smith Mountain, 1800 m. Rates for the dip-slip component of active range-bounding faults derived from offset Quaternary and Pliocene horizons in much of the eastern Sierra region lies between 0.2 and 0.5 mm/yr (Martel et al., 1987; Zhang et al., 1988; Depolo, 1989; Beanland and Clark, 1994; Klinger and Piety, 1996; Densmore and Anderson, 1997; Reheis and

Sawyer, 1997; Oswald and Wesnousky, 2001; Zehfuss et al., 2001; Vogel et al., 2002), although, rates as high as 0.9 and 1.0 mm/yr have also been reported along the Sierran Range crest (see references, Table 1).

There is no systematic variation between local dip-slip rates associated with active range-bounding faults and relief of the adjacent range evident from the available data (Table 1, Fig. 9). A preliminary minimum estimate for the dip-slip component along the Panamint fault in the central Panamint Range lies around 0.35 mm/yr (Vogel et al., 2002). Dip-slip rates on active range-bounding faults north of Panamint Valley in Fish Lake and Deep Springs Valleys range between 0.3 and 0.4 mm/yr (Reheis and Sawyer, 1997). There are fewer estimates of dip-slip rates for Death Valley proper, and estimated Holocene vertical rates of 3–5 mm/yr from the Mormon Point area (Klinger and Piety, 1996) are considerable greater than the rest of the region, thus may be suspect. In general, the Late Pleistocene to Holocene denudation rates from this study are about half to a third or less as large as the vertical component of the tectonic rate observed in the region, indicating relief is growing under present climatic conditions, a less than surprising conclusion.

The accumulation of sediment in the Death and Panamint basin depocenters is not considered in regard to denudation estimates in this study. Death Valley is an open system that presently receives runoff from the Amargosa River and has been linked to the Mojave River, as well as the Owens River system in the past. Thus, sediment influx is not restricted to the watersheds of the immediately adjacent ranges, and the larger watershed systems contain multiple subbasins of varying dimensions. Geomorphic evidence for eolian degradation and aggradation is widespread in the distal fan and playa fringe environments in both basins. The frequent strong wind events with accompanying sand and dust storms far out number precipitation events, thus are a major factor controlling playa and fringing playa degradation, as well as redistribution of fine grained sediment. Eolian processes remobilize playa sediments and re-transport them to the alluvial fan and adjacent environments as has been well established by previous studies of pavement and soil evolution (for example, McFadden and others, 1987).

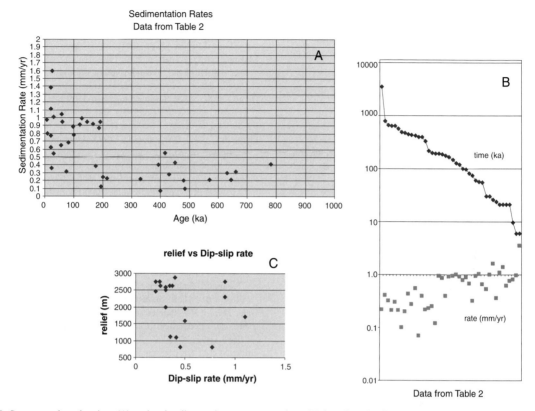

Fig. 9. Summary plots showing: (A) regional sedimentation rates versus time, (B) log plot of sedimentation rate versus time, and (C) vertical component of the faults-slip rates versus relief on the range front.

Factors, like climatic variation that influence sediment accumulation rates in the depocenters likely also influences denudation rates at the surface. The sedimentation rates determined from drill cores in Death, Searles and Owens Valleys show some systematic long term variation. The long-term sedimentation rates recorded in the basin depocenters are generally on the order of 0.2 to 0.4 mm/yr between about 800,000 and 200,000 yr (Liddicoat et al., 1980; Smith et al., 1983) and appear to increase from ~0.6 to 1.0 mm/yr during the last 200,000 yr (Hooke, 1972; Liddicoat et al., 1980; Smith et al., 1983; Bischoff, 1998; Bischoff et al., 1998; Lowenstein et al., 1999) (Fig. 9). The latest Pleistocene rates are generally higher, but also show extreme scatter which may be a consequence of the diagenetic and compaction processes rather than a major change in sedimentation rate, as thickness of the stratigraphic column rather than density is used to calculate the rate (Bischoff, 1998; Bischoff et al., 1998).

6. Conclusions

Estimated Late Pleistocene denudation rates for the Death and Panamint Valleys show variation which is most strongly correlated with mean watershed elevation, and thus is most likely in response to elevationally controlled mean annual precipitation. There is strong correlation between the fan morphology and watershed area, and similar properties have been described by Hooke (1968).

Acknowledgements

Comments on early drafts of the paper by Marith Reheis and Joe Smoot, U.S.G.S; Roger Hooke, U. Maine; Alan Gillespie, U. Washington, and Paul Bierman, U. Vermont markedly improved the paper and are greatly appreciated. Work was supported by U.S.G.S. National Cooperative Geologic Mapping

Project, Death Valley area. Daniel Pritchett, U.C. White Mountain Research Station assisted with acquisition of satellite imagery during preliminary phases of the work.

References

Ahnert, F., 1970. Functional relationships between denudation, relief, and uplift in large, mid-latitude drainage basins. American Journal of Science 268, 243–263.

Bachman, S.B., 1974. Depositional and structural history of the Waucobi Lake bed deposits. MSc, University of California, Los Angeles, Owens Valley, California, 129 pp.

Bachman, S.B., 1978. Pliocene-Pleistocene break-up of the Sierra Nevada-White-Inyo Mountains Block and formation of Owens Valley. Geology 6, 461–463.

Beanland, S., Clark, M.M., 1994. The Owens Valley fault zone, eastern California, and surface faulting associated with the 1872 earthquake. Bulletin, 29.

Bischoff, J.L., 1998. Introduction and rationale of study: the last interglaciation at Owens Lake, California; core OL-92. U.S. Geological Survey OF 98-0132, pp. 1–165.

Bischoff, J.L., Fitts, J.P., Fitzpatrick, J.A., 1997. Responses of sediment geochemistry to climate change in Owens Lake sediment; and 800-k.y. record of saline/fresh cycles in core OL-92. Special Paper, Geological Society of America, 37, 47 pp.

Bischoff, J.L., Chazan, D., Canavan, R.W.I.V., 1998. A high-resolution study of sediments from the last interglaciation at Owens Lake, California; geochemistry of sediments in core OL-92, 83–32 m depth. In: Smith, G.I., Bischoff, J.L. (Eds.), The Last Interglaciation at Owens Lake California; Core OL-92. U.S. Geological Survey OF 98-0132.

Blackwelder, E., 1933. Lake Manly, an extinct lake of Death Valley. Geographical Review 23, 464–471.

Brogan, G.E., Kellogg, K.S., Slemmons, D.B., Terhune, C.L., 1991. Late Quaternary Faulting Along the Death Valley–Furnace Creek Fault System, California and Nevada. U.S. Geological Survey, Reston, p. 23.

Bull, W.B., 1964. Geomorphology of segmented alluvial fans in western Fresno county, California. U.S. Geological Survey Professional Paper 352E, 79–129.

Burke, R.M., Birkeland, Peter W., 1979. Field guide to relative dating methods applied to glacial deposits in the third and fourth recesses and along the eastern Sierra Nevada, California, with supplementary notes on other Sierra Nevada localities. Friends of the Pleistocene, Pacific Cell, 131 pp.

Clark, M.M., Gillespie, A.R., 1981. Record of late Quaternary faulting along the Hilton Creek Fault in the Sierra Nevada, California. Earthquake Notes 52, 46.

Curry, R.R., 1971. Glacial and Pleistocene history of the Mammoth Lakes Sierra, a geologic guidebook. Geological Series, University of Montana, Department of Geology vol. 2, no. 11. 47 pp.

Davis, E.L., 1970. Archeology of the north basin of Panamint Valley, Inyo county, California. Nevada State Museum Anthropological Papers 15, 83–141.

Densmore, A.L., Anderson, R.S., 1997. Tectonic geomorphology of the Ash Hill Fault, Panamint Valley, California. Basin Research 9, 53–63.

Depolo, C.M., 1989. Seismotectonics of the White Mountains fault system, east-central California and west-central Nevada. MSc thesis, University of Nevada, Reno.

Fitzpatrick, J.A., Bischoff, J.L., 1993. Uranium-series dates on sediments of the high shoreline of Panamint Valley, California. U.S. Geological Survey OF 93-0232, p. 15.

Fitzpatrick, J.A., Bischoff, J.L., Smith, G.I., 1993. Uranium-Series Analyses of Evaporites from the 1000-Foot PAN-3 Core, Panamint Valley, California. U.S. Geological Survey, Reston, p. 22.

Frankel, K.L., Jayko, A.S., Glazner, A.F., 2001. Characteristics of Holocene fault scarp morphology, southern part of the Black Mountains fault zone, Death Valley. In: Machette, M.N., Johnson, M.L., Slate, J.L. (Eds.), Quaternary and Late Pliocene Geology of the Death Valley Region, Recent Observations On Tectonics, Stratigraphy and Lake Cycles. Volume Open-file Report 01-51U.S. Geological Survey, pp. 205–216.

Gale, H.S., 1914. Salines in the Owens, Searles, and Panamint basins, southeastern California. U.S. Geological Survey B 0580-L, 251–323.

Gilbert, C.M., Christensen, M.N., Al-Rawi, Yehya, Lajoie, K.R., 1968. Structural and volcanic history of Mono Basin. Memoir Geological Society of America, California-Nevada, pp. 275–329.

Granger, D.E., Kirchner, J.W., Finkel, R., 1996. Spatially averaged long-term erosion rates measured from in situ-produced cosmogenic nuclides in alluvial sediment. Journal of Geology 104, 249–257.

Hooke, R.L., 1968. Processes on arid-region alluvial fans. Journal of Geology 75, 438–460.

Hooke, R.L., 1972. Geomorphic evidence for Late-Wisconsin and Holocene tectonic deformation, Death Valley, California. Geological Society of America Bulletin 83, 2073–2097.

Hooke, R.L., Dorn, R.I., 1992. Segmentation of alluvial fans in Death Valley, California: new insights from surface exposure dating and laboratory modelling. Earth Surface Processes and Landforms 17, 557–574.

Hooke, R.L., Lively, R., Jacobsen, D., 1978. Guide for the Death Valley Leg of Friends of the Pleistocene trip, November 1978. Friends of the Pleistocene 1978, 6.

Hunt, C.B., Mabey, D.R., 1966. Stratigraphy and structure, Death Valley, California. U.S. Geological Survey Professional Paper 494-a, 1–162.

Jannik, N.O., Phillips, F.M., Smith, G.I., Elmore, D., 1991. A (super 36) Cl chronology of lacustrine sedimentation in the Pleistocene Owens River system. Geological Society of America Bulletin 103, 1146–1159.

Jannson, P., Jacobson, D., Hooke, R.L., 1993. Fan and playa areas in southern California and adjacent parts of Nevada. Earth Surface Processes and Landforms 18, 109–119.

Jayko, A.S., Forester, R.M., Sharpe, S.E., Smith, G.I., 2001. The last pluvial highstand in Panamint Valley, Southeast California. EOS, vol. 88. American Geophysical Union, p. 788.

Kistler, R.W., 1966. Structure and metamorphism in the Mono Craters Quadrangle. U. S. Geological Survey Bulletin, Sierra Nevada, California, pp. E1–E53 (Report: B 1221-E).

Klinger, R.E., 1996. Late Quaternary activity on the Furnace Creek Fault, Northern Death Valley, California. Geological Society of America, 28th Annual Meeting, vol. 28. Geological Society of America (GSA), Boulder, p. 193.

Klinger, R.E., 1999. Tectonic geomorphology along the Death Valley fault system; evidence for recurrent late Quaternary activity in Death Valley National Park. In: Slate, J.L. (Ed.), Proceedings of Conference on Status of Geologic Research and Mapping, Death Valley National Park. Volume Open-file Report 99-153. U.S. Geological Survey, Reston, pp. 132–140.

Klinger, R.E., Piety, L.A., 1996. Evaluation and Characterization of Quaternary Faulting on the Death Valley and Furnace Creek Faults, Death Valley, California, pp. 1–98.

Knott, J.R., Sarna-Wojcicki, A.M., Tinsley III, J.C., Wells, S.G., 2001. Late Quaternary tectonic–geomorphic development and pluvial lakes at Mormon Point. Open-File report United States Geological Survey 01-51, 89–115.

Ku, T.-L., Luo, S., Lowenstein, T.K., Li, J., Spencer, R.J., 1998. U-series chronology of lacustrine deposits in Death Valley, California. Quaternary Research 50, 261–275.

Liddicoat, J.C., Opdyke, N.D., Smith, G.I., 1980. Palaeomagnetic polarity in a 930-m core from Searles Valley, California. Nature 286, 22–25.

Lowenstein, T.K., Li, J., Brown, C., Roberts, S.M., Ku, T.-L., Luo, S., Yang, W., 1999. 200 k.y. paleoclimate record from Death Valley salt core. Geology 27, 3–6.

Lowenstein, T.K., Li, J., Brown, C.B., Spencer, R.J., Roberts, S.M., Yang, W., Ku, T.-L., Luo, S., Anonymous, 1994. Death Valley salt core. 200,000 Year Record of Closed-Basin Subenvironments and Climates. Geological Society of America, 1994 Annual Meeting, 26. Geological Society of America (GSA), Boulder, p. 169.

Lubetkin, L.K.C., Clark, M.M., 1988. Late Quaternary activity along the Lone Pine Fault, eastern California. Geological Society of America Bulletin 100, 755–766.

Lustig, L.K., 1965. Clastic sedimentation in Deep Springs Valley, California. U.S. Geological Survey Professional Paper 352-F, 192.

Martel, S.J., Harrison, T.M., Gillespie, A.R., 1987. Late Quaternary vertical displacement rate across the Fish Springs Fault, Owens Valley fault zone, California. Quaternary Research 27, 113–129.

Maxson, J.H., 1950. Physiographic features of the Panamint Range, California. Geological Society of America Bulletin 61, 99–114.

McFadden, L.D., Wells, S.G., Jercinovich, M.J., 1987. Influences of eolian and pedogenic processes on the origin and evolution of desert pavements. Geology 15, 504–508.

Ore, T.H., Warren, C.N., 1971. Late Pleistocene–Early Holocene Geomorphic History of Lake Mojave, California. Geological Society of America Bulletin 82, 2553–2562.

Oswald, J.A., Wesnousky, S.G., 2001. Neotectonics and Quaternary geology of the Hunter Mountain Fault Zone and Saline Valley Region, Southeastern California. Geomorphology 42, 255–278.

Phillips, Fred M., Zreda, Marek G., Smith, Stewart S., Elmore, David, Kubik, Peter W., Sharma, Pankaj, 1990. Cosmogenic chlorine-36 chronology for glacial deposits at Bloody Canyon, eastern Sierra Nevada. Science 248, 1529.

Reheis, M.C., Sawyer, T.L., 1997. Late Cenozoic history and slip rates of the Fish Lake Valley, Emigrant Peak, and Deep Springs fault zones, Nevada and California. Geological Society of America Bulletin 109, 280–299.

Schumm, S.A., 1963. The disparity between present rates of denudation and orogeny. U.S. Geological Survey P 0454-H, pp. 1–13.

Smith, G.I., 1968. Late-Quaternary geologic and climatic history of Searles Lake, southeastern California—means of correlation of Quaternary successions— 1965, Proc.,: Internat. Assoc. Quaternary Research, 7th Cong., USA, v. V. 8.

Smith, G.I., 1997. Stratigraphy, lithologies, and sedimentary structures of Owens Lake core OL-92, an 800,000-year paleoclimatic record from core OL-92, Owens Lake, Southeast California. Special Paper—Geological Society of America 317, 9–23.

Smith, G.I., Bischoff, J.L., 1997. Core OL-92 from Owens Lake; project rationale, geologic setting, drilling procedures, and summary. Special Paper, Geological Society of America 317, 1–8.

Smith, G.I., Pratt, W.P., 1957. Core logs from Owens, China, Searles, and Panamint basins, California. U.S. Geological Survey Bulletin 1045-A, 1–62.

Smith, G.I., Barczak, V.J., Moulton, G.F., Liddicoat, J.C., 1983. Core KM-3, a surface-to-bedrock record of late Cenozoic sedimentation in Searles Valley, California. U.S. Geological Survey PP-1256, pp. 1–24.

Smith, R.S.U., 1976. Late-Quaternary fluvial and tectonic history of Panamint Valley, Inyo and San Bernadino counties, California [Doctoral thesis]: Pasadena, California Institute of Technology. 295 p.

Smith, R.S.U., 1978. Pluvial history of Panamint Valley, California. Guidebook for the Friends of the Pleistocene, Pacific Cell, pp. 1–36.

Stuiver, M., 1964. Carbon isotopic distribution and correlated chronology of Searles Lake sediments. American Journal of Science 262, 377–392.

Summerfield, M.A., Hulton, N.J., 1994. Natural controls of fluvial denudation rates in major world drainage basins. Journal of Geophysical Research 99, 13871–13883.

Vogel, M.B., Jayko, A.S., Wooden, J.L., Smith, R.S.U., 2002. Quaternary exhumation rate Central Panamint Range, California from U–Pb Zircon Ages. Abstracts with Programs, Geological Society of America 34, 249.

Zehfuss, P.H., Bierman, P.R., Gillespie, A.R., Burke, R.M., Caffee, M.W., 2001. Slip rates on the Fish Springs Fault, Owens Valley, California, deduced from cosmogenic (super 10) Be and (super 26) Al and soil development on fan surfaces. Geological Society of America Bulletin 113, 241–255.

Zhang, P., Ellis, M.A., Slemmons, D.B., Anonymous, 1988. Holocene slip rate and earthquake recurrence interval on the southern Panamint Valley fault zone. AGU 1988 Fall Meeting, Abstracts, vol. 69. American Geophysical Union, Washington, pp. 1459–1460.

Available online at www.sciencedirect.com

Earth-Science Reviews 73 (2005) 291–307

www.elsevier.com/locate/earscirev

Holocene fluvial geomorphology of the Amargosa River through Amargosa Canyon, California

D.E. Anderson *

*Center for Environmental Science and Education and the Quaternary Sciences Program, P.O. Box 5694,
Northern Arizona University, Flagstaff, AZ 86011, USA*

Accepted 1 April 2005

Abstract

The 275-km-long, mostly ephemeral, Amargosa River flows through some of the driest terrain in the western U.S. and terminates below sea level at Badwater Basin in Death Valley. An understanding of the geomorphic history and climatic response of an 18-km reach of the Amargosa River through Amargosa Canyon was sought by: (1) the development of a fluvial chronology for Amargosa Canyon; (2) the analysis of the modern hydroclimatology of the Amargosa River; and (3) the comparison of the Amargosa River record with other local and regional paleohydrologic reconstructions. Late Holocene fluvial landforms preserved in the upper 11 km of Amargosa Canyon include a suite of five terraces ranging from the active floodplain level to approximately 12 m above modern stream. Terraces in this reach are both fill and fill-cut, deposits are generally fine-grained, and the stream is an incised, straight channel. Deposits underlying the dominant terrace in upper Amargosa Canyon represent two main aggradational periods, one around A.D. 646 to A.D. 760, and a later period of rapid aggradation around A.D. 1485 to A.D. 1663. These two aggradational periods were separated by an erosional period. Both aggradational periods are associated with wetter conditions. Hydroclimatologic analyses suggest that troughs or cut-off low-pressure systems over the west coast of the U.S. are the dominant synoptic patterns associated with modern flow events in the Amargosa River. The lower 7-km portion of the river through Amargosa Canyon is dramatically different. Here, the stream has a braided channel, and the coarse-grained deposits form a series of abandoned braid bars. Also present in Amargosa Canyon are pediments and truncated alluvial fan sediments interfingering with predominantly coarse-grained axial channel sediments that are topographically higher and older than preserved late Holocene fluvial deposits. The Amargosa River through Amargosa Canyon appears to be quite sensitive to the low-magnitude climatic fluctuations of the latest Holocene, but this record is periodically lost through higher magnitude climatic fluctuations that cause erosion.
© 2005 Elsevier B.V. All rights reserved.

Keywords: Holocene fluvial geomorphology; arid land geomorphology; Death Valley; Amargosa River

* Tel.: +1 928 523 1276; fax: +1 928 523 1276.
  E-mail address: sally.evans@nau.edu.

0012-8252/$ - see front matter © 2005 Elsevier B.V. All rights reserved.
doi:10.1016/j.earscirev.2005.04.010

1. Introduction

The starkly beautiful landscape of the Death Valley region has inspired much geomorphic study. Tectonism and dramatic fluctuations in paleoclimatic regimes through the Quaternary have contributed to a landscape that displays a vast array of desert landforms

(see Slate, 1999, for a comprehensive list of recent work). Death Valley and its major tributary, the Amargosa River, lie at the southernmost boundary of the Great Basin section of the Basin and Range Physiographic Province (Fig. 1; Dohrenwend, 1986). The Great Basin is characterized by internal drainage resulting in many basins that contained lakes during

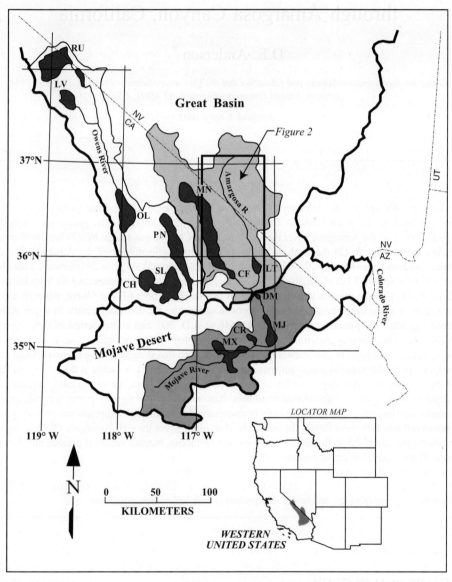

Fig. 1. Pluvial lakes in basins of the Owens, Mojave, and Amargosa Rivers. RU, Lake Russell; LV, Long Valley; OL, Owens Lake; CH, China Lake; SL, Searles Lake; PN, Panamint Lake; MN, Lake Manly; LT, Lake Tecopa; DM, Lake Dumont; MJ, Lake Mojave; CR, Cronese Lakes; MX, Lake Manix. CF denotes Confidence Flats sub-basin. Figure modified from Morrison (1991), and Anderson and Wells (2003a,b). Note inset square delineating the boundaries of Fig. 2.

the late Pleistocene pluvials (Dohrenwend, 1986; Mifflin and Wheat, 1979; Morrison, 1991). Understanding the link between climatic fluctuations and the response of fluvial and lacustrine systems is important in order to assess the degree of landscape sensitivity of various landscapes to climate change. The Amargosa River offers an excellent opportunity to explore linkages between climate and fluvial activity, in what is now the only major river to contribute to Death Valley, located in the most arid region of the United States.

The 275-km-long, mostly ephemeral, Amargosa River flows through some of the driest terrain in the western U.S. The headwaters of the Amargosa River include Pahute Mesa and the Timber Mountains (~2200 m elevation) north and northeast of Beatty, Nevada (Figs. 1 and 2). The upper Amargosa River drainage basin became integrated with the lower Amargosa River drainage basin about 195 ka, when Tecopa Basin was breached (Anderson et al., 1998; Morrison, 1991, 1999), providing through-flow from the headwaters to Death Valley (Figs. 1 and 2).

Hunt et al. (1966), Hunt and Mabey (1966), and Hunt (1975) found evidence for late Pleistocene and Holocene lakes in Death Valley based on identification of shoreline features, analyses of lake cores, and archaeological evidence. Hooke (1972) recovered cores from the basin depocenter at Badwater Basin and reported lake sediments dating from 26 to 10.5 ka. In 1993, Lowenstein et al. (1998, 1999), Li et al. (1996), and Ku et al. (1998) recovered a single, 185-m core from Badwater Basin and found evidence for lake stands at 186–128 ka and again from 35 to 10 ka. Anderson and Wells (2003a) reported the presence of fluctuating highstands in the basin during the 35–10-ka lake stand, where lake levels changed in response to climate and variable input from both the Amargosa and Mojave rivers (Anderson and Wells, 2003b). Identification of these highstands was based on four radiocarbon-dated lake cores from a 75-km transect down the axis of southern Death Valley; highstands occurred at ~26 ka, ~18 ka, and 12 ka (Anderson and Wells, 2003a,b). Both the Amargosa and Mojave rivers, at present, encompass broad drainage areas, 15,540 km^2 (Hunt et al., 1966) and 9500 km^2 (Wells et al., 1989), respectively.

The modern climate of the Amargosa River near Beatty is mesic to thermic and semiarid, with an annual mean temperature of 15 °C and average annual precipitation of 134.6 mm/year (Spaulding, 1985). The northwest-trending Funeral and Greenwater Ranges direct surface flow southeast in the vicinity of Tecopa, where the river enters Amargosa Canyon (Fig. 2). After emerging from Amargosa Canyon near the Dumont Sand Dunes, the river is deflected southwestward by the Salt Spring Hills, then west by the alluvial fans of the Avawatz Mountains, and finally flows northwest into southern Death Valley near Saratoga Springs. The modern terminus of the river is in Death Valley, CA at Badwater Basin (−86 m) which is hyperthermic and extremely arid with a mean annual temperature of 24 °C and mean precipitation of a scant 4 mm/year (Hunt, 1975).

The overarching goal of this research is to understand the fluvial geomorphic history of the Amargosa River and its relationship to low-amplitude climatic fluctuations during the Holocene in the context of late Quaternary landscape evolution. The three main components of this study presented in this paper include: (1) development of a fluvial chronology for Amargosa Canyon; (2) analysis of the modern hydroclimatology and erosion of the Amargosa River; and (3) comparison of the Amargosa River record with other local and regional paleohydrologic reconstructions.

The Amargosa Canyon reach of the Amargosa River provides a unique opportunity to investigate the fluvial history of the river where the generally broad and aggrading channel is constricted through Amargosa Canyon. This setting has provided the most complete sequence of exposed fluvial deposits found along the Amargosa River, although the canyon does appear to be periodically eroded. Exposed fluvial deposits and extensive terrace surfaces provide the means to resolve the sensitivity of the 8750-km^2 upper Amargosa drainage basin to climatic fluctuations during the late Holocene.

2. Pleistocene history of Amargosa Canyon: setting the stage

Amargosa Canyon is 18 km in length, beginning south of the town of Tecopa and emerging north of the Dumont Sand Dunes (Figs. 1–4). The canyon is 122 m wide in the northern portion and up to 610 m wide at mouth and is incised up to 275 m into predominantly China Ranch Beds, a Tertiary unit composed of debris flows, fanglomerates, pumice beds, and gypsi-

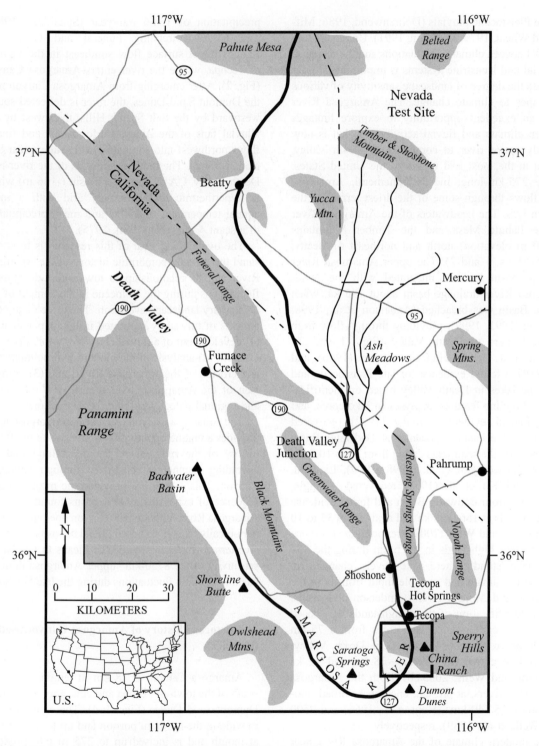

Fig. 2. Location map showing the course of the Amargosa River. The Amargosa Canyon reach of the river through the Sperry Hills is delineated by the black box, this area is shown at larger scale in Fig. 4. Circles indicate town sites, triangles show other significant geographic sites. Map modified from Elliott (1982).

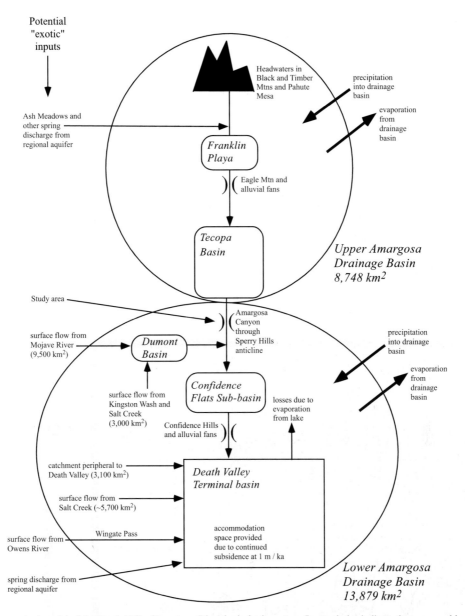

Fig. 3. Simple conceptual model of the Death Valley/Amargosa River hydrologic system. Large circles indicate the upper and lower Amargosa drainage basins; rounded squares show basins along the river, all are through-flowing currently except, of course, Death Valley; arrows show input as pointing toward the cascading system, output pointing away from the system; exotic inputs listed on left of page originate outside large circles; back-to-back parentheses show hydrologic discontinuities. Not to scale. Note: there are two distinct streams named "Salt Creek," one flowing into northern Death Valley and the other into Dumont Basin. The catchment peripheral to Death Valley includes springs that emanate at the alluvial fan playa interface on the western edge of the basin.

ferous mudstone (Hillhouse, 1987). The China Ranch Beds have been folded into an anticline that forms the Sperry Hills. The anticline is cored by Precambrian sedimentary and metamorphic rocks (Jennings et al., 1962).

The drainage basin of the Amargosa River as it enters the canyon is 8750 km² (U.S. Geological Survey, 1995). Flow in this reach is perennial, fed predominantly by the alluvial aquifer as it encounters shallow bedrock, and from groundwater emerging as

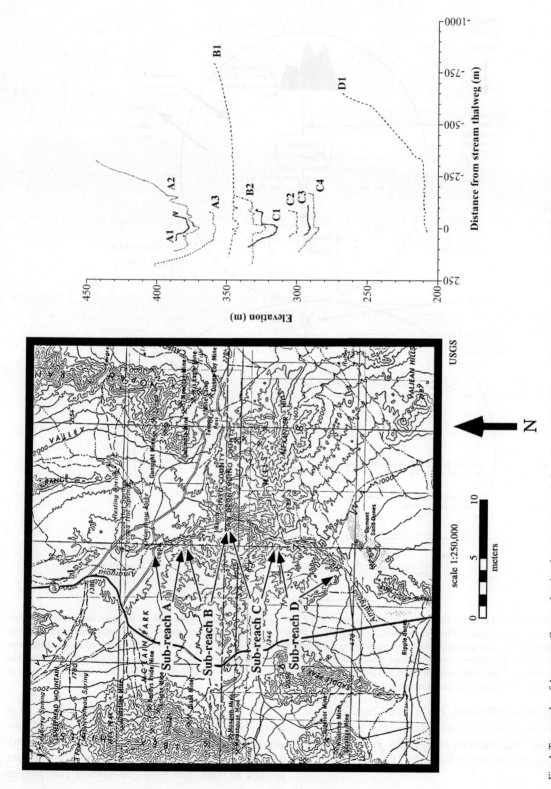

Fig. 4. Topography of Amargosa Canyon showing the upper canyon (subreaches A, B, and C), and the lower canyon (subreach D). Also shown are selected cross sections through the canyon plotted at their actual elevation. Cross sections A1, A2, and A3 are in subreach A, cross sections B1 and B2 are in subreach B, cross sections C1, C2, C3, and C4 are in subreach C, and cross section D1 is in subreach D.

springs at the contact between the Precambrian core of the anticline and the overlying Tertiary fanglomerates of the China Ranch Beds. Average daily discharge of the Amargosa River through Amargosa Canyon has been measured at 2.21 ft^3/s (U.S. Geological Survey, 1995; station number 10251300). Maximum instantaneous peak discharge was measured at 1500 ft^3/s during February 1969.

Three distinct pediments formed on the China Ranch Beds after the breach of the Tecopa basin (~195 ka) and prior to the late Holocene. Also present are truncated alluvial fan sediments interfingering with predominantly coarse-grained axial channel sediments that are topographically higher and substantially older, than the preserved late Holocene fluvial deposits. Both the pediments and alluvial fans are capped by desert pavement surfaces that appear, by degree of varnish, rubification of the undersides of clasts, and degree of clast interlocking, to be late Pleistocene in age (K. Anderson, personal communication, 1999). Three different surfaces were distinguished on the basis of characteristics listed above. They have been designated as Dumont (Qsd), Sperry (Qss), and Acme (Qsa) surfaces based on their type location.

The Dumont surface is present only in the lower section of the Amargosa River through Amargosa Canyon, where it locally underlies the spectacular Dumont Sand Dunes (Figs. 2 and 4). The Dumont Sand Dunes consist of a 20-km^2 field of star dunes up to 120 m in height surrounded by smaller crescentic-ridge, barchan, and linear dunes (Nielson and Kocurek, 1987). The presence of the Dumont surface on either side of the river indicates it was formed before the deep incision of Amargosa Canyon. The surface is underlain by a thin veneer of alluvial fan gravels that unconformably overlie the fanglomerates of the China Ranch Beds which compose the upper units in the Sperry Hills anticline. If the surface was formed before the deep incision of the canyon, it would predate the breach of Tecopa Basin at <195 ka (Morrison, 1991, 1999; Anderson, 1999), suggesting a middle Pleistocene age. The thin veneer of colluvial material underlying the Dumont surface is derived mainly from the China Ranch fanglomerates but is often covered with a veneer of eolian sand.

The best preserved Sperry surface is located near the Sperry site along the abandoned Tonapah and Tidewater Railroad. It is found in both the upper

and lower portions of the canyon and consists of topographically high surfaces with well-developed pavement that are underlain by alluvial fan or fluvial material deposited directly on pedimented China Ranch Beds. This surface has an inset relationship with the deposits underlying the Dumont surface. In the lower portion of the canyon, the Sperry surface is often underlain by large (<2 m) rounded boulders possibly related to high discharges following the breach of Tecopa Basin. This association suggests the age of the Sperry surface is <195 ka.

The Acme surface, the youngest surface, is found throughout the canyon. It is also coarse-grained and inset into the Sperry surface. The type locale of the Acme surface is near the Acme site at the confluence of China Ranch Wash and the Amargosa River.

Although these surfaces were not studied in detail, three observations were made. First, the Amargosa stream level to which the Sperry and Acme surfaces are graded was topographically higher (greater than 10 m) than the modern level, suggesting that significant incision has occurred since the formation of the surfaces. Second, the age range of the two surfaces (late Pleistocene) supports the hypothesis that establishment of the Sperry surface occurred after the breach of Tecopa Basin and following the oxygen isotope stage 6 moist climatic regime and the 186–128-ka Death Valley lake highstand (Lowenstein et al., 1999). The deposits underlying the Acme surface may have formed during OIS 2 when the Amargosa River contributed to Death Valley lake highstands. Third, the fluvial facies of deposits underlying both the Sperry and Acme surfaces include clasts whose size requires a moderate to high discharge for transport. To substantiate these hypotheses, however, further study and better age control are necessary.

3. Holocene deposits

A geomorphic map of Amargosa Canyon was constructed using 1:6000 B/W aerial photographs, chronologic control was provided by historic artifacts, and radiocarbon dating. Late Holocene fluvial landforms preserved in the upper 11 km of Amargosa Canyon include a suite of five fluvial terraces (Fig. 5) ranging from the active floodplain level up to approximately 12 m above modern stream level in some sections. The terraces in the upper reaches of

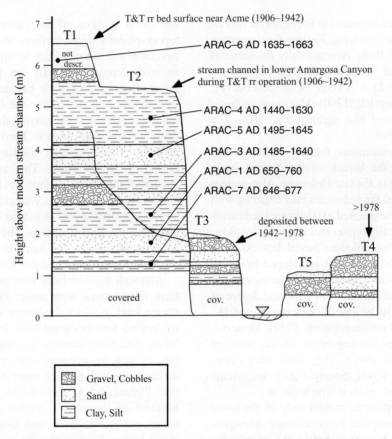

Fig. 5. Schematic composite diagram of the terraces and deposits preserved in the upper and middle portions of Amargosa Canyon with radiocarbon ages. Measured section described approximately 1 km upstream from the confluence of the China Ranch Wash and the Amargsoa River. Radiocarbon sample lab numbers, measured height above the modern stream level, conventional radiocarbon age and error are shown in Table 2.

Amargosa Canyon are represented by both fill and fill-cut terrace surfaces, and the stream may be characterized as an incised, straight channel. The lower 7-km portion of the river through Amargosa Canyon is dramatically different. Here, the stream has a braided channel, and the preserved deposits form three separate, abandoned braid bars (Table 1). In many cases, alluvial fan surfaces grade to various different terraces; these fans are usually dissected.

3.1. Holocene fluvial geomorphology of Upper Amargosa Canyon

Based on the character of the mapped geomorphic units and the modern stream morphology, upper Amargosa Canyon is divided into three sub-reaches (Fig. 4, Table 2). Sub-reach A is a relatively narrow segment, extending for 3.95 km; it is incised into the China Ranch Bed fanglomerate facies, and Precambrian

Table 1
General characteristics of mapped Pleistocene and Holocene units

Age	Dominant geomorphic unit	Relative texture	Location of dominant occurrence in the canyon	Apparent sensitivity to low-amplitude climatic change	Relative length of record
Pleistocene	Alluvial fans, pediments	Coarse	Lower canyon	Low	Long
Holocene	Fluvial terraces	Fine	Upper canyon	High	Short
Holocene	Braid bars	Coarse	Lower canyon	Low	Intermediate

Table 2
Sub-reach characteristics in Amargosa Canyon

Sub-reach	Length (km)	Gradient (m/km)	Sinuosity	Pattern	Surfaces present	Fluvial terraces or braid bars present (in addition to the modern floodplain)	Relative vertical stability
A	3.95	11.20	1.06	Straight	Qss	T1, T2, T3	Low
B	3.59	9.98	1.13	Straight	Qss, Qsa	T1, T3	Low
C	4.54	9.77	1.07	Straight	Qss, Qsa	T1, T3	Low
D	7.13	11.4	1.06	Braided	Qsd, Qss, Qsa	Tbr1, Tbr2	High

rocks. In sub-reach B, the channel is wide (up to 900 m) and the canyon is incised into the lacustrine facies of the China Ranch Beds. This segment is 2.59 km long with a gradient of 9.98 m/km. Sub-reach C is 4.54 km long and only 100 m wide from bedrock to bedrock. This sub-reach is incised through the core of the Sperry Hills anticline which exposes Precambrian bedrock. Fill and fill-cut terraces and abandoned braid bars are present in this sub-reach. The segment length is 4.54 km with a gradient of 9.77 m/km. At the lower end of the sub-reach, as the canyon leaves the Precambrian bedrock and enters predominantly fanglomerates of the China Ranch Beds, fluvial deposits are braided.

Five terrace levels are preserved in the upper Amargosa Canyon; these are designated T1 through T5, from oldest to the modern active floodplain (Fig. 5). T1 is the most widespread fill terrace, and the deposits underlying this terrace are comprised of light-colored, fine-grained material derived from the Tecopa lake beds immediately upstream of the canyon (Fig. 6). These deposits include cut/fill units, gravel lenses, and cienega deposits within a predominantly low-energy fluvial environment. Radiocarbon dating of organic material underlying the T1 terrace at a location approximately 1 km upstream from the confluence with China Ranch Wash suggests that deposi-

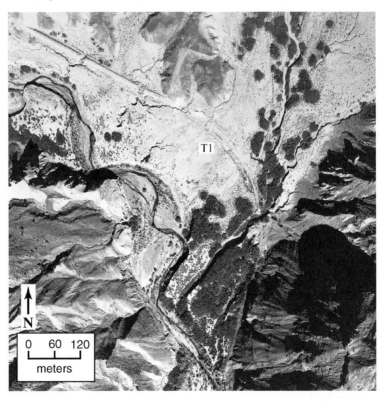

Fig. 6. Aerial photograph of deposits in upper Amargosa Canyon at the confluence of the Amargosa River and China Ranch Wash (Fig. 2). The main terrace is T1.

tion began before A.D. 646–760 and continued until at least A.D. 1635–1663 (Table 3; Fig. 5). This aggradational sequence was punctuated by one major erosional event that occurred sometime between A.D. 760 and A.D. 1440 (Table 3; Fig. 5). The age of the upper portion of the preserved deposits underlying the T1 terrace suggest rapid aggradation at between A.D. 1485–1663.

Historic photographs, taken during the time of railroad operation, show the now-abandoned Tonapah and Tidewater (T&T) Railroad (Myrick, 1991:569) and suggest that the T1 terrace was incised only about 1.5 m, a depth similar to present incision depths, at a site 2 km downstream of the Amargosa River/China Ranch confluence. The T&T Railroad was active through Amargosa Canyon between 1907 and 1940 (Myrick, 1991; Lingenfelter, 1986). Evidence supporting this shallow level of incision at other locations along the railroad bed are bridge abutments over fan tributaries to the Amargosa River. The shallow level of incision is preserved as the T2 fill-cut terrace (Fig. 5). The formation of the T2 terrace was contemporaneous with large-scale erosion on the T1 surface leaving a shallowly dissected, undulating surface morphology.

In upper Amargosa Canyon, the T1 terrace has been deeply incised, down to bedrock in some stretches. The timing of the incision is not well constrained, railroad abutments indicate some incision of the T1 terrace before 1907, but it is not clear that the incision seen today had occurred by that time. An assumption might be made that the railroad would not have been

constructed through the canyon if the deep incision had already occurred, due to the threat of lateral erosion to the roadbed. Documents related to valuation hearings do not specifically mention loss of railroad ties to floods, but rather due to decomposition (U.S. Interstate Commerce Hearing Transcripts, 1921).

Additional evidence suggesting that the deep incision of the T1 terrace occurred recently is the character of the Amargosa River floodplain through Tecopa, immediately upstream of the canyon. The nickpoints have not yet migrated upstream of the canyon, and the channel is still only shallowly incised; what is preserved as the T2 fill-cut terrace downstream here forms the modern floodplain level. A working hypothesis is that sometime after the construction of the railroad, but before 1953 (aerial photographs of that date show the arroyo), the T1 terrace was deeply incised during a major flood event.

Nickpoints are common in the upper Amargosa Canyon, creating a complex series of converging and diverging terrace surfaces. The formation of these surfaces is mainly due to variation in the size of local available sediment and the deposition and erosion of active tributary fans proximal to the Amargosa River. These fans occasionally deliver sediment that exceeds the carrying capacity of the axial channel, causing local gradient changes.

Fluvial activity after the major incision of the T1 terrace formed three fill terraces, T3, T4, and T5, at 2, 1.5, and 1 m above the modern channel, respectively (Fig. 5). The deposits underlying the T3 terrace are

Table 3
Location and result of radiocarbon dating of deposits underlying the T1 terrace in upper Amargosa Canyon

Sample number	Height above modern stream channel (m)	Conventional radiocarbon age and error[a]	Calibrated 1σ calendar age bracket[b]
ARAC-6/AA-19240	6.0	274 ± 44	A.D. 1635–1663
ARAC-4/BETA-72090	4.9	400 ± 70	A.D. 1440–1630
ARAC-5/BETA-93064	3.8	330 ± 40	A.D. 1495–1645
ARAC-3/BETA-93063	2.5	340 ± 40	A.D. 1485–1640
Cut/fill boundary			
ARAC-2/AA-19239	1.8	2374 ± 48	B.C. 415–392[c]
ARAC-1/BETA-70974	1.6	1340 ± 60	A.D. 650–760
ARAC-7/AA-25671	1.3	1375 ± 45	A.D. 646–670

[a] All results are AMS ages except BETA-72090, a conventional age with extended counting.

[b] Conversion to calendar ages for samples BETA-70974,-93063,-72090, and-93064 by Pretoria, all other conversions by Calib (A).

[c] This sample is considered to be transported and redeposited charcoal and was not included in the interpretation.

characterized by several coarse to fine couplets, suggesting multiple flooding events. Age constraint for deposits underlying T3 is provided by the discovery of a railroad tie buried at the base of the deposit that is assumed to have been deposited by a flood following the abandonment of the railroad, although the tie might have been relocated anytime following 1907. Aerial photographs show the presence of the T3 terrace by 1978. The deposits underlying the T4 terrace were emplaced and incised after 1978, and the T5 is considered to be the infrequently flooded, modern floodplain.

3.2. Holocene fluvial geomorphology of lower Amargosa Canyon

The Holocene deposits in lower Amargosa Canyon, sub-reach D (Fig. 4), are quite different from those in the upper canyon. Sub-reach D is characterized by a broad, open channel with abandoned braid bars and a modern braided system. These deposits include two abandoned braid bars, Tbr1 and Tbr2, and the modern deposits along the river (Fig. 7). The abandoned braid bars are composed of coarser material than most of the Holocene deposits in upper Amargosa Canyon. The surface of Tbr1, the older abandoned braid bar, includes dark varnish and shows no evidence of inundation by modern floods. The pavement characteristics suggest a mid- to late Holocene age (K. Anderson, personal communication, 1997), suggesting that the lower portion of the canyon has been relatively stable since the mid- to late Holocene. The younger abandoned braid bar Tbr2 is lithologically similar to Tbr1, but it has a better developed bar and swale topography, lighter varnish, is lower in elevation, and is occasionally inundated. Blocks ~1.5 m in size comprised of material from deposits underlying the upstream T1 terrace (i.e., transported blocks eroded from the fine-grained floodplain deposits) have been deposited on the Tbr2 abandoned braid bar. The blocks would probably not be preserved for more than a few decades due to the friable nature of

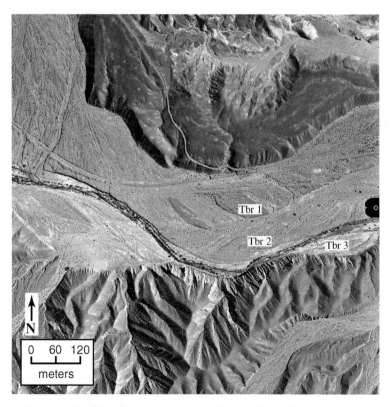

Fig. 7. Aerial photograph of deposits in lower Amargosa Canyon, all three braid bar deposits are shown.

the fine-grained sediment. Railroad debris (cans, boards, etc.) is present on the fans grading to the Tbr2 braid bars, suggesting that the deposits post-date 1907. A surveyed longitudinal profile shows nearly parallel gradients for all surfaces and suggests relative stability of the channel in sub-reach D at least since Tbr1 (mid- to late Holocene) emplacement.

4. Modern hydroclimatology and channel erosion

Stream gauge data from Tecopa, California (station number 10251300, U.S. Geological Survey, 1995), were analyzed for 1962 (A.D.) to 1983 (A.D.) and 1992 (A.D.) to 1994 (A.D.) (the gauging station was inoperative 1983–1991). Climate data were utilized from Death Valley, CA, Las Vegas, NV, and Beatty, NV, weather stations and the historical climatology network (HCN) data from Karl et al. (1990) and Quinlan et al. (1987). Climate analyses were conducted by R. Balling, Jr., and R. Vose from the Arizona State University Climatology Laboratory. Aerial photographs dating from 1953 (1:47,2000), 1978 (1:30,000), 1983 (1:58,000), and 1989 (1:40,000) and 1996 (1:6000) were obtained and analyzed for changes in channel morphology.

Modern stream flow events at the Tecopa stream gauge and concurrent meteorological data for the Amargosa River drainage basin were analyzed by month for the period of record between January 1963 and December 1983. Specifically, monthly mean 700 mb heights (28-grid network in the Pacific and North America, 20°N to 50°N and 110°W to 170°W, for each day from 1963 to 1983 at 12 Z) and event day 700-mb heights when $Q>50$ ft^3/s,

were analyzed for winter (November–April) and summer (May–October).

Winter event days show a large depression (>75 m) of the 700-mb heights in the eastern North Pacific when the circumpolar vortex expands southward with a large trough over the eastern North Pacific, causing upper level divergence over the study area promoting moisture advection and supporting ascending motion in the troposphere. During the summer, there is a depression of the 700-mb heights off the coast of Baja which encourages advection and upper-level divergence. The pattern also suggests that tropical cyclones may play a role in generating flow events.

Averaged, calculated mean monthly sea surface temperature anomalies (northeast Pacific, 20°N to 50°N and 110°W to 170°W, 5° latitude by 5° longitude cells, for each month from 1963 to 1983) and flow event months during which $Q>50$ ft^3/s were also analyzed. The resultant winter anomaly map includes cool water north of Hawaii and warm water south of Baja, indicating that warm water is available for advection south and west of the Amargosa River watershed. Summer is similar to winter, and the pattern is again suggestive of the role tropical cyclones play in generating flood events.

The 700-mb height data was used to group the 107 discharge events (Tecopa, California stream gauge), period of record 1962–1983 and 1992–1994, discharge ($Q>10$ ft^3/s) into a small number of synoptic weather types. A matrix was constructed using the 28-grid 700-mb value for each discharge event (107 events). The matrix was subjected to a principal component analysis and the results were grouped into seven categories using the "average linkage" clustering technique. The five most common of the seven

Table 4
Classification of the dominant flood (>10 ft^3/s) generating synoptic climatology at the Tecopa stream gauge on the Amargosa River

Type	Dominant characteristic	Dominant season of occurrence	Occurrence (%)	Significant discharge event	
				Month	Mean daily discharge (ft^3/s)
1	Ridge over north Pacific, trough over west coast of U.S.	Winter	20	February 1969	1500
2	Zonal flow over Pacific, trough over west coast	Winter	15	February 1983	448
3	Weak upper level flow, jet in Canada	Summer	10	August 1983	1280
4	Zonal flow over Pacific, "cut-off" low over west coast	Both	22	October 1976	757
5	Zonal flow over North Pacific, ridge over west coast	Winter	15	January 1969	396
6	"cut-off" low off west coast	Winter	Infrequent		
7	Strong west coast ridge	Winter	Infrequent		

categories, account for approximately 82% of all discharge events where $Q > 10$ ft^3/s (Table 4).

Changes in channel morphology interpreted from available aerial photos were compared to main flood events on the Amargosa River (Table 4). Analyses of the photographs revealed that the greatest incision and meander width expansion occurred after June 1983, probably during the August 1983 event which had the second highest daily mean discharge, at 1280 ft^3/s, for the period of record. The highest daily mean discharge occurred in February 1969 at 1500 ft^3/s, with an instantaneous peak discharge of 10,500 ft^3/s. The flood event of 1983 was of lower discharge than the 1969 event and was shorter in duration, with no antecedent precipitation, and occurred during the season when vegetative cover is typically low. This may have promoted more erosive potential. Long intervals between the sequence photography (6–25 years) rend the relationship between the flood events and erosion only preliminary.

5. Discussion

5.1. Latest Pleistocene

The Dumont surface pre-dates the breach of Tecopa Basin based on surface-weathering characteristics. Since the breach of Tecopa Basin at <195 ka (Morrison, 1999; Anderson, 1999), Amargosa Canyon has been in an overall incisional regime carving a canyon up to 275 m deep through the Sperry Hills. This incision has been punctuated at least twice by aggradational periods during which the materials underlying the Sperry and Acme surfaces were deposited. These deposits are not dated, but surface-weathering characteristics suggest a late Pleistocene age for the stabilization of the surfaces. Large boulders underlying the Sperry surface may be related to the high discharges associated with the breach of Tecopa Basin and the deposits underlying the both the Sperry and Acme surfaces probably resulted from large-scale climatic oscillations since 195 ka.

5.2. Late Holocene

The floodplain deposits underlying the dominant T1 terrace in Amargosa Canyon suggest that deposi-

tion began before A.D. 646–760 and continued until at least A.D. 1635–1663 (Table 3; Fig. 5). This aggradational sequence was punctuated by a major erosional event that occurred sometime between A.D. 760 and A.D. 1440 (Table 3; Fig. 5). The age of the upper portion of the deposits underlying the T1 terrace suggest rapid aggradation between A.D. 1485–1663. Fluvial activity since formation of the T1 terrace has included one fill-cut and three fill terraces. The oldest of the three fill terraces is assumed to post-date abandonment of the railroad due to the burial of a railroad tie near the bottom of the depositional unit underlying the T3 terrace, although the relocation of the tie could have occurred anytime since initiation of the railroad.

The first late Holocene aggradational period, represented by deposits underlying the T1 terrace date to ~646 to 760 A.D., is contemporaneous with lake sediments in Cronese Lakes (Fig. 1) dating at approximately 760 A.D. (1190 ± 90 BP) (Drover, 1979; Anderson and Wells, 1996). There is no record of a lake in Silver Lake playa (Fig. 1) at this time.

The later aggradational period (1450 A.D. to 1850 A.D.) is the time during which several regional proxy indicators suggest wetter conditions. This period approximates the "Little Ice Age" from the late A.D. 1500s to the early A.D. 1800s (Bradley, 1999). Wells et al. (1989), Enzel et al. (1989, 1992) and Enzel (1992) found evidence for a lake in Silver Lake playa (Fig. 1) between 1470 A.D. and 1650 A.D. Radiocarbon ages on lake sediments in Cronese Lake (Fig. 1) date at approximately 560 ± 110 BP (1390 A.D.) and 390 ± 140 BP (1560 A.D.) (Drover, 1979; Anderson and Wells, 1996).

Using *Pinus* and other pollen from Lower Pahranagat Lake (210 km northeast of Amargosa Canyon), Wigand (1996) interpreted a wetter period due to an increase in summer precipitation during the earlier aggradational period, from approximately 1600 BP to 1200 BP (350 A.D. to 750 A.D.). This increase in summer precipitation may have extended to the Amargosa River watershed as well.

Conifer/saltbush pollen indices from Lower Pahranagat Lake (210 km northeast of Amargosa Canyon) indicate a wet period due to increased winter precipitation (Wigand, 1996) during the later period of aggradation. *Juniperus* pollen, also from Lower Pahranagat Lake, shows a dramatic increase from 400 BP to 300 BP (1550 A.D.–1650 A.D), again suggesting greater effective winter precipitation, or reduced evaporation rates (Wigand, 1996).

The two major aggradational periods appear to occur during wetter climates and are separated by a substantial erosional event that may have occurred sometime between A.D. 760 and A.D. 1450. This interval brackets the Medieval Warm Epoch (MWE), A.D. 950 to A.D. 1100. (Bradley, 1999) A manifestation of the MWE in the local botanical records is a significant decrease in *Pinus* pollen from Lower Pahranagat Lake (Wigand, 1996). This period also brackets a period of especially high-magnitude floods that occurred around 1 ka during a cool-to-warm climate transition (Ely et al., 1993).

The arroyo-cutting event that deeply incised the T1 terrace is not well-constrained by this study. It is apparent that there was some entrenchment of the T1 terrace during railroad operation, but the extent and depth of the arroyo is unknown. Three subsequent fill terraces developed after incision of the T1 terrace and after abandonment of the railroad in 1940.

Inman and Jenkins (1999) analyzed changes in sediment flux of the 20 largest streams in a transect from Monterey Bay to south of the U.S.–Mexico border during the period 1940–1995. The authors found increased sediment flux during a wet phase of the Pacific North American (PNA) climate pattern that occurred from somewhere between 1968 and 1977 and lasting until at least 1995. The wet phase of the PNA climate pattern is associated with a period of strong and persistent El Niño events (Inman and Jenkins, 1999). The increase in sediment flux is largely believed to be due to large flood events (like those occurring in 1969 and 1983) that follow short, relatively dry periods during an overall wet phase PNA that occurred from 1944 to 1968. The sediment delivery was closely tied to the erodibility of the available sediment.

Hereford and Webb (2001) evaluated alluvium ponded by railroad beds in and around the Amargosa Canyon area and found evidence for increases in sediment accumulation during wet periods and suggested that the large floods that eroded alluvial washes occurred during the relatively wet periods such as those found during a warm phase PDO (Pacific Decadal Oscillation). The warm phase PDO, such as that which occurred from 1977 to the mid-1990s, is characterized by the predominance of wetter El Niño

events, among other variables (Mantua and Hare, 2002).

Enzel et al. (1989) have studied modern, extreme flood events (n=8, including 1969, 1978, 1980, and 1983) in the nearby Mojave River drainage. These floods resulted in short-lived lake-building events in Silver Lake playa north of Baker, California, and were a product of two dominant synoptic patterns: (1) a trough or cut-off low off the coast with a ridge in the North Pacific and a southerly displaced storm track; and (2) a strong southerly displacement of the westerlies. The five biggest, modern flood events (1969, 1976 [2], 1977, 1983) on the Amargosa River appear to be products of: (1) a ridge over the North Pacific and a trough over the west coast; (2) zonal flow over the Pacific and a possible trough or cut-off low over the west coast in winter; or (3) weak upper level flow in summer. Although different definitions of "flood" conditions and different river morphologies exist between the two rivers, similarities exist in synoptic conditions that led to the development of floods in the two arid drainage systems. Variability exists in 1976 and 1977 when the Amargosa flowed in response to zonal flow in the Pacific and a potential cut-off low off the coast (synoptic types 1 or 4, Table 4), and experienced floods in 1976 and 1983 during weak upper level flow; (synoptic type 2, summer "monsoon" conditions or tropical disturbances, Table 4; Anderson et al., 1994).

6. Conclusions

The older, latest Pleistocene deposits represent large-scale or high-amplitude changes in the hydro-climatologic regime whereby aggradation dominated during an overall erosional environment (the cutting of Amargosa Canyon). These deposits and overlying surfaces are the pre-195-ka Dumont surface, the younger Sperry surface, and the Acme surface. A dramatic period of erosion appears to occur after abandonment of the Acme surface and 646 A.D. The only deposits found that relate to this period are few, isolated, dissected tributary fan remnants. Subsequent fan surfaces are coincident with extant terrace surfaces. Exposed fluvial deposits underlying the terraces show evidence of a long history of interfingering of material from tributary fans and the axial channel.

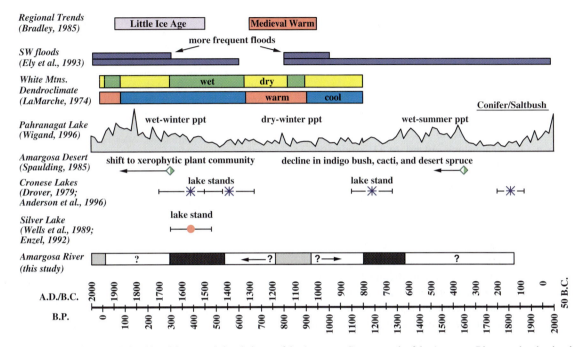

Fig. 8. Graph showing the relationship of the aggradational phases of the Amargosa Canyon reach of the Amargosa River as related to local and regional climatic variations for the late Holocene (LaMarche, 1974; Drover, 1979; Bradley, 1985; Spaulding, 1985; Wells et al., 1989; Enzel, 1992; Ely et al., 1993; Anderson et al., 1996; Waters and Haynes, 2001; Hereford, 2002; Wigand, 1996).

The distribution of radiocarbon ages from sediments underlying the T1 terrace in Amargosa Canyon indicates that the Amargosa River has had at least two major aggradational periods during the past ~2000 years; an early period within the range of 646 A.D. to 760 A.D. and a later period of rapid aggradation from 1450 A.D. to 1663 A.D. (Fig. 5), both occurring during wetter climates (Fig. 8).

The contemporaneity of the aggradational periods with wetter climates suggests that the canyon floor aggraded in response to relatively persistent wetter climatic intervals. The T1 terrace is underlain predominantly by fine-grained material probably derived from redeposition of eroded upstream playa deposits (Fig. 3; Franklin Playa, Tecopa Flats) in addition to sources from the local China Ranch Beds. In this transport-limited system, persistent discharge allows transport of the fine-grained material and vertical accretion.

Regional (<210 km) botanical proxy climate indicators suggest that portions of the earlier aggradational period may correlate with a wetter period dominated by summer precipitation, and that the more recent aggradational period may have correlated with a wetter period dominated by winter precipitation (Fig. 8).

While late Holocene aggradation of the Amargosa River appears to coincide with periods of sustained precipitation, low-frequency, high-magnitude floods may act to incise the existing floodplain. Modern incisional events are manifested as a series of nickpoints influenced by tributary fan deposition, and the formation of short-lived, time transgressive terraces within the canyon. Prehistoric incision, like that interrupted aggradation of the deposits underlying the T1 terrace between 760 and 1450 A.D., may have been in response to low-frequency, high-magnitude floods that occurred during an overall drier climatic interval (i.e., the MWE). If this model may be extrapolated further, the lack of sediments from the latest Pleistocene and the early to mid-Holocene may relate to incision during earlier relatively dry climatic intervals. The Amargosa River fluvial system appears to be quite sensitive to the low-magnitude climatic fluctuations of the latest Holocene, but this record is periodically lost through higher magnitude climatic fluctuations that may lead to dramatic erosion.

Acknowledgements

Funding was provided by an NSF EAR (Award Number 9316443) grant to Dr. Steve Wells and through a GSA Fahnstock Award to D. Anderson. Brian and Bonnie Brown of China Ranch kindly facilitated access to Amargosa Canyon and provided logistical support. I thank fellow graduate students Kirk Anderson Jeff Knott, and Yvonne Katzenstein-Wood for many interesting and insightful discussions in the field. Pete Wigand kindly provided guidance in understanding the local pollen record. Finally, I am very grateful to Lauren Wright and Bennie Troxel, for keeping track of who was doing what and where in the Death Valley region, for keeping us all in contact with one another, and for generously sharing their vast geologic knowledge of the region.

References

Anderson, K.C., 1999. Processes of vesicular horizon development and desert pavement formation on basalt flows of the Cima volacanic field and alluvial fans of the Avawatz Mountains piedmont, Mojave Desert, California. PhD thesis, Riverside, University of California, 191 pp.

Anderson, K.C., Wells, S.G., 1996. Late Pleistocene lacustrine record of Lake Dumont and the relationship to pluvial lakes Mojave and Manly. Abstract, Geological Society of America Annual Meeting, October, Denver, CO.

Anderson, D.E., Wells, S.G., 2003a. Latest Pleistocene lake high-stands in Death Valley, California. In: Enzel, Y., Wells, S.G., Lancaster, N. (Eds.), Paleoenvironments and Paleohydrology of the Mojave and Southern Great Basin Deserts. Geological Society of America Special Paper, 368, 115–128.

Anderson, K.C., Wells, S.G., 2003b. Latest Quaternary paleohydrology of Silurian Lake and Salt Spring basin, Silurian Valley, California. In: Enzel, Y., Wells, S.G., Lancaster, N. (Eds.), Paleoenvironments and paleohydrology of the Mojave and southern Great Basin Deserts. Geological Society of America Special Paper, 368, 129–141.

Anderson, D.E., Balling Jr., R., Wells, S.G., 1994. Modern Hydrology and Climatology of the Amargosa River near Tecopa, California. Abstract, Proceedings of the Sixth Annual Pacific Climate (PACLIM) Workshop, April, Asilomar, California.

Anderson, D.E., Wells, S.G., Balling Jr., R.C., Vose, R., 1996. Late Holocene hydroclimatology of the Amargosa River: geomorphic sensitivity of an arid watershed to low amplitude climate change.

Abstract, INQUA Commission on Global Continental Paleohydrology, 2nd International Meeting, September, Toledo, Spain.

Anderson, D.E., Poths, J., Wells, S.G., 1998. Cosmogenic ^{3}He age estimates for the terminal highstand of Plio-Pleistocene Lake Tecopa, California—Integration of the upper Amargosa River with the Death Valley drainage basin. Geological Society of America Abstracts with Programs, Rocky Mountain Section, 50th Annual Meeting, vol. 30, p. 2.

Bradley, R.S., 1985. Quaternary Paleoclimatology: methods of paleoclimatic reconstruction. Allen and Unwin, Boston, MA, 472 pp.

Bradley, R.S., 1999. Paleoclimatology, Reconstructing Climates of the Quaternary, 2nd ed. Academic Press, San Diego, 613 pp.

Dohrenwend, J.C., 1986. Basin and Range. In: Graf, W.L. (Ed.), Geomorphic Systems of North America, Centennial Special, vol. 2. Geological Society of America, Boulder, CO.

Drover, C.E., 1979. The late prehistoric human ecology of the Northern Mojave Sink San Bernardino County, California. Doctoral thesis, University of California at Riverside.

Elliott, B., 1982. An investigation of selected water quality parameters in the Amargosa drainage basin. Water Resources Center, Desert Research Institute, University of Las Vegas System. Prepared in cooperation with the U.S. Department of Energy, Nevada Operations Office, under contract DE-AC08-81NV10162.

Ely, L.L., Enzel, Y., Baker, V.R., Cayan, D.R., 1993. A 5000-year record of extreme floods and climate change in the Southwestern United States. Science 262, 410–412.

Enzel, Y., 1992. Flood frequency of the Mojave River and the formation of late Holocene Playa Lakes, Southern California, USA. The Holocene 2, 11–18.

Enzel, Y., Cayan, D.R., Anderson, R.Y., Wells, S.G., 1989. Atmospheric circulation during Holocene lake stands in the Mojave Desert: evidence of regional climate change. Nature 341, 44–48.

Enzel, Y., Brown, W.J., Anderson, R.Y., McFadden, L.D., Wells, S.G., 1992. Short-duration Holocene lakes in the Mojave River drainage basin, Southern California. Quaternary Research 38, 60–73.

Hereford, R.H., 2002. Valley-fill alluviation during the Little Ice Age (ca. A.D. 1400–1880), Paria River basin and southern Colorado Plateau, United States. Geological Society of America Bulletin 114 (12), 1550–1563.

Hereford, R., Webb, R.H., 2001. Climate variation since 1900 in the Mojave Desert region affects geomorphic processes and raises issues for land management. In: Reynolds, R.E. (Ed.), The Changing Face of the East Mojave Desert. Desert Studies Consortium, California State University, Fullerton, pp. 54–55.

Hillhouse, J., 1987. Late Tertiary and Quaternary geology of the Tecopa Basin, Southeastern California. Miscellaneous Investigations Series U.S. Geological Survey, Report: I-1728.

Hooke, R.L., 1972. Geomorphic evidence for late Wisconsin and Holocene tectonic deformation, Death Valley, California. Geological Society of America Bulletin 83, 2073–2098.

Hunt, C.B., 1975. Death Valley Geology Ecology Archaeology. University of California Press, Berkeley.

Hunt, C.B., Mabey, D.R., 1966. Stratigraphy and structure, Death Valley, California. U.S. Geological Survey Professional Paper 494-A, 162 pp.

Hunt, C.B., Robinson, T.W., Bowles, W.A., Washburn, A.L., 1966. Hydrologic basin, Death Valley, California. U.S. Geological Survey Professional Paper 494-B, 130 pp.

Inman, D.L, Jenkins, S.A., 1999. Climate change and the episodicity of sediment flux of small California rivers. Journal of Geology 107, 251–270.

Jennings, C.W., Burnett, J.L., Troxel, B.W., 1962. Geologic map of California Trona sheet, 1:250,000 scale.

Karl, T.R., Williams J., C.N., Quinlan, F.T., 1990. United States historical climatology network (HCN) serial temperature and precipitation data. Carbon Dioxide Information Analysis Center, Oak Ridge National Laboratory, Oak Ridge, TN.

Ku, T.-L., Luo, S., Lowenstein, T.K., Li, J., Spencer, R.J., 1998. U-Series Chronology of Lacustrine Deposits in Death Valley, California. Quaternary Research 50, 261–275.

LaMarche Jr., V.C., 1974. Paleoclimatic inferences from long tree-ring records. Science 183, 1043–1048.

Li, J., Lowenstein, T.K., Brown, C.B., Ku, T.-L., Luo, S., 1996. A 100 ka record of water tables and paleoclimates from Salt Cores, Death Valley, California. Palaeogeography, Palaeoclimatology, Palaeoecology 124, 1–4.

Lingenfelter, R.E., 1986. Death Valley and the Amargosa, Land of Illusion. University of California Press, 622 pp.

Lowenstein, T.K., Li, J., Brown, C.B., 1998. Paleotemperatures from fluid inclusions in halite: method verification and a 100,000 year paleotemperature record, Death Valley, CA. Chemical Geology 150, 223–245.

Lowenstein, T.K., Li, J., Brown, C., Roberts, S.M., Ku, T.-L., Luo, S., Yang, W., 1999. 200 k.y. paleoclimate record from Death Valley salt core. Geology 27, 3–6.

Mantua, N.J., Hare, S.R., 2002. The Pacific decadal oscillation. Journal of Oceanography 58, 35–44.

Mifflin, M.D., Wheat, M.M., 1979. Pluvial lakes and estimated pluvial climates of Nevada. Water Resources Center, Desert Research Institute, Reno, NV.

Morrison, R.B., 1991. Quaternary stratigraphic, hydrologic, and climatic history of the Great Basin, with emphasis on Lakes Lahontan, Bonneville, and Tecopa. In: Morrison, R. (Ed.), Quaternary Nonglacial Geology; Conterminous U.S. Geological Society of America, Boulder, pp. 283–320.

Morrison, R.B., 1999. Lake Tecopa: Quaternary Geology of Tecopa Valley, California, a multimillion-year record and its relevance to the proposed nuclear-waste repository at Yucca Mountain, Nevada. In: Wright, L.H., Troxel, B.W. (Eds.), Cenozoic Basins of the Death Valley Region, Special Paper Geological Society of America, 333, 301–344.

Myrick, D.F., 1991. Railroads of Nevada and Eastern California. University of Nevada Press, Reno.

Nielson, J., Kocurek, G., 1987. Surface processes, deposits, and development of star dunes: dumont dune field, California. Geological Society of America Bulletin 99, 177–186.

Quinlan, F.T., Karl, T.R., Williams J., C.N., 1987. United States historical climatology network (HCN) serial temperature and precipitation data. Carbon Dioxide Information Analysis Center, Oak Ridge National Laboratory, Oak Ridge, TN.

Slate, J. (Ed.), 1999. Proceedings of Conference on Status of Geologic Research and Mapping, Death Valley National Park, U.S. Geological Survey Open File Report, vol. 99-153, 177 pp.

Spaulding, W.G., 1985. Vegetation and climates of the last 45,000 years in the vicinity of the Nevada Test Site, South-Central Nevada. U.S. Geological Survey.

U.S. Geological Survey, 1995. Water Resources Data Nevada Water Year 1994. U.S. Geological Survey Water Data Report NV-94-1, Prepared in cooperation with the State of Nevada and with other agencies by S. L. Clary, D. R. McClary, R. Whitney, and D. D. Reeves, USGS, Carson City, NV.

Waters, M.R., Haynes, C.V., 2001. Late Quaternary arroyo formation and climate change in the American Southwest. Geology 29 (5), 399–402.

Wells, S.G., Anderson, R.Y., McFadden, L.D., Brown, W.J., Enzel, Y., Miossec, J., 1989. Late Quaternary Paleohydrology of the Eastern Mojave River Drainage, Southern California: Quantitative Assessment of the Late Quaternary Hydrologic Cycle in Large Arid Watersheds. New Mexico State University, Las Cruces.

Wigand, P.E., 1996. A late-Holocene pollen record from Lower Pahranagat Lake, Southern Nevada, USA: high resolution paleoclimatic records and analysis of environmental responses to climate change. In: Issacs, C.M., Tharp, V.L. (Eds.), Thirteenth Annual Pacific Climate (PACLIM) Workshop. California Department of Water Resources, Asilomar, CA.

Available online at www.sciencedirect.com

SCIENCE ⓓ DIRECT®

Earth-Science Reviews 73 (2005) 309–322

EARTH-SCIENCE
REVIEWS

www.elsevier.com/locate/earscirev

Macropolygon morphology, development, and classification on North Panamint and Eureka playas, Death Valley National Park CA

Paula Messina [a,*], Phil Stoffer [b], Ward C. Smith [c,1]

[a] San José State University, Geology Department, San José, CA 95192-0102, United States
[b] United States Geological Survey, 345 Middlefield Road, MS-975, Menlo Park, CA 94025, United States
[c] Stanford University, Department of Applied Earth Sciences, Stanford, CA 94305, United States

Accepted 1 April 2005

Abstract

Panamint and Eureka playas, both located within Death Valley National Park, exhibit a host of surficial features including fissures, pits, mounds, and plant-covered ridges, representing topographic highs and lows that vary up to 2 m of relief from the playa surface. Aerial photographs reveal that these linear strands often converge to form polygons, ranging in length from several meters to nearly a kilometer. These features stand out in generally dark contrast to the brighter intervening expanse of flat, plant-free, desiccated mud of the typical playa surface.

Ground-truth mapping of playa features with differential GPS (Global Positioning System) was conducted in 1999 (North Panamint Valley) and 2002 (Eureka Valley). High-resolution digital maps reveal that both playas possess macropolygons of similar scale and geometry, and that fissures may be categorized into one of two genetic groups: (1) shore-parallel or playa-interior desiccation and shrinkage; and (2) tectonic-induced cracks. Early investigations of these features in Eureka Valley concluded that their origin may have been related to agricultural activity by paleo-Indian communities. Although human artifacts are abundant at each locale, there is no evidence to support the inference that surface features reported on Eureka Playa are anthropogenic in origin.

Our assumptions into the genesis of polygons on playas is based on our fortuitous experience of witnessing a fissure in the process of formation on Panamint Playa after a flash flood (May 1999); our observations revealed a paradox that saturation of the upper playa crusts contributes to the establishment of some desiccation features. Follow-up visits to the same feature over 2 yrs' time are a foundation for insight into the evolution and possible longevity of these features.
© 2005 Elsevier B.V. All rights reserved.

Keywords: GIS; GPS; surficial processes; Quaternary; desiccation fissures; playa

1. Introduction

North Panamint and Eureka playas were annexed to Death Valley National Park in 1994, when the

* Corresponding author. Fax: +1 408 924 5053.
  E-mail address: pmessina@geosun.sjsu.edu (P. Messina).
[1] Manuscript incorporates research on bequest of Jenny Smith.

0012-8252/$ - see front matter © 2005 Elsevier B.V. All rights reserved.
doi:10.1016/j.earscirev.2005.04.011

park's boundaries were extended. North Panamint Playa lies within Panamint Basin, a north–south-striking valley separated from Death Valley by the Panamint Range to the east, from Owens Valley by the Argus and Coso Ranges to the west, and from Searles Valley by the Slate Range to the southwest (Smith, 1976). Eureka Playa lies within Eureka Valley, bounded by the Inyo Range to the west, and separated by Northern Death Valley by the Last Chance Range to the east. Both playas lie within structurally controlled basins that serve as local watershed termini. At comparable elevations (North Panamint Playa, 640 m; Eureka Playa, 877 m), their hard, smooth surfaces are composed of similar unremarkable clays, fine-grained silts, and minor sands (Motts and Carpenter, 1968; Smith, 1978). North Panamint Playa possesses more-plastic lacustrine sediments of Pleistocene Lake Panamint at depth (Neal, 1968b); there are no reports of Eureka Playa cores in the literature, and samples were not collected for this study, as per the limitations of our national park research permit.

North Panamint and Eureka playas are typical of numerous southwestern dry lake beds, in that they possess several common geomorphic features, including solution depressions, drain holes, mud volcanoes, phreatophyte and spring mounds.

Among the most striking surface features on these playas are long, deep cracks, vegetated lineaments, or tonal stripes. These features represent what were originally fissures in various forms of development; they may extend for hundreds of meters, and many are part of geometric networks, observable on aerial photographs.

2. Playa fissures and macropolygons

Extensive reconnaissance was conducted by the United States military in the second half of the twentieth century (Neal, 1968a,b; Motts, 1970) with an interest in identifying playa surfaces that would be suitable for landing aircraft. The research revealed desiccation cracks, often defining patterns radiating at roughly 90° or 120° angles (hence the term *macropolygon*) on numerous southwestern playas. This form of "patterned ground" had not been extensively studied, and some aspects of the origin of patterned

ground in warm arid climates remains puzzling (Hunt et al., 1966).

Neal and Motts (1967) reported that in the 15 yrs prior to their study, increasingly greater numbers of giant desiccation polygons had been noted on aerial photographs. While a small fraction of this increase may be due to enhanced remote sensing capabilities, much of the increase has been confirmed by ground-truthing. Over the last century, natural climatic and environmental changes (i.e., variations in the frequency and intensity of rainstorms), and changes induced by human activity (i.e., ground water pumping) have resulted in a lowering of ground water tables, piezometric surfaces, capillary zones, and overall soil moisture; it is likely that all of these factors are playing a role in the increasing development of giant desiccation features (Neal and Motts, 1967). However, drying and shrinking may not be the only phenomena that form fissures and macropolygon segments.

In tectonically active areas such as the Basin and Range Province, earthquakes may rupture playa sediments forming long fissure-like scarps. Such a rupture was noted on Lavic Lake resulting from the Hector Mine earthquake (magnitude 7.1) in October 1999 (Morrison, 2002). The Bullion Fault and at least four others ruptured coseismically, producing a spectacular surface rupture with a total length of 48 m on Lavic Lake (Treiman et al., 2002).

Another hypothesis proposes that some linear scars are "irrigation ditches," excavated by paleo-Indians (Patch, 1951). One such feature described as an irrigation ditch was exploited as evidence of an agricultural community in Eureka Valley; this classification was later used to support the existence of farming in Owens Valley (Lawton et al., 1976). The persistence of this inference led Ward C. Smith (1978) to visit Eureka Playa to investigate the "aqueduct." His observations laid the groundwork for our field investigations conducted on Eureka Playa in 2002.

3. Field methods

While many earlier surveys of southwestern playas were conducted remotely (most notably, Neal, 1965a,b), we found that it was important to investigate features on-site. While aerial photographs and satellite images reveal general patterns, they cannot provide information about relative ages or specific morphology of fissures and macropolygon networks.

A reconnaissance effort at North Panamint Playa in January 1999 revealed that its most common features were fissures, spring mounds, and drainage pits in varying stages of development, plant overgrowth, and deterioration. The northern part of the playa displayed a dense network of older-looking fissures. Our interest in the features was piqued during our January 1999 visit when we discovered a fresh network of fissures (Fig. 1). Our original questions were: (1) how old are they? and (2) how did they form? Alignment with mapped faults suggested that fissures may represent earthquake rupture scarps, but later review of recent seismic data for the area showed that orientation of the fissures was inconsistent with this hypothesis.

Fig. 1. A young fissure at the south end of North Panamint Playa, as photographed in January 1999.

It was clear that field mapping and classification of features needed to be conducted so that the entire playa could be analyzed for spatial and temporal clues to the genesis and morphological progression of desiccation features. Differential GPS mapping was chosen as the preferred field method, so that comparison of recent data to historic aerial photographs could be conducted in a GIS (Geographic Information System) environment.

Prior to field mapping of the entire playa in May 1999, a data dictionary, or feature/attribute template was constructed based on observations recorded during the reconnaissance visit. Trimble Pathfinder Office v. 2.51 software was used to create this template (similar to a table with only column headings and rows), which included point features (bush, generic point); line features (fissures, linear bushes, roads, and generic line); and areas (playa perimeter, mound, depression, Lake Hill, and generic area), for automated data collection. Each feature was associated with descriptive attributes which were also stored as part of the digital database. For example, fissures were described by average depth, relative moisture content (if any), and relative age (a subjective decision, based on degree of weathering/sediment fill, amount and condition of vegetation). Qualitative age classifications, based on a model for fissure development (Messina et al., 1999) are summarized in Table 1.

Mapping at North Panamint took place over a one-week period by a field crew of six using one Trimble Pro-XRS sub-meter receiver connected to a TSC1 GIS data logger, and one GeoExplorer II. Features mapped by the Pro-XRS made use of real-time differential correction (via Coast Guard beacon transmissions); positions are accurate to within 1 m; GeoExplorer II files were differentially corrected using post-processing routines (via a remote Bureau of Land Management base station in Las Vegas) which enhanced position accuracy to approximately 2–5 m. The GeoExplorer II was used in circumstances where coarser resolution was sufficient. One- to two-day return trips to North Panamint Playa have been conducted annually to monitor changes in previously mapped features using the sub-meter GPS receiver.

Eureka Playa mapping was conducted in April 2002, after a reconnaissance visit in January of the same year. As before, a data dictionary was

Table 1
Playa fissure relative age classifications

Relative age class	Characteristics
New (recent)	A crack appears fresh, deep (often over 1 m), and devoid of vegetation. Edges are sharp, with little if any slumping (see Fig. 4).
Young	Live bushes indicate a more recent origin; portions of the fissure may be filled with recent sediments, sides may have slumped, but its trace is topographically expressed, although likely uneven (see Fig. 9).
Mature	Aligned dead bushes are clearly visible; little to no original negative relief; sediments collected at the vegetated areas may produce mounds of as much as 0.5 m (see Fig. 6).
Old	The fissure appears as a healed scar, sometimes with remnants of vegetation indicating its existence; no topographic expression (see Fig. 10).

assembled after the preliminary visit, and it included many of the same features and attributes of the Panamint database. However, added to this digital spreadsheet were various anthropological elements, such as rock mounds and artifacts (points), and rock alignments (lines). Specific attention was paid to archaeological features on Eureka Playa in hopes of concluding whether irrigation networks had indeed been constructed. Rock alignments and artifacts are present on North Panamint Playa as well, but there has neither been mention in the literature, nor in-field observations that agricultural activities took place here in the past.

Field mapping took place in 1 day's time, using a field crew of 11. Three Pro-XRSs were used (although real-time differential correction was not possible 100% of the time); non-critical data were collected by a GeoExplorer III. Data files were post-processed to attain sub-meter (Pro-XRS), or 2–5 m accuracy (GeoExplorer III) using the CORS base station in Dyer, Nevada. All data were exported to GIS shape file format for use in ArcView GIS; all attribute data were exported to Microsoft Excel for analysis.

4. North Panamint Playa fissures

Fissures were defined as lineaments with discernable start and end points (some included bends); while others (perhaps not contemporaries) formed closed areas, defining macropolygons. A total of 305 fissures and macropolygon segments were mapped in May 1999 (Fig. 2). Features ranged from 4 to 911 m in length, with a mean of 107 m; the most deeply incised fissures were over 1 m deep, some were flush with the playa surface, while others were expressed as positive

topographic features (probably caused by the accumulation of wind-blown sediment beneath plants and around the roots of fissure-related vegetation). Age classification revealed that 143 fissures were old, 106 were mature, 39 were young, and 20 were new, although absolute ages of each category could not be determined.

The complete data set for North Panamint fissures' age/length characteristics are detailed in Fig. 3. There is a positive correlation between length and relative age; the mean linear extent of fissures is 114 m, 117 m, 75 m, and 66 m for old, mature, young, and new fissure categories, respectively. The slightly shorter average length for the oldest fissures is likely due to the high degree of degradation that occurred over time, making the true extent obscure, and difficult to define. These data suggest that fissures continue to grow over time; long after initial formation occurs, perhaps for decades. One such fissure that opened on Yucca Lake NV in 1969 showed episodic lengthening over a period of years. The original crack measured 1600 m but it extended 137 m in 1978, and 357 m in 1980 (Doty and Rush, 1985).

Motts and Carpenter (1968) reported that there were no giant polygons visible on North Panamint Playa in the region of two test holes (Fig. 2). The two cores drilled at these locations revealed that a highly fractured clay zone existed at shallow depth (6–10 m), although there was no surface expression of desiccation features. It was concluded that these zones resulted from playa desiccation, or that they were related to the formation of macropolygons (Motts and Carpenter, 1968).

During our survey 21 yrs later, there were well-defined consequently formed desiccation cracks in those areas. Of the 91 features investigated, 43%

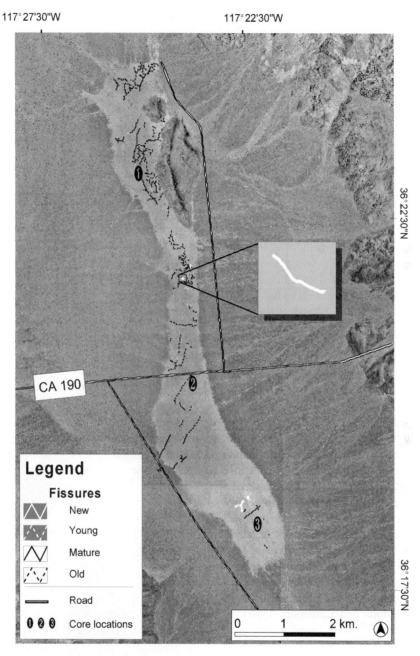

Fig. 2. Differential GPS/GIS map of North Panamint Playa layered on USGS Digital Orthophoto Quads. Cores bored by Motts and Carpenter (1968) are indicated by their original identifying numbers (1, 2, and 3). Fissures are classified by inferred relative age. The inset shows the location of the fissure that formed in May 1999, whose midpoint's coordinates are 36.3626932E North latitude; 117.4011947E West longitude.

were classified as old; 50% were mature; 7% were young. If surface lineaments were absent in 1968, it implies that these mapped features developed subsequent to the Motts and Carpenter research. It should be noted, however, that examination of aerial photographs taken in 1948 shows a network of macropolygons at the playa's north end. It is possible that either this older network was healed before the 1968 survey,

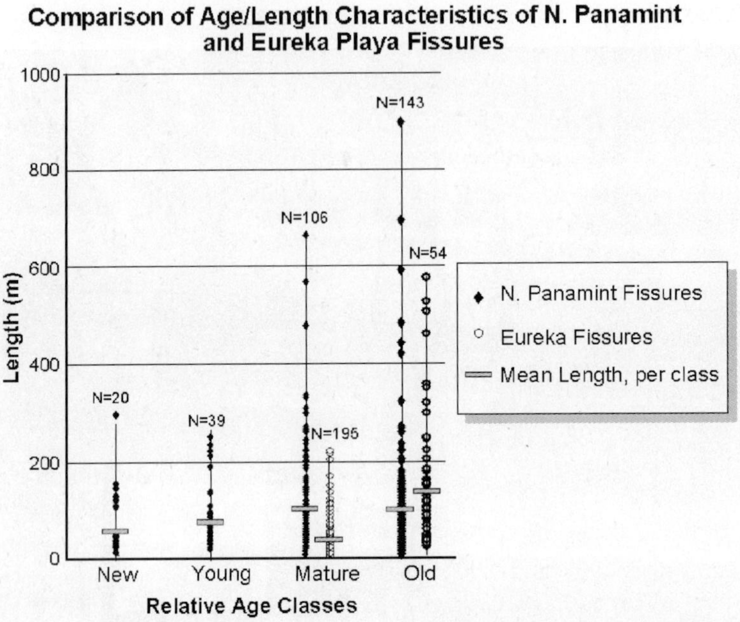

Comparison of Age/Length Characteristics of N. Panamint and Eureka Playa Fissures

Fig. 3. Graph showing a comparison of the relative age/length characteristics of North Panamint and Eureka Playa fissures.

or Motts and Carpenter did not include vegetated lineaments as indications of older macropolygons. In either case, the longevity of desiccation cracks may be measured in decades.

There were few "young" and no "new" fissures in Panamint's north end, which suggests that certain segments of the playa may experience periods of activity or inactivity, depending on general weather and localized soil moisture conditions. Supporting this hypothesis, a cluster of young and new fissures was mapped near the southern end of North Panamint Playa, not far from another test hole bored by Motts and Carpenter (1968). This core was drilled to a depth of 50 m, where standing water was encountered; it was concluded that the borehole was terminated close to the water table (Motts and Carpenter, 1968). In 1999, a total of 45 fissures were found in this area: 26% new; 55% young; 13% mature; 6% old. Some of these cracks' ages may be measured merely in years, or less. Their spatial grouping and proximity to the formerly saturated zone at depth (as confirmed by bore hole evidence) suggests that this region of the playa dehydrated since 1968. Sub-surface desiccation fractures later became expressed at the surface, rupturing before May 1999 after perhaps decades of upward migration. We fortuitously experienced the final phase

Fig. 4. New fissure on North Panamint Playa, as it formed the morning after a rain in May, 1999. The depth of the fissure was over 1.5 m at the time this photograph was taken.

of fissure development (i.e., surface expression) as it transpired at North Panamint Playa.

Paradoxically, a rare late-spring rain one evening in May 1999 provided the final requirement for the formation of giant desiccation fissures. The morning after precipitation, the upper surface of the playa was saturated although there was no standing water (except in topographic sinks); without the luxury of direct measurement, it was difficult to infer the total meteoric input of the event. While traversing the east-central quadrant of the playa, two field mappers observed a fissure in the act of "zipping open" (Fig. 4). The uppermost surface sediments were softened enough to were pull apart along a localized "rift zone." The new fissure lengthened over the course of several hours. Periodic mapping of the fissure while in development showed that it lengthened as much as 3 m in 30 min. Observing this process allowed for the construction of one plausible sequence of events in the evolution of macropolygons.

5. Eureka Playa fissures

Eureka Valley is the site of a playa that has not been extensively studied by the geologic community (Smith, 1978) as is evidenced by the scant mention it receives in the literature. The basin is also host to Eureka Dunes, which by contrast has captured the attention of geomorphologists, botanists (as it is the site of two endangered indigenous plants), and archaeologists (due to its prevalence of anthropogenic artifacts). The playa displays most of the same surface features as North Panamint Playa (fissures, spring mounds, and depressions), but it shows little evidence of recent surficial process activity.

In 2002, a total of 196 fissures were mapped (defined by negative-relief, vegetated, or "sutured" lineaments, Fig. 5), ranging in length from 0.5 to 224 m (mean=26.18 m). None of the fissures were characteristic of new or young features (Table 1); the age/length distribution of Eureka Playa fissures are detailed in Fig. 3. With the exception of 13 intercon-

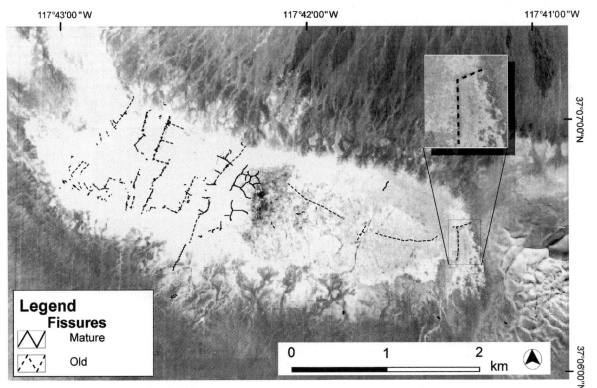

Fig. 5. Eureka Playa, as mapped by Differential GPS/GIS in April, 2002 layered on a digitized, georeferenced aerial photograph dating from 1947. The inset shows the location of an "irrigation ditch" as reported by Patch (1951).

nected fissures in the central part of the playa (which display roughly 120° intersections), all others are oriented in orthogonal or sub-orthogonal networks. Similar rectangular macropolygons have been reported on many other dry lakes, including Coyote Playa, CA (Hagar, 1966). Despite those who argue that cracks develop based on the laws of least energy, orthogonal patterns are far more prevalent than hexagonal or pentagonal networks (Neal, 1966).

Due to an advanced state of development (e.g., extensive deposition), some fissures can be detected only by examining the location of dead bushes on the playa surface. Since the perimeters of isolated bushes and vegetated mounds were mapped and added to the GIS as polygon features, inspection of these polygons revealed that they define an additional 54 lineaments (minimum length: 11 m; maximum length: 580 m; mean length: 148 m) which likely represent old fissures in the latest stages of their development. In some cases, individual vegetated patches were separated from others by as much as 100 m, and so it is uncertain whether these are vestiges of a contiguous feature; however, they were congruent with the 90° blueprint expressed by most younger, more distinct fissures.

One linear vegetation system of particular interest consists of two sub-orthogonal lines at the eastern margin of the playa in close proximity to the dune

field (Fig. 5). It was reported by Patch (1951) that this feature was an irrigation ditch constructed by an ancient (~1000 A.D.) indigenous community. It had been reported previously that the Paiute Indians constructed canals and sloughs in Owens Valley to divert water from Eastern Sierra streams to uncultivated plots of bushes and grasses that were desirable for their edible seeds (Steward, 1929). Patch visited Eureka Valley to observe the sand dune, but he was immediately struck by the rectangular pattern inscribed in the dry lakebed. At the time of his research, the furrow was approximately 15 cm deep; he described its longest segment as 375 yards (343 m) with three equidistant sub-parallel "laterals" branching from the main ditch.

The straightness of the gullies, the regularity of the pattern (and perhaps the absence of published geologic research of the playa) suggested that the system was laid out to plan, and hence was anthropogenic. Patch surmised that an aqueduct was excavated in the playa sediments at the easternmost edge of the playa; this "feeder canal" diverted water (Patch believed that the dunes collected rain that percolated through, reached a layer of impermeable lacustrine sediments, and emerged as a spring at the base of the dunes), which was used in the main irrigation ditch and lateral distributaries. Patch's hypothesis has persisted in the archaeological literature (Lawton et al.,

Fig. 6. Dried and dead vegetation are all that demarcate the inferred location of a fissure, as noted in Fig. 4 (inset).

1976); however, the current morphology and historic descriptions of this feature are consistent with playa fissure development found here and elsewhere, and so it is our conclusion that this feature is not man-made, confirming an earlier conclusion by Smith (1978).

In 2002, the "ditch" and "feeder canal" were no longer visible as recessed features, but instead were discernable by the existence of vegetation arranged in an orthogonal pattern (Fig. 6). The main extent measures 347 m, which is consistent with Patch's figure

(343 m). In places, aeolian deposits have accumulated to heights of up to 0.5 m at the base of the aligned bushes.

The description of this feature as 15 cm deep, as reported by Patch in 1951, contrasting to the positive relief noted in 2002, provides a general timetable of macropolygon evolution. The crack was vegetated with sage and creosote in Patch's report, yet it was recognizable and sunken, and would have therefore been classified as a "young" fissure at the time of his report, using a consistent nomenclature. Although this

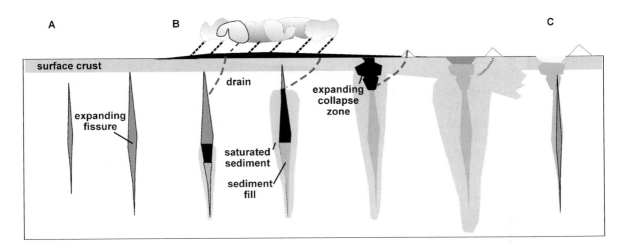

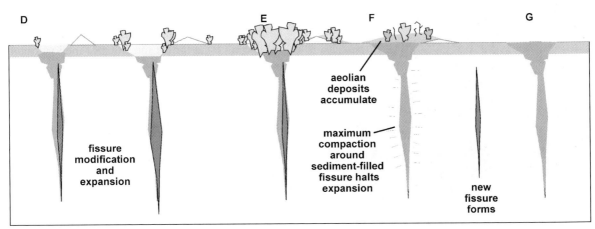

Fig. 7. Cross-section of inferred fissure development. (A) Desiccation at depth causes buried sediments to shrink, forming subsurface joints. (B) A flood event may trigger upward extension of the subsurface crack, forming a new fissure; water may flow into the joint, and if pressure increases, form a spring mound at the surface. (C) After further desiccation occurs, the spring mound may become inactive, and the fissure may drain. (D) Plants become established as the crevasse evolves into a reservoir for sediments and water. (E) Young fissures are characterized by a proliferation of vegetation. (F) At maturity and old age, former fissures may be recognized only by the lifeless bushes that once lined their extents. (G) Weathering processes erode the dead vegetation; erosion fills in the fissure; deflation scours the surface leaving a scar, or "healed fissure".

feature was clearly several years old in 1951, and it has not yet advanced to a "healed" state, these comparisons give further evidence to the decadal longevity of surface expressions of desiccation features on desert playas.

6. A general model for fissure development

Fig. 7 shows a schematic of the inferred evolution of fissures and related positive-relief features. Typically, new surface fissures and pits open and expand as floodwaters infiltrate into the playa sediments. The water-softened surficial layer collapses, exposing a linear pre-existing desiccation cavity below; sediment exposed in the fissure at depth may still be dry. Fissures likely extend for many meters in the subsurface, but they fill in with sediment freed from collapsing walls or carried in by flowing water. Water draining from the surface and along passages in the subsurface collects in pits. As water soaks in, clays expand; ultimately, sidewalls collapse when clays dry and shrink. The expansion of clays or hydrostatic

Fig. 9. The new North Panamint fissure pictured in Fig. 3, as it appeared in November, 2001.

head pressure from surrounding alluvial fans may be the driving force for the formation of mud volcanoes or spring mounds (Fig. 8) where fissures are not exposed to the surface. Some mounds may also represent sand boils: features commonly associated with earthquake liquefaction. However, not all sand boils form from earthquakes. In Fremont Valley, California, Holzer and Malcom Clark (1993) described sand boils associated with a 180-m-long sand dike that formed when sediment-laden surface runoff was intercepted along the upslope part of a 500-m-long pre-existing ground crack. Flood water flowed horizontally in the crack, and then flowed upward in the downslope part of the crack where it deposited sediment as a sand boil on the land surface. Hence, some mounds may form as secondary features, once a fissure has developed.

After several wetting and drying cycles (which may span many years), fissures become host to a

Fig. 8. A mud volcano on North Panamint Playa.

desert plants (salt bush, Russian thistle, grasses, etc.); the fissure is still considered to be in a youthful stage. As an example, Fig. 9 shows the condition of the fissure that formed in May 1999 two-and-one-half years later; incipient vegetation had become established, and surface characteristics were slightly degraded by subsequent inferred flash flood-related sheet wash erosion.

In a "mature" fissure (see Fig. 6), heavy brush contrasts the barren playa surface. The plant "ridge" buffers dust-bearing winds building up aeolian deposits. Thick bunches of brush mask both mounds and pits along the fissure. A fissure in this "healing" stage may act as a "cistern" following a flooding event for extended periods while the surrounding playa undergoes desiccation.

Old fissures are often recognized as an intermittent patchwork of dead bushes or salt-tolerant grasses. The mound-like aeolian deposits are highly degraded. Even later, deflation may degrade and expose passages of ancestral fissures. Healed fissures appear white in contrast to the darker patina of the surrounding crust (Fig. 10). All trace of plants and furrows are gone. Digital image processing methods can enhance and locate such vestiges.

Core records from North Panamint Playa obtained by Motts and Carpenter (1968) provide evidence that the piezometric surface underlying playa surfaces may be irregular, resulting in regions of preferred fissure development over time. The borehole that reached closest to the ground water table is in the location where a great concentration of young fissures was found three decades later. It is probable that desiccation took place at depth over the 30 subsequent years, ultimately expressing itself at the surface prior to 1999. Ground water recharge or differential meteoric infiltration may saturate areas of a playa's subsurface; after these areas dry, and as long as there is a triggering event, they can rupture years later.

7. Genetic fissure classes

The spatial orientation, morphology, and arrangement of macropolygons on North Panamint Playa suggests that there are at least two mechanisms by which fissures may form: one group of fissures is in close proximity to, and lie parallel or sub-parallel to the playa boundary. Other interior groups of fissures

Fig. 10. A healed fissure on North Panamint Playa.

are found in radiating patterns of roughly 120° angles. Both of these groups form from playa drying and shrinking, which may be caused by either natural desiccation (i.e., climate change), or lowering of ground water tables caused by aquifer pumping. A third, more anomalous type is found with neither of these characteristics. Two such fissures on North Panamint Playa, both striking east–west that measured 705 and 157 m, respectively—spanning the width of the playa, were found approximately 1.25 km north of State Highway 190. This rare orientation and absence of connectivity to a larger network suggests that these features may have been borne of another process. A witness to the opening of this fissure system was reported by Oreste Lombardi, who worked at the China Lake Naval Ordinance Test Station; it was coincident (or slightly after) a seismic event in the early 1920s (James T. Neal, personal communication). They are both characteristic of old fissures (i.e., decadal or older), which is consistent with this time frame of origin. While this particular fissure cannot be definitely attributed to an earthquake, it is likely (based on the positively correlated rupture) that a tremor can serve as a trigger, encouraging upward extension of cracking at depth (Neal, 1965c).

On Eureka Playa, macropolygon networks are strikingly rectangular in morphology; all fissures mapped in 2002 showed orientations and surface morphologies consistent with desiccation and contraction processes. There is no distinguishing sub-class of fissures that roughly parallel the playa boundary, and there were no "anomalous" fissures in suspect orientations. However, the three-category classification developed at North Panamint Playa may be applicable to many other lakebeds of the southwest. Furthermore, it is likely that linear playa features may be categorized by their spatial patterns. Scars generated by processes other than dehydration and attenuation may be easily distinguishable by their spatial discordance to the regular patterns of that playa's desiccation fissures (which usually intersect at uniform angles).

8. Conclusions

North Panamint and Eureka playas are similar in physical nature and feature morphology. Desiccation macropolygons display similar evolutionary phases, although angles of intersection are unique for each playa. There is a positive correlation between inferred fissure age and fissure length. Fissure extension may continue for decades, may be episodic, and typically occurs when a flood or tectonic event allows the less brittle surface to express tension at depth.

These cracks form from the "bottom up"; desiccation by water table reduction or climatic factors forms a fractured zone at depth. Surface rupture can occur after the surface becomes wet and pliable (as we witnessed on North Panamint Playa), or after a tectonic trigger allows tensional forces to be released toward the surface.

Tectonically induced surface ruptures on playas may be distinguished from contraction features in their unusual orientation, relative to a particular playa's "standard" or "expected" fissure pattern, although surface morphology may be indistinguishable. Differential GPS/GIS mapping can be used to analyze and classify specific lineaments toward this end.

Fissures persist on playas for decades, or even longer. Young fissures are characterized as distinct cracks; they later show evidence of gullying after sheet wash floodwaters cascade over their rims. After a year or so, plants become established, opportunistically availing themselves to the fissures' roles as moisture sinks. Plants continue to grow, roots collect aeolian sediment, eventually causing the original negative relief structures to be expressed by inverted topography. This progression continues for decades until the sediment mounds and remaining organic material is ultimately planed off, leaving only a scar of contrasting sediment along the original fissure lineament.

Linear anthropogenic features, in the form of straight rock alignments, are present on both playas, but there is no evidence on either North Panamint or Eureka playas that irrigation or agriculture took place on their surfaces in the past. Since a supposed "irrigation ditch" on Eureka Playa matches other natural features nearby (in spatial and morphological characteristics), it is undoubtedly of physical origin. Furthermore, although fissures may persist for decades, it is unlikely that such a crack could survive many centuries, as the case would have been if the

Eureka furrow were carved by paleo-Indians, unless new fissures are borne of old sutures. We saw no such evidence of such geographically constrained periodicity on either playa.

Acknowledgements

This research was partially funded by two California State University grants for Scholarly Research (1998–1999; 2001–2002), and was conducted under National Park permits DEVA199904598 and DEVA2002 SCI0003; we wish to thank Richard Anderson, Death Valley's Park Environmental Specialist, for his assistance and support.

Further, we would like to extend our gratitude to two colleagues, Lt. Col. James T. Neal (ret.), USAF and Dr. Kevin Schmidt (USGS, Earth Surfaces Processes Team, Menlo Park) who honored us by agreeing to review this manuscript: their comments and insights were invaluable.

Field work in Panamint Valley was conducted with the assistance of Eileen Brennan, Robin Dornfest, Jeannette Candau, and Jean Weekes. The Eureka Valley mapping project would have been impossible without the hard work of San José State University's spring 2002 Geology 137 class: Chanie Abuye, James Bonnie, Malia Burrows, Deanna Dawn, Linda Hambrick, Chris Jones, Chris Lay, Scott McPeek, Rebecca Promessi, and Christine Schelin. Stephanie MacDonald offered her tireless assistance to the mission.

Finally, the authors wish to thank Jenny Smith who brought the Eureka Playa controversy to our attention, and who generously gave us Ward C. Smith's extensive research notes.

References

Doty, G.C., Rush, F.E., 1985. Inflow to a crack in playa deposits of Yucca Lake, Nevada Test Site, Nye County, Nevada. U.S. Geological Survey Water Resources Investigations Report, pp. 84–4296.

Hagar, D.J., 1966. Geology and hydrology of Coyote Playa. In: Motts, W.S. (Ed.), Geology and Hydrology of Selected Playas in Western United States. Geology Department, University of Massachusetts, Amherst, 1970, pp. 66–107 (Final scientific report for the Air Force Cambridge Research Laboratories).

Holzer, T.L., Clark, M.M., 1993. Sand boils without earthquakes. Geology 21, 873–876.

Hunt, C.B., Robinson, T.W., Bowles, W.A., Washburn, A.L., 1966. Patterned ground. U.S. Geol. Survey Paper 494-B, 1–3, 104–108, 129–130. In: Neal, J.T. (Ed.), Playas and Dried Lakes. 1975. Benchmark Papers in Geology v. 20. Stroudsburg, Pennsylvania: Dowden, Hutchinson and Ross, Inc.

Lawton, H.W., Wilke, P.J., DeDecker, M., Mason, W.M., 1976. Agriculture among the Paiute of Owens Valley. Journal of California Anthropology 3 (1), 13–50.

Messina, P., Dornfest, R., Brennan, E., Stoffer, P., 1999. The paradoxical role of episodic flooding in the formation of giant desiccation polygons on North Panamint Playa, Death Valley, California. Abstracts with Programs-GSA 31, 7.

Morrison, H., 2002. Geophysical studies of the Lavic Lake Fault, Lavic Lake Playa, California. M.S. Thesis, Department of Geology, San José State University, San José, California.

Motts, W.S. (Ed.), 1970. Geology and hydrology of selected playas in Western United States. Final Scientific Report, prepared by the Geology Department, University of Massachusetts, Amherst for the Air Force–Cambridge Research Labs. 286 pp.

Motts W.S., Carpenter, D., 1968. Report of test drilling on Rogers, Coyote, Rosamond, and Panamint Playas in 1966. In: Neal, J.T. (Ed.), Playa Surface Morphology: Miscellaneous Investigations. Office of Aerospace Research, U.S. Air Force. AFCRL-68-0133, pp. 31–57.

Neal, J.T., 1965a. Airphoto characteristics of playas. In: Neal, J.T. (Ed.), Geology, Mineralogy, and Hydrology of U.S. Playas, 1965. Office of Aerospace Research, U.S. Air Force. AFCRL-65-266, pp. 149–176.

Neal, J.T. (Ed.), 1965b. Geology, mineralogy, and hydrology of U.S. playas. Office of Aerospace Research, U.S. Air Force. AFCRL-65-266. 176 pp.

Neal, J.T. (Ed.), 1965c. Tectonic influences on playa morphology. In: Neal, J.T. (Ed.), Geology, Mineralogy, and Hydrology of U.S. Playas, 1965. Office of Aerospace Research, U.S. Air Force. AFCRL-65-266, pp. 105–121.

Neal, J.T., 1966. Polygonal sandstone features in Boundary Butte Anticline area, San Juan County, Utah: discussion. Geological Society of America Bulletin 77, 1327–1330.

Neal, J.T. (Ed.), 1968a. Playa surface morphology: miscellaneous investigations. Office of Aerospace Research, U.S. Air Force. AFCRL-68-0133. 150 pp.

Neal, J.T., 1968b. Satellite monitoring of lakebed surfaces. In: Neal, J.T. (Ed.), Playa Surface Morphology: Miscellaneous Investigations, 1968. Office of Aerospace Research, U.S. Air Force. AFCRL-68-0133, pp. 131–149.

Neal, J.T., Motts, W.S., 1967. Recent geomorphic changes in playas of western United States. Journal of Geology 75 (5), 511–525.

Patch, R., 1951. Irrigation in east central California. American Antiquity 17, 50–52.

Smith, R.S.U., 1976. Late Quaternary pluvial and tectonic history of Panamint Valley, Inyo and San Bernardino Counties, California. Published dissertation; California Institute of Technology, University Microfilms, Ann Arbor, Michigan. 295 pp.

Smith, W.C., 1978. Contraction crack origin of "irrigation ditches" in Eureka Valley, California. Unpublished manuscript. 13 pp.

Steward, J.H., 1929. Irrigation without agriculture. Papers of the Michigan Academy of Arts, Sciences, and Letters 12, 149–156 (Ann Arbor).

Treiman, J.A., Kendrick, K.J., Bryant, W.A., Rockwell, T.K., McGill, S.F., 2002. Primary surface rupture associated with the Mw 7.1 16 October 1999 Hector Mine earthquake, San Bernardino County, California. Bulletin of the Seismological Society of America 92 (4), 1171–1191.

Available online at www.sciencedirect.com

Earth-Science Reviews 73 (2005) 323–346

www.elsevier.com/locate/earscirev

Base- and precious-metal deposits in the Basin and Range of Southern California and Southern Nevada—Metallogenic implications of lead isotope studies

Stanley E. Church [a,*], Dennis P. Cox [b,1], Joseph L. Wooden [b],
Joseph V. Tingley [c,1], Robert B. Vaughn [a,1]

[a] *U.S. Geological Survey, P.O. Box 25046, MS 973, Denver, CO 80225, United States*
[b] *U.S. Geological Survey, 345 Middlefield Rd., Menlo Park, CA 94025, United States*
[c] *Nevada Bureau of Mines and Geology, MS 178, University of Nevada, Reno, NV 89557, United States*

Accepted 1 April 2005

Abstract

Southern California and southern Nevada contain abundant lead–zinc deposits with strikingly different characteristics. On the west, the Darwin Terrane contains abundant Jurassic and Cretaceous intrusions surrounded by lead–zinc skarn and replacement deposits rich in pyrite and manganese minerals. The Tecopa Terrane is east of the Darwin Terrane and contains some lead deposits that are hosted by the Proterozoic Noonday Dolomite. These lead deposits have no consistent relation to igneous rocks; they contain mainly galena, and are devoid of pyrite and manganese minerals. Other skarn and vein deposits in the Ivanpah and Tecopa districts are more closely associated with igneous rocks. Mississippi Valley type lead–zinc deposits are present still farther to the east in the Goodsprings Terrane in Nevada. These deposits are hosted by breccias formed below the Mississippian–Pennsylvanian unconformity and are unrelated to igneous rocks.

Deposits in the Darwin Terrane have lead isotopic signatures that lie along a mantle–sediment mixing line indicating that they formed in a continental arc setting analogous to that for the plutons in the Sierra Nevada batholith [Chen, J.H. and Tilton, G.R., 1991. Application of lead and strontium isotopic relationships to the petrogenesis of granitoid rocks, central Sierra Nevada batholith, California. Geological Society of America Bulletin 103, 439–447]. Encroachment of this continental arc on the North American continent in the eastern part of the Darwin Terrane resulted in a lead isotopic signature that is like that of the strongly contaminated plutons [Chen, J.H. and Tilton, G.R., 1991. Application of lead and strontium isotopic relationships to the petrogenesis of granitoid rocks, central Sierra Nevada batholith, California. Geological Society of America Bulletin 103, 439–447]. Many deposits from the Inyo Mountains on east side of the Owens Valley have lead isotopic signatures that reflect this encroachment. To the east in the Tecopa Terrane, encroachment of the continental arc on the Mojave crust resulted in partial melting of 1.7 Ga amphibolite and granulite facies rocks to

---

* Corresponding author.
[1] Retired.

0012-8252/$ - see front matter. Published by Elsevier B.V.
doi:10.1016/j.earscirev.2005.04.012

produce the plutons and mineral deposits associated with plutons in this terrane. Lead from deposits in this terrane hosted in the Proterozoic Noonday Dolomite and associated rocks have a lead isotope signature that reflects hydrothermal circulation of fluids in the Mojave supracrustal rocks. The boundary between the Darwin Terrane and the Tecopa Terrane lies just west of the Ash Valley–Panamint Range Fault and is bounded by the Towne Fault on the north and the Garlock Fault on the south.

Lead isotopic data from the Goodsprings district in southwestern Nevada, east of the Tecopa Terrane, form a 1.45 Ga linear array that is indicative of Mississippi Valley type mineralization. Although we have no independent evidence of the timing of the tectonic events that formed these deposits, it is likely that all formed as a result of the Mesozoic collision of the Panthalassen crust with the North American continent.

Published by Elsevier B.V.

Keywords: lead isotope; lead–zinc deposits; Basin and Range; metallogeny

1. Introduction

Isotopic work in the Sierra Nevada Range (Chen and Tilton, 1991), the Mojave Desert (Wooden and Miller, 1990; Wooden and DeWitt, 1991; Rämö et al., 2002), and Nevada and Utah (Wooden et al., 1998, 1999) provided a new level of refinement of the geologic province boundaries identified in southern California by Zartman (1974), Kistler and Peterman (1973), and Kistler (1978, 1990). In conjunction with work conducted by the U.S. Geological Survey in the Death Valley area, we collected a suite of galena samples (Fig. 1) to evaluate how the isotopic composition of lead in mineral deposits reflected crustal structure in the Basin and Range east of the Sierra Nevada batholith. The deposit lead data are used to evaluate the boundaries between crustal blocks where lead isotope data from igneous rocks were not available. Work by John and Wooden (1990), Miller et al. (1990), Miller and Wooden (1994), Rämö et al. (2002), and Calzia and Rämö (2005—this volume) provided additional igneous lead and strontium isotopic data from this area. During a preliminary study of the Barstow-Ridgecrest Management Area of the Bureau of Land Management, we noted contrasting types of base- and precious-metal deposits in Inyo and San Bernardino Counties, California, and the study was expanded to evaluate lead isotope ratios from mineral deposits as tools for discriminating between deposit types and between tectonic terranes. In 1995, the study was extended to the east to include lead from Mississippi Valley type (MVT) deposits and occurrences in neighboring Clark County, Nevada.

The result is a 1300 km geologic–metallogenic–lead isotope transect extending from the Sierra Nevada batholith into southern Nevada. This transect reveals concomitant changes in geology, mineral deposit types, and lead isotope ratios. It also refines crustal boundaries previously described in Arizona (Wooden and DeWitt, 1991), Nevada (Wooden et al., 1998), and southern California (Wooden and Miller, 1990).

2. Methods

Samples of lead minerals and chalcopyrite were collected from 30 localities including mine dumps and prospect pits during 1994 and 1995. Joe V. Tingley collected galena from eight localities in Clark County, Nevada. Gary A. Nowlan and Sherman P. Marsh contributed additional samples from California and Nevada, collected as part of a geochemical study of the Mojave National Preserve. Lead isotope data from eight samples reported by Zartman (1974) also were used.

Galena samples were digested with concentrated ultrapure HNO_3 whereas mixed sulfide minerals were digested in aqua regia. Lead from the resulting solutions was isolated using anion exchange in bromide medium followed by a second step using anion exchange in chloride medium. The lead isotope compositions were determined at the U.S. Geo logical Survey in Denver using a 30.5-cm, 68° sector, solid-source, single-collector mass spectrometer of NBS design. Samples were analyzed using the single re-filament, silica-gel emitter tech-

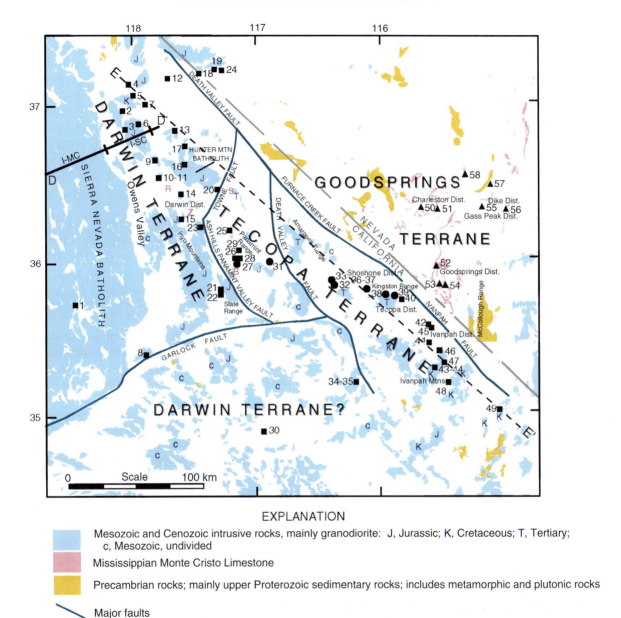

Fig. 1. Map of the geologic and metallogenic terranes in the Mojave region of California and Nevada showing selected geologic features and localities of samples and mine sites discussed in text. Geology is from Jennings (1961), Jennings et al. (1962), Smith (1964), Streitz and Stinson (1974), and Stewart and Carlson (1978) . Lines D–D′ and E–E′ show transects for data plotted in Figs. 2 and 6, respectively.

nique at 1250 ± 20 °C (Cameron et al., 1969). Analytical precision was estimated from internal data from individual analyses and from NBS standard reference materials. Reported lead isotope ratios are accurate to circa $\pm$ 0.05%/amu (e.g., $^{206}\text{Pb}/^{204}\text{Pb}$ of 18.xx $\pm$ 0.02). Corrections for isotopic fractionation were made from replicate analyses of the NBS 981 and 982 lead standards.

3. Regional geology

The study area extends from the eastern slopes of the Sierra Nevada Range to the Colorado River and encompasses three distinct geologic and metallogenic terranes. These are here named Darwin, Tecopa, and Goodsprings Terranes for the most prominent mining district in each (Fig. 1). These terranes are defined mainly by pre-Tertiary geology that are covered in large areas by Mesozoic and Tertiary volcanic rocks and strongly deformed by Tertiary extension.

Granitoid rocks mapped in the Tecopa Terrane give Proterozoic ages that range from 1.76 to 1.64 Ga and were intruded during the Ivanpah orogeny (Wooden and Miller, 1990). Supracrustal rocks contain Proterozoic and Archean zircons with Pb–Pb ages that range from 1.76 to 2.3 Ga. The rocks are a part of the Mojave crustal province.

Mesozoic metaluminous and peraluminous plutons have also been studied in this area and are characterized by elevated $^{87}Sr/^{86}Sr_i$ values that are generally between 0.710 and 0.726, but can be as high as 0.78–0.79 (Miller et al., 1990; Miller and Wooden, 1994; Rämö et al., 2002). These magmas were formed by partial melting of granulite facies Mojave crust at the base of the crust and contain neodymium isotopic signatures that reflect some mantle components in the magmas (Rämö and Calzia, 1998; Rämö et al., 2002). However, their lead and strontium signatures are totally dominated by the 1.7 Ga rocks of the Mojave crustal province (Miller et al., 1990; Chen and Tilton, 1991; Miller and Wooden, 1994; Rämö and Calzia, 1998; Rämö et al., 2002).

3.1. Darwin Terrane

This terrane is mainly underlain by folded and faulted Cambrian to Permian platform carbonate and clastic strata that are intruded by abundant Jurassic plutons ranging in composition from granodiorite to monzogranite and rarely to syenite. These plutons are responsible for skarn and replacement deposits like Darwin (Hall and MacKevett, 1962). To the south, the Slate Range is composed of Mesozoic sedimentary and volcanic rocks that are highly deformed, metamorphosed, and intruded by Jurassic granitic plutons and by dacite porphyry related to Tertiary volcanic rocks (Smith et al., 1968).

The Darwin Terrane is bounded on the west by the Jurassic–Cretaceous Sierra Nevada batholith and on the east is roughly coincident with faults in the Panamint Valley including the Panamint Valley, Ash Hills, and Towne Fault zones. The north and south boundaries are not defined in this paper and lie outside of the limits of Fig. 1. The Garlock Fault coincides with the southern limit of abundant skarn and replacement deposits typical of this terrane, but Paleozoic carbonates and Jurassic and undated Mesozoic plutons extend farther to the south and may contain buried, undiscovered base-metal resources.

The detailed study along a transverse cross section through the Sierra Nevada batholith by Chen and Moore (1982) and Chen and Tilton (1991) is indicted by the line D–D′ (Fig. 1).

3.2. Tecopa Terrane

In the Tecopa Terrane, Precambrian crystalline basement rocks are exposed and Proterozoic and lower Paleozoic sedimentary rocks are widespread. A 4500-m-thick section of unmetamorphosed Proterozoic rocks of the Pahrump Group form a small-scale aulacogen extending from the Kingston Range west to the southern Panamint Range (Wright et al., 1976). The Late Proterozoic Noonday Dolomite overlies this structure and hosts the base-metal deposits. The Mojave crustal rocks were intruded by Precambrian diabase dated at 1.1 Ga (Hammond and Wooden, 1990). Wooden and Miller (1990) dated granulite facies rocks from the Ivanpah and New York Mountains south of the Kingston Range that give $^{207}Pb/^{206}Pb$ ages of 1.76 to 1.64 Ga and suggest a protolith as old as 2.3 Ga. They named this crustal block the Mojave crustal province.

Jurassic plutons are present in the southern Panamint Range and Tertiary calcalkaline and peraluminous granite plutons are present along the east side of the terrane. Cretaceous granitic plutons including the Teutonia batholith are abundant in the southeastern part (Miller and Wooden, 1994; Rämö et al., 2002).

The terrane boundary on the west follows faults in the Panamint Valley and Towne Pass and on the east the boundary follows the Death Valley—Furnace Creek and Ivanpah Faults. The terrane extends north to the point where these faults converge. To the south, the western boundary is apparently offset by the left-lateral Garlock Fault. The southern boundary lies outside of the limits of our study area. To the southeast, the terrane probably extends into the McCullough Range area in southern Nevada. The aforementioned faults are much younger than the terranes and probably represent reactivation of deep-seated Precambrian structures.

3.3. Goodsprings Terrane

The Goodsprings Terrane extends from the Death Valley—Furnace Creek–Ivanpah Faults east into southern Nevada. It is underlain by Lower Proterozoic to Permian platform carbonate rocks, folded during the Cretaceous Sevier orogeny (Longwell et al., 1965). Metallogenically most significant of these rocks is the Mississippian Monte Cristo Limestone which hosts the mineral deposits of the Goodsprings district (Hewett, 1931) and other lead–zinc occurrences in the terrane. The terrane is marked by the nearly complete absence of igneous rocks. Dikes and sills of lower Mesozoic feldspar porphyry in the Goodsprings district and a small Tertiary rhyolite plug are the sole exceptions. The terrane extends north to the first appearance of Tertiary volcanic rocks at about latitude N36°42′ and southeast to the McCullogh Range where Precambrian metamorphic rocks and Cretaceous plutons are exposed.

4. Ore deposits and mineral occurrences

4.1. Darwin Terrane

Skarn and replacement deposits of the Darwin Terrane are described from north to south in Table 1, which lists deposits with recorded production. These are mainly lead mines with subsidiary silver, zinc and copper. Galena, sphalerite, chalcopyrite, and tetrahedrite are the main ore minerals. Tungsten was mined at Darwin (loc. 15, Fig. 1) from an early stage of skarn formation (Newberry, 1987). Pyrite and manganese minerals are abundant in the Modoc district. Some skarn deposits such as Darwin contain garnet and pyroxene with pyrrhotite and magnetite (Hall and MacKevett, 1962). Other deposits are simple replacement deposits with alteration shells of ocher-brown-

weathering ferroan dolomite exemplified by relations at the Lemoigne mine (loc. 20, Fig. 1), or jasperoid as noted at Zinc Hill mine (loc. Z, Fig. 1).

The deposits tend to be lenticular and subparallel to bedding but may form steeply plunging chimneys as at Cerro Gordo (loc. 10–11, Fig. 1) and Lemoigne mines (loc. 20, Fig. 1). At Santa Rosa (loc. R, Fig. 1), veins are the dominant deposit form (MacKevett, 1953). Host rocks range from Cambrian to Permian age. Deposits tend to form in limestone rather than dolostone beds, but do not show a preference for any particular age or stratigraphic interval. They are all believed to have been formed by hydrothermal fluids from Jurassic granodiorite plutons. At Ubehebe mine (loc. 17, Fig. 1), ores are associated with syenodiorite, monzonite and syenite intrusions (McAllister, 1955).

Ore deposits of the Slate Range (loc. 21, 22, Fig. 1) are in Precambrian, Paleozoic, and Mesozoic sedimentary and volcanic rocks that are highly deformed, metamorphosed, and intruded by granitic plutons of Cretaceous and older age and by dacite porphyry related to Tertiary volcanic rocks (Smith et al., 1968).

4.2. Tecopa Terrane

The Tecopa Terrane is characterized by a wide variety of small- to medium-sized gold and base-metal deposits. Lead–zinc–silver mineralization is present in two types of deposits: (1) replacement bodies hosted in the late Proterozoic Noonday Dolomite in the Panamint (loc. 27, 31, Fig. 1) and Amargosa Ranges (32, 33, 36–39, Fig. 1), and (2) skarn deposits of Pb, Zn, Cu, and W and polymetallic quartz veins (loc. 23, 25, 26, 28, 29, 40–44, 46–49, Fig. 1; Nowlan and Theodore, in press). Of these, only the Noonday, Grant, War Eagle, and Columbia mines in the Shoshone district (loc. 36, 37, Fig. 1), and the Queen of Sheba mine (loc. 31, Fig. 1) had significant base-metal production or resources (Table 1). These deposits had only about one third the production of the Darwin district.

Deposits hosted by the Noonday Dolomite share the following characteristics: (1) all are localized within the lower part of the Late Proterozoic Noonday Dolomite, (2) they have no consistent relation with igneous rocks of any specific age, and (3) they have much lower pyrite and manganese content compared to the ores related to Jurassic plutons in the Darwin

Table 1
Description of mineral deposits with recorded production by terrane [sample localities are in Fig. 1 and locations given in Table 2; NA, not available]

Map no.	Name	Description	Host rock	Nearest igneous rock	Production	References
Darwin Terrane						
10, 11	Cerro Gordo district	Galena, and minor sphalerite, tetrahedrite, and pyrite in lenses up to 14 m thick concordant with bedding in the host limestone, and in chimneys cutting across bedding.	Devonian Lost Burro Fm.	Granodiorite porphyry dike	21,000 metric tons of lead–silver bullion	Knopf (1918), Meriam (1963)
15	Darwin mines	Skarns contain abundant grandite garnet, minor diopside, sphalerite, galena, pyrite, pyrrhotite, and magnetite. Silver, antimony, bismuth and selenium are important trace elements. Scheelite ores are 20 Ma earlier than base-metal ores	Pennsylvanian and Permian Keeler Canyon Fm.	Jurassic batholith, Darwin stock dated at 174 Ma	1.6 Mmt ore containing 9% Pb, 5% Zn, 0.2% Cu, 390 g/t Ag and recoverable Au. Plus 100,000 mt of W ore	Anaconda files, Hall and MacKevett (1962), Newberry et al. (1991)
17	Ubehebe Mine	Oxidized lead silver replacement	Ordovician Ely Springs Dolomite	Jurassic	NA	McAllister (1955)
20	Lemoigne mine	Bedding-plane replacement and steeply plunging pipes of lead–silver ore	Cambrian Racetrack Fm.	Jurassic Hunter Mtn batholith	NA	Hall and Stephens (1963)
21	Ophir mine	Galena-bearing fissure zone	Precambrian limestone	Cretaceous, Jurassic	NA	Smith et al. (1968)
22	Gold Bottom (Copper Queen)	Replacement lenses of galena, sphalerite and chalcopyrite	Precambrian limestone	Cretaceous, Jurassic granodiorite	NA	Smith et al. (1968)
23	Modoc district, Minietta mine	Pb Zn replacement bodies with minor Zn, Cu, Au. Abundant Mn oxides (coronadite)	Devonian Lost Burro Fm.	Andesite dike swarm	NA	Hall and Stephens (1963)
R	Santa Rosa mine	Veins from 30 to 200 m long, 1 m thick.	Permian Owens Valley Fm.	Dikes of syenodiorite	33,000 mt of ore containing 5400 mt Pb and 13 mt Ag	MacKevett (1953)
Z	Zinc Hill mine	Sphalerite (light brown color) with minor galena, pyrite, and chalcopyrite in a gangue of calcite, jaspar, gypsum, and quartz in replacement bodies parallel to bedding and in veins following steep faults	Mississippian Perdido Fm.	Jurassic stock	NA	Hall and MacKevett (1962)
Tecopa Terrane						
27	Honolulu mine	Small lead ore bodies	Upper Proterozoic Noonday Dolomite		NA	
31	Queen of Sheba mine	Two galena replacement bodies, one of which lies on the hanging wall of a low-angle fault	Upper Proterozoic Noonday Dolomite	None	4000 mt ore containing 15.5% Pb, 0.5% Cu, 214 g/t Ag and 3 g/t Au	Morton (1965)

No.	Mine	Description	Host rock	Igneous rock	Production/resource	References
32	Ibex and Paddy's Pride	Fracture-controlled galena veins	Upper Proterozoic Noonday Dolomite	None	NA	
33	Gladstone	Copper oxide minerals and quartz veinlets in arkose and dolomite breccia	Possibly Kingston Peak Fm. underlying Noonday Dolomite	None	NA	
36, 37	Noonday mine in Shoshone district which includes Grant, War Eagle, and Columbia mines	Fracture-controlled bodies of galena and minor sphalerite subparallel to bedding	Upper Proterozoic Noonday Dolomite	None	540,000 mt Pb–Zn ore	Anaconda files, Carlisle et al. (1954)
39, 40	Silver Rule and Blackwater-Pioneer mines	Fracture-controlled galena veins	Upper Proterozoic Noonday Dolomite	Tertiary granodiorite	NA	
38	Chambers mine	Fracture-controlled galena veins	Upper Proterozoic Noonday Dolomite	Tertiary granodiorite	NA	
41	Mohawk mine	Copper-rich skarn	Precambrian dolomite	Cretaceous Teutonia batholith	NA	Hewett (1956)
45	Colosseum gold mine	Breccia pipe with clasts of silicified limestone in a fine-grained matrix with chalcopyrite and sphalerite		Cretaceous Teutonia batholith	NA	Sharp (1984), Hodges (in press)
46	Teutonia mine	Quartz vein with galena			NA	Hewett (1956)
Goodsprings Terrane						
50	Ada and Edith claims	Lenses and pods of galena associated with white carbonate veins and banded "zebra rock"	Cambrian Nopah Fm.	None	8 mt Pb and 342 g Ag	
51	Lucky Strike mine		Mississippian Monte Cristo Lms.		NA	
52	Potosi	Oxidized Zn and Pb in irregular, roughly stratabound, breccias	Mississippian Monte Cristo Lms.	None	NA	Hewett (1931), Bray (1983)
53, 54	Yellowpine mine	Oxidized Zn and Pb in irregular, roughly stratabound, breccias	Mississippian Monte Cristo Lms.	Lower Mesozoic granite porphyry	Goodsprings dist produced 98,900 mt Zn, 43,600 mt Pb plus Cu, Ag, Co, V, Pd, and Pt	Hewett (1931), Bray (1983)
55	June Bug	Lenticular pods of zinc and lead minerals	Mississippian Monte Cristo Lms.	None	907 mt ore containing 31% Zn and 0.8% Pb	
57	White Caps prospect	Thin lenses of galena in shattered quartzite along a steep reverse fault	Cambrian Wood Canyon Fm., Precambrian Sterling Quartzite	None	NA	
58	Joe May prospect	Thin stringers of galena cementing carbonate fault-breccia	Mississippian Monte Cristo Lms.		NA	Guth (1981), Guth et al. (1988)

Terrane. Stope walls and mine waste piles lack the intense iron staining typical of the Jurassic polymetallic vein deposits. Ores hosted by the Noonday Dolomite contain galena, less abundant sphalerite, and minor chalcopyrite and pyrite in lenses and tabular bodies strongly controlled by faults; gangue minerals are chiefly coarsely crystalline dolomite and minor quartz. In the areas surrounding mines and prospects, the Noonday Dolomite contains coarse dolomite crystals in abundant irregular masses and short, curved veinlets that are unrelated to faulting. Less commonly, abundant small (5–10 cm wide) masses of quartz, shaped like tubes with a filling of dolomite, are arranged perpendicular to bedding and about 10 to 30 cm apart. The tubes are well developed around the Queen of Sheba deposit (loc. 31, Fig. 1; Morton, 1965; Cloud et al., 1974). These features suggest that a pre-ore-stage permeability of diagenetic origin may have contributed to fluid movement and ore deposition. Hall (1984) determined that the isotopic composition of sulfur from these deposits had a small, limited range in δ^{34}S from + 4 to + 7.

The Noonday Dolomite overlies an east–northeast-trending aulacogen that includes red sandstone and conglomerate in the Upper Proterozoic Pahrump Group (Wright et al., 1976). Saline fluids circulating through them may have transported the metals (Hall, 1984). Rocks of the Pahrump Group are considerably older than the Noonday Dolomite however, and were folded and faulted before Noonday deposition (J.P. Calzia, oral communication, 2003). This suggests that permeability of the sediment of the Pahrump Group was low during Noonday time and that mineralizing fluids probably followed faults. J.P. Calzia (oral communication, 1995) observed that sulfide-bearing fractures in the Chambers mine (loc. 38, Fig. 1) extend into faults that cut dikes that are apparently related to Tertiary volcanic rocks. These observations plus lead isotopic data (see below) cast doubt on the Precambrian diagenetic origin of deposits hosted in the Noonday Dolomite.

Two types of polymetallic veins cutting schistose rocks in the Panamint district (loc. 26, 28–29, Fig. 1) were described by Murphy (1930): (1) quartz–galena–pyrite–chalcopyrite–tetrahedrite veins, rich in silver, in which quartz is abundant and milky white in color, and (2) pyrrhotite–pyrite–quartz veins in which the quartz is granular and glassy. Siderite, chalcopyrite and mar-

casite also are present in the latter. There is no recorded production from prospects in the Panamint district.

The Skidoo district (loc. S, Fig. 1) contains low-sulfide auriferous quartz veins that appear to be genetically related to a peraluminous Skidoo granodiorite pluton of Cretaceous age (70 Ma, Rämö et al., 2002). These veins are similar to veins in major producing deposits at the Mineral Ridge (Silver Peak) district, 160 km to the north in Nevada.

The Briggs mine (loc. B, Fig. 1) and nearby associated deposits are apparently related to Miocene low-angle thrust (detachment) faults, and consist of gold, quartz, pyrite, Ca–Mg–Fe carbonate minerals, sericite, and chlorite disseminated in felsic gneiss. A Miocene basalt dike that intrudes the fault also is mineralized (S.D. Luddington, written communication, 1998). No samples from Panamint or Skidoo districts or Briggs mine were obtained for this study.

The Ivanpah district, a cluster of skarn, vein and replacement base- and precious-metal deposits (loc. 43, 45, Fig. 1) in the Ivanpah Mountains related to the Cretaceous Teutonia batholith was described by Hewett (1956). Rämö et al. (2002) report both Jurassic and Cretaceous ages for igneous rocks in the Ivanpah Mountains. Additional geochemical studies of mineral occurrences in the Ivanpah Mountains are described by Nowlan and Theodore (in press). These deposits are present in Precambrian metamorphic rocks, Paleozoic carbonate rocks, or as veins in the batholith itself.

4.3. Goodsprings Terrane

The Goodsprings Terrane contains mainly zinc–lead deposits (loc. 52–54, Fig. 1), but copper and gold deposits are also important. This terrane includes the Goodsprings, Gass Peak, Charleston, and Dike mining districts, located within a 60- to 75-km radius of Las Vegas, Nevada. Zinc and lead deposits, here classified as MVT, have been mined in each of these districts.

At the Goodsprings (loc. 53–54, Fig. 1), Gass Peak (loc. 55, Fig. 1) and Dike (loc. 56, Fig. 1) districts, and in Lucky Strike Canyon in the eastern Charleston district (loc. 50, 51, Fig. 1), mineralization is present mainly in the Anchor, Bullion, and Yellowpine members of the Mississippian Monte Cristo Limestone. Most zinc–lead ore bodies in these districts are localized below the unconformable contact between Monte Cristo Limestone and the overlying Pennsylvanian–Permian Bird Spring Formation. The pre-Pennsylvanian unconformity has been interpreted as an interval of erosion and karstification by Bray (1983).

Other zinc–lead prospects north of the Goodsprings district are localized in various stratigraphic and structural positions (Smith and Tingley, 1983a,b; Tingley et al., 1993). No igneous rocks are present in the vicinity of these occurrences. In the Gass Peak district, the deposits are within a zone of brecciation near the base of the massive, thick-bedded Bullion member of the Monte Cristo Limestone within envelopes of banded light and dark dolomite (zebra rock) and both the zebra banding and replacement mineralization conform to bedding in the host carbonate rock. Zebra rock is also present at the Ada and Edith claims (loc. 50, Fig. 1) in the Charleston district where lead–zinc ores are hosted in the Cambrian Nopah Formation. Quartzite of the Upper Proterozoic Wood Canyon Formation hosts lead veins in the White Caps prospect in the Dike district (loc. 57, Fig. 1; Sylvester, F.J., 1941, White Caps claims, Clark County, Nevada: unpublished report, U.S. Bureau of Mines files, 3 pp.).

According to Bray (1983), the zinc–lead deposits at Goodsprings formed within irregular roughly stratabound breccias which are commonly vertically zoned. The basal zones are characterized by abundant insoluble residue probably inherited from pre-mineral karstification. The upper parts of the breccias are open-textured and are cemented with ore and gangue. Ore mined was mainly hydrozincite with associated hemimorphite, cerussite, and other oxide zinc and lead minerals. Original unoxidized ores are presumed to have consisted predominantly of sphalerite and galena and these minerals occurred in parts of the mined ore bodies.

Although dikes of granite porphyry have intruded rocks close to the zinc–lead ore bodies, Hewett (1931) did not present data on the relative ages of dikes and lead–zinc ore. It is probable that the zinc–lead ores are older than the dikes and their related gold–copper mineralization. The gold, silver, and copper deposits are present within brecciated fault zones and in association with these granite porphyry dikes (Hewett, 1931). Gold production has come from vein-like deposits cutting hydrothermally altered dikes of granite porphyry. The copper deposits, which also contain minor amounts of cobalt, vanadium, platinum, and

palladium, are in veins, and all but two lie along high-angle shear zones which cut across bedding (Bray, 1983). The porphyritic intrusive rocks, which crop out at several locations in the central Goodsprings district, have been dated at 200 to 190 Ma (Carr et al., 1986; Garside et al., 1993).

5. Understanding and interpreting lead isotopic data

Four isotopes of lead exist in nature, three of which change as a function of time through the radioactive decay of uranium and thorium: ^{206}Pb is the daughter product of decay of ^{238}U, ^{207}Pb is the daughter product of decay of ^{235}U, and ^{208}Pb is the daughter product of decay of ^{232}Th. However, ^{204}Pb has no radioactive parent. Thus, the isotopic composition of lead in rocks in the earth's crust (e.g., ^{206}Pb/^{204}Pb, ^{207}Pb/^{204}Pb, and ^{208}Pb/^{204}Pb) changes regularly with time as uranium and thorium undergo radioactive decay. Crustal lead is characterized by elevated ^{238}U/^{204}Pb (mu) and ^{232}Th/^{204}Pb (kappa) values that reflect crustal evolution (Stacey and Kramers, 1975). During a mineral deposit-forming event, lead is separated from the parent uranium and thorium isotopes and the lead isotopic composition of the hydrothermal fluid is "frozen" into the sulfide minerals, usually in galena, within the mineral deposit. This mineral-deposit lead isotopic signature will eventually differ from that of the host rocks because the lead in the host rocks continues to change with time whereas the lead in the mineral deposit remains fixed. Lead in the continental crust differs substantially from that in the oceanic crust, which is characterized by lower ^{238}U/^{204}Pb (mu) and ^{232}Th/^{204}Pb (kappa) values and much lower lead concentrations (e.g., Church and Tatsumoto, 1975).

6. Implications of the lead isotope studies of the Sierra Nevada batholith on metallogenic terranes

Studies of the mineralogy and petrography of mafic minerals in plutons of the Sierra Nevada batholith lead Ague and Brimhall (1988) to propose a classification of plutons ranging from weakly contaminated to strongly contaminated plutons from west to east. They showed that the chemistry of the mafic minerals biotite and amphibole changed systematically across the batholith and proposed that these changes correlated with processes of magma formation. They correlated their work with previous work by Kistler and Peterman (1973) and Kistler (1978) who showed that the $^{87}Sr/^{86}Sr_i$ value in rocks increased from west to east across the Sierra Nevada batholith. Chen and Tilton (1991) subsequently used lead isotope data from feldspar and whole rock samples used in earlier geochronologic studies (Chen and Moore, 1982) to evaluate this pluton classification. They found a positive correlation between the initial lead isotopic data and the strontium isotopic data, however there was no correlation of the initial lead isotope values with the ages of the plutons. Furthermore, the data showed no positive correlation with the four-fold classification of plutons in the Sierra Nevada batholith proposed by Ague and Brimhall (1988) if the lead isotope data are plotted as groups of plutons by class. However, the lead isotope data, if plotted along a transect through the Sierra Nevada batholith (line D–D′, Fig. 1), show a strong correlation with distance (Fig. 2). Plutons showing the least effect of incorporation of crust on their mafic mineral chemistry and lead isotopic signature lie to the west whereas those plutons most affected by crust are on the east. The boundary between the WC and the SCR plutons at 35 km (Fig. 2) marks the transition from metaluminous to strongly peraliminous plutons (Fig. 5C, Ague and Brimhall, 1988) that contain elevated lead isotopic values (Fig. 2, Chen and Tilton, 1991) but $^{87}Sr/^{86}Sr_i < 0.706$. There is a sharp break in the lead isotope data between those plutons that lie on either side of the major crustal shear zone defined by Kistler (1990). This shear zone is nearly coincident with the boundary between the strongly contaminated and reduced (SCR) plutons and the moderately contaminated plutons (MC) of Ague and Brimhall (1988) at a distance of 55 km (Fig. 2). This boundary is approximately coincident with the west edge of Fig. 1 (Kistler, 1990, Fig. 3) and with the $^{87}Sr/^{86}Sr_i > 0.706$ isopleth that includes all moderately contaminated plutons in the Sierra Nevada batholith (Fig. 1). Plutons in the moderately contaminated suite (MC) are weakly metaluminous (Fig. 5C, Ague and Brimhall, 1988) and have lower lead isotopic values (Fig. 2, Chen and Tilton, 1991) and $^{87}Sr/^{86}Sr_i > 0.706$. Plutons

from the strongly contaminated (SC) group lie east of the Owens Valley (Fig. 1), are peraliminous (Fig. 5C, Ague and Brimhall, 1988), and have intermediate lead isotopic values and $^{87}Sr/^{86}Sr_i > 0.706$.

When plotted in conventional lead isotope diagrams, the initial lead isotope data from plutons show several important features:

(1) The lead isotope data from plutons of the Sierra Nevada have lead isotope ratios that plot along a mixing array between oceanic mantle values and sediment lead isotope values shown as the Sierra Nevada trend in Fig. 3. The strongly contaminated and reduced (SCR) plutons lie at the right and top of the lead-isotope mixing diagram indicating sediment contamination during partial melting to form these plutons. These changes are systematic along the transect D–D′ (Fig. 2). In this sense, the Jurassic–Cretaceous arc volcanism they represent is analogous to the Cascade volcanic arc forming today (Church, 1973; Church and Tilton, 1973).

(2) For plutons of the Sierra Nevada that have $^{87}Sr/^{86}Sr_i > 0.706$, the $^{208}Pb/^{204}Pb$ vs. $^{206}Pb/^{204}Pb$ values have a shallow slope (0.36 ± 0.06) whereas lead isotope data for the feldspars from these plutons have a slightly lower slope (0.34 ± 0.06).

(3) The lead isotope data for the strongly contaminated plutons containing $^{87}Sr/^{86}Sr_i > 0.706$ have lower values of $^{206}Pb/^{204}Pb$ which Chen and Tilton (1991) argue is evidence of assimilation of lead from granulite facies rocks in the lower crust during partial melting.

The previous studies by Ague and Brimhall (1988) and by Chen and Tilton (1991) provide telling insight into the processes active in the formation of magma in continental-margin arcs. The processes controlling the F/OH ratios, which Ague and Brimhall (1988) mapped in biotite and amphibole of the plutons in the strongly contaminated and reduced group appear to represent assimilation or partial melting of subducted pelitic sediment derived from the older Precambrian crust as supported by the high $\delta^{18}O$ values from these plutons (Kistler, 1990). Lead and strontium isotope signatures of the plutons are also indicative of processes in the source of the magmas, not of shallow contamination events during pluton emplacement in the shallow

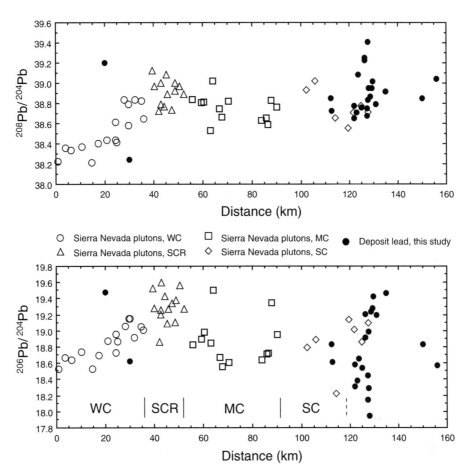

Fig. 2. Crustal transect along line D–D′ showing lead isotope data (Chen and Tilton, 1991) from different classes of plutons discussed in text. Sierra Nevada plutons divided into four groups following work by Ague and Brimhall (1988), from west to east: WC, weakly contaminated plutons; SCR, strongly contaminated and reduced plutons; MC, moderately contaminated plutons; and SC, strongly contaminated plutons. Boundary between SCR and MC rocks is the major crustal shear zone defined by Kistler (1990). The western boundary of Fig. 1 is at about 55 km, coincident with the SCR–MC transition. Deposit lead from Darwin Terrane is from this study (Table 2).

crust. Kistler (1990) argues that the western Panthalassen crust, through which the weakly and some moderately contaminated plutons were intruded, is thin (< 5 km) whereas the western Mojave crust is thick (> 34 km). Lead concentrations in the Sierra Nevada plutons increase from west to east and remain high in the strongly contaminated plutons (Chen and Tilton, 1991, their Fig. 2). Thorium concentrations also remain high in these plutons, but uranium concentrations are depleted, which results in low observed $^{206}Pb/^{204}Pb$ values and moderate to high $^{208}Pb/^{204}Pb$ values reflected in the SC plutons (Fig. 3).

Subsequent to these studies, additional lead and strontium isotopic data have become available from

Mesozoic granitic plutons from the Mojave crustal province (Miller et al., 1990; Miller and Wooden, 1994; Rämö et al., 2002; J.L. Wooden, unpublished data, 2003). These data are shown along with the data from mineral deposits in the figures below. Miller et al. (1992) documents the presence of amphibolite to granulite facies xenoliths in plutonic rocks from the region. The protoliths of these xenoliths were pelite, quartz-rich arenite, and greywacke as well as mafic and felsic intrusive rocks. Cretaceous plutons contain zircons with cores that have 1.4 to 1.8 Ga inheritance, which indicates that the rocks were formed by partial melting of old crust. Studies of the lead isotopic compositions of supracrustal rocks also are shown

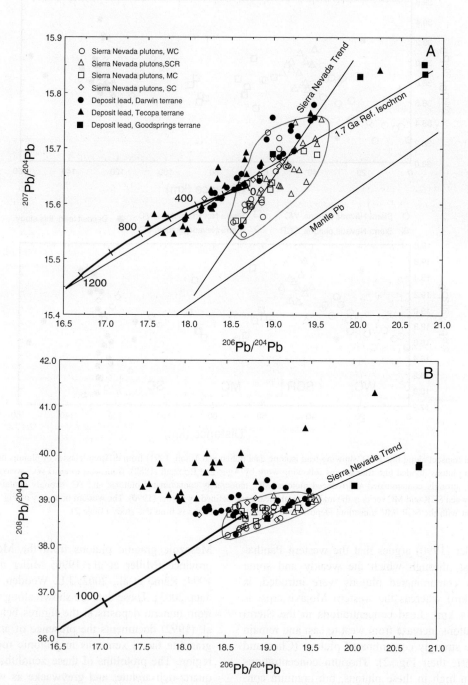

Fig. 3. $^{207}Pb/^{204}Pb$ vs. $^{206}Pb/^{204}Pb$ (A) and $^{208}Pb/^{204}Pb$ vs. $^{206}Pb/^{204}Pb$ (B) diagrams showing the lead isotopic signatures of plutons from the Sierra Nevada batholith (Chen and Tilton, 1991) and mineral deposits (Table 2) from the Darwin, Tecopa, and Goodsprings Terranes. The Stacy–Kramers model geochrons for evolution of crustal lead and the mantle lead isochron from the N.E. Pacific Ocean are shown for reference (Stacey and Kramers, 1975; Church and Tatsumoto, 1975). Sierra Nevada plutons divided into four groups following work by Ague and Brimhall (1988): WC, weakly contaminated plutons; SCR, strongly contaminated and reduced plutons; MC, moderately contaminated plutons; and SC, strongly contaminated plutons. Boundary between SCR and MC rocks is the major crustal shear zone defined by Kistler (1990).

where appropriate and are from J.L. Wooden (unpublished data, 2003).

7. Lead isotope results from mineral deposits

The lead isotope data from 58 samples from 53 mineral deposits are in Table 2 and Figs. 2 and 3. The lead isotope data generally follow Doe and Stacey's (1974) premise that deposits within a mining district have a common lead isotope signature within some small range of variation. Deposits in the Darwin, Tecopa, and Goodsprings Terranes have $^{208}Pb/^{204}Pb$ ratios that fall into distinct groups and arrays (Fig. 3). The isotope data from deposit leads from the Darwin Terrane indicates the influence of the encroachment of the Sierra Nevada continental arc on older crust whereas the deposit lead from the Tecopa Terrane shows the influence of the Mojave crustal province. None of the deposits have lead isotope signatures that would indicate we have sampled any of the western Great Basin crust as defined by Wooden et al. (1998). Although this terrane extends into eastern California, the southern boundary of the Great Basin province is north of the study area.

7.1. Darwin Terrane

Deposits in the Darwin Terrane exhibit some lead isotope ratios that can be correlated with the Sierra Nevada igneous lead signatures (Figs. 2 and 4). Data from samples within the area of the MC and SC plutons (Figs. 1 and 2) plot within the outline of the Sierra Nevada lead isotope field (Fig. 4). Lead from mineral deposits located within the Sierra Nevada batholith–the Kernville deposit (loc. 1, Fig. 1) and the Copper Basin Group (loc. 8, Fig. 1)–reflect the lead isotope compositions of WC and SCR plutons in the western segment of the Sierra Nevada batholith. However, the long distances that both have been projected to line D–D′ may have distorted their spatial magmatic association so they do not match well with the pluton data on the distance axis (Fig. 2). The character of the lead isotope array defined by lead from mineral deposits within the Sierra Nevada is like that observed in the Cascade Range (Church et al., 1986) where deposit leads matched with those measured in the volcanic rocks. The paucity of base-metal

deposits in the Sierra Nevada batholith may be one of the level of erosion. Undoubtedly, porphyry copper and base-metal vein deposits were present in the batholith after formation, but the deep level of erosion has preserved only a few deposits hosted in roof pendants.

The plutons from the Inyo Mountains east of the Owens Valley, on the east side of the Darwin Terrane (Fig. 1), were classified as strongly contaminated by both Chen and Tilton (1991) and Aguc and Brimhall (1988). It is apparent when examining all the lead isotopic data from deposits east of the Owens Valley that there are additional factors that have affected the lead isotopic signature in some of these deposits (Figs. 2 and 4). As shown in Fig. 2, the deposit lead data have a wide spread in $^{206}Pb/^{204}Pb$ values over a short distance (120–140 km). As indicated in Fig. 4, lead from some of the mineral deposits and the plutons (Fig. 2) in the Darwin Terrane east of the Owens Valley indicate that they have been derived from older sources. The data plot along the 1.7 Ga isochron (Fig. 4A) and have slightly elevated $^{208}Pb/^{204}Pb$ values relative to $^{206}Pb/^{204}Pb$ (Fig. 4B) reflecting the presence of the Mojave crustal province at depth. Lead isotope data from mineral deposits east of the study area have been identified from this crustal province (Wooden and Miller, 1990; Wooden and DeWitt, 1991), and are characterized by a pronounced enrichment of thorogenic lead. Samples from deposits in the east side of the Inyo Mountains and the Slate Range (loc. 4–6, 9–11, 14, 15, 21–23, Fig. 1) show the effects of the Mojave crust whereas deposits from the west side of the Inyo Mountains (loc. 2, 3, Fig. 1), from deposits associated with the Hunter Mountain batholith (loc. 13, 16. 17, 20, Fig. 1), and from the cluster of prospects in the Gold Mountain area east of the extension of line D–D′ on the California–Nevada border (12, 18, 19, Fig. 1) apparently were not generally affected by the Mojave crust. The discontinuous nature of the lead isotope data from this area of the Inyo Mountains we interpret to reflect the strike-slip rifting of old crust during the pluton emplacement process as discussed by Kistler (1990).

South of the Garlock Fault, lead from mineral deposits (loc. 30, 34–35, Fig. 1) also have a $^{208}Pb/^{204}Pb$ vs.$^{206}Pb/^{204}Pb$ signature that reflects this older thorogenic lead-rich Mojave crustal province and are consistent with the $^{87}Sr/^{86}Sr_i > 0.706$ (Kistler, 1990).

Table 2

Lead isotopic data from the Basin and Range of California and southern Nevada [gn, galena; sp, spalerite; cp, chalcopyrite; dol, dolominte; lms, limestone; cal, calcite; qtz, quartz]

Map no.	Name	Comments	Method	$\frac{^{206}Pb}{^{204}Pb}$	$\frac{^{276}Pb}{^{204}Pb}$	$\frac{^{208}Pb}{^{204}Pb}$	Latitude	Longitude
1	Kernville	Gn	3FI-N[a]	19.479	15.755	39.197	35 44	118 27
2	S. Badger Flat	Gn in W skarn	Gel-N	18.614	15.654	38.724	36 57 55	118 05 20
3	Whiteside mine	Gn in jasperoid	Gel-N	18.835	15.667	38.849	36 50 05	118 04 55
4	Overholtz mine	Gn in brecciated qtz vein parallel to bedding in argillite	Gel-N	19.203	15.741	39.246	37 08 29	118 02 38
5	Opal mine	Gn and sp in calcite from mine dump	Gel-N	18.915	15.716	39.225	37 04 07	117 59 49
6	Green Monster mine	qtz vein with pyrite	Gel-N	18.381	15.609	38.712	36 53 37	117 57 22
7	Morning Star mine # 2	Gn in brecciated qtz vein	Gel-N	18.987	15.755	39.406	37 00 48	117 54 43
8	Copper Basin Group	Cp in qtz vein in granodiorite	Gel-N	18.621	15.559	38.238	35 24 50	117 53 00
9	Big Silver mine	Gn in qtz vein	Gel-N	18.649	15.635	39.087	36 39 20	117 49 20
10	Cerro Gordo	Gn and qtz veins in lms	Gel-N	18.307	15.619	38.777	36 32 30	117 47 30
11	Cerro Gordo	Gn and qtz veins in lms	3FI-N	18.584	15.630	38.647	36 32 24	117 47 24
12	Unnamed prospect	Gn in qtz veins in breciated zone in Zebrisky qtzite	Gel-N	19.466	15.779	38.916	37 10 36	117 43 10
13	Unnamed prospect	Disseminated oxidized sulfides, Last Chance Range, Calif.	Gel-N	19.279	15.722	38.947	36 51 14	117 39 35
14	Silver Ried	Gn, cerargyrite, and anglesite in replacement deposits in Pz lms	3FI-N[a]	18.446	15.629	38.746	36 26 30	117 36 30
15	Darwin	Gn in silicified lms; intrusive nearby	3FI-N[a]	18.535	15.626	38.759	36 17 00	117 36 00
16	Lippencott mine	Gn in limestone	Gel-N	19.235	15.695	38.865	36 38 03	117 34 55
17	Ubehebe mine	Gn in limestone	Gel-N	19.426	15.751	39.015	36 45 10	117 34 50
18	Unnamed prospect	Oxidized sulfides in qtz veins; west of Gold Mtn. Area, Nev.	Gel-N	18.828	15.642	38.853	37 12 17	117 28 02
19	Unnamed prospect	Oxidized sulfides in qtz veins; Gold Mtn. Area, Nev.	Gel-N	18.566	15.646	39.043	37 15 02	117 20 33
20	Lemoigne mine	Gn in red-brown dolomite	Gel-N	19.194	15.732	38.793	36 28 10	117 18 30
21	Ophir mine	Gn and qtz in dolomite	Gel-N	18.143	15.587	38.677	35 50 00	117 17 00
22	Copper Queen	Gn and qtz in skarn	Gel-N	18.294	15.597	38.832	35 48 00	117 16 40
23	Minnietta mine	Gn in Mn-stained dolomite	Gel-N	17.942	15.572	38.951	36 14 10	117 26 20
24	Unnamed prospect	Oxidized sulfides in qtz veins; Gold Mtn. Area, Nev.	Gel-N	19.110	15.691	38.976	37 14 32	117 17 03
25	OBJ prospect		Gel-N	18.283	15.590	39.965	36 12 29	117 12 41
26	Gold Bug mine	Cp in qtz veins	Gel-N	18.129	15.569	39.764	36 01 40	117 09 40
27	Honolulu mine	Gn, sp in bleached Noonday Dol.	Gel-N	20.260	15.840	41.294	35 59 30	117 09 15
28	Clair Camp	Limonite, mill tailings	Gel-N	18.688	15.652	39.753	36 02 00	117 08 00
29	Sentinel Peak	Gn in qtz veins in Precambrian lms	3FI-N[a]	19.434	15.763	40.528	36 05 00	117 07 30
30	Lead Mountain	Gn in brecciated Tertiary andesite	3FI-N[a]	19.059	15.684	39.275	34 55 18	116 55 54
31	Queen of Sheba	Gn in Noonday Dol.	Gel-N	18.859	15.673	39.370	35 59 50	116 53 30
32	Paddys Pride mine	Gn in Noonday Dol.	Gel-N	18.346	15.644	39.669	35 51 00	116 22 30
33	Gladstone mine	Cu oxides and limonite in ss	Gel-N	18.667	15.680	39.724	35 52 20	116 23 20
34	Blue Bell mine	Py and azurite in andesite	Gel-N	18.263	15.619	38.735	35 14 25	116 12 10
35	Blue Bell mine	Garnet skarn	Gel-N	18.247	15.593	38.787	35 14 30	116 12 09
36	Noonday mine	Gn in Noonday Dol.	Gel-N	18.355	15.610	39.358	35 49 45	116 05 45
37	Noonday mine	Gn in Noonday Dol.	3FI-N[a]	18.268	15.624	39.553	35 49 42	116 05 48
38	Chambers mine	Gn in Noonday Dol.	Gel-N	18.348	15.665	39.780	35 48 10	115 57 40
39	Blackwater mine	Gn in Noonday Dol.	Gel-N	18.677	15.692	39.989	35 47 02	115 52 30
40	Blackwater mine	Oxidized dump material with high Pb conc.	Gel-N	18.432	15.632	39.774	35 45 22	115 49 41
43	Ada-Edith mines	Gn in Cambrian dol.	Gel-N	22.039	16.027	40.482	36 20 38	115 39 26
44	Mohawk mine	Gn and limonite in silicified dol.	Gel-N	17.586	15.580	39.314	35 28 45	115 36 40

Table 2 (*continued*)

Map no.	Name	Comments	Method	$\frac{^{206}\text{Pb}}{^{204}\text{Pb}}$	$\frac{^{276}\text{Pb}}{^{204}\text{Pb}}$	$\frac{^{208}\text{Pb}}{^{204}\text{Pb}}$	Latitude	Longitude
45	Unnamed prospect	Oxidized dump material with high Pb conc., adit in lms	Gel-N	17.695	15.546	39.172	35 35 43	115 36 40
46	Teutonia mine	Gn in qtz calcite vein in granite	Gel-N	17.819	15.545	39.076	35 18 25	115 34 05
47	Teutonia Pluton	K-feldspar in granite	Gel-N	17.788	15.564	39.146	35 18 26	115 34 06
48	Colosseum mine	Cp in breccia with calc-silicate clasts	Gel-N	17.984	15.553	38.982	35 34 10	115 34 10
49	Colosseum mine	Sp in breccia with calc-silicate clasts	Gel-N	17.933	15.558	39.037	35 34 10	115 34 10
50	Potosi mine	Gn in Monte Cristo Lms.	Gel-N	21.290	15.869	40.164	35 57 39	115 32 16
51	Lucky Strike mine	Gn in Monte Cristo Lms.	Gel-N	21.130	15.896	40.089	36 20 05	115 32 20
52	Iron Horse	Oxidized dump material with high Pb conc., adit in lms	Gel-N	17.708	15.580	39.750	35 25 50	115 31 53
54	Yellowpine mine	Gn, Pb and Zn Oxides in Monte Cristo Lms	Gel-N	21.343	15.899	40.270	35 51 00	115 29 43
55	Yellowpine mine	Gn, Pb and Zn Oxides in Monte Cristo Lms	Gel-N	21.448	15.899	40.335	35 51 00	115 29 39
56	Kewanee mine		Gel-N	17.506	15.563	39.252	35 20 52	115 29 08
57	Dolly Varden	Gn in qtz veins in Teutonia qtz monzonite	3FI-N[a]	17.965	15.578	39.135	35 13 12	115 27 30
58	Joe May prospect	Gn in carbonate fault breccia	Gel-N	18.987	15.691	38.587	36 31 56	115 17 57
59	June Bug	Gn, Zn oxides in Monte Cristo Lms.	Gel-N	20.774	15.852	39.723	36 22 11	115 10 14
60	Quartzite Hill (White Caps prospect)	Gn in Sterling qtzite	Gel-N	20.009	15.830	39.302	36 29 01	115 06 02
61	Unnamed prospect	Oxidized dump material with high Pb conc., adit in granitic rks	Gel-N	18.158	15.595	39.006	35 02 18	115 03 41
62	Lead King	Gn, Zn oxide, in Monte Cristo Lms.	Gel-N	20.781	15.835	39.758	36 19 11	114 58 38

Analytical methods: 3FI-N, thermal ionization mass spectrometry using the triple filament method (Zartman, 1974); Gel-N, thermal ionization using the silica gel method (Cameron et al., 1969).

[a] Data from Zartman (1974).

7.2. Tecopa Terrane

This terrane is characterized by distinctly higher $^{208}\text{Pb}/^{204}\text{Pb}$ values indicating much higher kappa ($^{232}\text{Th}/^{204}\text{Pb}$) and different mu ($^{238}\text{U}/^{204}\text{Pb}$) values than most of the plutons of the Sierra Nevada batholith (Fig. 5). Wooden and Miller (1990) and Wooden and DeWitt (1991) have described the lead isotopic characteristics of the crust from southern California and Arizona and mapped them using their isotopic character. They define three provinces: central Arizona which has kappa values near 2, southeastern Arizona which has kappa values near 4, and the Mojave crustal province which has kappa values between about 4 and 15 (Fig. 4, Wooden and DeWitt, 1991). In Fig. 5B, the deposit lead isotope data plot above the Sierra Nevada batholith field indicating a higher kappa value for lead in these deposits. In Fig. 6, deposit and pluton lead isotope data from west of the Death Valley—Furnace Creek–Ivanpah Fault zone from the Darwin and Tecopa Terranes are plotted against distance along the transect E–E′ (Fig. 1).

Deposit lead from the Ivanpah and Tecopa districts closely matches the igneous lead from plutons in the same vicinity within the Tecopa Terrane. Deposits from the Ivanpah district (loc. 41–43, 45–48, Fig. 1 and Table 2) are characterized by $^{206}\text{Pb}/^{204}\text{Pb}$ from 17.5 to 18.0 and $^{208}\text{Pb}/^{204}\text{Pb}$ from circa 39.0–39.3 (Fig. 5). Lead from a large feldspar phenocryst from the Teutonia pluton gave a $^{206}\text{Pb}/^{204}\text{Pb}$ value of 17.79 (loc. 44, Fig. 1 and Table 2) whereas the $^{206}\text{Pb}/^{204}\text{Pb}$ from the galena deposited in the pluton gave a value of 17.82 (loc. 43, Fig. 1 and Table 2). Agreement between the two values is excellent given that no corrections for radiogenic growth were made to the feldspar lead isotope data. Lead isotopes from three plutons in the Ivanpah Mountains were also analyzed by Rämö et al. (2002). Data from the Cretaceous Kessler Springs monzogranite gave a $^{206}\text{Pb}/^{204}\text{Pb}$ value of 17.83 in excellent agreement with the galena (loc. 43, Fig. 1 and Table 2) whereas samples from the Teutonia granite and the Ivanpah monzonogranite gave lower $^{206}\text{Pb}/^{204}\text{Pb}$ values of 17.40 and 17.46. The mineral deposit sampled (loc. 43, Fig. 1 and Table

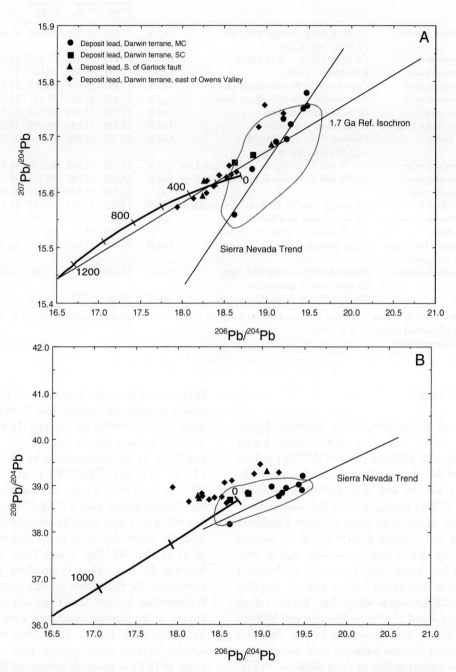

Fig. 4. $^{207}Pb/^{204}Pb$ vs. $^{206}Pb/^{204}Pb$ (A) and $^{208}Pb/^{204}Pb$ vs. $^{206}Pb/^{204}Pb$ (B) diagrams showing the lead isotopic signatures (Table 2) of mineral deposits from the Darwin Terrane. Symbols shown are for deposits within the same area of the moderately contaminated (MC) and the strongly contaminated (SC) plutons (Ague and Brimhall, 1988) and by the lead isotope data of Chen and Tilton (1991). The Stacy–Kramers model geochrons, the field of Sierra Nevada lead data from plutons (Chen and Tilton, 1991) from Fig. 3, and the 1.7 Ga reference isochron for the Mojave crustal province (Wooden and Miller, 1990; Wooden and DeWitt, 1991; J.L. Wooden, unpublished data, 2003) are shown for reference.

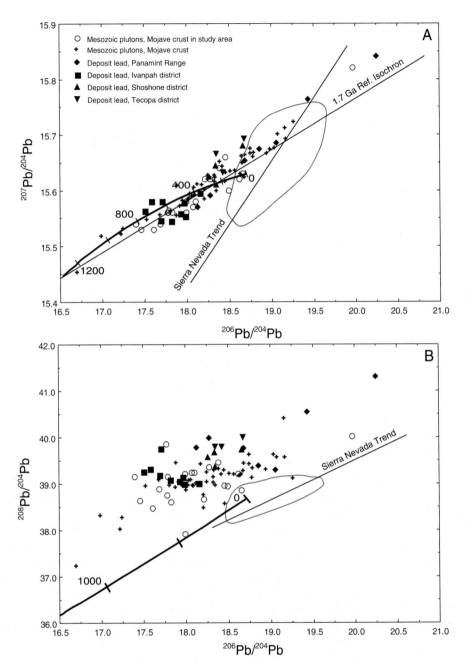

Fig. 5. $^{207}Pb/^{204}Pb$ vs. $^{206}Pb/^{204}Pb$ (A) and $^{208}Pb/^{204}Pb$ vs. $^{206}Pb/^{204}Pb$ (B) diagrams showing the lead isotopic signatures (Table 2) of mineral deposits from the Tecopa Terrane. The Stacy–Kramers model geochrons, the field of Sierra Nevada lead data from plutons (Chen and Tilton, 1991) from Fig. 3, and the 1.7 Ga reference isochron for the Mojave crustal province are shown for reference. Data from Mesozoic plutons are from Miller and Wooden (1994), Rämö et al. (2002), and J.L. Wooden (unpublished data, 2003).

2) may be related to the porphyritic phase of the Cretaceous Kessler Springs pluton. The lead isotopic signature from the unnamed prospect at loc. 49 (Fig. 1

and Table 2) south of the Ivanpah district in the New York Mountains also falls in this same range. Three analyses of lead from plutons in this area (Rämö et al.,

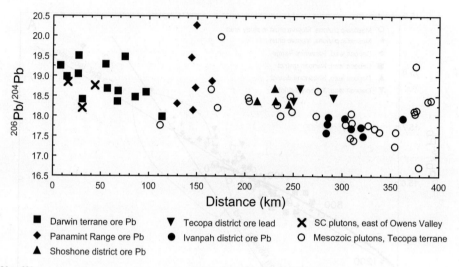

Fig. 6. Plot of $^{206}Pb/^{204}Pb$ versus distance along transect E–E' (Fig. 1). Deposit data are from the Darwin Terrane west of the Furnace Creek Fault and north of the Garlock Fault and from districts in the Tecopa Terrane (Table 2). Pluton data are from Chen and Tilton (1991), Miller and Wooden (1994), Rämö and Calzia (1998), and Rämö et al. (2002).

2002) gave lower lead isotope ratios. Although the leads in the exposed plutonic rocks generally do not give values that correspond closely with that of the deposit leads, the values of both sets of data tend to cluster in the same general area in Fig. 5.

In contrast, mineral deposit lead from the Tecopa district (loc. 38–40, Fig. 1 and Table 2) provides a different signature ($^{206}Pb/^{204}Pb$ from 18.35–18.68, Fig. 5A, and $^{208}Pb/^{204}Pb$ from circa 39.7–40.0, Fig. 5B). Lead isotope data from the Tertiary Granite of Kingston Peak (Calzia and Rämö, 2005—this volume) have a higher $^{206}Pb/^{204}Pb$ value than mineral deposits in the district. Thus, the source of ore fluids for rocks in the Ivanpah and New York Mountains and in the Kingston Range are distinct, but all reflect the lead isotopic signature of the granulite facies lower crustal rocks of the Mojave crustal province which formed during the Ivanpah orogeny (Wooden and Miller, 1990). The Skidoo granodiorite (loc. S, Fig. 1) in the northern part of the Panamint Range also has a lead isotopic signature ($^{206}Pb/^{204}Pb$ of 17.77 and $^{208}Pb/^{204}Pb$ of 39.83, Fig. 5; Rämö et al., 2002), which indicates that Mojave crust affected by the Ivanpah orogeny also underlies this area. The progressive decrease in $^{206}Pb/^{204}Pb$ values from the Tecopa to the Ivanpah districts (Fig. 6) is indicative of partial melting deeper in the lower crust. The readers are referred to Rämö et al. (2002) for a thorough discus-

sion of the origin of the plutons. In contrast, deposit lead from the Panamint Range is generally radiogenic and is interpreted to reflect circulation of hydrothermal fluids in supracrustal rocks as discussed below.

7.3. Deposit lead hosted in supracrustal rocks of the Mojave crustal province

Deposits sampled in the Shoshone district (loc. 36, 37, Fig. 1) as well as the rest of the deposits from the Amargosa and Panamint Ranges are generally hosted in Precambrian rocks, usually in the Late Proterozoic Noonday Dolomite (Figs. 5–7). These rocks also reflect the lead isotopic signature of the Mojave supracrustal rocks (Wooden and Miller, 1990; J.L. Wooden, unpublished data, 2003), but do not have the reduced $^{206}Pb/^{204}Pb$ signature indicative of a source in the granulite facies metamorphosed lower crust. They have elevated mu and kappa values indicative of supracrustal rocks and plot coincident with the 1.7 Ga reference isochron from the Mojave crustal province (Fig. 5A). The large producing deposits in the Shoshone district (Table 1) as well as smaller occurrences hosted by Noonday Dolomite east and west of the Shoshone district (loc. 27, 31–33, 38–40, Fig. 1 and Table 2) have the more radiogenic lead isotopic signature (Figs. 5 and 7). They plot coincident with the 1.7 Ga isochron from Mojave supra-

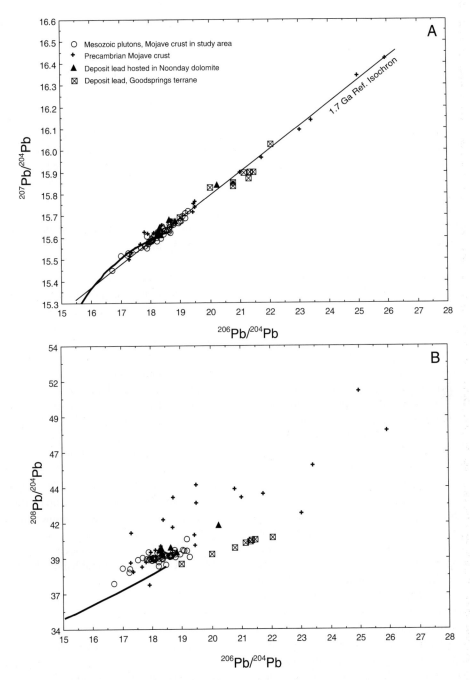

Fig. 7. ^{207}Pb/^{204}Pb vs. ^{206}Pb/^{204}Pb (A) and ^{208}Pb/^{204}Pb vs. ^{206}Pb/^{204}Pb (B) diagrams showing the lead isotopic signatures (Table 2) of mineral deposits hosted in supracrustal rocks in the Tecopa and Goodsprings Terranes. The Stacy–Kramers model geochrons and the 1.7 Ga reference isochron for the Mojave crustal province are shown for reference. Data from supracrustal rocks are from study by Wooden and Miller (1990), Wooden and DeWitt (1991), Calzia and Rämö (2005—this volume), and J.L. Wooden (unpublished data, 2003).

crustal rocks (Fig. 7A) and have elevated and $^{208}Pb/^{204}Pb$ values (Fig. 7B). We interpret the lead from this group of deposits to have been leached from the supracrustal rocks relatively recently in geologic time to allow for the buildup of radiogenic lead seen in the deposit at the Honolulu mine (loc. 27, Fig. 1 and Table 2). Samples from the Noonday mine have the least radiogenic lead isotope signature (Table 2) of the deposits in the district and the largest known resources of lead and zinc (Table 1, Carlisle et al., 1954). Textural evidence suggests that these lead ores were formed during diagenesis of the Noonday Dolomite, but the radiogenic character of the lead isotopic data argues that the deposits are young relative to the age of the host rocks. The age of this mineralization is unresolved.

7.4. Goodsprings Terrane

Lead isotope ratios from samples in the Goodsprings Terrane form a linear array typical of large zinc–lead deposits hosted in carbonate rocks. The lead isotope data generally plot below the 1.7 Ga reference isochron for the Mojave crustal rocks (Fig. 7A) and have a lower slope in the $^{206}Pb/^{204}Pb$ vs. $^{208}Pb/^{204}Pb$ diagram (Fig. 7B). The age calculated from the lead isotope data from these deposits gives a value of 1450 ± 620 Ma, which is nominally younger than but lies within the range of error of the 1.7 Ga age of the Mojave crustal province (Wooden and Miller, 1990; Wooden and DeWitt, 1991). Similar lead isotopic results from the carbonate-hosted Pb–Zn deposits in the district and values of $\delta^{34}S$ of $+ 6 \pm 2$ are reported by Vikre (2001). These differ from the lead and sulfur isotopic signatures of copper and gold pluton-related deposits from the district (Vikre, 2001), which we did not sample. The lead isotope arrays are typical of MVT carbonate-hosted zinc–lead deposits that overlie a Precambrian crystalline terrane and show no influence by igneous processes. The lead isotope data and field observations strongly indicate that these are MVT deposits that were formed by hydrothermal fluids leaching lead from an orogenic belt containing uranium-rich coarse-grained fluvial sedimentary deposits derived from old crystalline rocks. Studies of lead isotopic data from the MVT deposits of SE Missouri (Goldhaber et al., 1995; Goldhaber and Taylor, 2001)

showed very similar lead isotope arrays. Preferential leaching of ^{206}Pb relative to ^{208}Pb results in an apparent low kappa value. Deposition of metals from the hydrothermal solutions occurs in karst cavities in carbonate host rock when chloride brines containing metals mixes with groundwater containing H_2S (Goldhaber and Mosier, 1990). The younger age and large error reflect an isotopically heterogeneous source, probably resulting from weathering of detritus derived from the Mojave crustal province through which hydrothermal MVT fluids flowed. We suggest that these are MVT deposits formed in a foreland bulge and associated subaerial karst formation (Bradley and Leach, 2003).

8. The western boundary of the Mojave crust in the study area

Our studies, in conjunction with other published lead isotope data, suggest that the boundary between the Darwin Terrane and the Tecopa Terrane is constrained to lie near or just west of the Ash Hills–Panamint Valley Fault (Fig. 1). Lead isotopic data from three deposits west of the fault (loc. 21–23, Fig. 1) all show contributions from the Mojave crust. North of the Towne Fault, most mineral deposit lead was not affected by Mojave crust although isolated deposits have elevated $^{208}Pb/^{204}Pb$ values (Fig. 4; loc. 4, 5, 7, 9, Fig. 1). Lead from minerals deposits and plutons south of the Garlock Fault (loc. 34, 35, Fig. 1) show the effects of Mojave crust and have elevated $^{87}Sr/^{86}Sr_i > 0.706$ (Kistler, 1990). All deposits and plutons from the Tecopa and Goodsprings Terranes show the lead isotopic signature of the Mojave crust (John and Wooden, 1990; Miller et al., 1990; Miller and Wooden, 1994; Rämö et al., 2002; J.L. Wooden, unpublished data).

9. Conclusions

The formation of lead and zinc deposits in southern California and southern Nevada began in the Panamint Range and the Shoshone district with formation of mineral deposits in Proterozoic supracrustal rocks of the Tecopa terrane. The time of deposition of lead in veins in the Noonday Dolomite and the processes of fluid-flow are uncertain because of contradictory evidence resulting from this and previous studies. On the one hand, lead isotope data indicated that mineral deposits formed as a result of circulation of hydrothermal fluids in the 1.7 Ga supracrustal rocks of the Mojave crust (Wooden and Miller, 1990). On the other hand, diagenetic features in Noonday rocks associated with the veins suggest that the deposits were formed soon after Noonday deposition. Sulfur isotope data from galena, however, do not indicate that sulfur was derived from biogenic reduction of sedimentary sulfate as would be expected in a deposit that formed during diagenesis.

Many polymetallic and gold vein deposits in the Tecopa Terrane are related to Mesozoic and Tertiary plutonic activity; however, the Noonday-hosted deposits are clearly not, owing to their lack of consistent relation to igneous rocks and their low pyrite content. Thus, even though one of the original objectives of this study was to throw light on the origin of the Noonday-hosted deposits, we have not been able to provide sufficient constraints on the timing of this mineralization event to resolve when and how these deposits might have formed.

A second period of lead and zinc deposition occurred during development of the Paleozoic depositional platform on the western margin of the North American continent. The occurrence in the Goodsprings district of lead and zinc ore in karst breccias which formed below the Devonian–Pennsylvanian unconformity indicates that these are MVT deposits. Lead isotopic data reported here show that the lead isotope array does have the character of MVT deposits formed in carbonate rocks that overlie old Precambrian basement. The lead isotopic data lie along a linear array that has a slope, which is slightly younger than that of the basement rocks and shows significant depletion of ^{208}Pb relative to the source rocks. The source of fluids and the timing of deposition and relationship to orogeny, however, are uncertain.

The third period of lead and zinc deposition resulted from subduction, accretion, and magmatism in a continental arc setting along the western margin of the continent during the Jurassic. Numerous plutons within the Sierra Nevada batholith have lead–zinc skarn and replacement deposits clustered near their contacts and may have had porphyry copper and related deposits in higher levels, now removed by erosion. Lead isotope ratios measured in these deposits conform closely to those obtained from feldspars in the plutonic rocks although matches between lead from individual plutons and mineral deposits are not common. The lead isotope data from the deposits indicate the same mixing process that resulted in the formation of the Sierra Nevada pluton array defined by Chen and Tilton (1991) caused by mixing of mantle lead with crustal lead from subducted sediment and by anatexis at the base of the crust. Rocks and mineral deposits with these lead isotope ratios make up the Darwin Terrane and extend to just west of the Ash Hills–Panamint Valley Fault. Here, they give way to mineral deposits of the Tecopa Terrane with higher ^{208}Pb/^{204}Pb ratios typical of Mojave crust. Melting of granulite to amphibolite facies rocks at the base of the crust has been proposed for plutons of the Tecopa Terrane (John and Wooden, 1990, Miller et al., 1990; Miller and Wooden, 1994; Rämö and Calzia, 1998; Rämö et al., 2002) and for the strongly contaminated plutons on the east side of the Owens Valley in the Inyo Mountains (Chen and Tilton, 1991). Mineral deposits from the Tecopa Terrane also have this same deep crustal melting signature indicating that the hydrothermal fluids separated from magmas that originated at the source region, presumably the result of this same partial melting process, rather than in a hydrothermal cell that circulated in the supracrustal rocks of the Mojave crust.

Isotopic studies of mineral deposit lead have provided data on the structure of the crust in the area that compliments results obtained from studies of initial lead and strontium in plutons. As a result and based upon the sampling of deposits and plutons to date, the boundary of the Mojave crust with the Panthalassen crust is constrained just to the west of the Ash Hills–Panamint Valley Fault, south of the Towne Fault, and south of the Garlock Fault.

Acknowledgments

The authors wish to thank Ben Troxel, James Calzia, and Robert Zierenberg for helpful discussions in the field. We also thank Boris Kotlyar, Barbara Ramsey, and Barry Moring for help in creating Fig. 1, and Gary

Nowlan, Sherman Marsh, and Ken Bishop for contributing samples. Joe Wooden provided unpublished data and made many useful suggestions while the authors were writing this paper. Reviews by Ted Theodore and Dan Unruh are greatly appreciated.

References

Ague, J.J., Brimhall, G.H., 1988. Regional variation in bulk chemistry, mineralogy, and the compositions of mafic and accessory minerals in the batholiths of California. Geological Society of America Bulletin 100, 891–911.

Bradley, D.C., Leach, D.L., 2003. Tectonic controls of Mississippi Valley-type lead–zinc mineralization in orogenic forelands. Mineralium Deposita 38, 652–667.

Bray, T.D., 1983. Stratabound zinc–lead deposits in the Monte Cristo Limestone, Goodsprings, Nevada. M.A. thesis, Dartmouth College, Hanover, New Hampshire, 235 pp.

Calzia, J.P., Rämö, O.T., 2005—this volume. Miocene rapaviki granites in the southern Death Valley region, California, USA. Earth-Science Reviews 73, 221–243. doi:10.1016/j.earscirev. 2005.07.006.

Cameron, A.E., Smith, D.H., Walker, R.L., 1969. Mass spectrometer analysis of nanogram quantities of lead. Analytical Chemistry 41, 525–526.

Carlisle, Donald, Davis, D.L., Kildale, M.B., Stewart, R.M., 1954. Base metal and iron deposits of southern California. Bulletin-California. Division of Mines and Geology 170, 41–49 (Chapter 8).

Carr, M.D., Evans, K.V., Fleck, R.J., Frizzell Jr., V.A., Ort, K.M., Zartman, R.E., 1986. Early Middle Jurassic upper limit for movement on the Keystone Thrust, southern Nevada [abs.]. Abstracts with Programs, Geological Society of America 18 (5), 345.

Chen, J.H., Moore, J.G., 1982. Uranium–lead isotopic ages from the Sierra Nevada batholith, California. Journal of Geophysical Research 87, 4761–4784.

Chen, J.H., Tilton, G.R., 1991. Application of lead and strontium isotopic relationships to the petrogenesis of granitoid rocks, central Sierra Nevada batholith, California. Geological Society of America Bulletin 103, 439–447.

Church, S.E., 1973. Limits of sediment involvement in the genesis of orogenic volcanic rocks. Contributions to Mineralogy and Petrology 39, 17–32.

Church, S.E., Tatsumoto, M., 1975. Lead isotope relations in oceanic basalts from the Juan de Fuca-Gorda ridge area, N.E. Pacific Ocean. Contributions to Mineralogy and Petrology 53, 253–279.

Church, S.E., Tilton, G.R., 1973. The bearing of lead and strontium isotope geochemistry of the Cascade Mountains on andesite genesis. Geological Society of America Bulletin 84, 431–454.

Church, S.E., LeHuray, A.P., Grant, A.R., Delevaux, M.H., Gray, J.E., 1986. Lead-isotopic data from sulfide minerals from the Cascade Range, Oregon and Washington. Geochimica et Cosmochimica Acta 50, 317–328.

Cloud, Preston, Wright, L.A., Williams, E.G., Diehl, Paul, Walter, M.R., 1974. Giant stromatolites and associated vertical tubes from the upper Proterozoic Noonday Dolomite, Death Valley Region, Eastern California. Geological Society of America Bulletin 85 (12), 1869–1882.

Doe, B.R., Stacey, J.S., 1974. The application of lead isotopes to the problems of ore genesis and ore prospect evaluation: a review. Economic Geology 69 (6), 757–776.

Garside, L.J., Bonham Jr., H.F., Tingley, J.V., McKee, E.H., 1993. Potassium–argon ages of igneous rocks and alteration minerals associated with mineral deposits, western and southern Nevada and eastern California. Isochron-West (59), 17–23.

Goldhaber, M.B., Mosier, E.L., 1990. Sulfur sources for southeast Missouri, MVT ores—implications for ore genesis. In: Pratt, W.P., Goldhaber, M.B. (Eds.), U.S. Geological Survey Symposium—Mineral Resource Potential of the Midcontinent, Program and Abstracts, St. Louis, Missouri, April 11–12, 1989. U.S. Geological Survey Circular, vol. 1043, pp. 8–9.

Goldhaber, M.B., Taylor, C.D., 2001. Subcontinental scale fluid transport of the sulfide component of Mississippi Valley-type ores [Abst.]. Abstracts with Programs, Geological Society of America 33 (6), A-96.

Goldhaber, M.B., Church, S.E., Doe, B.R., Aleinikoff, J.N., Brannon, J.C., Podosek, F.A., Mosier, E.L., Taylor, C.D., Gent, C.A., 1995. Lead- and sulfur-isotope investigation of Paleozoic sedimentary rocks from the southern midcontinent of the United States: implications for paleohydrology and ore genesis of the southeast Missouri lead belts. Economic Geology 90, 1875–1910.

Guth, P.L., 1981. Tertiary extension north of the Las Vegas Valley shear zone, Sheep and Desert Ranges, Clark County, Nevada. Geological Society of America Bulletin 92, 763–771.

Guth, P.L., Schmidt, D.L., Deibert, J., Yount, J.C., 1988. Tertiary extensional basins of northwestern Clark County, Nevada. In: Weide, D.L., Faber, M.L. (Eds.), This Extended Land, Geologic Journeys in Southern Basin and Range. Geological Society of America. University of Nevada Printing Service, Cordilleran Section, Las Vegas, pp. 239–253.

Hall, S.M., 1984. Metallogenesis and stable isotope systematics of stratabound lead/silver deposits in the Precambrian Noonday dolomite, Death Valley, California. MS thesis, University of California, Davis, 133 pp.

Hall, W.E., MacKevett Jr., E.M., 1962. Geology and ore deposits of the Darwin quadrangle, Inyo County California. U.S. Geological Survey Professional Paper 368, 87 pp.

Hall, W.E., Stephens, H.G., 1963. Economic geology of the Panamint Butte Quadrangle and Modoc District, Inyo County, California. Special Report-California Division of Mines and Geology 73, 39 pp.

Hammond, J.G., Wooden, J.L., 1990. Isotopic constraints on the petrogenesis of Proterozoic diabase in the southwestern USA. In: Parker, A.J., Rickwood, P.C., Tucker, D.H. (Eds.), Mafic Dykes and Emplacement Mechanisms. A.A. Balkerma, Rotterdam, Netherlands, pp. 145–146.

Hewett, D.F., 1931. Geology and ore deposits of the Goodsprings quadrangle, Nevada. U.S. Geological Survey Professional Paper 162, 171 pp.

Hewett, D.F., 1956. Geology and ore deposits of the Ivanpah quadrangle, Nevada. U.S. Geological Survey Professional Paper 275, 172 pp.

Hodges, C.A., in press. Breccia pipe and related deposits. In: Theodore, T.G. (Ed.), Geology and Mineral Resources of the East Mojave National Scenic Area, San Bernardino County, California. U.S. Geological Survey Bulletin 2160.

Jennings, C.W., 1961. Geologic map of California, Kingman sheet. California Division of Mines and Geology, 1 sheet, scale 1:250,000.

Jennings, C.W., Burnett, L.L., Troxel, B.W., 1962. Geologic map of California, Trona sheet. California Division of Mines and Geology, 1 sheet, scale 1:250,000.

John, B.E., Wooden, J.L., 1990. Petrology and geochemistry of the metaluminous to peraluminous Chemehuevi Mountains plutonic suite, southeastern California. In: Anderson, J.L. (Ed.), The Nature and Origin of Cordilleran Magmatism. Memoir Geological Society of America 174, 71–98.

Kistler, R.W., 1978. Reconstruction of crustal blocks of California on the basis of initial strontium isotopic compositions of Mesozoic granitic rocks. U.S. Geological Survey Professional Paper 1071, 17 pp.

Kistler, R.W., 1990. Two different lithosphere types in the Sierra Nevada, California. In: Anderson, J.L. (Ed.), The Nature and Origin of Cordilleran Magmatism. Memoir Geological Society of America 174, 271–281.

Kistler, R.W., Peterman, Z.E., 1973. Variations in Sr, Rb, K, Na, and initial $^{87}Sr/^{86}Sr$ in Mesozoic granitic rocks and intruded wall rocks, central California. Geological Society of America Bulletin 84, 3489–3511.

Knopf, Adolph, 1918. A geologic reconnaissance of the Inyo Range and the slope of the southern Sierra Nevada. U.S. Geological Survey Professional Paper 110, 130 pp.

Longwell, C.R., Pampeyan, E.H., Bowyer, B., Roberts, R.J., 1965. Geology and mineral deposits of Clark County, Nevada. Bulletin Nevada Bureau of Mines and Geology 62, 218 pp.

MacKevett Jr., E.M., 1953. Geology of the Santa Rosa lead mine, Inyo County, California. Special Report California Division of Mines and Geology 34, 9 pp.

McAllister, J.F., 1955. Geology and mineral deposits of the Ubehebe Peak quadrangle. Special Report California Division of Mines and Geology 42, 63 pp.

Meriam, C.W., 1963. Geology of the Cerro Gordo mining district, Inyo County, California. U.S. Geological Survey Professional Paper 408, 83 pp.

Miller, C.F., Wooden, J.L., 1994. Anatexis, hybridization and modification of ancient crust: Mesozoic plutonism in the old woman mountains area, California. Lithos 32, 111–133.

Miller, C.F., Wooden, J.L., Bennett, V.C., Wright, J.E., Solomon, G.C., Hurst, R.W., 1990. Petrogenesis of the composite peraluminous–metaluminous Old Woman-Piute Range batholith, southeastern California; isotopic constraints. In: Anderson, J.L. (Ed.), The Nature and Origin of Cordilleran Magmatism. Memoir Geological Society of America 174, 99–110.

Miller, C.F., Hanchar, J.M. Wooden, J.L., Bennett, V.C., Harison, T.M., Wark, D.A., Foster, D.A., 1992. Source region of a batholith: evidence from lower crustal xenoliths and inherited

accessory minerals. Transactions of the Royal Society of Edinburgh. Earth Sciences 53, 49–62.

Morton, P.K., 1965. Geology of the Queen of Sheba lead mine, Death Valley, California. Special Report California Division of Mines and Geology 88, 18 pp.

Murphy, F.M., 1930. Geology of the Panamint silver district, California. Economic Geology 25, 305–325.

Newberry, R.J., 1987. Use of intrusive and calc-silicate compositional data to distinguish skarn types in the Darwin polymetallic skarn district, California. Mineralium Deposita 22, 207–215.

Newberry, R.J., Einaudi, M.T., Eastman, H.S., 1991. Zoning and genesis of the Darwin Pb–Zn–Ag skarn deposit, California: a reinterpretation based on new data. Economic Geology 86 (5), 960–982.

Nowlan, G.A., Theodore, T.G., in press. Geochemistry. In: Theodore, T.G. (Ed.), Geology and Mineral Resources of the East Mojave National Scenic Area, San Bernardino County, California. U.S. Geological Survey Bulletin 2160.

Rämö, O.T., Calzia, J.P., 1998. Nd isotopic composition of cratonic rocks in the southern Death Valley region: evidence for a substantial Archean source component in Mojavia. Geology 26, 891–894.

Rämö, O.T., Calzia, J.P., Kosunen, P.J., 2002. Geochemistry of Mesozoic plutons, southern Death Valley region, California: Insights into the origin of Cordilleran interior magmatism. Contributions to Mineralogy and Petrology 143, 416–437.

Sharp, J.E., 1984. A gold mineralized breccia pipe complex in the Clark Mountains, San Bernardino County, California. In: Wilkins Jr., Joe (Ed.), Gold and Silver Deposits of the Basin and Range Province Western U.S.A. Arizona Geological Society Digest 15, 119–139.

Smith, A.R., 1964. Geologic map of California, Bakersfield sheet. California Division of Mines and Geology, 1 sheet, scale 1:250,000.

Smith, P.L., Tingley, J.V., 1983a. A mineral inventory of the Esmeralda-Stateline Resource Area, Las Vegas District, Nevada. Nevada Bureau of Mines and Geology Open-File Report 83-11, 175 pp.

Smith, P.L., Tingley, J.V., 1983b. Results of geochemical sampling within the Esmeralda-Stateline Resource Area, Esmeralda, Clark, and southern Nye Counties, Nevada (portions of the Death Valley, Goldfield, Kingman, Las Vegas, Mariposa, and Tonopah 2° sheets). Nevada Bureau of Mines and Geology Open-File Report 83-12, 172 pp.

Smith, G.I., Troxel, B.W., Gray Jr., C.H., von Huene, Roland, 1968. Geologic reconnaissance of the Slate Range, San Bernardino and Inyo Counties, California. Special Report California Division of Mines and Geology 96, 33 pp.

Stacey, J.S., Kramers, J.D., 1975. Approximation of terrestrial lead isotope evolution by a two-stage model. Earth and Planetary Science Letters 26, 207–221.

Stewart, J.H., Carlson, J.E., 1978. Geological map of Nevada. U.S. Geological Survey and Nevada Bureau of Mines and Geology, 1 sheet, scale 1:500,000.

Streitz, Robert, Stinson, W.C., 1974. Geologic map of California, Death Valley sheet. California Division of Mines and Geology, 1 sheet, scale 1:250,000.

Tingley, J.V., Castor, S.B., Garside, L.J., Bonham Jr., H.F., Lugaski, T.P., Lechler, P.J., 1993. Energy and mineral resources of the Desert National Wildlife Range, eastern section, Clark and Lincoln Counties, Nevada. Nevada Bureau of Mines and Geology Open-File Report 93-2, 372 pp.

Vikre, Peter, 2001. Diverse styles and ages of base, precious metal, and PGE mineralization at Goodsprings, Nevada [Abst.]. Abstracts with Programs, Geological Society America 33 (6), A-98.

Wooden, J.L., DeWitt, Ed., 1991. Pb isotope evidence for the boundary between the Early Proterozoic Mojave and Central Arizona Provinces in Western Arizona. In: Karlstrom, K.E. (Ed.), Proterozoic Geology and Ore Deposits of Arizona. Arizona Geological Digest 19, 27–50.

Wooden, J.L., Miller, D.M., 1990. Chronologic and isotopic framework for early Proterozoic crustal evolution in the eastern Mojave Desert region, S.E. California. Journal of Geophysical Research 95, 20133–20146.

Wooden, J.L., Kistler, R.W., Tosdal, R.M., 1998. Pb isotopic mapping of crustal structure in the northern Great Basin and relationships to Au deposit trends. In: Tosdal, R.M. (Ed.), Contributions to the Gold Metallogeny of Northern Nevada. United States Geological Survey, Open-File Report 98-338, pp. 20–33.

Wooden, J.L., Kistler, R.W., Tosdal, R.M., 1999. Strontium, lead, and oxygen isotopic data for Granitoid and volcanic rocks from the northern Great Basin and Sierra Nevada, California, Nevada, and Utah. United States Geological Survey, Open-File Report 99–569, 20 pp.

Wright, L.A., Troxel, B.W., Williams, E.G., Roberts, M.T., Diehl, P.E., 1976. Precambrian sedimentary environments in the Death Valley region, eastern California. Special Report-California Division of Mines and Geology 106, 7–17.

Zartman, R.E., 1974. Lead isotope provinces in the Cordillera of the Western United States and their geologic significance. Economic Geology 69 (6), 792–805.

Available online at www.sciencedirect.com

Earth-Science Reviews 73 (2005) 347–348

www.elsevier.com/locate/earscirev

Contents Volume 73, 2005

Special Issue
Fifty Years of Death Valley Research
A volume in honor of Lauren A. Wright and Bennie Troxel

Edited by: J.P. Calzia

Fifty years of Death Valley research: A volume in honor of Lauren Wright and Bennie Troxel
J.P. Calzia 1
Acknowledgements of a professional lifetime
L.A. Wright 3
My geological career in Death Valley
B.W. Troxel 13
Geological landscapes of the Death Valley region
M.B. Miller 17
Bibliography L.A. Wright and B.W. Troxel 1944–present 31
Eolian deposits in the Neoproterozoic Big Bear Group, San Bernardino Mountains, California, USA
J.H. Stewart 47
The relationship between the Neoproterozoic Noonday Dolomite and the Ibex Formation: New observations and
their bearing on 'snowball Earth'
F.A. Corsetti and A.J. Kaufman 63
Interpretation of the Last Chance thrust, Death Valley region, California, as an Early Permian décollement in a
previously undeformed shale basin
C.H. Stevens and P. Stone 79
Structure and regional significance of the Late Permian(?) Sierra Nevada–Death Valley thrust system, east-central
California
C.H. Stevens and P. Stone 103
The Black Mountains turtlebacks: Rosetta stones of Death Valley tectonics
M.B. Miller and T.L. Pavlis 115
Are turtleback fault surfaces common structural elements of highly extended terranes?
I. Çemen, O. Tekeli, G. Seyitoğlu and V. Isik 139
Large-scale gravity sliding in the Miocene Shadow Valley Supradetachment Basin, Eastern Mojave Desert, California
G.A. Davis and S.J. Friedmann 149
Late Cenozoic sedimentation and volcanism during transtensional deformation in Wingate Wash and the
Owlshead Mountains, Death Valley
H.G. Luckow, T.L. Pavlis, L.F. Serpa, B. Guest, D.L. Wagner, L. Snee, T.M. Hensley and A. Korjenkov 177
Miocene rapakivi granites in the southern Death Valley region, California, USA
J.P. Calzia and O.T. Rämö 221
Upper Neogene stratigraphy and tectonics of Death Valley — a review
J.R. Knott, A.M. Sarna-Wojcicki, M.N. Machette and R.E. Klinger 245

Late Quaternary denudation, Death and Panamint Valleys, eastern California
   A.S. Jayko   271
Holocene fluvial geomorphology of the Amargosa River through Amargosa Canyon, California
   D.E. Anderson   291
Macropolygon morphology, development, and classification on North Panamint and Eureka playas, Death Valley
   National Park CA
   P. Messina, P. Stoffer and W.C. Smith   309
Base- and precious-metal deposits in the Basin and Range of Southern California and Southern
   Nevada—Metallogenic implications of lead isotope studies
   S.E. Church, D.P. Cox, J.L. Wooden, J.V. Tingley and R.B. Vaughn   323

Contents Volume 73, 2005   347